Nanoparticles in the Lung

Environmental Exposure and Drug Delivery

Nanoparticles in the Lung

Environmental Exposure and Drug Delivery

Edited by
Akira Tsuda
Harvard University
Cambridge Massachusetts, USA
Peter Gehr
University of Bern
Switzerland

CRC Press
Taylor & Francis Group
Boca Raton London New York

CRC Press is an imprint of the
Taylor & Francis Group, an **informa** business

CRC Press
Taylor & Francis Group
6000 Broken Sound Parkway NW, Suite 300
Boca Raton, FL 33487-2742

First issued in paperback 2021

ISBN 13: 978-1-138-74920-7 (pbk)
ISBN 13: 978-1-4398-9279-4 (hbk)

Publisher's Note
The publisher has gone to great lengths to ensure the quality of this reprint but points out that some imperfections in the original copies may be apparent.

Visit the Taylor & Francis Web site at
http://www.taylorandfrancis.com

and the CRC Press Web site at
http://www.crcpress.com

Contents

Section IV Translocation

Section V Drug Delivery to the Respiratory Tract

Section VI Special Issues

Foreword

There is no doubt that the topics of nanotechnology and the associated fields of nanomedicine and nanotoxicology have grown dramatically during the past decades. Corporations, universities, and governments are supporting a rapid growth in research and development. We also see substantial numbers of nano-enabled products. Of particular interest are highly engineered nanomaterials, such as quantum dots, carbon nanotubes, and fullerenes.

At the same time, the words of Solomon in the Book of Ecclesiastes seem appropriate, "There is nothing new under the sun." India ink is a stable colloid of carbon nanoparticles (NPs). Artisans in the ninth century used silver and copper NPs in pottery glazes. Other familiar examples are tattoo inks, which have many NPs, and especially tobacco smoke. Intravenously injected radioactive NPs of ^{198}Au colloid were used decades ago to treat liver cancer. All contain polydisperse particles that have a substantial fraction whose diameter is less than 100 nm, especially as a function of number.

One can go back even further to when humans discovered fire. Smokes are polydisperse aerosols that include NPs, usually defined as being less than 0.1 μm in at least one dimension. The air pollution field has long recognized the existence of nanosized or ultrafine particles, sometimes referred to as $PM_{0.1}$. Important sources of such "unintentionally produced NPs" include cars, trucks, airplanes, other combustion processes, metal mining, welding, and smelting, as well as those produced by natural processes such as forest fires, volcanoes, sea spray, or wind erosion. Atmospheric chemical processes acting on air pollutants produce NPs following nucleation of extremely small particles, which then grow by condensation and coagulation. When ingested by alveolar macrophages, respirable particles, e.g., $PM_{2.5}$, if they completely dissolve, inevitably progress through a NP size range.

This excellent book focuses primarily on the intentional production of NPs, which are sometimes highly engineered. The universe of technologies, new materials, and products is vast. There is a pressing need to ask and answer the questions articulated in the many chapters of this comprehensive volume. As is true for the existence of NPs, there are both old and new concepts in this book. The familiar saying, "the dose makes the poison," continues to be true. We need to understand the temporal relationships of NP concentrations in relevant organs during and after exposure. This in turn depends on initial deposition, clearance processes, solubility of the particles, translocation, and overall persistence. For organs and tissues of interest, we need to know the area under the curve as we describe concentration versus time.

In contrast, there are many chapters in this volume that discuss topics hitherto ignored or even unimagined. For example, we are just beginning to describe how the size, shape, and surface of NPs determine the protein and phospholipid corona, which rapidly envelops nanomaterials in the lungs as molecules from the alveolar lining layer and airway mucus bind reversibly and irreversibly to NP surfaces. We also need to better understand how that corona then affects the fate of NPs. For example, the protein corona likely defines the particle affinity for different specific particle receptors on macrophage surfaces. Other topics in this book that focus on important unexplored areas are the extent to which NPs enter the pulmonary lymphatics and the likelihood of NP transport to the brain via the nose and the olfactory system. Finally, we are just beginning to characterize the potential of NP drug delivery to and through the respiratory tract, another important theme in this book.

Should we be concerned about the hazards of nanomaterials? Is there a future "asbestos" that will do great harm? We need to be concerned about the potential toxicity of nanomaterials because they are novel and poorly studied. Moreover, nanomaterials have been observed in unexpected places, e.g., in mitochondria or even in the nucleus. When NPs become intracellular, they sometimes enter by non-endocytic pathways and appear not membrane-bound. It is also true that because of their small size, there are greater numbers of particles and greater surface area per unit weight than larger respirable particles. Increasing evidence suggests that toxicity is better correlated with particle number or surface area than total mass, yet exposures are typically given in terms of mass per unit volume. In toto, NPs may have unexpected biologic consequences.

It is somewhat reassuring that there are no current examples of nanodiseases in humans. We know of no epidemic of diseases or conditions elicited by NP exposures. Moreover, the most exotic and unusual NPs, such as fullerenes and quantum dots, are produced in small quantities, and exposures of workers or the public are limited. Finally, since dissolution of particles is proportional to surface area, one would expect that NPs would dissolve faster than their larger brothers. An exception are carbon nanotubes. They have raised concerns because they can be similar in size, shape, and durability to asbestos. They also can translocate in ways reminiscent to asbestos. At high doses in animals, lesions similar to mesothelioma have been reported.

In conclusion, there are legitimate concerns about nanosafety given the widespread and increasing use of novel nanomaterials. Data and risk analysis are needed. We need to discover the laws of nanotoxicology. We should not and cannot evaluate just one material at a time. It is also important to compare the risks of NPs to other health hazards from inhaled particles, such as fossil fuel combustion, tobacco, and diverse occupational exposures. Like other new technologies and materials, nanotechnology has great potential benefits. We need to quantify both benefits and risks in order to make informed and rational decisions as individuals and societies.

Readers of this book will be treated to a diverse and delicious menu of new ideas and data. Inevitably, in addition to providing important answers, new questions are raised. Readers from many disciplines will find this book valuable. I express my appreciation to the book editors, who also authored many chapters, Professors Akira Tsuda and Peter Gehr. They had the vision and persistence to bring this ambitious project to a successful conclusion.

Joseph D. Brain, SD

Preface

The lungs are exposed to approximately 100 billion to 10 trillion particles per day. To cope with this, the lungs are equipped with sophisticated mechanisms to handle the large particle load. The lungs efficiently eliminate particles; however, some particles, especially those of nanosize (nanoparticles [NPs]), manage to escape from the lung clearance mechanisms and may enter the bloodstream through the air–blood tissue barrier. Only nanosized particles are able to do this.

The rapid development of nanotechnology has resulted in an increased engineering of a variety of different NP types that may be released into the environment, and, in medicine, nanosize particles are designed to deliver drugs and/or genes. Therefore, it became imperative to obtain a better understanding of how inhaled NPs behave in our lungs and body, to protect humans from their harmful effects and also to explore their utility as drug delivery carriers. This book serves this purpose; it covers our current knowledge on both the possibly toxic and/or pathologic nature and the promising beneficial aspects of NPs applied by inhalation.

Our intention with this book is to present to the reader the events that NPs encounter in the respiratory system when moving from the air to the bloodstream. We invited experts for each topic to cover the sequence of events, and we are very pleased that most of the scientists we have contacted accepted our invitation. All chapters have been thoroughly reviewed by external reviewers (other than the book contributors) to ensure the quality of each chapter with the most updated and comprehensive knowledge.

We are indebted to many people for producing this book. First of all, we are very grateful to the chapter authors for their contributions to the book. We would also like to thank the external reviewers. Finally, we would like to express our appreciation to the staff of Taylor & Francis for their patient and invaluable professional assistance in producing this book.

Akira Tsuda
Peter Gehr

Editors

Akira Tsuda is a principal research scientist at the Molecular and Integrative Physiological Sciences Program, Harvard School of Public Health, working on aerosol physics in the respiratory system. His research focuses on the intersection of lung biology and particle exposure, in particular on the effects of lung structure on particle behavior. Most notably, he has discovered the existence of chaotic mixing (i.e., irreversible acinar flow kinematics due to the recirculating nature of alveolar flow) and demonstrated its importance on aerosol behavior in the gas-exchange region of the lungs. Currently, he is working on the effects of inhaled nanoparticles on postnatal lung development and lung remodeling. He is the author of about 100 peer-reviewed articles and mentor of postdoctors (Harvard School of Public Health and Harvard Medical School). He serves as a reviewer for many international scientific journals, as well as research grants. He was a member of the National Institutes of Health (NIH) Respiratory Physiology, Respiratory Integrative Biology and Translational Research Study Section, and is currently on the editorial board of the *Journal of Applied Physiology* and *Journal of Aerosol Medicine and Pulmonary Drug Delivery*.

Peter Gehr is a professor emeritus from the University of Bern. He was chair of the Institute of Anatomy. He was a visiting assistant professor at the Harvard School of Public Health and a guest professor at the University of Nairobi; he spent sabbaticals at the University of Western Australia, at National Jewish Health in Denver, and the Harvard School of Public Health. His field of research was first the structure–function correlation of the gas-exchange apparatus of the lungs of mammals and humans. Later he focused his studies on the interaction of particles with lung structures. His main interest was the mechanisms of interaction of particles with the surfactant, and later with cells and subcellular structures. His studies finally concentrated on the interaction of nanoparticles with cells and intracellular trafficking. A specialty of his research was the investigation of the cellular interplay upon exposure to nanoparticles. He is currently the president of the Steering Committee of the National Research Program on Opportunities and Risks of Nanomaterials of the Swiss National Science Foundation, and he is a member of the Swiss Federal Commission for Air Hygiene.

Contributors

Dominique Balharry
School of Life Sciences
Heriot-Watt University
Edinburgh, United Kingdom

Vianney Bernau
Powder Technology Laboratory
Institute of Materials
Ecole Polytechnique Fédérale de Lausanne
Lausanne, Switzerland

Marie-Gabrielle Beuzelin
Powder Technology Laboratory
Institute of Materials
Ecole Polytechnique Fédérale de Lausanne
Lausanne, Switzerland

Fabian Blank
Respiratory Medicine
Bern University Hospital
Bern, Switzerland

Zea Borok
Will Rogers Institute Pulmonary Research Center
Departments of Medicine and Biochemistry and Molecular Biology
University of Southern California
Los Angeles, California

Joseph D. Brain
Molecular and Integrative Physiological Sciences Program
Department of Environmental Health
Harvard School of Public Health
Boston, Massachusetts

Vincent Castranova
Department of Basic Pharmaceutical Sciences
West Virginia University
Morgantown, West Virginia

Martin J.D. Clift
Adolphe Merkle Institute
Université de Fribourg
Fribourg, Switzerland

Joel M. Cohen
Center for Nanotechnology and Nanotoxicology
Department of Environmental Health
Harvard School of Public Health
Boston, Massachusetts

Edward D. Crandall
Will Rogers Institute Pulmonary Research Center
Departments of Medicine, Pathology, and Chemical Engineering and Materials Science
University of Southern California
Los Angeles, California

Philip Demokritou
Center for Nanotechnology and Nanotoxicology
Department of Environmental Health
Harvard School of Public Health
Boston, Massachusetts

Farnoosh Fazlollahi
Will Rogers Institute Pulmonary Research Center
Department of Chemical Engineering and Materials Science
University of Southern California
Los Angeles, California

Peter Gehr
Institute of Anatomy
University of Bern
Bern, Switzerland

Eva Gubbins
School of Life Sciences
Heriot-Watt University
Edinburgh, United Kingdom

Frank S. Henry
Department of Mechanical Engineering
Manhattan College
Riverdale, New York

Fabian Herzog
Adolphe Merkle Institute
Université de Fribourg
Fribourg, Switzerland

Heinrich Hofmann
Powder Technology Laboratory
Institute of Materials
Ecole Polytechnique Fédérale de Lausanne
Lausanne, Switzerland

Lisbeth Illum
IDentity
Nottingham, United Kingdom

Keiji Itaka
Division of Clinical Biotechnology
Center for Disease Biology and Integrative Medicine
Graduate School of Medicine
University of Tokyo
Tokyo, Japan

Xiue Jiang
State Key Laboratory of Electroanalytical Chemistry
Changchun Institute of Applied Chemistry
Chinese Academy of Sciences
Changchun, China

Helinor Johnston
School of Life Sciences
Heriot-Watt University
Edinburgh, United Kingdom

Kazunori Kataoka
Division of Clinical Biotechnology
Center for Disease Biology and Integrative Medicine
Graduate School of Medicine
and
Department of Materials Engineering
Graduate School of Engineering
University of Tokyo
Tokyo, Japan

Ali Kermanizadeh
School of Life Sciences
Heriot-Watt University
Edinburgh, United Kingdom

Kwang-Jin Kim
Will Rogers Institute Pulmonary Research Center
Departments of Medicine, Physiology and Biophysics, Pharmacology and Pharmaceutical Sciences, and Biomedical Engineering
University of Southern California
Los Angeles, California

Yong Ho Kim
Will Rogers Institute Pulmonary Research Center
Department of Medicine
University of Southern California
Los Angeles, California

Wolfgang G. Kreyling
Institute for Lung Biology and Disease, Focus Network: Nanoparticles and Health
Helmholtz Center Munich – Research Center for Environmental Health
Neuherberg/Munich, Germany

Dagmar A. Kuhn
Adolphe Merkle Institute
Université de Fribourg
Fribourg, Switzerland

Shun'ichi Kuroda
Department of Bioengineering Sciences
Graduate School of Bioagricultural Sciences
Nagoya University
Nagoya, Japan

Debra L. Laskin
Department of Pharmacology and Toxicology
Ernest Mario School of Pharmacy
Rutgers University
Piscataway, New Jersey

Jeffrey D. Laskin
Department of Environmental and Occupational Medicine
Rutgers Robert Wood Johnson Medical School
Piscataway, New Jersey

Iseult Lynch
School of Geography, Earth and Environmental Science
University of Birmingham
Birmingham, United Kingdom

Lionel Maurizi
Powder Technology Laboratory
Institute of Materials
Ecole Polytechnique Fédérale de Lausanne
Lausanne, Switzerland

Robert R. Mercer
National Institute for Occupational Safety and Health
Morgantown, West Virginia

Gerd Ulrich Nienhaus
Institute of Applied Physics and Center for Functional Nanostructures
Karlsruhe Institute of Technology
Karlsruhe, Germany

and

Department of Physics
University of Illinois at Urbana-Champaign
Urbana, Illinois

John S. Patton
Dance Biopharma Inc.
San Francisco, California

Alke Petri-Fink
Adolphe Merkle Institute
Department of Chemistry
Université de Fribourg
Fribourg, Switzerland

Dale W. Porter
National Institute for Occupational Safety and Health
Morgantown, West Virginia

Barbara Rothen-Rutishauser
Adolphe Merkle Institute
Université de Fribourg
Fribourg, Switzerland

and

Respiratory Medicine
Bern University Hospital
Bern, Switzerland

Usawadee Sakulkhu
Powder Technology Laboratory
Institute of Materials
Ecole Polytechnique Fédérale de Lausanne
Lausanne, Switzerland

Patrick J. Sinko
Department of Pharmaceutics
Ernest Mario School of Pharmacy
Rutgers University
Piscataway, New Jersey

Arnold Sipos
Will Rogers Institute Pulmonary Research Center
Department of Medicine
University of Southern California
Los Angeles, California

Vicki Stone
School of Life Sciences
Heriot-Watt University
Edinburgh, United Kingdom

Lennart Treuel
Institute of Applied Physics and Center for Functional Nanostructures
Karlsruhe Institute of Technology
Karlsruhe, Germany

and

Institute of Physical Chemistry
University of Duisburg-Essen
Essen, Germany

Akira Tsuda
Molecular and Integrative Physiological Sciences Program
Department of Environmental Health
Harvard School of Public Health
Boston, Massachusetts

Satoshi Uchida
Division of Clinical Biotechnology
Center for Disease Biology and Integrative Medicine
Graduate School of Medicine
University of Tokyo
Tokyo, Japan

Dimitri Vanhecke
Adolphe Merkle Institute
Université de Fribourg
Fribourg, Switzerland

Christophe von Garnier
Respiratory Medicine
Bern University Hospital
Bern, Switzerland

Barry Weinberger
Department of Pediatrics
Division of Neonatology
Rutgers Robert Wood Johnson Medical School
New Brunswick, New Jersey

1

Introduction

Akira Tsuda and Peter Gehr

We will sequentially follow the events that nanoparticles (NPs) encounter in the respiratory system from the air to the bloodstream after being deposited. Thus, the book starts with deposition (Section I, Chapter 2). First of all, owing to the physically small size of NPs, inertia-driven transport (e.g., mass-mediated deviation from the carrier airflow streamlines) and inertia-driven deposition (e.g., inertial impaction or gravitational sedimentation) of NPs are negligible. Whereas NPs eventually deposit on the internal surface of the lung in the gas-exchange region by diffusion, what brings them close to the alveolar surface structures is still largely a mystery. Since the diffusivity of NPs is small—several orders of magnitude smaller than those of the typical respiratory gases (oxygen and carbon dioxide)—sole diffusion cannot explain the transport of NPs for a long distance (more than several microns); the main transport mechanism of NPs must be due to convection. Exploring these problems, a large portion of Chapter 2 is devoted to the basic physics of acinar flow patterns.

Section II (Airway/Alveolar Surface) focuses on the events that NPs encounter on the lung surface. We start by giving the readers a broad picture on this topic (Chapter 3), followed by details. In Chapter 4, we discuss how NPs are coated by proteins ("protein corona") in the alveoli, a topic extensively studied currently. As surfactant proteins are the first proteins NPs encounter after deposition on the alveolar surface, a large portion of Chapter 4 is devoted to this topic. In Chapter 5, we discuss the interactions of NPs with the hypophase (i.e., alveolar lining fluid), with emphasis on carbon nanofibers. The section ends with a discussion of lung defense mechanisms (Chapter 6); these are essential for the host because the lungs are exposed to an astronomically large number of particles constantly.

The next topic is the encounter of NPs and the epithelial cell layer (Section III). This important topic is examined in three different chapters (Chapters 7 through 9). How NPs cross the lipid bilayer of the cell membranes has a great impact on their ability to induce inflammation, and therefore also affects NP-based therapeutic drug delivery.

Section IV (Translocation) focuses on the translocation events and the subsequent fate of NPs. We start with NP translocation across the air–blood barrier (Chapter 10), followed by discussing NPs in the pulmonary lymphatic system (Chapter 11). The section will end with examining how NPs reach the secondary organs via the bloodstream (Chapter 12).

Switching gears, NP-based drug delivery will be discussed in Section V. We start with a general overview of this topic, followed by specific technical applications, nanomicelles (Chapter 14), nanobiocapsules, and multifunctional envelope-type nanodevices (Chapter 15), as examples. The field (nanomedicine) is rapidly moving and is an active area of research, with new studies performed and new papers published in ever-larger numbers.

Section VI covers additional issues of special concern regarding NPs in the lungs. The section starts by describing the unique physicochemical characteristics of NPs. The surface per volume ratio increases with decreasing particle size, but also quantum effects come into the picture at the nanometer size range (Chapter 16). The section continues by addressing the issue of accurate dose calculation in *in vitro* cell culture systems. Since the size of NPs covers both diffusion-dominated and gravity-dominated size ranges in the culture media, accurate dose estimation is not simple (Chapter 17). Chapter 18 discusses NP-based drug delivery from the nose to the brain; this topic attracts a lot of attention. Children are one of the most susceptible and vulnerable age groups for exposure to air pollution. Chapter 19 describes how the developing lungs of young children are structurally and functionally different from the fully developed lung; data obtained from an animal model demonstrate how the deposition of NPs can be influenced by the non-alveolated lung structure of the developing lung. Nanotoxicology—the study of the toxic effects caused by nanomaterials—is the subject of Chapter 20.

Finally, in Chapter 21, we summarize the unique physicochemical features of NPs and outline our views on the future direction of the field.

Section I

Gas Phase

2

Deposition

Akira Tsuda and Frank S. Henry

CONTENTS

2.1 Introduction

The transport and deposition of aerosol particles in the respiratory tract is primarily determined by their physical characteristics (e.g., size, shape, and density) and the surrounding carrier airflow patterns. Large and/or dense aerosol particles (e.g., micron-size and/or heavy particles) tend not to follow the patterns of the carrier airflows owing to a combination of their large inertia and the effect of gravity; thus, they deviate from the airflow streamlines and deposit onto the airway surface. Inertia-driven deposition is called inertial impaction, whereas deposition due to gravity is called gravitational sedimentation. Smaller and/or lighter particles, such as nano-size particles (NPs) or ultrafine particles (UFPs), have negligible inertia; thus, they cannot deposit by inertial impaction or gravitational sedimentation. It is, however, known that such particles do deposit; hence, they must do so by some other means.

What are these other mechanisms? Ultimately, NPs deposit by diffusion; however, they first have to be brought close to the alveolar surface by convection. Since the diffusivity of NPs is larger than that of micron-size particles, it is often thought that diffusion is the major deposition mechanism for NPs. However, this is only true for the smaller NPs ($\lesssim$10 nm). Indeed, for those particles that reach the acinus, cross-stream transport by diffusion cannot be significant; otherwise, they would deposit before reaching the acinus. We note that although the diffusivity of NPs is 2–3 orders of magnitude larger than that of micron-size particles (see Table 1 in Tsuda et al. 2013), the diffusivity of NPs is 5–6 orders of magnitude smaller than the molecular diffusivity of typical respiratory gases (oxygen and carbon dioxide). Hence, while respiratory gases can migrate from the central airway to the alveolar surface in a fraction of a breathing cycle, this is not the case for NPs.

We see that particles enter the respiratory system from the environment (i.e., atmosphere) and reach the lung (alveolar) surface to deposit through a combination of convective and diffusive transport. For most of its journey (e.g., from airway opening to the gas-exchange region of the lungs), the trajectory of an NP follows closely that of the surrounding airflow in a piggy-back fashion (due to fluid viscosity). That is, the time taken for an NP to be convected to the acinus is short compared with the time scale of cross-stream diffusion. We note that convective transport largely depends on local flow patterns, and that the flow patterns in the expanding alveoli are much different from those in the central airways. Once an NP enters an alveolus, the convective velocity is much reduced and diffusion becomes more effective; the trajectory of an NP near the alveolar surface can then deviate from the convective airflow streamline by diffusion and deposit on the lung surface. Because convection and diffusion are the main transport mechanisms, NP diffusivity should be compared with airflow convection—the only other main mechanism. In this way, it will be possible to tell when NP diffusivity becomes effective.

In fluid mechanics theory, there is a dimensionless parameter called the Péclet number (Pe), which exactly expresses the relative importance of convective transport to diffusive transport:

$$\mathrm{Pe} = \frac{uL}{D}$$

where u is the characteristic airflow velocity, L is the characteristic length scale of airways (e.g., airway diameter), and D is the particle diffusivity.

In Figure 2.1, the Pe of NPs (<100 nm in diameter) is given for human acinar flow under normal breathing conditions (Tsuda et al. 2013) and is plotted against particle size. The

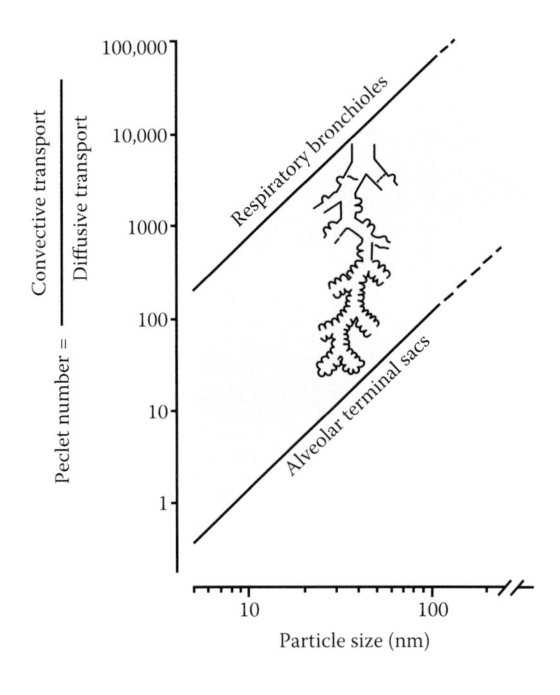

FIGURE 2.1
Péclet number (Pe) of NPs in the pulmonary acinus under normal breathing conditions plotted versus the size of NPs. Note that Pe of NPs is normally much more than unity, indicating that the airflow convective transport dominates over the diffusive transport throughout the acinus. (Adapted from Semmler-Behnke M et al., *Proc. Natl. Acad. Sci. U. S. A.*, 109, 5092, 2012.)

two lines in Figure 2.1 represent the conditions in the central airways at the entrance of the acinus and those in the alveolar terminal sacs, respectively; all NPs within the acinus have Pe between these two limiting lines. For instance, for 100-nm NPs (the upper size limit of NPs), $100 < Pe_{NP=100\ nm} < 60{,}000$, demonstrating that for this particle size, convective transport greatly outweighs diffusive transport everywhere in the acinus. For much smaller NPs (e.g., NP = 8 nm), $1 < Pe_{NP=8\ nm} < 500$, showing that convection still dominates in most of the acinus except at the very end of the acinus, where convection and diffusion are comparable (Pe ~1). Diffusive transport ultimately dominates for 8-nm particles but only at the very end of the acinar tree. We note that NPs entering the alveoli in the proximal region of the acinus, where the central channel flow is still relatively high, will experience a much less vigorous flow within the alveoli (the average velocity in the proximal alveoli being approximately 100th of that in the duct). Hence, even in the proximal alveoli, diffusion is an important transport mechanism and, ultimately, is the means by which the NPs deposit.

The most important message to be learned from Figure 2.1 is that although it is often thought that NPs deposit in the lung solely by diffusion, in fact convective transport dominates over diffusive transport by far (for all generations but the terminal sacs). The question arises: What kind of airflow is in the lungs? As predictions using the ICRP (International Commission on Radiological Protection) model show (e.g., Figure 6 in Tsuda et al. 2013), NPs dominantly deposit deep in the lungs where gas exchange occurs. Therefore, in the rest of this chapter, we will focus our discussion on airflow patterns and NP behavior in that region (i.e., gas-exchange region) of the lungs.

2.2 Pulmonary Acinus and Its Geometry

The chief function of the lungs is gas exchange, and this mainly takes place deep in the organ. In the respiratory zone of the lungs, the distance between air and blood is extremely thin (0.62 μm [Gehr et al. 1978]; 100 times thinner than human hair); however, the surface area is enormously large (130 m^2 [Gehr et al. 1978]; comparable to the size of a tennis court). To achieve this large surface area, the airways in the gas-exchange region are alveolated and form a dichotomously bifurcating structure (10 generations on average; Haefeli-Bleuer and Weibel 1988). Because of its appearance as a grapelike cluster, the gas-exchange region of the lungs is called the pulmonary acinus.

Weibel (1963) defines the pulmonary acinus starting from the airway on which the first alveolus appears (i.e., after 14–15 airway generations from the trachea), just beyond the terminal bronchioles. In the case of humans and primates, the first few generations of the acinus are partially alveolated, forming a transitional zone between the conducting airways and the gas-exchange region. The airways in this zone are called respiratory bronchioles. In the case of rodents, the transition from non-alveolated airways (i.e., terminal bronchiole) to alveolated airways is more abrupt with none or only one generation of partially alveolated bronchioles (Bastacky et al. 1983; Tyler 1983; Rodriguez et al. 1987). Because of the abrupt increase in the cross-sectional area at the entrance of the acinus, there is a sudden change in fluid mechanics in this region of the bronchopulmonary tree (see in the following).

Beyond the transitional zone, the airways are fully alveolated, forming a unit called the subacinus (Haefeli-Bleuer and Weibel 1988). There are, on average, six generations in the subacinus. According to Haefeli-Bleuer and Weibel (1988), the volume of each acinus is, on average, about 187 mm^3 at the total lung capacity (TLC); there are roughly 30,000 acini in the human lung, with >10,000 alveoli in each acinus; the diameter of the central channel in the human acinus falls from 500 to 270 μm; the depth and width of a typical alveolus is approximately 250 μm; and the longitudinal path length of the acinus (i.e., the distance along the ducts from the transitional bronchiole to the alveolar sacs) averages 8.8 mm.

Regarding the shape of each alveolus, the (truncated) spherical shape is the most commonly used model geometry for alveoli (e.g., Haber et al. 2000; Sznitman et al. 2007; Ma and Darquenne 2011; Semmler-Behnke et al. 2012). However, owing to space-filling considerations, alveoli cannot be treated as spherical. A polyhedral shape (i.e., a three-dimensional [3D] geometric object with flat faces and straight edges) is often proposed as space-filling alveolar model geometry (Fung 1988; Denny and Schroter 1995, 2000, 2006; Sznitman et al. 2009; Kumar et al. 2009, 2011). As far as basic acinar flow characteristics and behavior of particles therein are concerned, however, there is some evidence (Henry and Tsuda 2010) showing that the global geometric arrangement of acinar airway architecture (i.e., the central channel vs. surrounding alveoli; explained in detail later in this chapter) is more important than the precise shape of each individual alveolus.

The pulmonary acinus was recently visualized in 3D (Figure 2.2) by Tsuda et al. (2008). Using high-resolution synchrotron radiation-based x-ray tomographic microscopy (HR-SRXTM, beamline-TOMCAT; Swiss Light Source, Paul Scherrer Institut, Villigen, Switzerland), stereologically well-characterized rat lung samples (Tschanz et al. 2003) were imaged with a resolution of 1.4 μm^3 per voxel (at least, a resolution of a few micron or less is necessary because inter-airspace septal thickness is approximately 10 μm; Gehr et al. 1978); a part of the acinus was reconstructed by using the 3D finite element shell technique (Tsuda et al. 2008). The size of an acinus is relatively large (e.g., rat acinus is ~2.33 mm^3 at TLC; Rodriguez et al. 1987); however, a high resolution (<1 μm) is required to image the fine details of acinar microstructure. These somewhat conflicting requirements were addressed recently by developing a new imaging protocol called "wide field scanning SRXTM" protocol (Haberthür et al. 2010); with this new protocol, a volume of up to 3.6 mm

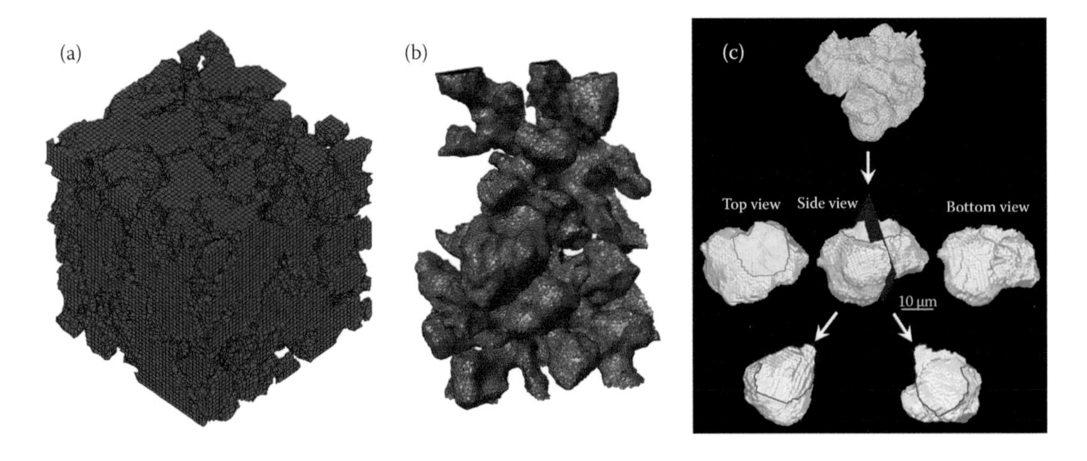

FIGURE 2.2
Using synchrotron radiation-based x-ray tomographic microscopy, a part of rat acinus (a and b) was reconstructed in 3D. (c) Once alveoli are reconstructed in 3D, they can be viewed from any arbitrary angle (c, middle). Also, the alveolus can be cut into half (c, bottom) and also sliced to make 2D sections in any orientation with any slab thickness (not shown here), and there is no limitation to the number of times the sectioning process can be repeated electronically. (Adapted from Tsuda A et al., *J. Appl. Physiol.*, 105, 964, 2008.)

in diameter (which can contain multiple acini—easily accommodates an entire rat acinus) can be imaged with high resolution (0.7 μm).

2.3 Acinar Fluid Mechanics

As mentioned previously, owing to the abrupt increase in the cross-sectional area at the entrance of the acinus and the unique branching structure of the bronchial tree, the magnitude of the linear air velocity (u) decreases as inspiration air reaches the pulmonary acinus and the diameter of the acinar airways (L) is <0.5 mm. In the acinus, therefore, the airflow Reynolds number (i.e., $\mathrm{Re} = uL/\nu$, where ν denotes the kinematic fluid viscosity) is normally much less than unity. Such a flow is called a low Reynolds number flow, the airflow momentum of which is largely governed by the interaction of pressure drops and viscous forces. In the acinus, therefore, the configuration of alveolated airway walls and their tidal motion become ultimately very important in determining the airflow patterns.

In classical fluid mechanics theory, a low Reynolds number flow is considered kinematically reversible (Taylor 1960). On the basis of this classical theory, Davies (1972) stated that "...In the alveolar regions, there is no mixing...," and therefore concluded that any deposition in the acinus, if any, must be due to the intrinsic motion of aerosol particles. However, as discussed previously, the diffusivity of submicron particles (i.e., one of the intrinsic motions of aerosol particles) is too small to account for the amount of deposition observed in the real lungs (Heyder et al. 1988); there was a gap between Davies' theoretical explanation and experimental realities. Despite this discrepancy, Davies' classical paper was influential; the paper dominated the field for several decades; thus, aerosol deposition in the pulmonary acinus was excluded on theoretical grounds for a long time from classical deposition analyses. In 1988, however, Heyder et al. challenged Davies' argument, experimentally demonstrating that submicron particles indeed deposit in the acinus by a nondiffusive but unknown deposition mechanism. In 1995, Tsuda et al. came up with the idea of "chaotic mixing" to explain Heyder's experimental findings.

The key to understanding the basic physics behind acinar chaotic airflow and its effects on NP deposition in the acinus is to recognize that the unique geometric feature of acinar airways—characterized as a central thoroughfare channel and surrounding dead-end air pockets—is indeed an important factor in determining acinar fluid mechanics. As the air in the central channels washes back and forth with tidal motion, it drags, via viscous interaction, otherwise quiescent fluid residing inside the dead-end side pockets along the alveolar opening. As a result, the air in the alveoli rotates slowly with a family of different frequencies (f_1) due to a purely fluid mechanical mechanism: alveolar air rotates faster at the center of recirculation and more slowly near the walls owing to the high viscous resistance from the walls (Figure 2.3).

It is critical to realize that the family of alveolar rotation frequencies (f_1) is not governed by the breathing frequency (f_2); the latter is determined by the diaphragm and rib cage motion. Since the alveolar rotation frequency (f_1) involves a full frequency spectrum, whereas the breathing frequency (f_2) is fixed, the frequencies of some alveolar recirculation orbits can be an integer multiple of f_2. It turned out that some of those alveolar recirculation orbits, which resonate with tidal breathing, can break into chaos when the flow system is under perturbation (Tsuda et al. 2011). While kinematical reversibility is ensured in a strictly "unperturbed," idealized alveolar flow system, to which Davies (1972) referred,

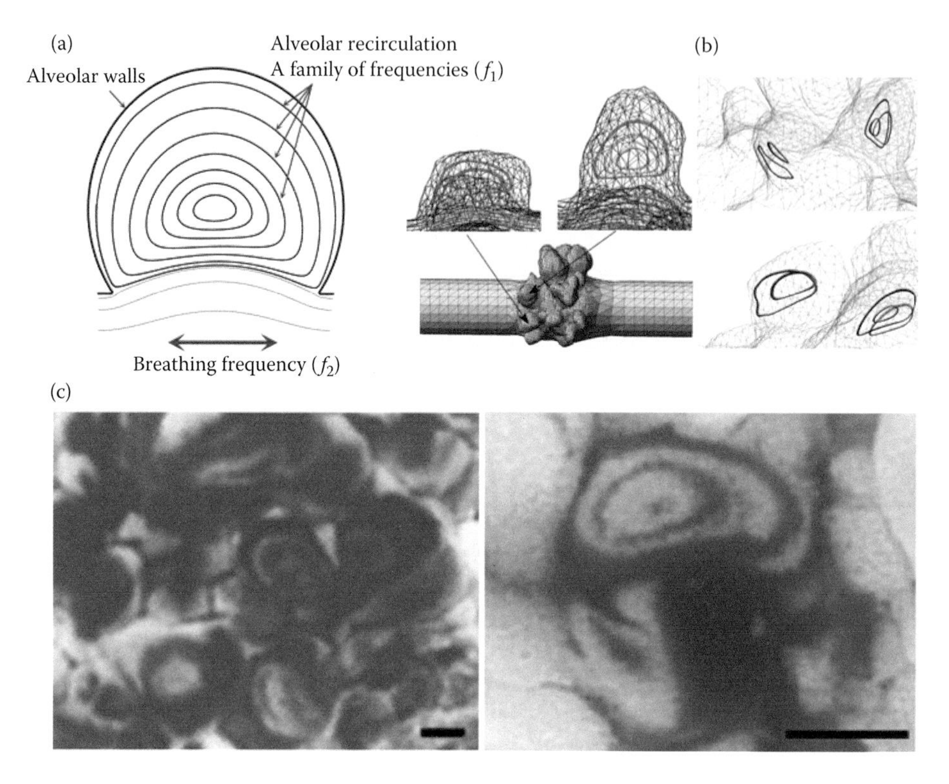

FIGURE 2.3
Air inside an alveolus typically forms rotational motion. Because of unique geometric feature of acinar airways—characterized as a central thoroughfare channel and surrounding dead-end air pockets—quiescent fluid residing inside an alveolus is dragged by the ductal flow, slashing back and forth along the central channel owing to tidal motion, with a breathing frequency of f_2. This viscous drag causes alveolar fluid to rotate with a family of different frequencies (f_1). Top: computational predictions. (a) Idealized model. (Modified from Tsuda A et al., *J. Appl. Physiol.*, 79, 1055, 1995. With permission.) (b) Realistic alveolar duct was reconstructed on the basis of SRXTM, and finite element computational fluid dynamics techniques were performed to demonstrate alveolar recirculation flows. (Adapted from Filipovic N et al., Recirculation identified in a 3D alveolar duct reconstructed using synchrotron radiation based X-ray tomographic microscopy. ATS meeting, New Orleans, 2010. [abstr.]. With permission.) (c) Experimental observation of rotational flow patterns in alveoli of excised rat lung. Bar, 100 μm. (Adapted from Tsuda A et al., *Proc. Natl. Acad. Sci. U. S. A.*, 99, 10173, 2002.)

in the more realistic situation of tidally breathing lungs there are many sources of perturbations. These include small wall movement (Tsuda et al. 1995; Henry et al. 2002; Laine-Pearson and Hydon 2006), small but persistent geometric hysteresis (Haber et al. 2000; Haber and Tsuda 2006), small diffusion effects (Laine-Pearson and Hydon 2008), and nonzero Reynolds number effects (Tsuda et al. 1995; Henry et al. 2002, 2009). Henry and Tsuda (2010) showed that rotational flow due to nonzero Reynolds number effects occurs to some degree in all but the terminal alveolar ducts. Therefore, although the extent of chaotic mixing due to nonzero Reynolds number effects might be larger at the entrance of the acinus than deeper in the acinus (Henry et al. 2009), the potential of chaotic mixing persists in most of the alveoli in the acinus. Recently, numerous experimental (van Ertbruggen et al. 2008; Berg et al. 2010; Chhabra and Prasad 2010a,b; Oakes et al. 2010; Berg and Robinson 2011) and numerical studies (Lee and Lee 2003; Dailey and Ghadiali 2007; Sznitman et al. 2007, 2009; Darquenne et al. 2009; Kumar et al. 2009, 2011; Harding and Robinson 2010; Kleinstreuer and

Zhang 2010; Ma and Darquenne 2011) of acinar flows have appeared in the literature, and many of these have confirmed the presence of recirculating alveolar flow.

Not only theoretical investigations but also the occurrence of chaotic mixing between inhaled tidal fluid and alveolar residual fluid was demonstrated experimentally (Tsuda et al. 2002). Figure 2.4 shows a stretch-and-fold mixing pattern—a hallmark of chaos (Aref 1984; Ottino 1989; Tabor 1989)—observed on transverse cross sections of the airways after only one breathing cycle. As these complicated mixing patterns indicate a history of what the inhaled tidal fluid experienced in the pulmonary acinus, each tidal fluid element must sample alveolar recirculation flow, which can produce chaotic dynamics. It is well known that the iteration of chaotic trajectories is often manifested in fractal geometries (Mandelbrot 1982; Sommerer 1994). Indeed, spatial correlation analysis revealed that these observed stretch-and-fold patterns were fractal, with the same fractal dimension persisting over many airway generations, suggesting that the chaotic mixing originates deep in the lung—in the pulmonary acinus.

Figure 2.5 shows the mixing pattern between the alveolar residual fluid and the inhaled tidal fluid on the cross sections of the acinar airways. After one cycle ($n = 1$), most of the acinar airways appear to be predominantly the color of the alveolar residual (white) fluid. After two or three cycles, however, a large amount of tidal (blue) fluid appears on the cross sections of acinar airways ($n = 2, 3$), showing clearly delineated interface patterns with both blue and white fluids being stretched and folded. After four cycles, the clarity of the interface patterns had largely disappeared, and the clear color patterns observed at $n = 2$, 3 were smeared to uniformity ($n = 4$). This phenomenon—an abrupt increase in mixing

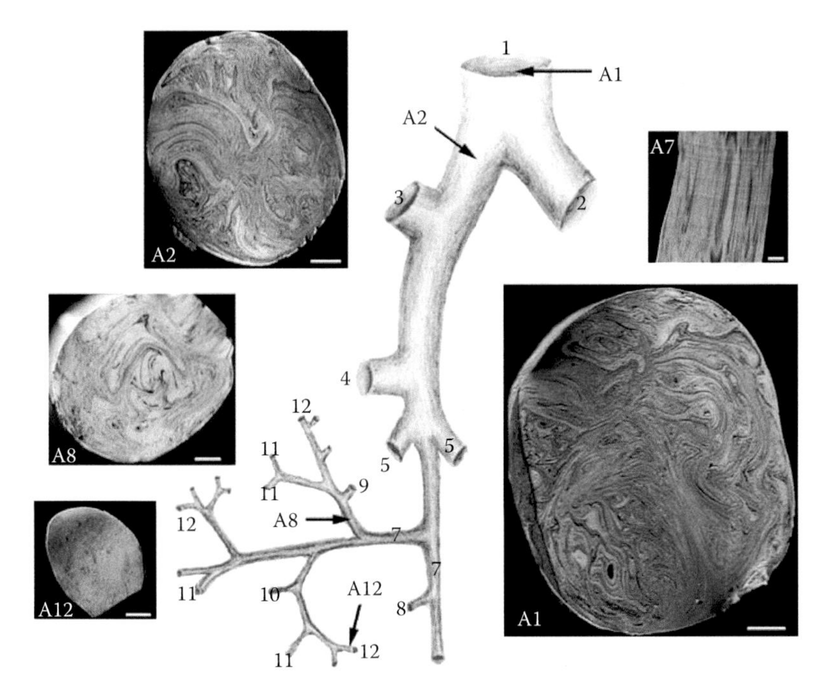

FIGURE 2.4
Stretch-and-fold mixing patterns—a hallmark of chaos (Aref 1984; Ottino 1989; Tabor 1989)—observed on airway cross sections at different locations in the tracheobronchial tree after one breathing cycle. Scale bar, 500 μm in A1 and A2, 200 μm in A8, and 100 μm in A7 and A12. (Adapted from Tsuda A et al., *Proc. Natl. Acad. Sci. U. S. A.*, 99, 10173, 2002.)

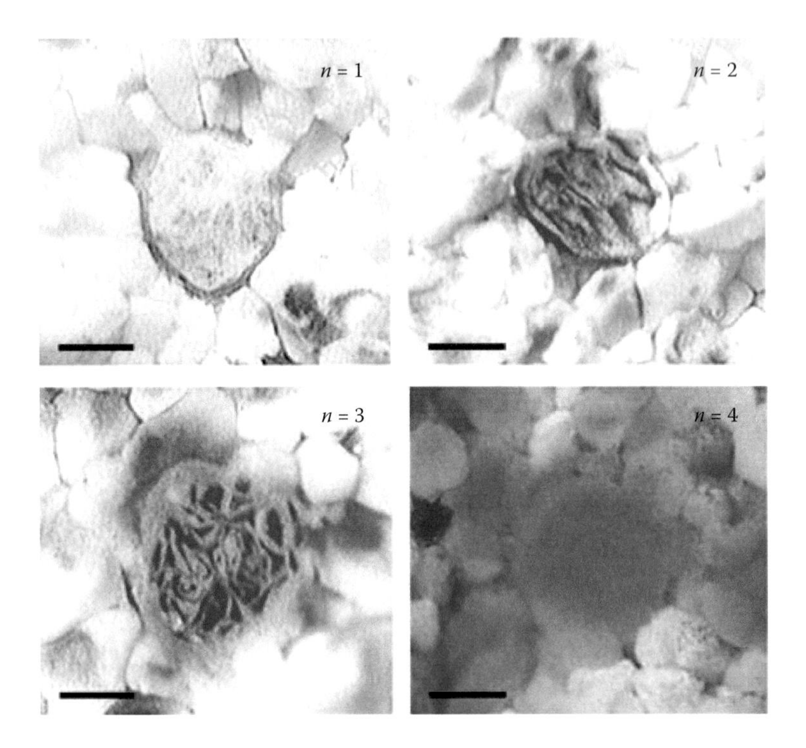

FIGURE 2.5
Typical mixing pattern between the alveolar residual fluid and inhaled tidal fluid on the cross sections of the rat acinar airways. After ventilatory cycles of one ($n = 1$), two ($n = 2$), three ($n = 3$), and four ($n = 4$). Scale bar, 100 μm. (Adapted from Tsuda A et al., *Proc. Natl. Acad. Sci. U. S. A.*, 99, 10173, 2002.)

(from $n = 2$, 3 to $n = 4$)—is a characteristic feature of chaotic mixing (Aref 1984; Ottino 1989; Tabor 1989), and is entirely different from temporally smooth diffusive mixing.

2.4 Nanoparticle Deposition in the Pulmonary Acinus

Despite Davies' classical prediction to the contrary, NPs are known to deposit in the pulmonary acinus at a rate greater than is possible by simple diffusion (Heyder et al. 1988; Schulz et al. 1992). This clearly indicates that a mechanism other than diffusion must be operating for NP deposition. Recently, the importance of alveolated wall structure and the role played by the resulting chaotic mixing on NP deposition were experimentally demonstrated (Semmler-Behnke et al. 2012). Briefly (details in Section VI, Chapter 19), because a dramatic structural remodeling takes place postnatally, from a saccular stage (i.e., few and shallow alveoli) to fully developed alveoli (i.e., deep alveoli), in the developing infant lungs, Semmler-Behnke et al. (2012) hypothesized that there is a critical developmental stage at which the shape of the developing alveolus becomes deep enough to cause the alveolar flow to start rotating. At that stage of development, the characteristics of the alveolar flow will change qualitatively and start to exhibit chaotic dynamics. This radical shift in the fluid dynamics of the acinus must lead to significant mixing of inhaled and residual air in the lungs, enhancing NP deposition. Semmler-Behnke et al. (2012) tested this hypothesis in a rat animal model because the remodeling process associated with development and the resulting structural change is essentially the same in humans and rodents, the only difference being one of time scales (2–3 years in humans, 3 weeks in rodents).

Remarkably, Semmler-Behnke et al. (2012) found that NP deposition was strongly age dependent (see Figure 19.6 in Chapter 19); NP deposition was initially very low (in 7-day-old rats) but later significantly increased and peaked at the end of the bulk alveolation stage of postnatal lung development (21-day-old rats). These findings clearly demonstrate that postnatal structural changes fundamentally alter the airflow in the lungs, and thus NP deposition in the acinus. In other words, the alveolated structure has a crucial role in convective transport and deposition of NPs in the pulmonary acinus.

What is still largely unknown is the detailed patterns of NP deposition distribution along the acinar airways. Is this important to consider? Since 90% of the lung volume belongs to the pulmonary acinus (Weibel 1963), and most of the inhaled NPs deposit in the acinus (except very small NPs [<8 nm], which deposit in the nose), the distribution patterns of NP deposition in the acinus must be ultimately important for the onset of NP-mediated lung diseases as well as NP-based therapeutic drug delivery. Pinkerton et al. (2000) visualized deposition patterns along the airway paths, from the trachea down to the acinus, in human autopsy specimens. Although they found little evidence of particle deposition in the larger conducting airways (generations 2–6), they observed great accumulations of particles in the acinus, especially in the entrance region (the highest retention in the first, second, and third respiratory bronchioles, in that order). It is noteworthy that it has been repeatedly demonstrated that the proximal area of the acinus—a few generations beyond the terminal bronchioles—is the primary site of lung injury after exposure to airborne pollutants (Niewoehner et al. 1974; Kleinerman et al. 1979; Craighead et al. 1982, 1988; Wright et al. 1992; Harkema et al. 1993; Green and Churg 1998; Pinkerton et al. 2000; Saldiva et al. 2002; Churg et al. 2003). Churg and Brauer (2000) reported that long-term accumulation of submicron-size particles were found at the entrance of the acinus, at typically 25–100 times higher concentrations than in the main stem bronchi. Lung tissues in the transitional zone are likely to respond to particle insults in a more complex fashion than in other parts of the lung because bronchiolar and alveolar epithelial cells interdigitate in this zone (Plopper 1990; Plopper and Ten Have-Opbroek 1995) and are of different phenotypes.

Although the study of Pinkerton et al. (2000), mentioned previously, revealed the important consequence of particle deposition at the entrance of the acinus, the study involved a combination of elaborate, time-consuming tissue sampling and careful tissue histology. Indeed, the acinus is, in general, difficult to access because of its distant location from the airway opening, its small size, and its complex and tortuous structure. To circumvent these difficulties and to access more general and comprehensive information about the deposition distribution in the entire acinus, there are currently two lines of research undertaken: one is theoretical and the other experimental.

In the theoretical study, let us first note that flow patterns in individual alveoli are determined by the ratio QA/QD (Tsuda et al. 1995), where QA is the rate of flow entering the alveolus and QD is the ductal flow passing by the alveolar opening. Since each alveolus expands at approximately the same rate regardless of its position in the acinus, QA is roughly the same throughout the acinus, while QD is a function of the expanding lung volume distal to the alveolus of interest. Therefore, the ratio QA/QD, that is, the flow pattern in an individual alveolus, depends on its generational position in the acinar tree. Since NP deposition is strongly influenced by convection (i.e., alveolar flow pattern), the heterogeneous distribution of NP deposition in the level of the entire acinar tree is likely a result of the change in QA/QD in the acinus.

On the basis of the longitudinal change in QA/QD along the acinar airway pathway, the preferentially high deposition in the proximal region of the acinus (PRA) can be predicted by the chaotic mixing theory described previously. As QD is larger in the PRA than that in the distal region of the acinus (DRA), the ratio QA/QD in the PRA is smaller than that in the DRA. Because a smaller QA/QD ratio is associated with a stronger and larger alveolar recirculation, the flow in the PRA is subjected to more chaotic mixing than the flow in the DRA, resulting in higher deposition. This prediction is consistent with the results of Pinkerton et al. (2000).

At each acinar bifurcation, the volume distal of the daughter ducts will, in general, be unequal in magnitude. This heterogeneity of the acinar tree results in differences in deposition between parallel pathways. The daughter pathway leading to larger lung volume distally will have a smaller QA/QD ratio compared with that of the other pathway. Accordingly, the daughter pathway with the larger distal volume is more strongly influenced by chaotic mixing, and thus NPs deposit more on that side of the bifurcation. This explains the heterogeneity in deposition in parallel pathways in the acinus.

In experimental studies, attempts have been made to visualize the sites of particle deposition on real alveolar walls. As mentioned previously, a wide field scanning protocol developed for SRXTM by Haberthür et al. (2010) was used to image and reconstruct a section of rat lung tissue containing two to three acini in 3D. The branching pattern of an acinus can be visualized by skeletonizing the acinar tree (Haberthür 2010), and the global location of each individual alveolus within the acinus can be registered using Horsfield's airway ordering system (e.g., Horsfield and Cumming 1968). With this technique, therefore, the exact location of each alveolus in the acinus can be identified regardless of the complexity of acinar tree structure, once the acinar tree is skeletonized.

Using SRXTM with a combination of absorption-contrast and phase-contrast modes, deposition sites of 200 nm and 700 nm gold particles were visualized on the unstained rat alveolar tissue (Figure 2.6). The lung tissue sample containing deposited gold NPs was scanned twice. In the first scan, in which the absorption-contrast mode was used, small, very bright spots, representing the gold particles, were detected; the contrast between the gold grains and the alveolar septal tissue allowed for a segmentation of gold particles

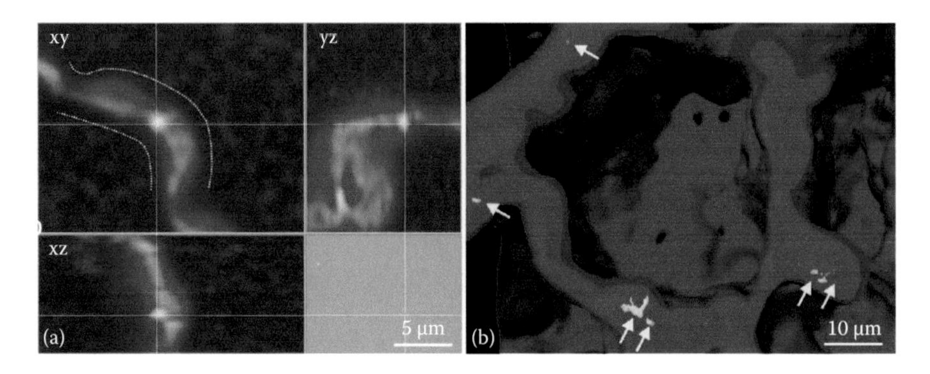

FIGURE 2.6
Submicron gold particles visualized in the alveolar septa using SRXTM with a combination of absorption- and phase-contrast modes. Absorption-contrast mode was used to detect gold particles in the unstained rat alveolar tissue, and phase-contrast mode was used to segment tissues from airspace. (a) Orthogonal slices showing a gold particle (bright spots in the center of the crosshair) inside the alveolar septa. (b) Submicron gold particles (arrows) in the alveolar septa. (Adapted from Schittny JC et al., *AIP Conf. Proc.*, 1365, 384, 2011.)

using a global threshold of black-and-white levels. With a voxel size of approximately 370 nm, the segmentation of even a single 200 nm gold particle was able to be detected. In the second scan, in which the phase-contrast mode was used, the difference of the black-and-white levels between the tissue and the airspace was noticeably larger, allowing the segmentation of the alveolar septa distinct from airspace in 3D with a voxel side length of 370 nm. Combining the absorption-contrast data with the phase-contrast data, therefore, the location of the submicron gold particles can be visualized in the 3D space of the pulmonary acinus.

To visualize NPs—smaller than the resolution of synchrotron—on the alveolar surface requires an electron microscope. We have developed a technique called "multimodal imaging" (Haberthür et al. 2009), which combines synchrotron imaging and electron microscopy (EM). In this technique, high-resolution 3D reconstruction of the acinus based on the synchrotron imaging followed by acinar tree mapping through skeletonization allows us to locate and visualize individual alveoli within an acinus in 3D. Once the global position of the alveolus of interest within the acinar tree is identified, NPs can be visualized on the alveolar surface by taking a series of transmission EM images. Thus, combining the global (acinar) level and the local (alveolar) level (i.e., multimodal imaging), the site of NP deposition can be visualized on the surface of alveolus of interest within the acinar tree.

2.5 Conclusion

There is no doubt that NPs deposit in the lung and that the major location of NP deposition is the acinus—the gas-exchange region of the lung. It is, however, often incorrectly considered that NPs deposit in the acinus solely by diffusion. This misunderstanding is commonly held because the diffusivity of NPs is larger than that of micron-size particles. However, we have shown that this is an incorrect comparison; in fact, the diffusivity of NPs should be compared with the extent of its convective transport (due to airflow). The Péclet number of NPs in the central airways of the acinus is much larger than unity under normal breathing conditions, which clearly indicates that the extent and complexity of the airflow patterns in the acinus are much more important than the diffusivity of NPs when considering the axial transport of NPs along the airways. It is, however, important to note that the balance of convection and diffusion in the alveoli is not well defined by the Péclet number of the central channel flow, and that NPs ultimately need to deviate from the airflow streamlines to deposit on the lung surface; that is, diffusion of NPs for a short distance is crucial for the deposition process.

An acinar flow is typically a low Reynolds number flow, governed by the interaction of pressure drops and viscous forces. Therefore, the unique geometric feature of acinar airways—characterized by a central thoroughfare channel and surrounding dead-end air pockets—is important in determining acinar fluid mechanics. Alveolar flow, which can be characterized by the ratio of QA/QD, is often rotational, causing chaotic mixing. Deposition of NPs associated with chaotic mixing due to the alveolated geometry likely explains experimental observation, which has been classically excluded on theoretical grounds. In addition, facilitated by the recent rapid technological advancements, the visualization of NP deposition on the alveolar walls is now possible by combining HR-SRXTM and EM.

Acknowledgment

This work was supported in part by research grants from the National Institutes of Health (HL054885, HL070542, HL074022, HL094567, and ES000002).

References

Aref H. 1984. Stirring by chaotic advection. *J. Fluid Mech.* 143:1–21.

Bastacky J, Hayes TL, von Schmidt B. 1983. Lung structure as revealed by microdissection. *Am. Rev. Resp. Dis.* 128:S7–S13.

Berg EJ, Weisman JL, Oldham MJ, Robinson RJ. 2010. Flow field analysis in a compliant acinus replica model using particle image velocimetry (PIV). *J. Biomech.* 43(6):1039–1047.

Berg EJ, Robinson RJ. 2011. Stereoscopic particle image velocimetry analysis of healthy and emphysemic alveolar sac models. *J. Biomech. Eng.* 133(6):061004.

Chhabra S, Prasad AK. 2010a. Flow and particle dispersion in a pulmonary alveolus—Part I: Velocity measurements and convective particle transport. *J. Biomech. Eng.* 132(5):051009.

Chhabra S, Prasad AK. 2010b. Flow and particle dispersion in a pulmonary alveolus—Part II: Effect of gravity on particle transport. *J. Biomech. Eng.* 132(5):051010.

Churg A, Brauer M. 2000. Ambient atmospheric particles in the airways of human lungs. *Ultrastruct. Pathol.* 24(6):353–361.

Churg A, Brauer M, del Carmen Avila-Casado M, Fortoul TI, Wright JL. 2003. Chronic exposure to high levels of particulate air pollution and small airway remodeling. *Environ. Health Perspect.* 111:714–718.

Craighead JE, Abraham JL, Churg A, Green FHY, Kleinerman J, Pratt PC, Seemayer TA, Vallyathan V, Weill H. 1982. The pathology of asbestos-associated diseases of the lungs and pleural cavities: Diagnostic criteria and proposed grading scheme. *Arch. Pathol. Lab. Med.* 106(1):544–596.

Craighead JE, Kleinerman J, Abraham JL, Gibbs AR, Green FHY, Harley RA, Ruttner JR, Vallyathan V, Juliano EB. 1988. Diseases associated with exposure to silica and non-fibrous silicate minerals. *Arch. Pathol. Lab. Med.* 112:673–720.

Dailey HL, Ghadiali SN. 2007. Fluid-structure analysis of microparticle transport in deformable pulmonary alveoli. *J. Aerosol Sci.* 38(3):269–288.

Darquenne C, Harrington L, Prisk G. 2009. Alveolar duct expansion greatly enhances aerosol deposition: A three-dimensional computational fluid dynamics study. *Philos. Trans. R. Soc. A* 367:2333–2346.

Davies CN. 1972. Breathing of half-micron aerosols. II. Interpretation of experimental results. *J. Appl. Physiol.* 32:601–611.

Denny E, Schroter RC. 1995. The mechanical behavior of mammalian lung alveolar duct model. *J. Biomech. Eng. Trans. ASME* 117:254–261.

Denny E, Schroter RC. 2000. Viscolelastic behavior of a lung alveolar duct model. *J. Biomech. Eng. Trans. ASME* 122:143–151.

Denny E, Schroter RC. 2006. A model of non-uniform lung parenchyma distortion. *J. Biomech.* 39:652–663.

Filipovic N, Haberthür D, Henry FS, Milasinovic D, Nikolic D, Schittny J, Tsuda A. 2010. Recirculation identified in a 3D alveolar duct reconstructed using synchrotron radiation based X-ray tomographic microscopy (abstr.). ATS Meeting, New Orleans.

Fung YC. 1988. A model of the lung structure and its validation. *J. Appl. Physiol.* 64:2132–2141.

Gehr P, Bachofen M, Weibel ER. 1978. The normal human lung: Ultrastructure and morphometric estimation of diffusion capacity. *Respir. Physiol.* 32:121–140.

Green FHY, Churg A. 1998. Occupational asthma, byssinosis, extrinsic allergic alveolitis and related conditions. In: *Pathology of Occupational Lung Disease* (Churg A, Green FHY, eds.) 2nd ed. Baltimore: Williams & Wilkins, 403–450.
Haber S, Butler JP, Brenner H, Emanuel I, Tsuda A. 2000. Flow field in self-similar expansion on a pulmonary alveolus during rhythmical breathing. *J. Fluid Mech.* 405:243–268.
Haber S, Tsuda A. 2006. Cyclic model for particle motion in the pulmonary acinus. *J. Fluid Mech.* 567:157–184.
Haberthür D, Semmler-Behnke M, Takenaka S, Kreyling WG, Stampanoni M, Tsuda A, Schittny JC. 2009. Multimodal imaging for the detection of sub-micron particles in the gas-exchange region of the mammalian lung. *J. Phys. Conf. Ser.* 186:012040.
Haberthür D. 2010. High resolution to tomographic imaging of the alveolar region of the mammalian lung: A close look deep into the lung. Graduate School for Cellular and Biomedical Sciences. PhD Thesis. Bern, Switzerland: University of Bern.
Haberthür D, Hintermüller C, Marone F, Schittny JC, Stampanoni M. 2010. Radiation dose optimized lateral expansion of the field of view in synchrotron radiation X-ray tomographic microscopy. *J. Synchrotron Radiat.* 17:1–10.
Haefeli-Bleuer B, Weibel ER. 1988. Morphometry of the human pulmonary acinus. *Anat. Rec.* 220(4):401–414.
Harding EM Jr, Robinson RJ. 2010. Flow in a terminal alveolar sac model with expanding walls using computational fluid dynamics. *Inhal. Toxicol.* 22(8):669–678.
Harkema J, Plopper C, Hyde D, St. George J, Wilson D, Dungworth D. 1993. Response of macaque bronchiolar epithelium to ambient concentrations of ozone. *Am. J. Pathol.* 143:857–866.
Henry FS, Butler JP, Tsuda A. 2002. Kinematically irreversible flow and aerosol transport in the pulmonary acinus: A departure from classical dispersive transport. *J. Appl. Physiol.* 92:835–845.
Henry FS, Laine-Pearson FE, Tsuda A. 2009. Hamiltonian chaos in a model alveolus. *ASME J. Biomech. Eng.* 131(1):011006.
Henry FS, Tsuda A. 2010. Flow and particle tracks in model acini. *ASME J. Biomech. Eng.* 132(10):101001.
Heyder J, Blanchard D, Feldman HA, Brain JD. 1988. Convective mixing in human respiratory tract: Estimates with aerosol boli. *J. Appl. Physiol.* 64(3):1273–1278.
Horsfield K, Cumming G. 1968. Morphology of the bronchial tree in man. *J. Appl. Physiol.* 24(3):373–383.
Kleinerman J, Green F, Harley RA, Lapp L, Laqueur L, Laqueur W, Naeye RL, Taylor G, Wiot J, Wyatt J. 1979. Pathology standards for coal worker's pneumoconiosis. (A report of the Pneumoconiosis Committee of the College of American Pathologists.) *Arch. Pathol. Lab. Med.* 101:375–431.
Kleinstreuer C, Zhang Z. 2010. Airflow and particle transport in the human respiratory system. *Annu. Rev. Fluid Mech.* 42:301–334.
Kumar H, Tawhai MH, Hoffman EA, Lin CL. 2009. The effects of geometry on airflow in the acinar region of the human lung. *J. Biomech.* 42(11):1635–1642.
Kumar H, Tawhai MH, Hoffman EA, Lin CL. 2011. Steady streaming: A key mixing mechanism in low-Reynolds-number acinar flows. *Phys. Fluids* 23(4):41902.
Laine-Pearson FE, Hydon PE. 2006. Particle transport in a moving corner. *J. Fluid Mech.* 559:379–390.
Laine-Pearson FE, Hydon PE. 2008. Carousel effect in alveolar models. *J. Biomech. Eng.* 130(2):1–6.
Lee D, Lee J. 2003. Characteristics of particle transport in an expanding or contracting alveolated tube. *J. Aerosol Sci.* 34:1193–1215.
Ma B, Darquenne C. 2011. Aerosol deposition characteristics in distal acinar airways under cyclic breathing conditions. *J. Appl. Physiol.* 110(5):1271–1282.
Mandelbrot BB. 1982. *The Fractal Geometry of Nature.* San Francisco: Freeman.
Niewoehner DE, Kleinerman J, Rice DB. 1974. Pathologic changes in the peripheral airways of young cigarette smokers. *N. Engl. J. Med.* 291(15):755–758.
Oakes JM, Day S, Weinstein SJ, Robinson RJ. 2010. Flow field analysis in expanding healthy and emphysematous alveolar models using particle image velocimetry. *J. Biomech. Eng.* 132(2):021008.
Ottino JM. 1989. *The Kinematics of Mixing: Stretching, Chaos, and Transport.* Cambridge, UK: Cambridge University Press.

Pinkerton KE, Green FH, Saiki C, Vallyathan V, Plopper CG, Gopal V, Hung D, Bahne EB, Lin SS, Ménache MG, Schenker MB. 2000. Distribution of particulate matter and tissue remodeling in the human lung. *Environ. Health Perspect.* 108(11):1063–1069.

Plopper C. 1990. Structural methods for studying bronchiolar epithelial cells. In: *Models of Lung Disease: Microscopy and Structural Methods* (Gil J, ed.). New York: Marcel Dekker, Inc., 537–559.

Plopper CG, Ten Have-Opbroek A. 1995. Anatomical and histological classification of the bronchioles. In: *Diseases of the Bronchioles* (Epler G, ed.). New York: Raven Press, Ltd., 15–25.

Rodriguez M, Bur S, Favre A, Weibel ER. 1987. Pulmonary acinus: Geometry and morphometry of the peripheral airway system in rat and rabbit. *Am. J. Anat.* 180(2):143–155.

Saldiva PH, Clarke RW, Coull BA, Stearns RC, Lawrence J, Murthy GG, Diaz E, Koutrakis P, Suh H, Tsuda A, Godleski JJ. 2002. Lung inflammation induced by concentrated ambient air particles is related to particle composition. *Am. J. Respir. Crit. Care Med.* 165(12):1610–1617.

Schittny JC, Barré SF, Mokso R, Haberthür D, Semmler-Behnke M, Kreyling WG, Tsuda A, Stampanoni M. 2011. High resolution phase-contrast imaging of submicron particles in unstained lung tissue. *AIP Conf. Proc.* 1365:384–387.

Schulz H, Heilmann P, Hillebrecht A, Gebhart J, Meyer M, Piiper J, Heyder J. 1992. Convective and diffusive gas transport in canine intrapulmonary airways. *J. Appl. Physiol.* 72:1557–1562.

Semmler-Behnke M, Kreyling WG, Schulz H, Takenaka S, Butler JP, Henry FS, Tsuda A. 2012. Nanoparticle delivery in infant lungs. *Proc. Natl. Acad. Sci. U.S.A.* 109(13):5092–5097.

Sommerer JC. 1994. Fractal tracer distributions in complicated surface flows: An application of random maps to fluid dynamics. *Physica D* 76:85–98.

Sznitman J, Heimsch F, Heimsch T, Rusch D, Rosgen T. 2007. Three-dimensional convective alveolar flow induced by rhythmic breathing motion of the pulmonary acinus. *ASME J. Biomech. Eng.* 129:658–665.

Sznitman J, Heimsch T, Wildhaber JH, Tsuda A, Rösgen T. 2009. Respiratory flow phenomena and gravitational sedimentation in a three-dimensional space-filling model of the pulmonary acinar tree. *ASME J. Biomech. Eng.* 131(3):031010-1-16.

Tabor M. 1989. *Chaos and Integrability in Nonlinear Dynamics*. New York: Wiley.

Taylor GI. 1960. *Low Reynolds Number Flow. (16mmfilm)*. Newton, MA: Educational Services Inc.

Tschanz SA, Makanya AN, Haenni B, Burri PH. 2003. Effects of neonatal high-dose short-term glucocorticoid treatment on the lung: A morphologic and morphometric study in the rat. *Pediatr. Res.* 53:72–80.

Tsuda A, Henry FS, Butler JP. 1995. Chaotic mixing of alveolated duct flow in rhythmically expanding pulmonary acinus. *J. Appl. Physiol.* 79:1055–1063.

Tsuda A, Rogers RA, Hydon PE, Butler JP. 2002. Chaotic mixing deep in the lung. *Proc. Natl. Acad. Sci. U.S.A.* 99:10173–10178.

Tsuda A, Filipovic N, Haberthür D, Dickie R, Stampanoni M, Matsui Y, Schittny JC. 2008. The finite element 3D reconstruction of the pulmonary acinus imaged by synchrotron X-ray tomography. *J. Appl. Physiol.* 105:964–976.

Tsuda A, Laine-Pearson FE, Hydon PE. 2011. Why chaotic mixing of particles is inevitable in the deep lung. *J. Theor. Biol.* 286:57–66.

Tsuda A, Henry FS, Butler JP. 2013. Particle transport and deposition. *Compr. Physiol.* 3(4):1437–1471.

Tyler WS. 1983. Comparative subgross anatomy of lungs. *Am. Rev. Resp. Dis.* 128:S32–S36.

van Ertbruggen C, Corierib P, Theunissenb R, Riethmuller M, Darquenne C. 2008. Validation of CFD predictions of flow in a 3D alveolated bend with experimental data. *J. Biomech.* 41:399–405.

Weibel ER. 1963. *Morphometry of the Human Lung*. Heidelberg/New York: Springer Verlag/Academic Press.

Wright JL, Cagle P, Churg A, Colby TV, Myers J. 1992. State of the art: Diseases of the small airways. *Am. Rev. Respir. Dis.* 146:240–262.

Section II

Airway/Alveolar Surface

3

Interaction with the Lung Surface

Peter Gehr and Akira Tsuda

CONTENTS

3.1 Introduction

The major target of ambient particulate air pollution is the lung. The lung is the portal of entry for any particulate matter (PM). The relation between increased concentrations of particulate air pollution and adverse health effects in general, as well as in sensitized subjects such as children, the elderly, and diseased people in particular, is documented well and in great detail (Brunekreef and Holgate 2002; Gehr et al. 2009; Latzin et al. 2009; Pope et al. 2009; Perez et al. 2010). Many experimental studies have thus far been performed to investigate how PM can enter the body, organs, tissues, and cells, to determine what effect they may have (Gehr et al. 2009). It is not surprising that it is the lung as an organ, the lung tissue, and lung cells that have been studied with preference to find out why particles can have adverse health effects. In contrast to fine (i.e., micron-sized) particles that are taken up by cells by phagocytosis, nanosized particles (NPs; i.e., particles ≤0.1 μm in diameter) enter cells through an active endocytotic pathway or through a passive mechanism called adhesive interaction (Rothen-Rutishauser et al. 2007b; Wang et al. 2012). Evidence is growing that it is the nanosized fraction of PM that might be considered a particular health risk, on the one hand because of its high content of organic compounds and its pro-oxidative potential due to the high surface-to-volume ratio of the particles as compared with its bulk material (Peters et al. 1997; Oberdörster et al. 2005) and, on the other hand, because NPs could penetrate through tissues and cells and translocate in the lung periphery into capillary blood (Gehr et al. 1990; Schürch et al. 1990; Rothen-Rutishauser et al. 2007b). NPs could then be transported to all organs (Kreyling et al. 2002; Semmler et al. 2004). Many reports have shown that the main cellular effect of exposure to combustion-derived NPs is the production of reactive oxygen species (ROS) causing oxidative stress, which is a major factor in cellular inflammatory reaction and toxicity (Donaldson et al. 2001; Brook et al. 2003).

It was already Leonardo da Vinci (1452–1519), a universal genius, among others a sculptor and anatomist, who added in his *Anatomical Atlas* the short but, particularly for that time, incisive and most meaningful short comment above the picture of a trachea, "dust is harmful." Today, we know that Leonardo da Vinci was correct. There is a growing body of literature that has recently provided evidence that inhalation of airborne PM, particularly in nanosize, contributes substantially to causing adverse health effects (Pope et al. 2009; Puett et al. 2009; Bruske et al. 2010; He et al. 2010).

Many epidemiological studies provide evidence that PM, and particularly its NP fraction, cause adverse health effects associated with increased pulmonary and cardiovascular morbidity and mortality (Pope et al. 1995; Peters et al. 1997, 2001; Wichmann et al. 2000; Oberdörster 2001; Zanobetti et al. 2003; Künzli and Tager 2005; Schulz et al. 2005; Analitis et al. 2006; Dominici et al. 2006; Peel et al. 2011; Peters 2011; Shah et al. 2013). Exposure to elevated levels of PM leads to an increased heart rate and a decreased heart rate variability in the elderly and in more susceptible patients (Luttmann-Gibson et al. 2006). There are recent studies indicating a specific toxicological role of NPs (Borm and Kreyling 2004), and there is an increased body of information available today on their effects on cells, tissues, and organs. It has been described that inhaled combustion-derived NPs provoke oxidative stress, which can cause inflammation as well as oxidative adducts in the epithelium that can contribute to carcinogenesis (for a review, see Donaldson et al. 2005). Particle-induced pulmonary and systemic inflammation, accelerated atherosclerosis, and altered cardiac autonomic function may be part of the pathophysiological pathways, linking particulate air pollution to cardiovascular mortality (Künzli and Tager 2005). Oxidative stress-related pathways may mediate the effect of air pollutants to the cardiovascular system through the efflux of inflammatory cytokines and other mediators of oxidative stress into the blood system (Donaldson et al. 2001; Brook et al. 2003). Experimental investigations with cellular and animal models suggest a variety of possible mechanisms, including direct effects of particle components on the intracellular sources of ROS, indirect effects due to pro-inflammatory mediators released from macrophages stimulated by PM, and neural stimulation after particle deposition in the lungs (reviewed in Gonzàles-Flecha 2004).

In addition to NPs from combustion processes released in large amounts into the air, water, and soil, there are progressively more so-called engineered NPs from nanotechnology released into these systems (Mazzola 2003; Paull et al. 2003). All these engineered NPs have unique chemical and physical properties. They do not behave like solids, liquids, or gases of the same materials. New properties emerge that are not exhibited by the bulk of particles of the same chemical composition. These properties include different colors and different electronic, magnetic, and mechanical properties, any or all of which may be altered at the nanoscale. They are used in the energy, electronics, pharmaceutical, medicine, cosmetics, surface treatment, insulation, construction, automobile, tire, textile, food, and many other industries.

3.2 Structure of the Lung

Understanding how inhaled PM can affect health begins with understanding the structure and function of the respiratory system, which is the main portal of entry for PM, including NPs.

Since gas exchange is the major function of the lung, its morphology is particularly designed for this function, that is, for the exchange of oxygen and carbon dioxide. To let these two gases exchange efficiently by diffusion, the surface over which diffusion has to take place needs to be large and the tissue barrier between air and blood, the diffusion barrier, small (Weibel 1963; Gehr et al. 1978). In the lung, this is accomplished by bringing blood and air in very close proximity over a dense network of capillaries in the wall of a large number of alveoli in the gas-exchange parenchyma. In an adult human lung, owing to approximately 500 million alveoli (Ochs et al. 2004), the surface of the alveolar wall (i.e., the air–blood tissue barrier) is approximately 150 m^2 (Gehr et al. 1978; Stone et al. 1992). The thin air–blood tissue barrier is, on average, 2 μm in thickness; however, more than approximately half of the surface area is only a few hundred nanometers. It consists of the alveolar epithelium, an interstitial connective tissue space, and the capillary endothelium (Figure 3.1). An uneven alveolar surface is smoothed out by the aqueous hypophase (see Figure 3.2a). At the liquid–air interface, that is, between the liquid phase (hypophase) and the air phase, there is a continuous surfactant film.

The structural and functional barriers (Figure 3.3) that protect the respiratory system against innocuous and harmful particulate material are discussed here in more detail (for a review, see Gehr et al. 1978; Geiser et al. 2005; Nicod 2005; Rothen-Rutishauser et al. 2007b). The vast internal surface area of the lungs and the thin air–blood tissue barrier are facilitating an easy access to the lung tissue, including cells of the defense system and inhaled PM that reaches the alveoli. NPs may enter the blood capillaries and then translocate to other organs. The series of these barriers includes (from the alveolar air to the capillary blood)

1. The surfactant film (Gil and Weibel 1971; Gehr et al. 1990; Schürch et al. 1990)
2. The aqueous surface lining layers in the airways, including the mucociliary escalator, and in the alveoli where it is called hypophase (Kilburn 1968)
3. A population of macrophages (professional phagocytes) on the epithelial cell layer in the airways and in the alveoli (Brain 1988; Lehnert 1992)
4. The epithelial cell layer endowed with tight junctions between the cells (Schneeberger and Lynch 1984; Godfrey 1997)
5. A population of dendritic cells (professional antigen-presenting cells) mainly in the upper airways and in the alveoli at the base of the epithelial cell layer and between the epithelial cells (Holt et al. 1990; review by Blank et al. 2008)
6. The basal lamina (basement membrane) under the epithelial and endothelial cells, which may fuse to a single thicker basement lamina in certain areas of the alveoli, approximately more than half of the gas-exchange surface area (Timpl and Dzladek 1986; Yurchenco et al. 1986; Maina and West 2005)
7. The capillary endothelium (Schneeberger 1977; Dudek and Garcia 2001)

3.3 Epithelial Defense System

Despite the existence of these barriers, respiratory diseases are frequent and increasing (Peters et al. 1997; Wichmann et al. 2000; Schulz et al. 2005). More attention has been

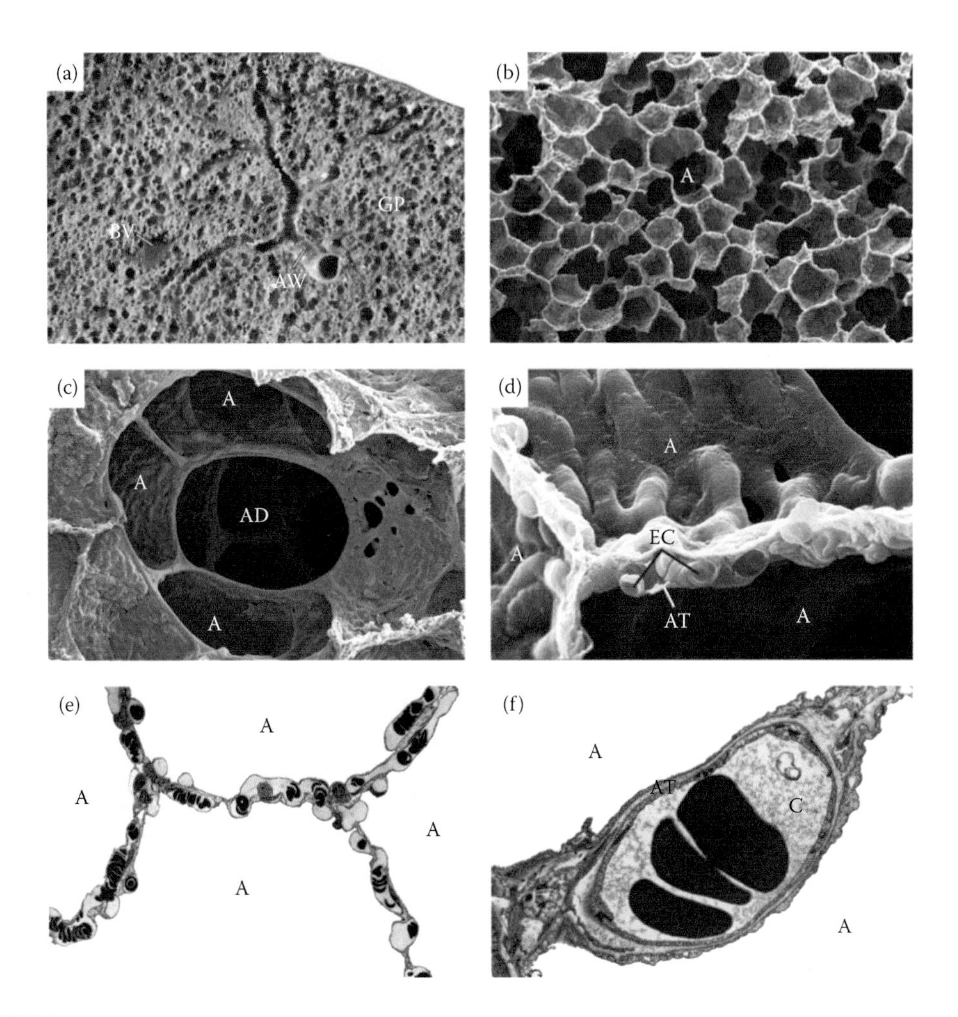

FIGURE 3.1
Structure of human lung. (a) Lung slice at low magnification (15×), showing gas-exchange parenchyma (GP), blood vessels (BV), and airways (AW). (Reprinted with permission from the Institute of Anatomy, University of Bern, Bern, Switzerland.) (b) Scanning electron micrograph of the gas-exchange parenchyma with alveoli (A); magnification, 60×. (Reprinted from *Comprehensive Toxicology*, ed. C.A. McQueen, 2nd edition, L.-Y. Chang, J.D. Crapo, P. Gehr, B. Rothen-Rutishauser, C. Mühlfeld, and F. Blank. Alveolar epithelium in lung toxicology, 59–91, Copyright 2010, with permission from Elsevier.) (c) Scanning electron micrograph of an alveolar duct (AD) with concentrically arranged alveoli (A) around it; magnification, 180×. (Reprinted from *Respir. Physiol.*, 32, P. Gehr, M. Bachofen, and E.R. Weibel, 121, Copyright 1978, with permission from Elsevier.) (d) Scanning electron micrograph of an interalveolar septum showing erythrocytes (EC) in a capillary and the thin air–blood tissue barrier (AT); A, alveoli; magnification, 550×. (Reprinted from *Respir. Physiol.*, 32, P. Gehr, M. Bachofen, and E.R. Weibel, 121, Copyright 1978, with permission from Elsevier.) (e) Transmission electron micrograph of interalveolar septa, showing capillaries with erythrocytes (black) meandering around the connective tissue frame; A, alveolar air; magnification, 500×. (Reprinted from *Respir. Physiol.*, 32, P. Gehr, M. Bachofen, and E.R. Weibel, 121, Copyright 1978, with permission from Elsevier.) (f) Transmission electron micrograph of a capillary (C) with erythrocytes (black) in an interalveolar septum; the three layers of the air–blood tissue barrier (AT) can be clearly seen; A, alveolar air; magnification, 4500×. (Reprinted from *Comprehensive Toxicology*, ed. C.A. McQueen, 2nd edition, L.-Y. Chang, J.D. Crapo, P. Gehr, B. Rothen-Rutishauser, C. Mühlfeld, and F. Blank. Alveolar epithelium in lung toxicology, 59–91, Copyright 2010, with permission from Elsevier.)

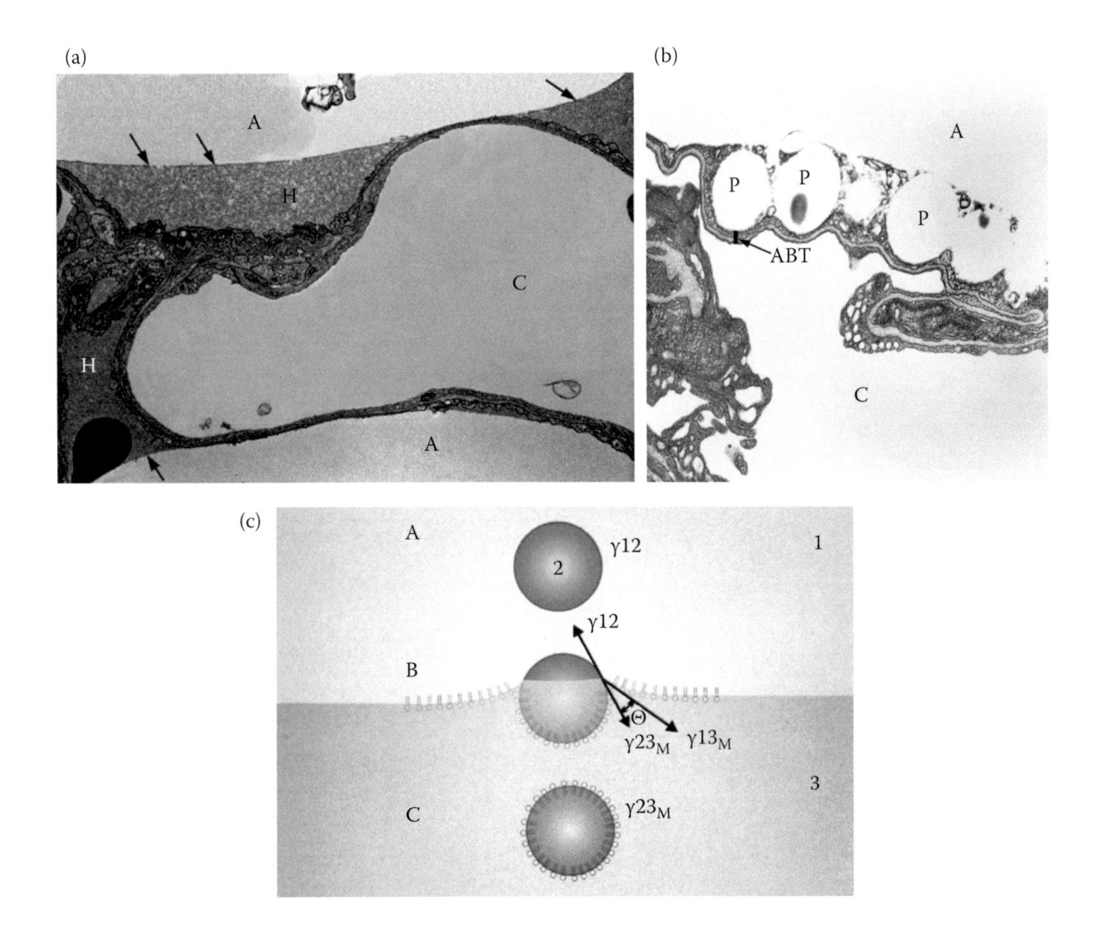

FIGURE 3.2
Particle–surfactant film interaction. (a) Electron micrograph of interalveolar septum of human lung showing aqueous hypophase (H) and surfactant (arrows). A lung lobe was fixed by intravascular perfusion with glutaraldehyde. A, alveolar air; C, capillary lumen; magnification, 7000×. (Reprinted from P. Gehr, Institute of Anatomy, University of Bern, Bern, Switzerland, 2014. With permission.) (b) Electron micrograph of 1-μm polystyrene particles deposited on the surface of the alveolar wall, after inhalation by hamster. The lung was fixed by intravascular perfusion with glutaraldehyde. They had been deposited on surfactant and displaced into the depth by surface and line tension forces, which were so strong that the particles caused indentations into the capillary. P, particle; A, alveolar air; ABT, air–blood tissue barrier; C, capillary lumen. Magnification, 11,000×. (Reprinted from *Respir. Physiol.*, 80, S. Schürch, P. Gehr, V. Im Hof, M. Geiser, and F. Green, 17, Copyright 1990, with permission from Elsevier.) (c) Schematic drawing of the interaction of a particle with the surfactant film at the air–liquid interface (internal lung surface) and the subsequent displacement of the particle by surface (and line tension) forces exerted on it by surfactant (wetting); during the displacement, the particle might be coated with a film of surfactant or surfactant components; phases: 1, air; 2, solid (particle); 3, liquid; forces: $\gamma 12$, surface tension between air phase and solid phase; $\gamma 13_M$, surface tension between air phase and liquid phase; $\gamma 23_M$, surface tension between solid phase (particle coated with a DPPC monolayer) and liquid phase; Θ, contact angle at three-phase point; A through C, steps of thermodynamically driven particle displacement. (Reprinted from *Comprehensive Toxicology*, ed. C.A. McQueen, 2nd edition, L.-Y. Chang, J.D. Crapo, P. Gehr, B. Rothen-Rutishauser, C. Mühlfeld, and F. Blank. Alveolar epithelium in lung toxicology, 59–91, Copyright 2010, with permission from Elsevier.)

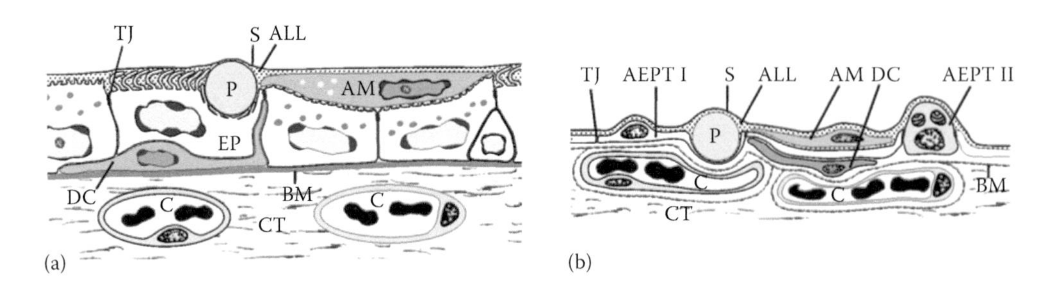

FIGURE 3.3
Air–blood tissue barrier system in the airways (a) and in the alveoli (b). Alveolar epithelial cells of type I (AEPT I), alveolar epithelial cells of type II (AEPT II), aqueous lining layer (ALL), airway/alveolar macrophage (AM), basement membrane (BM), capillary (C), connective tissue (CT), dendritic cell (DC), epithelial cell (EP), particle (P), surfactant (S), tight junction (TJ). (Reprinted from B. Rothen-Rutishauser, S. Schürch, and P. Gehr. Interaction of particles with membranes. In: *The Toxicology of Particles*, eds. K. Donaldson and P.J. Borm, 139–160. Taylor & Francis Group, LLC, CRC Press, Boca Raton, FL, 2007. With permission.)

directed toward elucidating how and when the antigens evade these barriers. Insoluble particles deposited in the airways are largely cleared by the mucociliary activity, but not all deposited particles are removed by this mechanism. The fate of these particles depends on their physicochemical characteristics and the nature of their interaction with the surfactant film at the air–liquid interface. All inhalable particles (PM_{10}: PM with a diameter ≤10 μm), including NPs deposited on the internal surface of the lung, are displaced into the aqueous subphase below the surfactant film and may be coated with surfactant or surfactant components during the displacement process and hence perhaps modified (Gehr et al. 1990, 1996; Schürch et al. 1990; Geiser et al. 2003a,b; Mijailovich et al. 2010). As a result of the displacement, particles may come into close association with the epithelium and the epithelial defense system (i.e., important cells in close association with the epithelium), such as alveolar macrophages and dendritic cells (Gehr et al. 1990, 1996; Schürch et al. 1990; Geiser et al. 2003a, 2005). Dendritic cells are the most competent antigen-presenting cells in the lung (Holt et al. 1990; Nicod 1997; reviewed by Blank et al. 2008). One of the most crucial specialized functions of dendritic cells is capturing antigens and delivering them to local lymphoid tissues (Blank et al. 2007). Their unique responsibility is to decide whether to present a sampled antigen in an immunogenic or tolerogenic way (reviewed by Vermaelen and Pauwels 2005).

In vivo, alveolar macrophages occupy the luminal aspect, while immature dendritic cells occupy the basal aspect of the epithelium (Brain et al. 1984; Geiser et al. 1994, 2005; Blank et al. 2007). Dendritic cells are located within the basement lamina and reach the tight junctions of the epithelium with slender cytoplasmic processes pushed up between the epithelial cells. They may, eventually, penetrate through the tight junctions into the luminal space of airways or alveoli to catch antigen material (Figure 3.3). If antigen is taken up, the dendritic cells mature; they process the antigen material (proteins chopped into peptides), transport the peptides to the MHC molecule at the outer cell surface, then detach from the epithelial layer and move as mature dendritic cells to T-lymphocytes to which they present the antigen material (peptides) that may trigger an immune reaction (Holt and Schon-Hegrad 1987; McWilliam et al. 2000; Blank et al. 2008).

By using confocal laser scanning light microscopy and transmission electron microscopy, it has been demonstrated that dendritic cells are efficiently phagocytic for a variety of particles, such as polystyrene particles (Matsuno et al. 1996; Kiama et al. 2001; Thiele et al. 2001), puffball spores (Geiser et al. 2000), biodegradable microspheres (Walter et al. 2001), and *Salmonella typhimurium* (Dreher et al. 2001). Although both macrophages and dendritic cells are derived from circulating blood monocytes, macrophages are twice as phagocytic as immature dendritic cells (Kiama et al. 2001). Whereas much is known on the interaction of particles with dendritic cells and with macrophages, hardly anything is known on how the antigens reach the dendritic cells and if macrophages and epithelial cells are involved. The transport of particles to the dendritic cells presupposes their passage probably between the epithelial cells, through the tight junctions. However, it is not known yet how dendritic cells open the tight junctions, which "seal" the airway epithelium at the apical side (Holt 2005). It could be shown in gut mucosa that subepithelial dendritic cells were capturing antigens outside the epithelium by extending fine cytoplasmic processes through the tight junctions (Rescigno et al. 2001; Niess et al. 2005). Dendritic cells in the lungs may behave in the same way (Geiser et al. 2005; Rothen-Rutishauser et al. 2005, 2007a; Blank et al. 2007, 2008). Vermaelen and colleagues (2001) found in an *in vivo* study that fluorescein isothiocyanate–conjugated macromolecules are transported to the tracheal lymph nodes by airway dendritic cells after intratracheal instillation. The mechanism by which the macromolecules pass through the epithelium to reach the dendritic cells was, however, not addressed. Takano and colleagues (2005) showed that dendritic cells can easily collect antigens beyond the epithelial tight junctions in human nasal mucosa, but in allergic rhinitis only. Another *in vitro* model using mouse tracheal epithelial cells and mouse bone marrow dendritic cells showed impaired migration of metalloproteinase-9-deficient dendritic cells through tracheal epithelial tight junctions (Ichiyasu et al. 2004); however, the mechanism of whole dendritic cell migration through the lung epithelium remains a mystery. Since they are found in bronchoalveolar lavage fluid, they seem to be able to translocate into the luminal space as whole cells. There is evidence, however, from *in vitro* studies with the triple-cell coculture model that dendritic cells collect particles on the luminal side of the epithelium, eventually also from macrophages, and transport them through the cytoplasmic processes to the basal side of the epithelium (Rothen-Rutishauser et al. 2005; Blank et al. 2007). No publications describe this kind of dendritic cell behavior in diseases. There is evidence that dendritic cells play an important role in the pathogenesis of allergic asthma, and an increased number of dendritic cells are found in the airway mucosa of patients with chronic obstructive pulmonary disease (COPD) (Vermaelen and Pauwels 2005).

After the antigens have passed freely through the epithelial barrier, they may pass through the basement membrane and subsequently through the subepithelial connective tissue layer and eventually come into contact with endothelial cells lining the capillaries. Since endothelial cells play an important role in inflammation processes (Michiels 2003), particles can have some effect on endothelial cell function and viability, inducing pro-inflammatory stimuli. It has been proposed that the permeability of the lung tissue barrier to NPs is controlled at the epithelial and endothelial levels (Meiring et al. 2005). Particulate antigens of nanosize are able to cross the air–blood tissue barrier of the lung, and thus can enter the circulatory system (Nemmar et al. 2002; Geiser et al. 2005; Rothen-Rutishauser et al. 2007a,b). Antigens that have passed through the epithelial and endothelial barriers are transported by the blood circulation and reach other organs such as the liver, heart, digestive tract, spleen, kidney (Brown et al. 2002; Kreyling et al. 2002), and brain (Oberdörster et al. 2004), or, as a matter of fact, any organ.

The series of tissue barriers may not be as effective in protecting the body against NPs. It was shown that these minuscule particles penetrate through tissues and cells and may eventually reach the capillaries as already mentioned, and nerves embedded in the subepithelial connective tissue (Figure 3.3). Once NPs reach the capillaries, they translocate into other organs where they may leave the blood and penetrate into the tissue and cells (Kreyling et al. 2002; Oberdörster et al. 2002). In the brain, they may clear the blood–brain barrier since they can obviously pass through endothelial cells and astrocytes and enter neurons of the brain (Oberdörster et al. 2004).

NPs inhaled through the nose can directly enter the endings of nerves exposed to air in the olfactory region of the upper nasal cavity, and may travel to the brain along the axons of the olfactory nerves by their very efficient axonal microtubular transport system (Oberdörster et al. 2004; Elder et al. 2006). Deposition on the internal nose surface, translocation into olfactory nerve endings, and transport to the brain along the axons depend on the size of the NPs. Nasal deposition is relatively small (Garcia and Kimbell 2009). On the one hand, NPs have been considered for nose-to-brain drug delivery; this, however, needs a lot of additional careful research (Mistry et al. 2009). On the other hand, it has been shown in animal experiments that diesel exhaust NPs deposited in the nose may travel to the olfactory bulb and there change gene expression (Yokota et al. 2013).

3.4 Air–Blood Tissue Barrier

The air–blood tissue barrier protects the blood circulating in capillaries in the lung alveoli from direct exposure to inhaled and deposited airborne toxicants and pathogens. Two types of cells, type I and type II pneumocytes, form the alveolar epithelium. The wall of each of the 500×10^6 alveoli of a human lung (Ochs et al. 2004) is lined, on average, by 40 alveolar type I cells and 67 alveolar type II cells (Stone et al. 1992; Mercer et al. 1994). The alveolar type I epithelial cells are the major component of the barrier. They are large squamous cells covering >95% of the total alveolar surface area of the lung. Its average thickness is approximately a few hundred nanometers (Gehr et al. 1978; Geiser et al. 2005; Blank et al. 2007). The functions of type I epithelial cells, other than forming a low permeability barrier, are poorly defined. Owing to the difficulty of isolating type I cells for *in vitro* studies, the metabolic capacity of these cells is unknown. Type II cells are cuboidal in shape, having approximately 1/5 of the mean cell size and 1/40 of the mean alveolar surface area of a type I cell. Type II cells are described to be preferentially located at so-called alveolar corners (Mensah et al. 1996). A distinguishing ultrastructural feature of type II cells, the producers of surfactants, is their lamellar bodies containing stacks of tightly packed surfactant phospholipids and surfactant proteins in a lamellar configuration (Williams 1977).

An aqueous surface lining layer, the hypophase, covers the whole alveolar surface area, smoothing out all unevenness of the epithelial surface of the alveoli. The aqueous phase–air phase interface consists of a continuous film of surfactant (from surface active agent) that consists of phospholipids (mostly dipalmitoylphosphatidylcholine [DPPC], other phospholipids, and neutral lipids) and proteins (i.e., four surfactant proteins: SP-A, SP-B, SP-C, and SP-D) (Possmayer 1988; review by Whitsett 2010). The hydrophobic SP-B and SP-C are involved in the formation of the surfactant film at the air–liquid interface (Schürch et al. 2010; Parra et al. 2011; Chavarha et al. 2013), whereas the hydrophilic SP-A and SP-D are involved in defense processes (Lawson and Reid 2000; Wright 2005; Jakel et al. 2013).

PM that is deposited on the internal lung surface first encounters the surfactant film (Figure 3.2a). The surfactant displaces the particles deposited on it into the surface lining layer of the airways and the hypophase of the alveoli (Figure 3.2b and c). These particles are displaced by wetting forces, that is, by surface and line tension forces exerted on them by the surfactant film. Displacement due to line tension is much stronger for smaller particles. Surface and line tension forces are so strong that the particles may cause indentations into the capillaries (Figure 3.2b) (Gehr et al. 1990; Schürch et al. 1990). By this displacement process, the particles may be coated with surfactant or surfactant components (Figure 3.2c) and brought into close association with epithelial cells and cells of the defense system, macrophages, and dendritic cells. The surfactant coating may alter the physical properties of the particles, reduce particle toxicity, and enhance phagocytosis by opsonization (Gehr et al. 1996). NPs may then penetrate through the epithelial cells and enter the connective tissue (Figure 3.4) from where they may penetrate through the endothelial cell layer and enter the capillaries (Figure 3.5). By blood circulation, the particles can be distributed throughout the whole organism (Gehr et al. 1990, 1996; Schürch et al. 1990; Kreyling et al. 2002; Oberdörster et al. 2002; Semmler et al. 2004; Choi et al. 2010).

The interaction of NPs with lipid and protein components of the pulmonary surfactant and the physicochemical properties of the surfactant film are important in understanding how inhaled and deposited NPs affect pulmonary function (Tatur and Badia 2012). SP-A and SP-D have been shown to enhance the binding, uptake, and clearance of NPs by macrophages and dendritic cells (Ruge et al. 2011, 2012; Kendall et al. 2013). However, it was shown that SP-B, along with SP-C, is vital to surfactant function (Bakshi et al. 2008).

The interaction of PM and, in particular, NPs with biological systems in general may also pose a point of nanosafety (Ahluwalia et al. 2013). The fate and behavior of NPs after interacting with biological systems (i.e., at the bio–nano interface) depends on coating, translocation, signaling, and kinetics. It is the coating, that is, the creation of a corona on the NP surface (discussed in Chapters 4 and 5), which will be mentioned here in particular. NPs that are inhaled may have already been coated in the environment. They will be further coated when encountering the first structure after inhalation and deposition, the surfactant film, by surfactant or surfactant components. Furthermore, they may be even further coated, or the biomolecule corona exchanged or altered, when penetrating

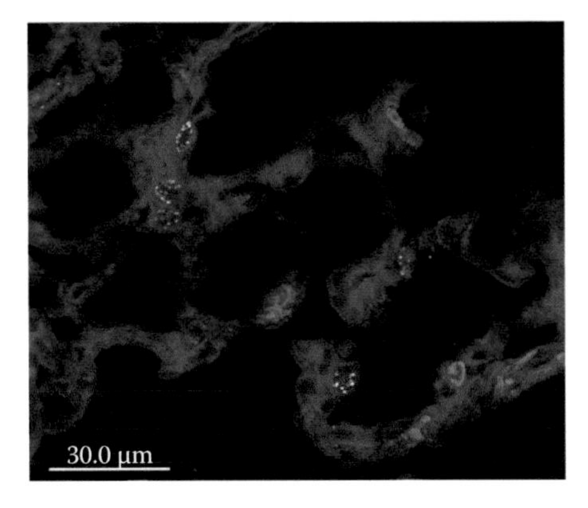

FIGURE 3.4
Lung parenchyma of a BALB/c mouse that inhaled 100-nm polystyrene NPs (yellow). The interalveolar septa (red, actin network) show polystyrene NPs that may have entered cells and connective tissue after displacement by surfactant. Macrophages and dendritic cells are stained blue. (Reprinted from F. Blank, Department of Clinical Research, University of Bern, Bern, Switzerland, 2014. With permission.)

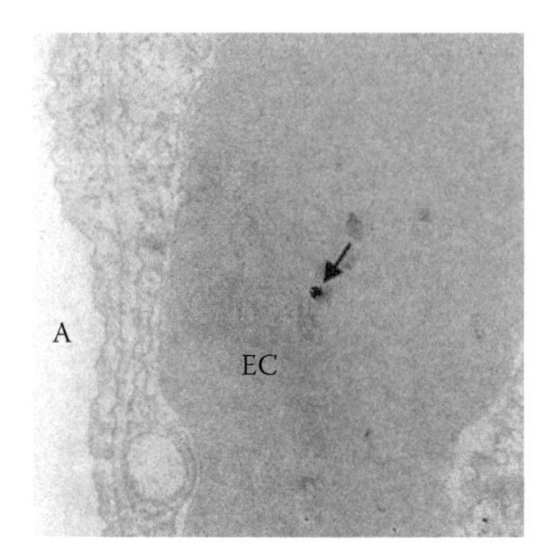

FIGURE 3.5
Energy-filtering transmission electron micrograph of an erythrocyte (EC) in a capillary of a rat lung after inhalation of a TiO_2 aerosol, containing a 22-nm TiO_2 particle (arrow); A, alveolar air. Magnification, 25.000×. (Reprinted and modified from M. Geiser, B. Rothen-Rutishauser, N. Kapp et al., *Environ. Health Perspect.*, 113, 1555, The National Institute of Environmental Health Sciences, National Institutes of Health, U.S. Department of Health and Human Services, 2005. With permission.)

deeper and deeper into the body, by interaction with proteins and other biomolecules during each step of penetration. It has been increasingly recognized that NPs could absorb proteins and other biomolecules (e.g., DPPC of the surfactant upon deposition on it) from the microenvironment they are interacting with, in order to lower their surface free energy (Rivera Gil et al. 2010), with important consequences for the interaction with biological systems in general. The corona composition may have an influence on the interaction or the interaction mechanism of NPs with lung cells. It is, however, not yet clear whether the biomolecule layer of the NPs really mediates the binding to cells, that is, to the cell surface (Ahluwalia et al. 2013). Understanding the biointerface is the key to understanding the mechanisms of uptake of NPs when penetrating through pulmonary tissue after deposition on the surfactant film on the internal lung surface.

The type I epithelium is both a primary target during any insults to the lung and a component in the front line of lung defense for inhaled pathogens and toxicants, including PM. Inhaled particulates that are deposited on the gas-exchange surface (i.e., on the surfactant film) are separated from the blood by only the hypophase with the surfactant film, the epithelial layer, the interstitium, and the endothelial layer. The interstitium may be reduced to the merged basement laminae of the endothelial and epithelial layers, which makes the air–blood tissue barrier as thin as a few hundred nanometers only. The hypophase is very variable in thickness, smoothing out even deep indentations in the interalveolar septa. Hence, it is of very variable thickness between several microns (in indentations) and a few nanometers only (over capillaries bulging into the alveolar space: personal observation; see also Figures 3.6 and 3.2a).

The aqueous hypophase also contributes to the protection of the alveolar epithelium. The surfactant film is capable of suppressing the oxidative burst and effector functions of alveolar macrophages (Toews et al. 1984; Hayakawa et al. 1989; Winkler and Hohlfeld 2013). The ability of fibers to activate alveolar macrophages is reduced by coating them with surfactant (Jabbour et al. 1991; Gasser et al. 2012; Mehra et al. 2014). Two of the surfactant proteins, SP-A and SP-D, as already mentioned, also have potential functions in host defense against invading pathogens (Lawson and Reid 2000; Jakel et al. 2013). The presence of nitric oxide ($NO^{\bullet}$) radical, extracellular superoxide dismutase (EC-SOD) (Kwon et

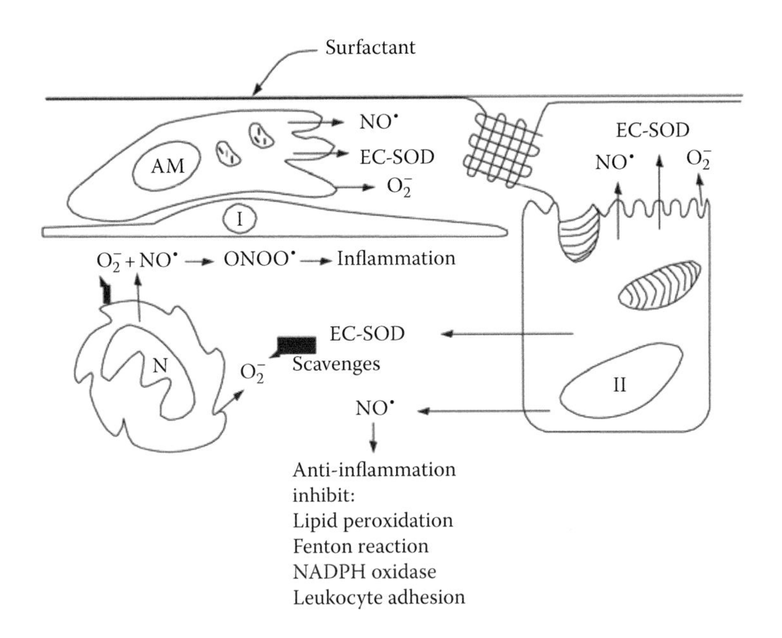

FIGURE 3.6
Proposed scheme of type II cell (II)-mediated anti-inflammatory situation on the alveolar surface. Superoxide (O_2^-) is generated by membrane oxidases on either epithelial or inflammatory cells. Nitric oxide (NO•) is known to be released by alveolar type II cells, alveolar macrophages, and neutrophils (N). The reaction between NO• and O_2^- leads to the formation of peroxynitrite ($ONOO^-$), a potent pro-inflammatory mediator. However, EC-SOD secreted by type II cells and alveolar macrophages will scavenge extracellular O_2^-. Nitric oxide, if protected from reaction with O_2^-, can inhibit inflammation. In addition to EC-SOD and NO•, surfactant plays an anti-inflammatory role by inhibiting leukocyte activation and particle-induced O_2^- production. (Reprinted from *Comprehensive Toxicology*, ed. C.A. McQueen, 2nd edition, L.-Y. Chang, J.D. Crapo, P. Gehr, B. Rothen-Rutishauser, C. Mühlfeld, and F. Blank. Alveolar epithelium in lung toxicology, 59–91, Copyright 2010, with permission from Elsevier.)

al. 2012), and surfactant creates a strong anti-inflammatory milieu on the alveolar surface. The proposed mechanisms for the type II cell–regulated anti-inflammatory milieu on alveolar surfaces are illustrated in Figures 3.3 and 3.6 (Rothen-Rutishauser et al. 2007b; Chang et al. 2010).

Excessive accumulation of alveolar fluid leads to alveolar flooding and impaired gas transport. Active absorption of alveolar fluid is therefore critical in preventing respiratory failure. The alveolar epithelium actively transports sodium and water from the surface fluids to the interstitial space (Voelker and Mason 1989). The ability of the alveolar epithelium to engage in transepithelial transport was first suggested by the observation of domes that were formed in primary cultures of type II epithelial cells (Mason et al. 1982).

Alveolar epithelial cell injury can derive initially from physical and chemical interactions of the epithelial cells with an injurious substance. Most of the reactions involved in lung epithelial cell toxicity caused by reagents are not specific to alveolar epithelial cells. The reactive oxidant gases ozone (O_3) and nitrogen dioxide (NO_2) exhibit specificity for the alveolar epithelium because of their reactivity. In the case of O_3, the major cellular target was found to be the plasma membrane (Pryor 1992). Deposited fine particles from dust, combustion-derived ultrafine particles, or manufactured NPs and pathogens may penetrate into cells and initiate cytotoxic, immunological, or pathogenic reactions (Schins 2002; Donaldson et al. 2005, 2010; Unfried et al. 2007; Han et al. 2012).

3.5 Interaction of Particulates with the Internal Pulmonary Surface: Consideration of Health Effects and Nanotoxicological Aspects

It is very important to collect as many risk data as possible, in particular health risk data, to address problems and answer questions early in the stage of the development of new technologies (Hoet et al. 2004). The rapid proliferation of many different engineered nanomaterials (NMs) as well as NPs generated by combustion processes requires defined screening strategies for the characterization of the potential human health effects from exposure to NM (Oberdörster et al. 2005). Most of the concerns regarding NM stem from the experiences with NPs being substantially more inflammatory and toxic than fine particles (Ferin et al. 1992; Li et al. 2003). Even within the NPs, the smaller ones may be more toxic than the larger ones, as was very recently demonstrated (Shang et al. 2014).

Since the lung has a large surface area (approximately 150 m^2; Gehr et al. 1978) that is in constant contact with inhaled air, it is vulnerable to airborne pollutants (i.e., particulates and chemicals). Complex environmental pollutants, such as car exhaust, cigarette smoke, pollutant gases (NO_2, O_3, and others), herbicides and pesticides (e.g., paraquat), organic chemicals (e.g., formaldehyde) and solvents (e.g., hexane), minerals (e.g., silica), and metals and metal oxides (e.g., Cd, Co_3O_4), have been shown to produce alveolar epithelial injuries. In addition, the entire cardiac output flows through pulmonary capillaries that are in close proximity to the epithelium. Parenteral administration of compounds, for example, ingested herbicides (paraquat), food additives (butylated hydroxytoluene), drugs (bleomycin), alkaloids (monocrotaline), and radiation, could also cause lung injuries (Kehrer and Kacew 1985; Shimabukuro et al. 2003; Federal Office for the Environment [FOEN] 2010, 2013). On the other hand, epithelial injuries as a consequence of exposures to drugs or radiation may mediate the epithelial–mesenchymal transition and initiate the development of interstitial lung disease.

The lung is one of the major sites of interaction with environmental particulates. PM consists of three fractions: a coarse fraction, $PM_{10-2.5}$ (2.5–10 μm); a fine fraction, $PM_{2.5-0.1}$ (2.5–0.1 μm); and a nanosize fraction, $PM_{0.1}$ (≤0.1 μm). The coarse fraction is considered responsible for hazards in the larger airways, whereas $PM_{2.5}$ (i.e., particles ≤2.5 μm) reaches the gas-exchange region. However, only the smallest fraction, the nanosize particles (i.e., $PM_{0.1}$), could penetrate in the alveoli through the air–blood tissue barrier into the capillaries, after being deposited on the alveolar wall (Geiser et al. 2005; Rothen-Rutishauser et al. 2005, 2006). Besides entering tissues and cells in the lung, facilitated by the minute size, these particles may translocate through the blood circulation to all other organs (Kreyling et al. 2002; Oberdörster et al. 2002; Semmler et al. 2004). Therefore, it is this fraction of PM that should be given particular attention as far as adverse health effects are concerned. Those NPs that are inhaled through the nose may be transported via the olfactory nerves directly into the brain (Oberdörster et al. 2004). By number, NPs make the biggest part of the fine particles. They have a longer lifetime in the atmosphere and can be transported over thousands of kilometers (Hinds 1999).

The concentration of fine particles and NPs in the air is estimated to be 5000–10,000 particles cm^{-3} in rural atmospheres; however, it can be as high as 150,000 particles cm^{-3} or even more in urban or heavy-traffic areas (highways) (Seaton et al. 1995; Seipenbusch et al. 2008). Assuming a tidal volume of 700 ml, 12–16 breaths min^{-1}, and a 10% deposition of particles in alveoli, the alveolar region of a human lung could, therefore, be exposed to as many as approximately 200 billion particles over 24 h. These particles may carry different kinds of chemicals on their surface. To maintain the integrity of the delicate alveolar epithelium, a potent defense mechanism is required. Particle clearance by alveolar macrophages is one of the protective mechanisms for the alveolar epithelium. The ability of alveolar type II cells to renew injured type I epithelium is another important protective mechanism (Marsh et al. 2009). Alveolar type II cells appear also to be responsible for maintaining an anti-inflammatory environment on the alveolar surface that is important in limiting neutrophil-mediated amplification of epithelial injury (Qian et al. 2006). There is evidence that the alveolar surface lining fluid and the type II epithelial cells play a primary role in pulmonary defense. The antioxidant enzyme MnSOD is found in all lung cells (Chang et al. 1995; Gehr et al. 1996; Caraballo et al. 2013). As already mentioned, SP-A and SP-D might also be involved in defense processes (Lawson and Reid 2000; Jakel et al. 2013).

The tissue barriers of the respiratory system may not be able to prevent NPs from entering the body. The fate of these minute particles once deposited is different from that of larger particles. It has been shown that TiO_2 NPs of only about 20–30 nm in diameter are able to cross cellular membranes in a rat lung exposure model that did not usually involve commonly known phagocytic or endocytic mechanisms (Geiser et al. 2005; Rothen-Rutishauser et al. 2007b). As a number of these particles were not membrane bound in the cytoplasm of all lung cells, and since they were found in erythrocytes (Figure 3.5), other mechanisms of translocating into cells were considered for these particles. An active process, a diffusion or signal-mediated transport through membrane pores, or a passive process, a translocation by electrostatic, van der Waals forces, or steric interactions subsumed under the expression "adhesive interaction" is possible. Furthermore, wetting forces, as it was suggested for the particle displacement by surfactant (Gehr et al. 1990; Schürch et al. 1990; Mijailovich et al. 2010), or thermal capillary fluctuations that eventually enhance particle transport through cell membranes (Rothen-Rutishauser et al. 2007a) may be considered as well. Most recently, a mechanism has been described by which NPs were shown to penetrate through cellular membranes passively. Through mechanical NP–cell membrane interaction, the cell membrane could become more flexible, allowing NPs to sneak through temporal spaces between the phospholipids (Wang et al. 2012).

A small fraction of displaced NPs is transported by one or the other of these mechanisms from the air side to the connective tissue (Figure 3.4) and released into the systemic circulation within less than 1 h (Mühlfeld et al. 2007). As NPs were also found in erythrocytes of the capillaries (Figure 3.5), it is not surprising that in other studies, NPs could be found in a number of other organs within a few hours after deposition in the lungs (Kreyling et al. 2002; Oberdörster et al. 2002, 2004; Semmler et al. 2004; Choi et al. 2010).

A substantial number of publications demonstrated the molecular toxicity of combustion-derived NPs, which pose a hazard to the lung through their potential to provoke an oxidative stress reaction. Reactive mechanisms mediated by oxygen species in low concentration not only may cause inflammation but also increase calcium concentrations, which can activate the transcription factors nuclear factor-κB and activator protein-1 after exposure of cells to NPs (Brown et al. 2004). Black and brown carbons are important constituents of ambient NPs. It may be the organic constituents of PM that are mainly associated with the effects of NPs. NPs could act as carriers that bring chemical and gaseous components to areas that they would ordinarily not reach. Oxidative stress may be followed by inflammation, apoptosis, and eventually cancer.

The key properties associated with these particles are, as already mentioned, the large surface area and the presence of metals and organics attached to the particle's surface that all have the potential to produce oxidative stress. They may eventually activate the inflammasome, which has been described to have the potential to cause cancer (Yazdi et al. 2010). The same particles may also have genotoxic effects, depending on their composition and the materials attached to the surface (Schins 2002; Donaldson et al. 2005; Unfried et al. 2007). Therefore, all these particulates taken together may be considered a group of particulate toxins that have the potential to mediate a range of adverse health effects not only in the lungs but also in other organs.

However, the lung remains the major target of ambient air pollution. The relation between increased ambient air pollution and adverse health effects in sensitive subjects such as children, the elderly, vulnerable persons, and diseased adults is well documented (Nel et al. 2006; Kelly et al. 2012; Schüep and Sly 2012; Cassee et al. 2013). Some of the major adverse acute health effects are increased respiratory symptoms, decreased lung function, and increased hospitalization, but also altered mucociliary clearance, asthma, COPD, and eventually increased mortality (Beeson et al. 1998; Pietropaoli et al. 2004; Gong et al. 2005; Koenig et al. 2005; Silkoff et al. 2005). Clinical studies have revealed high deposition efficiencies of nanosized particulates in the total respiratory tract of healthy subjects (Jacques and Kim 2000; Pekkanen et al. 2002). However, deposition was even higher in subjects with asthma and COPD (Anderson et al. 1990; Brown et al. 2002; Chalupa et al. 2004). Moreover, animal studies showed that NPs could induce pulmonary inflammatory responses (Ferin et al. 1991; Ferin and Oberdörster 1992; Li et al. 1999; Nemmar et al. 1999; Oberdörster et al. 2000; Zhou et al. 2003; Shvedova et al. 2005). There is increasing evidence that the NPs pose the greatest problems because of their large surface area relative to their volume, and hence their high content of a variety of toxic materials with a pro-oxidative potential attached to their surface (Araujo et al. 2008).

Undoubtedly, it is of particular concern that NPs deposited in the lung periphery have the potential to translocate to the brain via the vascular system coming from the pulmonary capillaries passing the blood–brain barrier (Kreyling et al. 2002; Semmler et al. 2004) or directly, deposited in the nose, via the olfactory nerve (Oberdörster et al. 2004; Matsui et al. 2009). In brains of persons exposed to severe air pollution, neurodegenerative pathology has been reported, as is known of diseases such as Parkinson and Alzheimer (Calderon-Carcidueñas et al. 2004). In human subjects exposed to diesel exhaust, consisting mainly of NPs, changes in electroencephalography have been measured (Cruts et al. 2008). Diesel exhaust has been called cancerogenic by the International Agency for Research on Cancer (IARC) of the World Health Organization (WHO) in June 2012, and categorized into class 1 carcinogenic agents. Very recently, in October 2013, the IARC of WHO declared air pollutants in general as carcinogenic (class 1).

3.6 *In Vitro* Investigations

A number of *in vitro* studies have been performed to investigate how fine particles and NPs enter the tissue and cells of the lung, and what effect they may have. In contrast to fine particles that are taken up by cells by an active, phagocytic mechanism, the NPs could enter cells by an active endocytic, energy-driven mechanism, or by a passive mechanism as mentioned before (Geiser et al. 2005; Rothen-Rutishauser et al. 2006; Wang et al. 2012). Very recently, it has been shown that even among NPs, size might matter as far as their interaction with cells are concerned. Smaller NPs were found to be more toxic than larger ones (Shang et al. 2014). In general, the main effect of NPs that entered cells is the production of ROS, and this is a major factor in mediating inflammation and toxicity. Oxidative stress, activation of signaling pathways, and apoptosis may be the responses and might lead to pulmonary and other diseases (Brown et al. 2000; Donaldson et al. 2001; Donaldson and Tran 2002; Schins 2002; Donaldson and Stone 2003; Li et al. 2003; Nel et al. 2006).

With a triple-cell co-culture model mimicking the airway epithelial layer (Figures 3.7 through 3.9), it could be shown that fine particles and NPs may interact with cells in a different manner (Rothen-Rutishauser et al. 2005, 2007b).

Micron-sized particles could be taken up by epithelial cells at a low number, but avidly by macrophages and dendritic cells (Figure 3.9). They could be shown to translocate from the apex to the base of a layer of A549 epithelial cells by transepithelial interdigitating slender cytoplasmic processes of macrophages and dendritic cells (Figure 3.8). If the particles are membrane bound, they will be moved by intracellular cytoskeletal transport mechanisms. If they are not membrane bound (i.e., free in the cytoplasm), as it can be observed

(a)

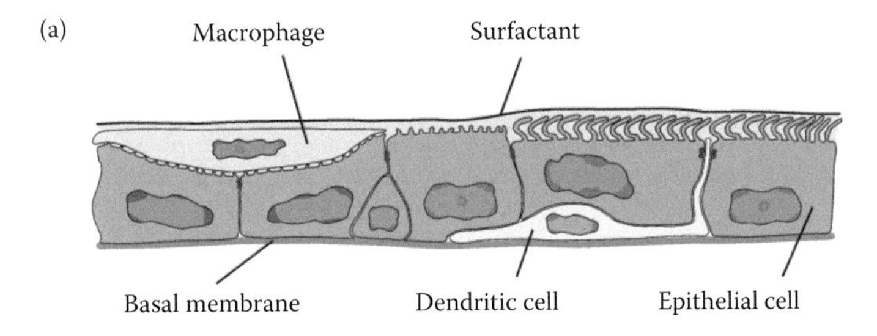

(b)

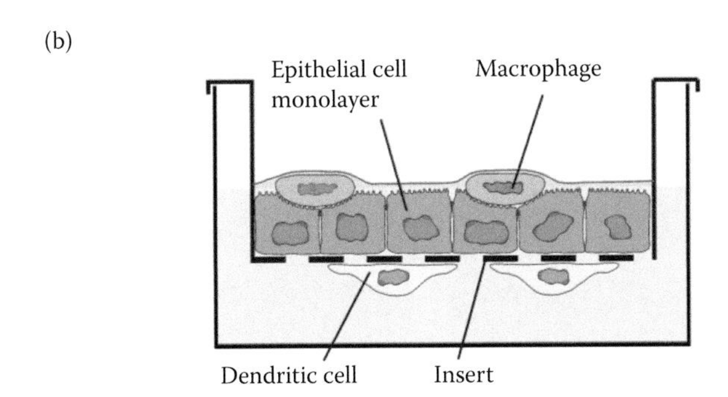

FIGURE 3.7
Schematic drawings of the epithelial layer of the airway wall (a) and the triple-cell coculture model designed to mimic the epithelial airway wall (b). Cells involved are human epithelial cells (A549 alveolar type II epithelial cell line in the model), macrophages on top, and dendritic cells at the base of the epithelial layer (both are human blood monocyte–derived cells in the model). (Reprinted from F. Blank et al., *Am. J .Respir. Cell. Mol. Biol.*, 36, 669, ATS Journals, New York, 2007. With permission.)

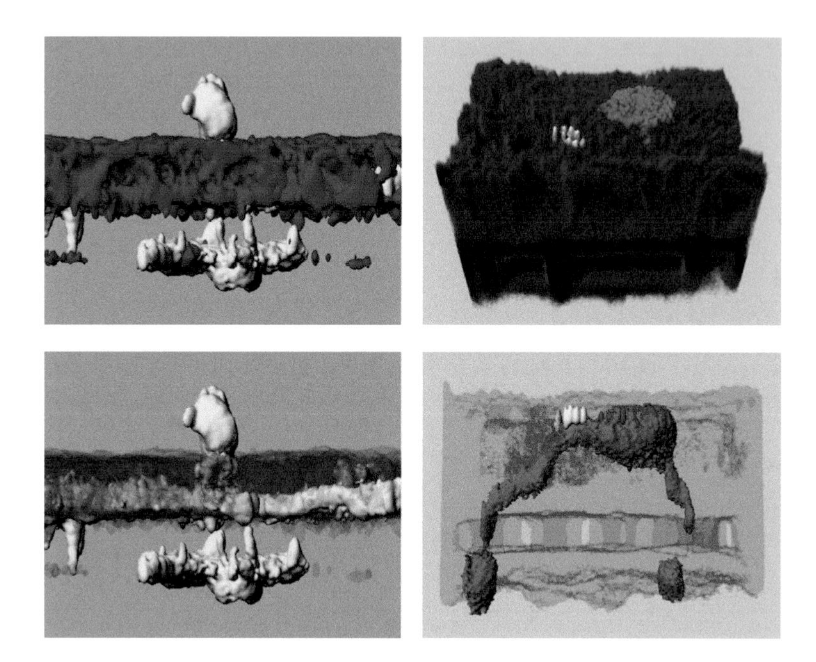

FIGURE 3.8
Confocal laser scanning light micrographs, showing the transepithelial network of slender cytoplasmic processes of macrophages (right, blue, toward the base) and of dendritic cells (left, yellow, toward the apex and into the luminal space), eventually interacting with each other; it is postulated that fine particles deposited on the epithelial layer (red) are transported through the epithelial layer by these structures. Image processing and 3D presentation by IMARIS (Bitplane AG, Zurich, Switzerland): upper pictures by volume rendering (shadow projection) and lower pictures by surface rendering; magnification, 1200×. (Right pictures reprinted from F. Blank et al., *Am. J. Respir. Cell. Mol. Biol.*, 36, 669, ATS Journals, New York, 2007. With permission.)

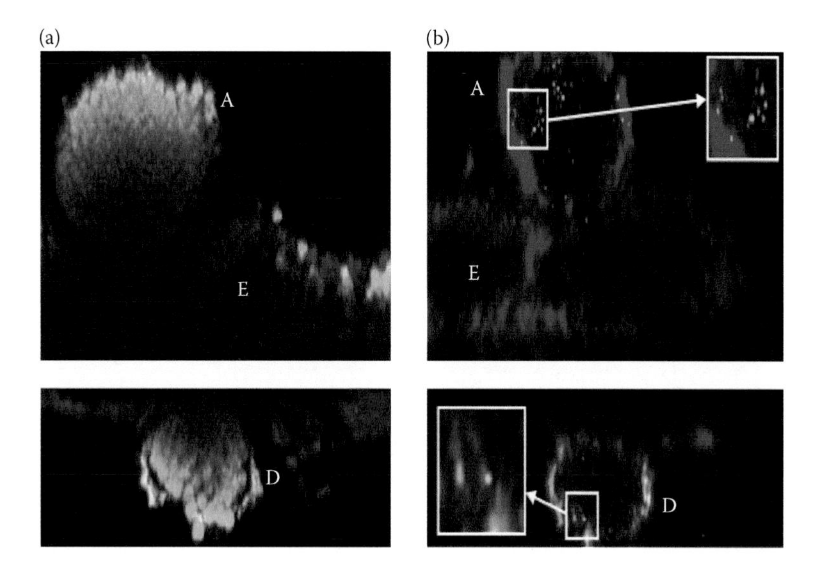

FIGURE 3.9
Confocal laser scanning light micrographs of the triple-cell coculture exposed to 1-μm (a) and 0.078-mm (b) polystyrene particles; presentation of *xz* axis; macrophages (A) and epithelial cells (E), dendritic cells (D); magnification, 1600×. (Reprinted from B. Rothen-Rutishauser et al., *Part. Fibre Toxicol.*, 4, 9, BioMed Central, 2007. With permission.)

with NPs (Brandenberger et al. 2010), they might be transported by Brownian motion or by cytoplasmic movements. In the case of cytoplasmic movements, again, cytoskeletal mechanisms might be involved.

3.7 Conclusions

While nanotechnology has been recognized as having the potential to advance science, a number of reports suggest that the potential toxicity of these particles should be considered as well (Donaldson and Tran 2002; Warheit et al. 2004; Limbach et al. 2005; Brunner et al. 2006; Stone et al. 2007). However, only a limited number of NP compositions and structures have been tested thus far. They all seem to generate ROS once they enter cells. Moreover, considerable work is needed to identify whether the conclusions drawn for combustion-derived particles can be extrapolated to engineered NPs (Gwinn and Vallyathan 2006).

NPs of various materials can cross any cellular membrane, either by endocytosis, which is based on vesicle formation, or an actin-based mechanism; these are active mechanisms. The results from the inhalation experiments with TiO_2 particles (Geiser et al. 2005) suggest, in addition, a penetration mechanism that includes a nonspecific, passive entering of particles by adhesive interactions (Rothen-Rutishauser et al. 2007b; Wang et al. 2012). After particle deposition on the internal surface of the lung, NPs can enter the tissue, cross cellular barriers, and eventually reach different tissue compartments with various cell types. They could also enter capillaries and ultimately end up in many other organs, although only in small amounts (Kreyling et al. 2002). It might be impossible to prevent, influence, or direct their entering process on a cellular level. Moreover, the adverse potential of NPs is greatly enhanced by their free location and movement within cells, which can promote interactions with intracellular proteins and organelles and even the nuclear DNA.

References

Ahluwalia, A., D. Boraschi, H.J. Byrne et al. 2013. The bio–nano-interface as a basis for predicting nanoparticle fate and behavior in living organisms: Towards grouping and categorising of nanomaterials and nanosafety by design. *BioNanoMaterials* 14: 195–216.

Analitis, A., K. Katsouyanni, K. Dimakopoulou et al. 2006. Short-term effects of ambient particles on cardiovascular and respiratory mortality. *Epidemiology* 17: 230–233.

Anderson, P.J., J.D. Wilson, and F.C. Hiller. 1990. Respiratory tract deposition of ultrafine particles in subjects with obstructive or restrictive lung disease. *Chest* 97: 1115–1120.

Araujo, J.A., B. Barajas, M. Kleinman et al. 2008. Ambient particulate pollutants in the ultrafine range promote early atherosclerosis and systemic oxidative stress. *Circ. Res.* 102: 589–596.

Bakshi, M.S., L. Zhao, R. Smith, F. Possmayer, and N.O. Petersen. 2008. Metal nanoparticle pollutants interfere with pulmonary surfactant function *in vitro*. *Biophys. J.* 94: 855–868.

Beeson, W.L., D.E. Abbey, and S.F. Knutsen. 1998. Long-term concentrations of ambient air pollutants and incident lung cancer in California adults: Results from the AHSMOG study Adventist Health Study on Smog. *Environ. Health Perspect.* 106: 813–822.

Blank, F., B. Rothen-Rutishauser, and P. Gehr. 2007. Dendritic cells and macrophages form a transepithelial network against foreign particulate antigens. *Am. J. Respir. Cell Mol. Biol.* 36: 669–677.

Blank, F., C. Von Garnier, B. Rothen-Rutishauser, C. Obregon, P. Gehr, and L. Nicod. 2008. The role of dendritic cells in the lung: What do we know from *in vitro* models, animal models and human studies? *Exp. Rev. Respir. Med.* 2: 215–233.

Borm, P.J., and W. Kreyling. 2004. Toxicological hazards of inhaled nanoparticles—Potential implications for drug delivery. *J. Nanosci. Nanotechnol.* 4: 521–531.

Brain, J.D. 1988. Lung macrophages: How many kinds are there? What do they do? *Am. Rev. Respir. Dis.* 137: 507–509.

Brain, J.D., P. Gehr, and R.I. Kavet. 1984. Airway macrophages. The importance of the fixation method. *Am. Rev. Respir. Dis.* 129: 823–826.

Brandenberger, C., C. Mühlfeld, Z. Ali et al. 2010. Quantitative evaluation of cellular uptake and trafficking of plain and polyethylene glycol-coated gold nanoparticles. *Small* 6: 1669–1678.

Brook, R.D., J.R. Brook, and S. Rajagopalan. 2003. Air pollution: the "heart" of the problem. *Curr. Hypertens. Rep.* 5: 32–39.

Brown, D.M., K. Donaldson, P.J. Borm et al. 2004. Calcium and ROS-mediated activation of transcription factors and TNF-alpha cytokine gene expression in macrophages exposed to ultrafine particles. *Am. J. Physiol. Lung Cell Mol. Physiol.* 286: L344–L353.

Brown, D.M., V. Stone, P. Findlay, W. MacNee, and K. Donaldson. 2000. Increased inflammation and intracellular calcium caused by ultrafine carbon black is independent of transition metals or other soluble components. *Occup. Environ. Med.* 57: 685–691.

Brown, J.S., K.L. Zeman, and W.D. Bennett. 2002. Ultrafine particle deposition and clearance in the healthy and obstructed lung. *Am. J. Respir. Crit. Care Med.* 166: 1240–1247.

Brunekreef, B., and T. Holgate. 2002. Air pollution and health. *Lancet* 360: 1233–1242.

Brunner, T.J., P. Wick, P. Manser et al. 2006. *In vitro* cytotoxicity of oxide nanoparticles: Comparison to asbestos, silica, and the effect of particle solubility. *J. Environ. Sci. Technol.* 44: 4374–4381.

Bruske, I., R. Hampel, M.M. Socher et al. 2010. Impact of ambient air pollution on the differential white blood cell count in patients with chronic pulmonary disease. *Inhal. Toxicol.* 22: 245–252.

Calderon-Carcidueñas, L., W. Reed, and R.R. Maronpot. 2004. Brain inflammation and Alzheimer's-like pathology in individuals exposed to severe air pollution. *Toxicol. Pathol.* 32: 650–658.

Caraballo, J.C., J. Borcherding, P.S. Thorne, and A.P. Comellas. 2013. Protein kinase C-ζ mediates lung injury induced by diesel exhaust particles. *Am. J. Respir. Cell Mol. Biol.* 48: 306–313.

Cassee, F.R., M.E. Héroux, M.E. Gerlofs-Nijland, and F.J. Kelly. 2013. Particulate matter beyond mass: Recent health evidence on the role of fractions, chemical constituents and sources of emission. *Inhal. Toxicol.* 14: 802–812.

Chalupa, D.C., P.E. Morrow, G. Oberdörster, M.J. Utell, and M.W. Frampton. 2004. Ultrafine particle deposition in subjects with asthma. *Environ. Health Perspect.* 112: 879–882.

Chang, L.Y., B.H. Kang, J.W. Slot, R. Vincent, and J.D. Crapo. 1995. Immunocytochemical localization of the sites of superoxide dismutase induction by hyperoxia in rat lungs. *Lab. Invest.* 73: 29–39.

Chang, L.-Y., J.D. Crapo, P. Gehr, B. Rothen-Rutishauser, C. Mühlfeld, and F. Blank. 2010. Alveolar epithelium in lung toxicology. In: *Comprehensive Toxicology*, 2nd edition, ed. C.A. McQueen, 59–91. Elsevier Limited, Oxford, UK.

Chavarha, M., R.W. Loney, S.B. Rananavare, and S.B. Hall. 2013. An anionic phospholipid enables the hydrophobic surfactant proteins to alter spontaneous curvature. *Biophys. J.* 104: 594–603.

Choi, H.S., Y. Ashitate, J.H. Lee et al. 2010. Rapid translocation of nanoparticles from the lung airspaces to the body. *Nat. Biotechnol.* 28: 1300–1303.

Cruts, B., L. van Etten, and H. Tornqvist. 2008. Exposure to diesel exhaust induces changes in EEG in human volunteers. *Part. Fibre Toxicol.* 5: 4.

Dominici, F., R.D. Peng, M.L. Bell et al. 2006. Fine particulate air pollution and hospital admission for cardiovascular and respiratory diseases. *JAMA* 295: 1127–1134.

Donaldson, K., C.A. Polandand, and R.P. Schins. 2010. Possible genotoxic mechanisms of nanoparticles: Criteria for improved test strategies. *Nanotoxicology* 4: 414–420.

Donaldson, K., and V. Stone. 2003. Current hypotheses on the mechanisms of toxicity of ultrafine particles. *Ann. Ist. Super. Sanita* 39: 405–410.

Donaldson, K., V. Stone, A. Seaton, and W. MacNee. 2001. Ambient particle inhalation and the cardiovascular system: Potential mechanisms. *Environ. Health Perspect.* 109: 523–527.
Donaldson, K., and C.L. Tran. 2002. Inflammation caused by particles and fibers. *Inhal. Toxicol.* 14: 5–27.
Donaldson, K., L. Tran, L. Jimenez et al. 2005. Combustion-derived nanoparticles: A review of their toxicology following inhalation exposure. *Part. Fibre Toxicol.* 2: 10.
Dreher, D., M. Kok, L. Cochand et al. 2001. Genetic background of attenuated *Salmonella typhimurium* has profound influence on infection and cytokine patterns in human dendritic cells. *J. Leukoc. Biol.* 69: 583–589.
Dudek, S.M., and J.G. Garcia. 2001. Cytoskeletal regulation of pulmonary vascular permeability. *J. Appl. Physiol.* 91: 1487–1500.
Elder, A., R. Gelein, T. Feikert et al. 2006. Translocation of inhaled ultrafine manganese oxide particles to the central nervous system. *Environ. Health Perspect.* 114: 1172–1178.
Federal Office for the Environment (FOEN). 2010. *Report*: 25 years of clean air on the basis of the Swiss Environmental Protection Act (in German), FOEN, Switzerland.
Federal Office for the Environment (FOEN). 2013. *Report*: PM10 and PM 2.5 ambient concentrations in Switzerland, FOEN, Switzerland.
Ferin, J., and G. Oberdörster. 1992. Translocation of particles from pulmonary alveoli into the interstitium. *J. Aerosol Med.* 5: 179–187.
Ferin, J., G. Oberdörster, and D.P. Penney. 1992. Pulmonary retention of ultrafine and fine particles in rats. *Am. J. Respir. Cell Mol. Biol.* 6: 535–542.
Ferin, J., G. Oberdörster, S.C. Soderholm, and R.J. Gelein. 1991. Pulmonary tissue access of ultrafine particles. *J. Aerosol. Med.* 4: 57–68.
Garcia, G.J., and J.S. Kimbell. 2009. Deposition of inhaled nanoparticles in the rat nasal passages: Dose to the olfactory region. *Inhal. Toxicol.* 21: 1165–1175.
Gasser, M., P. Wick, M.J. Clift et al. 2012. Pulmonary surfactant coating of multi-walled carbon nanotubes (MWCNTs) influences their oxidative and pro-inflammatory potential *in vitro*. *Part. Fibre Toxicol.* 9: 17.
Gehr, P., M. Bachofen, and E.R. Weibel. 1978. The normal human lung: Ultrastructure and morphometric estimation of diffusion capacity. *Respir. Physiol.* 32: 121–140.
Gehr, P., F.H. Green, M. Geiser, V. Im Hof, M.M. Lee, and S. Schürch. 1996. Airway surfactant, a primary defense barrier: Mechanical and immunological aspects. *J. Aerosol Med.* 9: 163–181.
Gehr, P., C. Muhlfeld, B. Rothen-Rutishauser, and F. Blank. 2009. *Particle–Lung Interactions*. Informa Healthcare USA Inc., New York.
Gehr, P., S. Schürch, Y. Berthiaume, V. Im Hof, and M. Geiser. 1990. Particle retention in airways by surfactant. *J. Aerosol Med.* 3: 27–43.
Geiser, M., M. Baumann, L.M. Cruz Orive, V. Im Hof, U. Waber, and P. Gehr. 1994. The effect of particle inhalation on macrophage number and phagocytic activity in the intrapulmonary conducting airways of hamsters. *Am. J. Respir. Cell Mol. Biol.* 10: 594–603.
Geiser, M., N. Leupin, I. Maye, V. Im Hof, and P. Gehr. 2000. Interaction of fungal spores with the lungs: Distribution and retention of inhaled puffball (*Calvatia excipuliformis*) spores. *J. Allergy Clin. Immunol.* 106: 92–100.
Geiser, M., M. Matter, I. Maye, V. Im Hof, P. Gehr, and S. Schürch. 2003b. Influence of airspace geometry and surfactant on the retention of man-made vitreous fibers (MMVF 10a). *Environ. Health Perspect.* 111: 895–901.
Geiser, M., B. Rothen-Rutishauser, N. Kapp et al. 2005. Ultrafine particles cross cellular membranes by non-phagocytic mechanisms in lungs and in cultured cells. *Environ. Health Perspect.* 113: 1555–1560.
Geiser, M., S. Schürch, and P. Gehr. 2003a. Influence of surface chemistry and topography of particles on their immersion into the lung's surface-lining layer. *J. Appl. Physiol.* 94: 1793–1801.
Gil, J., and E.R. Weibel. 1971. Extracellular lining of bronchioles after perfusion–fixation of rat lungs for electron microscopy. *Anat. Rec.* 169: 185–200.
Godfrey, R.W. 1997. Human airway epithelial tight junctions. *Microsc. Res. Tech.* 38: 488–499.

Gong, H. Jr., W.S. Linn, K.W. Clark, K.R. Andereson, M.D. Geller, and C. Sioutas. 2005. Respiratory responses to exposures with fine particulates and nitrogen dioxide. *Inhal. Toxicol.* 17: 123–132.
Gonzàlez-Flecha, B. 2004. Oxidant mechanisms in response to ambient air particles. *Mol. Aspects Med.* 25: 169–182.
Gwinn, M.R., and V. Vallyathan. 2006. Nanoparticles health effects—Pros and cons. *Environ. Health Perspect.* 114: 1818–1825.
Han, X., N. Corson, P. Wade-Mercer et al. 2012. Assessing the relevance of *in vitro* studies in nanotoxicology by examining correlations between *in vitro* and *in vivo* data. *Toxicology* 297: 1–9.
Hayakawa, H., Q.N. Myrvik, and R.W. St. Clair. 1989. Pulmonary surfactant inhibits priming of rabbit alveolar macrophage. Evidence that surfactant suppresses the oxidative burst of alveolar macrophage in infant rabbits. *Am. Rev. Respir. Dis.* 140: 1390–1397.
He, F., M.L. Shaffer, X. Li et al. 2010. Individual-level PM(2.5) exposure and the time course of impaired heart rate variability: The APACR study. *J. Expos. Sci. Environ. Epidemiol.* 21: 65–73.
Hinds, W.C. 1999. *Aerosol Technology: Properties, Behavior, and Measurement of Airborne Particles*, 2nd edition. Wiley-Interscience, New York.
Hoet, P.H., I. Bruske-Hohlfeld, and O.V. Salata. 2004. Nanoparticles—Known and unknown health risks. *J. Nanobiotechnol.* 2: 12.
Holt, P.G. 2005. Pulmonary dendritic cells in local immunity to inert and pathogenic antigens in the respiratory tract. *Proc. Am. Thorac. Soc.* 2: 116–120.
Holt, P.G., and M.A. Schon-Hegrad. 1987. Localization of T cells, macrophages and dendritic cells in rat respiratory tract tissue: Implications for immune function studies. *Immunology* 62: 349–356.
Holt, P.G., M.A. Schon-Hegrad, and P.G. McMenamin. 1990. Dendritic cells in the respiratory tract. *Int. Rev. Immunol.* 6: 139–149.
Ichiyasu, H., J.M. McCormack, K.M. McCarthy, D. Dombkowski, F.I. Preffer, and E.E. Schneeberger. 2004. Matrix metalloproteinase-9-deficient dendritic cells have impaired migration through tracheal epithelial tight junctions. *Am. J. Respir. Cell Mol. Biol.* 30: 761–770.
Jabbour, A.J., A. Holian, and R.K. Scheule. 1991. Lung lining fluid modification of asbestos bioactivity for the alveolar macrophage. *Toxicol. Appl. Pharmacol.* 110: 283–294.
Jacques, P.A., and C.S. Kim. 2000. Measurement of total lung deposition of inhaled ultrafine particles in healthy men and women. *Inhal. Toxicol.* 12: 715–731.
Jakel, A., A.S. Oaseem, U. Kishore, and R.B. Sim. 2013. Ligands and receptors of lung surfactant proteins SP-A and SP-D. *Front. Biosci. (Landmark Ed.)* 18: 1129–1140.
Kehrer, J.P., and S. Kacew. 1985. Systematically applied chemicals that damage lung tissue. *Toxicology* 35: 251–293.
Kelly, F.J., G.W. Fuller, H.A. Walton, and J.C. Fussell. 2012. Monitoring air pollution: Use of early warning systems for public health. *Respirology* 17: 7–19.
Kendall, M., P. Ding, R.M. Mackay et al. 2013. Surfactant protein D (SP-D) alters cellular uptake of particles and nanoparticles. *Nanotoxicology* 7: 963–973.
Kiama, S.G., L. Cochand, L. Karlsson, L.P. Nicod, and P. Gehr. 2001. Evaluation of phagocytic activity in human monocyte-derived dendritic cells. *J. Aerosol Med.* 14: 289–299.
Kilburn, K.H. 1968. A hypothesis for pulmonary clearance and its implications. *Am. Rev. Respir. Dis.* 98: 449–463.
Koenig, J.Q., T.F. Mar, R.W. Allen et al. 2005. Pulmonary effects of indoor and outdoor generated particles in children with asthma. *Environ. Health Perspect.* 113: 499–503.
Kreyling, W.G., M. Semmler, F. Erbe et al. 2002. Translocation of ultrafine insoluble iridium particles from lung epithelium to extrapulmonary organs is size dependent but very low. *J. Toxicol. Environ. Health A* 65: 1513–1530.
Künzli, N., and I.B. Tager. 2005. Air pollution: From lung to heart. *Swiss Med. Wkly.* 135: 697–702.
Kwon, M.J., Y.I. Jeon, K.Y. Lee, and T.V. Kim. 2012. Superoxide dismutase 3 controls adaptive immune responses and contributes to the inhibition of ovalbumin-induced allergic airway inflammation in mice. *Antioxid. Redox Signal.* 17: 1376–1392.
Latzin, P., M. Röösli, A. Huss, C.E. Kuehni, and U. Frey. 2009. Air pollution during pregnancy and lung function in newborns: A birth cohort study. *Eur. Respir. J.* 33: 594–603.

Lawson, P.R., and K.B. Reid. 2000. The roles of surfactant proteins A and D in innate immunity. *Immunol. Rev.* 173: 66–78.
Lehnert, B.E. 1992. Pulmonary and thoracic macrophage subpopulations and clearance of particles from the lung. *Environ. Health Perspect.* 97: 17–46.
Li, N., C. Sioutas, A. Cho et al. 2003. Ultrafine particulate pollutants induce oxidative stress and mitochondrial damage. *Environ. Health. Perspect.* 111: 455–460.
Li, X., D. Brown, S. Smith, J.T. Brunner, and K. Donaldson. 1999. Short-term inflammatory responses following intratracheal instillation of fine and ultrafine carbon black in rats. *Inhal. Toxicol.* 11: 709–731.
Limbach, L.K., Y. Li, R.N. Grass et al. 2005. Oxide nanoparticle uptake in human lung fibroblasts: Effects of particle size, agglomeration, and diffusion at low concentrations. *Environ. Sci. Technol.* 39: 9370–9376.
Luttmann-Gibson, H., H.H. Suh, B.A. Coull et al. 2006. Short-term effects of air pollution on heart rate variability in senior adults in Steubenville, Ohio. *J. Occup. Environ. Med.* 48: 780–788.
Maina, J.N., and J.B. West. 2005. Thin and strong! The bioengineering dilemma in the structural and functional design of the blood–gas barrier. *Physiol. Rev.* 85: 811–844.
Marsh, L.M., L. Cakarova, G. Kwapiszewska et al. 2009. Surface expression of CD74 by type II alveolar epithelial cells: A potential mechanism for macrophage migration inhibitory factor-induced epithelial repair. *Am. J. Physiol. Lung Cell Mol. Physiol.* 296: L442–L452.
Mason, R.J., M.C. Williams, J.H. Widdicombe, M.J. Sander, D.S. Misfeldt, and L.C. Berry Jr. 1982. Transepithelial transport by pulmonary alveolar type II cells in primary culture. *Proc. Natl. Acad. Sci. U.S.A.* 79: 6033–6037.
Matsui, Y., N. Sakai, A. Tsuda et al. 2009. Tracking the pathway of diesel exhaust particles from the nose to the brain by X-ray florescence analysis. *Spectrochim. Acta Part B.* 64: 796–801.
Matsuno, K., T. Ezaki, S. Kudo, and Y. Uehara. 1996. A life stage of particle–laden rat dendritic cells *in vivo*: Their terminal division, active phagocytosis, and translocation from the liver to the draining lymph. *J. Exp. Med.* 183: 1865–1878.
Mazzola, L. 2003. Commercializing nanotechnology. *Nat. Biotechnol.* 21: 1137–1143.
McWilliam, A.S., P.G. Holt, and P. Gehr. 2000. Dendritic cells as sentinels of immune surveillance in the airways. In: *Particle–Lung Interaction*, eds. P. Gehr, and J. Heyder. *Lung Biology in Health and Disease Series*, vol. 143, exec. ed. C. Lenfant, 473–489. Marcel Dekker, New York.
Mehra, N.K., V. Mishra, and N.K. Jain. 2014. A review of ligand tethered surface engineered carbon nanotubes. *Biomaterials* 35: 1267–1283.
Meiring, J.J., P.J. Borm, K. Bagate et al. 2005. The influence of hydrogen peroxide and histamine on lung permeability and translocation of iridium nanoparticles in the isolated perfused rat lung. *Part. Fibre Toxicol.* 2: 3.
Mensah, E.A., N.M. Kumar, L. Nirelsen, and J.S. Lwebuga-Mukasa. 1996. Distribution of alveolar type II cells in neonatal and adult rat lung revealed by RT-PCR *in situ*. *Am. J. Physiol.* 271: L178–L185.
Mercer, R.R., M.L. Russell, and J.D. Crapo. 1994. Alveolar septal structure in different species. *J. Appl. Physiol.* 77: 1060–1066.
Michiels, C. 2003. Endothelial cell functions. *J. Cell Physiol.* 196: 430–443.
Mijailovich, S., M. Kojic, and A. Tsuda. 2010. Particle-induced indentation of the alveolar epithelium caused by surface tension forces. *J. Appl. Physiol.* 109: 1179–1194.
Mistry, A., S. Stoinik, and L. Illum. 2009. Nanoparticles for direct nose-to-brain delivery of drugs. *Int. J. Pharm.* 8: 146–157.
Mühlfeld, C., T.M. Mayhew, P. Gehr, and B. Rothen-Rutishauser. 2007. A novel quantitative method for analyzing the distributions of nanoparticles between different tissue and intracellular compartments. *J. Aerosol Med.* 20: 395–407.
Nel, A.E., T. Xia, L. Madler, and N. Li. 2006. Toxic potential of materials at the nanolevel. *Science* 311: 622–627.
Nemmar, A., A. Delaunois, B. Nemery et al. 1999. Inflammatory effect of intratracheal instillation of ultrafine particles in the rabbit: Role of C-fiber and mast cells. *Toxicol. Appl. Pharmacol.* 160: 250–261.

Nemmar, A., P.H. Hoet, B. Vanquickenborne et al. 2002. Passage of inhaled particles into the blood circulation in humans. *Circulation* 105: 411–414.
Nicod, L.P. 1997. Function of human lung dendritic cells. In: *Particle Toxicology*, eds. K. Donaldson, and P. Borm, 311–334. Marcel Dekker Inc, New York.
Nicod, L.P. 2005. Lung defenses: An overview. *Eur. Respir. Rev.* 95: 45–50.
Niess, J.H., S. Brand, X. Gu et al. 2005. CX3CR1-mediated dendritic cell access to the intestinal lumen and bacterial clearance. *Science* 307: 254–258.
Oberdörster, G. 2001. Pulmonary effects of inhaled ultrafine particles. *Int. Arch. Occup. Environ. Health* 74: 1–8.
Oberdörster, G., J.N. Finkelstein, C. Johnston et al. 2000. Acute pulmonary effects of ultrafine particles in rats and mice. *Res. Rep. Health Eff. Inst.* 96: 5–74.
Oberdörster, G., E. Oberdörster, and J. Oberdörster. 2005. Nanotoxicology: An emerging discipline evolving from studies of ultrafine particles. *Environ. Health Perspect.* 113: 823–839.
Oberdörster, G., Z. Sharp, V. Atudorei et al. 2002. Extrapulmonary translocation of ultrafine carbon particles following whole body inhalation exposure of rats. *J. Toxicol. Environ. Health A* 65: 1531–1543.
Oberdörster, G., Z. Sharp, V. Atudorei et al. 2004. Translocation of inhaled ultrafine particles to the brain. *Inhal.* Toxicol. 16: 437–445.
Ochs, M., J.R. Nyengard, A. Jung et al. 2004. The number of alveoli in the human lung. *Am. J. Respir. Crit. Care Med.* 169: 120–124.
Parra, E., L.H. Moleiro, I. Lopez-Montero, A. Cruz, F. Monroy, and J. Pérez-Gil. 2011. A combined action of pulmonary surfactant proteins SP-B and SP-C modulates permeability and dynamics of phospholipid membranes. *Biochem. J.* 438: 555–564.
Paull, R., J. Wolfe, P. Hebert, and M. Sinkula. 2003. Investing in nanotechnology. *Nat. Biotechnol.* 21: 1144–1147.
Peel, J.L., W.D. Flanders, J.A. Mulholland, G. Freed, and P.E. Tolbert. 2011. Ambient air pollution and apnea and bradycardia in high-risk infants on home monitors. *Environ. Health Perspect.* 119: 1321–1327.
Pekkanen, J., A. Peters, G. Hoek et al. 2002. Particulate air pollution and risk of ST-segment depression during repeated submaximal exercise tests among subjects with coronary heart disease: The exposure and risk assessment for fine and ultrafine particles in ambient air (ULTRA) study. *Circulation* 106: 933–938.
Perez, L., R. Rapp, and N. Künli. 2010. The year of the lung: Outdoor air pollution and lung health. *Swiss Med. Wkly.* 140: w13129.
Peters, A. 2011. Ambient particulate matter and the risk for cardiovascular disease. *Prog. Cardiovasc. Dis.* 53: 327–333.
Peters, A., D.W. Dockery, J.E. Muller, and M.A. Mittleman. 2001. Increased particulate air pollution and the triggering of myocardial infarction. *Circulation* 103: 2810–2815.
Peters, A., H.E. Wichmann, T. Tuch, J. Heinrich, and J. Heyder. 1997. Respiratory effects are associated with the number of ultrafine particles. *Am. J. Respir. Crit. Care Med.* 155: 1376–1383.
Pietropaoli, A.P., M.W. Frampton, R.W. Hyde et al. 2004. Pulmonary function, diffusing capacity, and inflammation in healthy and asthmatic subjects exposed to ultrafine particles. *Inhal. Toxicol.* 16: 59–72.
Pope, C.A. III, D.W. Dockery, and J. Schwartz. 1995. Review of epidemiological evidence of health effects of particulate air pollution. *Inhal. Toxicol.* 7: 1–18.
Pope, C.A. III, M. Ezzati, and D.W. Dockery. 2009. Fine-particulate air pollution and life expectancy in the United States. *N. Engl. J. Med.* 360: 376–386.
Possmayer, F. 1988. Pulmonary perspective: A proposed nomenclature for pulmonary surfactant-associated proteins. *Am. Rev. Respir. Dis.* 138: 990–998.
Pryor, W.A. 1992. How far does ozone penetrate into the pulmonary air/tissue boundary before it reacts? *Free Radic. Biol. Med.* 12: 83–88.
Puett, R.C., J.E. Hart, J.D. Yanosky et al. 2009. Chronic fine and coarse particulate exposure, mortality, and coronary heart disease in the Nurses' Health Study. *Environ. Health Perspect.* 117: 1697–1701.

Qian, X., K. Agematsu, G.J. Freeman, Y. Tagawa, K. Sugane, and T. Hayashi. 2006. The ICOS-ligand B7-H2, expressed on human type II alveolar epithelial cells, plays a role in the pulmonary host in the pulmonary host defense system. *Eur. J. Immunol.* 36: 906–918.

Rescigno, M., M. Urbano, B. Valzasina et al. 2001. Dendritic cells express tight junction proteins and penetrate gut epithelial monolayers to sample bacteria. *Nat. Immunol.* 2: 361–367.

Rivera Gil, P., G. Oberdörster, A. Elder, V. Puntes, and W.J. Parak. 2010. Correlating physico-chemical with toxicological properties of nanoparticles: The present and the future. *ACS Nano* 4: 5527–5531.

Rothen-Rutishauser, B.M., S.G. Kiama, and P. Gehr. 2005. A 3D cellular model of the human respiratory tract to study the interaction with particles. *Am. J. Respir. Cell Mol. Biol.* 32: 281–289.

Rothen-Rutishauser, B., C. Muhlfeld, F. Blank, C. Musso, and P. Gehr. 2007a. Translocation of particles and inflammatory responses after exposure to fine particles and nanoparticles in an epithelial airway model. *Part. Fibre Toxicol.* 4: 9.

Rothen-Rutishauser, B., S. Schürch, and P. Gehr. 2007b. Interaction of particles with membranes. In: *The Toxicology of Particles*, eds. K. Donaldson, and P.J. Borm, 139–160. Taylor & Francis Group, LLC, CRC Press, Boca Raton, FL.

Rothen-Rutishauser, B., S. Schürch, B. Haenni, N. Kapp, and P. Gehr. 2006. Interaction of fine particles and nanoparticles with red blood cells visualized with advanced microscopic techniques. *Environ. Sci. Technol.* 40: 4353–4359.

Ruge, C.A., J. Kirch, O. Cañadas et al. 2011. Uptake of nanoparticles by alveolar macrophages is triggered by surfactant protein A. *Nanomedicine* 7: 690–693.

Ruge, C., U.F. Schaefer, J. Herrmann et al. 2012. The interplay of lung surfactant proteins and lipids assimilates the macrophage clearance of nanoparticles. *PLoS One* 7: e40775.

Schins, R.P. 2002. Mechanism of genotoxicity of particles and fibers. *Inhal. Toxicol.* 14: 57–78.

Schneeberger, E.E. 1977. Ultrastructure of intercellular junctions in the freeze fractured alveolar-capillary membrane of mouse lung. *Chest* 71: 299–300.

Schneeberger, E.E., and R.D. Lynch. 1984. Tight junctions. Their structure, composition, and function. *Circ. Res.* 55: 723–733.

Schüepp, K., and P.D. Sly. 2012. The developing respiratory tract and its specific needs in regard to ultrafine particulate matter exposure. *Paediatr. Respir. Rev.* 13: 95–99.

Schulz, H., V. Harder, A. Ibald-Mulli et al. 2005. Cardiovascular effects of fine and ultrafine particles. *J. Aerosol Med.* 18: 1–22.

Schürch, D., O.L. Ospina, A. Cruz, and J. Pérez-Gil. 2010. Combined and independent action of proteins SP-B and SP-C in the surface behavior and mechanical stability of pulmonary surfactant films. *Biophys. J.* 99: 3290–3299.

Schürch, S., P. Gehr, V. Im Hof, M. Geiser, and F. Green. 1990. Surfactant displaces particles toward the epithelium in airways and alveoli. *Respir. Physiol.* 80: 17–32.

Seaton, A., W. MacNee, K. Donaldson, and D. Godden. 1995. Particulate air pollution and acute health effects. *Lancet* 345(8943): 176–178.

Seipenbusch, M., A. Binder, and G. Kasper. 2008. Temporal evolution of nanoparticle aerosols in workplace exposure. *Ann. Occup. Hyg.* 53: 707–716.

Semmler, M., J. Seitz, F. Erbe et al. 2004. Long-term clearance kinetics of inhaled ultrafine insoluble iridium particles from the rat lung, including transient translocation into secondary organs. *Inhal. Toxicol.* 16(6–7): 453–459.

Shah, A.S., J.P. Langrish, H. Nair et al. 2013. Global association of air pollution and heart failure: A systematic review and meta-analysis. *Lancet* 382: 1039–1048.

Shang, L., K. Nienhaus, and G.U. Nienhaus. 2014. Engineered nanoparticles interacting with cells: Size matters. *J. Nanobiotechnol.* 12: 5.

Shimabukuro, D.W., T. Sawa, and M.A. Gropper. 2003. Injury and repair in lung and airways. *Crit. Care Med.* 31: S524–S531.

Shvedova, A.A., E.R. Kisin, R. Mercer et al. 2005. Unusual inflammatory and fibrogenic pulmonary responses to single-walled carbon nanotubes in mice. *Am. J. Physiol. Lung* 289: L698–L708.

Silkoff, P. E., L. Zhang, S. Dutton et al. 2005. Winter air pollution and disease parameters in advanced chronic obstructive pulmonary disease panels residing in Denver, Colorado. *Allergy Clin. Immunol.* 115: 337–344.
Stone, K.C., R.R. Mercer, P. Gehr, B. Stockstill, and J.D. Crapo. 1992. Allometric relationships of cell numbers and size in the mammalian lung. *Am. J. Respir. Cell Mol. Biol.* 6: 235–243.
Stone, V., H. Johnston, and M.J.D. Clif. 2007. Air pollution, ultrafine and nanoparticle toxicology, cellular and molecular interactions. *IEEE Trans. Nanobioscience* 4: 331–340.
Takano, K., T. Kojima, M. Go et al. 2005. HLA-DR- and CD11c-positive dendritic cells penetrate beyond well-developed epithelial tight junctions in human nasal mucosa of allergic rhinitis. *J. Histochem. Cytochem.* 53: 611–619.
Tatur, S., and A. Badia. 2012. Influence of hydrophobic alkylated gold nanoparticles on the phase behavior of monolayers of DPPC and clinical lung surfactant. *Langmuir* 28: 628–639.
Thiele, L., B. Rothen-Rutishauser, S. Jilek, H. Wunderli-Allenspach, H.P. Merkle, and E. Walter. 2001. Evaluation of particle uptake in human blood monocyte-derived cells *in vitro*. Does phagocytosis activity of dendritic cells measure up with macrophages? *J. Control Rel.* 76: 59–71.
Timpl, R., and M. Dzladek. 1986. Structure, development and molecular pathology of basement membranes. *Int. Rev. Exp. Pathol.* 29: 1–112.
Toews, G.B., W.C. Vial, M.M. Dunn, P. Guzzetta, G. Nunez, P. Stastny, and M.F.J. Lipscomb. 1984. The accessory cell function of human alveolar macrophages in specific T cell proliferation. *J. Immunol.* 132: 181–186.
Unfried, K., C. Albrecht, L.-O. Klotz, S. Grether-Beck, and R.P.F. Schins. 2007. Cellular responses to nanoparticles: Target structures and mechanisms. *Nanotoxicology* 1: 52–71.
Vermaelen, K.Y., I. Carro-Muino, B.N. Lambrecht, and R.A. Pauwels. 2001. Specific migratory dendritic cells rapidly transport antigen from the airways to the thoracic lymph nodes. *J. Exp. Med.* 193: 51–60.
Vermaelen, K., and R. Pauwels. 2005. Pulmonary dendritic cells. *Am. J. Respir. Crit. Care Med.* 172: 530–551.
Voelker, D.R., and R.J. Mason. 1989. Alveolar type II cells. In: *Lung Cell Biology*, ed. D. Massaro. *Lung Biology in Health and Disease Series*, vol. 4, exec. ed. C. Lenfant, 487–538. Marcel Dekker, New York.
Walter, E., D. Dreher, M. Kok et al. 2001. Hydrophilic poly(DL-lactide-co-glycolide) microspheres for the delivery of DNA to human-derived macrophages and dendritic cells. *J. Control Rel.* 76: 149–168.
Wang, T., J. Bai, X. Jiang, and G.U. Nienhaus. 2012. Cellular uptake of nanoparticles by membrane penetration: A study combining confocal microscopy with RTIR spectroelectrochemistry. *ACS Nano* 6: 1251–1259.
Warheit, D.B., B.R. Laurence, K.L. Reed, D.H. Roach, G.A.M. Reynolds, and T.R. Webb. 2004. Comparative pulmonary toxicity assessment of single-wall carbon nanotubes in rats. *Toxicol. Sci.* 77: 117–125.
Weibel, E.R. 1963. *Morphometry of the Human Lung*. Academic Press, New York.
Whitsett, J.A. 2010. Review: The intersection of surfactant homeostasis and innate host defense of the lung: Lessons from newborn infants. *Innate Immun.* 16: 138–142.
Wichmann, H.E., C. Spix, T. Tuch et al. 2000. Daily mortality and fine and ultrafine particles in Erfurt, Germany part I: Role of particle number and particle mass. *Res. Rep. Health Eff. Inst.* 98: 5–86.
Williams, M.C. 1977. Conversion of lamellar body membranes into tubular myelin in alveoli of fetal rat lungs. *J. Cell Biol.* 72: 260–277.
Winkler, C., and J.M. Hohlfeld. 2013. Surfactant and allergic airway inflammation. *Swiss Med. Wkly.* 143: w13818.
Wright, J.R. 2005. Immunoregulatory functions of surfactant proteins. *Nat. Immunol.* 5: 58–68.
Yazdi, A.S., G. Guarda, N. Riteau et al. 2010. Nanoparticles activate the NLR pyrin domain containing 3 (Nlrp3) inflammasome and cause pulmonary inflammation through release of IL-1α and IL-1β. *Proc. Natl. Acad. Sci. U.S.A.* 107: 19449–19454.

Yokota, S., H. Hori, M. Umezawa et al. 2013. Gene expression changes in the olfactory bulb of mice induced by exposure to diesel exhaust are dependent on animal rearing environment. *PLoS One* 8: e70145.

Yurchenco, P.D., E.C. Tsilibary, A.S. Charonis, and H. Furthmayr. 1986. Models for the self-assembly of basement membrane. *J. Histochem. Cytochem.* 34: 93–102.

Zanobetti, A., J. Schwartz, E. Samoli et al. 2003. The temporal pattern of respiratory and heart disease mortality in response to air pollution. *Environ. Health Perspect.* 111: 1188–1193.

Zhou, Y.M., C.Y. Zhong, I.M. Kennedy, V.J. Leppert, and K.E. Pinkerton. 2003. Oxidative stress and NFkappaB activation in the lungs of rats: A synergistic interaction between soot and iron particles. *Toxicol. Appl. Pharmacol.* 190: 157–169.

4

Role of the Biomolecule Corona in Nanoparticle Fate and Behavior

Iseult Lynch

CONTENTS

4.1 Introduction

Understanding and predicting the fate and behavior of engineered nanoparticles (NPs) as they interact with living systems has become an international research priority during recent years, as researchers and regulators struggle to ensure the safe and responsible implementation of nanotechnologies in a wide variety of applications from information technology to construction to medicine. The key features of nanomaterials that make them attractive for a range of applications include their small size and their consequent large surface area-to-volume ratio and high surface energy, which can be exploited for a range of effects, such as catalysis or drug transport and delivery. However, there are persistent concerns that this increased reactivity could lead to toxicity (Abbott et al. 2006; Borm et al. 2006), via, for example, the well-established oxidative stress paradigm (Nel et al. 2006), in parallel with long-standing research into respirable air pollution particles (PM_{10}). Epidemiological studies have repeatedly found a positive correlation between the level of particulate air pollution and increased morbidity and mortality rates in both adults and children. Such studies have also identified a link between respiratory ill health and the number of ambient ultrafine particles (Stone et al. 2007).

An important consequence of the large surface area-to-volume ratio of NPs, and their high surface energy, is the tendency of NPs to reduce their surface energy via adsorption of biomolecules from their surroundings, forming the protein or biomolecule corona (Walczyk et al. 2010; Monopoli et al. 2011). This adsorbed biomolecule layer has emerged as one of the leading approaches to understanding NP interactions with living systems, and is based on the concept that the nature and identity of the absorbed biomolecules play a vital role in determining the fate and behavior of particles, as these are what engage with biological membranes and cellular receptors. However, it is clear that the core particle composition and specific surface characteristics determine which proteins bind, and likely also influence the protein/biomolecule orientation and conformation upon binding. While most research to date has focused on the proteins in the NP corona (Abbott et al. 2006), there have also been some studies on lipid interactions, and combination of lipid and protein interactions (Abbott et al. 2006; Hellstrand et al. 2009). Among the key findings to date are the fact that the corona is remarkably stable once formed, allowing it to be isolated and studied (Walczyk et al. 2010), but that it does evolve as NPs move from one biological environment (compartment) to another (Lundqvist et al. 2011). Additionally, it has emerged that the composition of the corona may be different under *in vivo* conditions (i.e., at high protein concentrations) than under *in vitro* conditions (typically 3%–10% proteins) (Monopoli et al. 2011). This has important consequences for the assessment of NP interactions with the lungs, due to their complex arrangement of cells, lipids, and proteins.

Upon reaching the alveolar region, inhaled NPs must first interact with the pulmonary surfactant (PS) lining layer of alveoli (Peters et al. 2006; Schleh and Hohlfeld 2009). PS is a phospholipid–protein mixture synthesized by type II alveolar epithelial cells. It covers the entire internal surface of the respiratory tract as a thin film and plays a crucial role in surface tension reduction and host defense (Zuo et al. 2008; Possmayer et al. 2010). By reducing alveolar surface tension to near zero, the PS film stabilizes alveoli against collapse, thereby maintaining a large surface area for gas exchange. Thus, these outer noncellular pulmonary structures of the lung (i.e., mucus or PS) are the first point of contact for inhaled NPs, and are likely to alter the surface of the NPs via adsorption. Interaction with the PS film determines the subsequent clearance, retention, and translocation of inhaled NPs and hence their potential toxicity, such as interference with PS metabolism; interaction with lung epithelial cells, alveolar macrophages, and neutrophils; and translocation to the bloodstream and other organs (Peters et al. 2006; Schleh and Hohlfeld 2009).

Regardless of whether the conducting airways or the alveolar epithelium is the site of NP deposition, soluble compounds (i.e., proteins, glycoproteins, lipids) are secreted by respective cells in both compartments of the lung (e.g., goblet cells in the airways or AT2 cells in the alveoli) and can bind to particles once they have landed (Ruge 2012). Among these biomolecules, proteins are probably the dominant components, whereas lipids or salts may play a mediating role in such particle–protein interactions.

Research into the mechanism of NP interactions with the lungs, both the cellular and biomolecular components such as the lipids and proteins, and the consequences of such interactions, has become increasingly vibrant during recent years. As knowledge increased, the focus of research (as in other areas of nanosafety) has moved increasingly toward understanding the role of the bio–nano interface. For example, an early review by Mühlfeld et al. focused on the interaction of inhaled NPs with the structures of the respiratory tract, including surfactant, alveolar macrophages, and epithelial cells, and placed special emphasis on methodological differences between experimental studies and the caveats associated with the dose metrics and pointed out ways to overcome inherent methodological problems (Mühlfeld et al. 2008). Building on that, an increased understanding that NP interactions with lung surfactant can have a significant effect on PS metabolism was used as the basis for a review of the effect of NPs on biophysical surfactant function, surfactant metabolism, particle clearance, and particle-induced toxicity by Schleh et al. (2009). More recently, the significant body of work and the approaches developed to understand and characterize the NP–protein corona formed upon contact with blood plasma are now being translated to the study of NP interactions with lung surfactant proteins, and the role that these proteins play in determining the fate and impacts of the NPs. This chapter attempts to frame some of these studies, highlight significant advances, and suggest some directions for future studies.

Thus, the interaction of NPs with the lungs leads to adsorption of PS components onto the NP surface as a consequence of particle displacement into the surfactant layer, and likely leads to a PS corona, which influences the further biological fate of NPs, in parallel to the role/impact of the corona formed in blood (Ruge 2012).

4.2 Key Players: Phospholipids and Pulmonary Surfactant Proteins A and D

The lungs represent the largest surface contact that most air-breathing vertebrates have with their environment (Pérez-Gil 2008), and are thus potentially the most vulnerable organ for unintended contact with pollutants, including particles released from combustion processes and engineered NPs. A complex macromolecular system has evolved at the environmental interface, to provide optimal properties in terms of structural stability and accessibility to the air phase while raising an efficient barrier against environmental insults, including the entrance of pathogens (Pérez-Gil 2008).

PS, a membrane-based lipid–protein complex, which is assembled and secreted onto the respiratory surface by specialized cells of the alveolar epithelium, contains molecular components simultaneously and inseparably responsible for biophysical stabilizing activities and innate defense mechanisms (Pérez-Gil 2008). In general terms, PS is composed of around 80% phospholipids, 5%–10% neutral lipids (mainly cholesterol), and 8%–10% proteins, with 5%–6% of total surfactant mass being constituted by specific surfactant proteins (Pérez-Gil 2008).

The phospholipid fraction of surfactant is mainly responsible for forming surface-active films at the respiratory air–liquid interface, but it also provides the scaffold or matrix on which the different surfactant structures are assembled. In most mammals, half of the surfactant phospholipids (by mass) are composed of disaturated species, mainly dipalmitoylphosphatidylcholine (DPPC), a phospholipid that is scarce in other tissues. Evolution has probably selected DPPC as the main phospholipid species in surfactant because at physiological temperature, the saturated chains of DPPC can be packed to a very high density at the air–water interface, providing the large reduction of surface tension required to stabilize the lung at the end of expiration (Hawco et al. 1981a,b; Pérez-Gil 2008). By contrast, the kinked chains of unsaturated phospholipid species—constituting roughly the other half mass of surfactant phospholipids—cannot be interfacially packed beyond a certain threshold and cannot therefore sustain low enough tensions (Hawco et al. 1981a,b). Packing propensity is then a main feature to define surface activity and organization of surfactant structures, while the actual packing of surfactant phospholipids depends not only on the acyl chain composition but also on temperature, the presence of other lipids such as cholesterol, and, at the interface, the state of compression, as summarized in Figure 4.1.

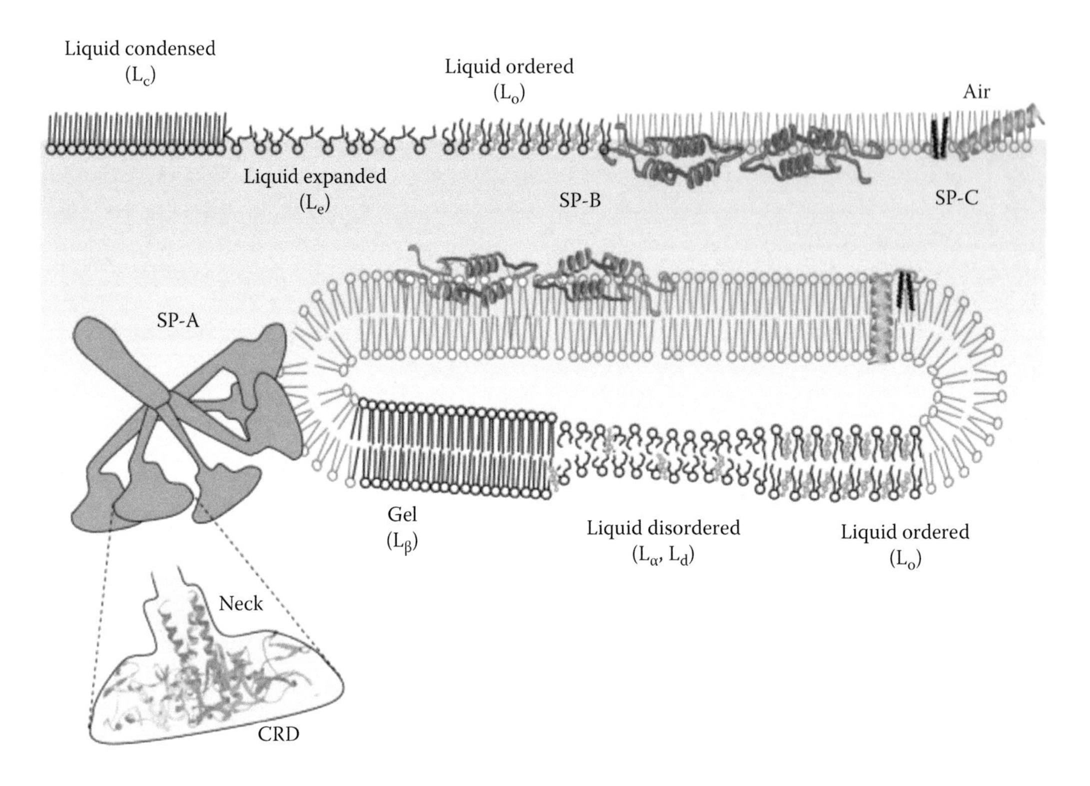

FIGURE 4.1
Structure of lipid phases and membrane-associated proteins in PS. The cartoon summarizes current models of the structure and orientation of the three proteins usually obtained associated with PS membranes: SP-A, SP-B, and SP-C. Differences in organization and packing of phospholipids (indicated as variations of the circular head group and the double tail, in the bilayer structure or as a monolayer at the liquid–air interface) have been illustrated in gel, liquid-disordered, and liquid-ordered phases in bilayers, and in liquid-expanded and liquid-condensed two-dimensional phases in interfacial films. (From Pérez-Gil, J., *Biochim. Biophys. Acta [BBA] Biomembr.*, *1778*, 1676, 2008.)

Figure 4.1 also includes a representation of the three main surfactant proteins associated with PS membranes: SP-A, SP-B, and SP-C. A fourth surfactant protein, SP-D, is not usually associated with membranes but is involved in immune response (Crouch and Wright 2001). Proteins SP-A and SP-D are hydrophilic in nature and consist of large macromolecular assemblies belonging to the family of collectins, Ca^{2+}-dependent C-type lectins possessing both collagen-like and carbohydrate-recognition domains (CRD) (Hawgood and Poulain 2001).

The quaternary structure of SP-A consists of a hexamer of trimers, as shown in Figure 4.2. Each trimer contains a long triple-helical collagenous stem, interrupted by a flexible hinge, a helical bundle connector, and a globular head, which contains the CRD (Haagsman 2002), and enables SP-A to bind multiple ligands, including sugars, Ca^{2+}, and phospholipids in a cooperative manner (Casals 2001). Recognition of SP-A by specific receptors in alveolar macrophages stimulates phagocytosis and other pathogenicidal events; however, the binding of SP-A to bacteria, although confirmed to occur, was not necessary for SP-A-induced enhancement of phagocytosis (Sano et al. 2006). It has been proposed that SP-A recognizes ordered lipid patterns, and that this is the rationale behind a certain preference for the protein to interact with ordered membranes such as those in gel-like or in liquid-ordered phases.

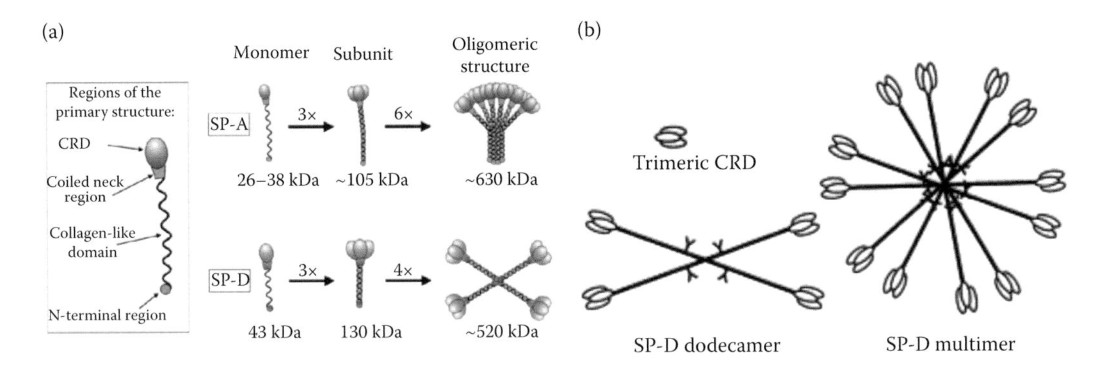

FIGURE 4.2
(a) Overall structure of SP-A and SP-D. SP-A (top) has a bouquet-like structure. SP-D (bottom) has a cruciform-shaped structure. (From Ruge, C.A. Bio–nano interactions in the peripheral lungs: Role of pulmonary surfactant components in alveolar macrophage clearance of NPs. Dissertation, Saarbrücken, 2012.) (b) Alternative molecular forms of SP-D. Trimeric CRDs, SP-D multimers and SP-D dodecamers. The minimum structure required for high-affinity binding is the trimeric CRD. (From Crouch, E.C., *Biochim. Biophys. Acta [BBA] Mol. Basis Dis., 1408*, 278, 1998.)

SP-D (43 kDa) is predominantly assembled as dodecamers consisting of four homotrimeric subunits (4 × 3 = 12 chains) with relatively long triple helical arms (Crouch 1998). Each trimeric subunit contains four major domains: a short cysteine-containing NH_2-terminal cross-linking domain (N), a triple-helical collagen domain of variable length, a trimeric coiled-coil linking domain or neck (L), and a carboxy-terminal, C-type lectin CRD. Interactions between the amino-terminal domains of SP-D subunits are stabilized by interchain disulfide bonds (Crouch et al. 1994). SP-D dodecamers can self-associate at their amino termini to form highly ordered multimers with complex arrays of up to 32 (or more) trimeric CRDs (Figure 4.2) (Crouch et al. 1994).

Both SP-A and SP-D interact with lipids: SP-A binds to DPPC, the major surfactant phospholipid, as well as to the lipid A domain of gram-negative lipopolysaccharide, and to several glycolipids and neutral glycosphingolipids. SP-D interacts with the inositol and lipid moieties of phosphatidylinositol and with glucosylceramide (Crouch et al. 1994). The interactions of collectins with lipid ligands could contribute to surfactant lipid reorganization and/or the interactions of these molecules with host cells.

Generally, the pulmonary collectins act by three main mechanisms:

1. Opsonization of inhaled pathogens or particles;
2. Activation of immunocompetent cells such as alveolar macrophages, neutrophils, or dendritic cells (DC);
3. Regulation of cellular responses, such as release of cytokines or expression of surface receptors.

Owing to their large oligomeric structures, SP-A and SP-D can be considered as broadly selective opsonins with high avidity, allowing them to tightly bind biological structures and patterns present on bacteria, viruses, fungi, or yeast (Seaton et al. 2010; Ruge 2012).

McCormack and Whitsett (2002) collected evidence from *in vitro* and *in vivo* studies to support roles for the pulmonary collectins as opsonins that enhance bacterial clearance and killing, and as immunomodulators that regulate cellular recruitment and activation as part of the host response, although the mechanisms involved in these functions are

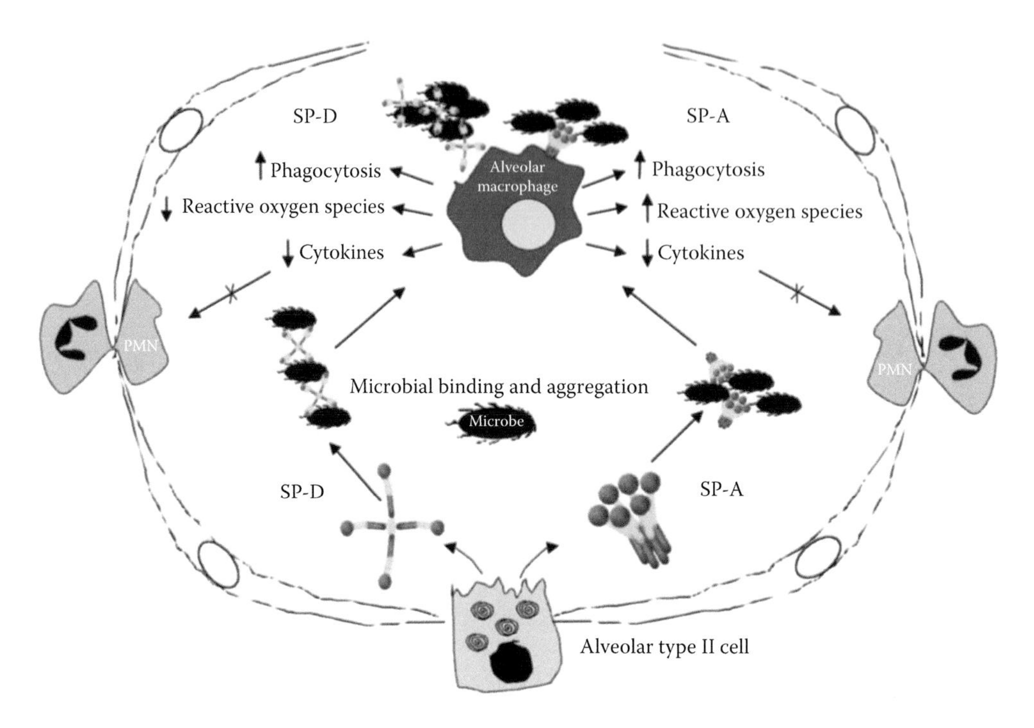

FIGURE 4.3
Pulmonary collectins orchestrate host defense, inflammation, and oxidant production. Schematic of alveolar defense roles for SP-A and SP-D suggested by *in vitro* studies and genetically engineered animal models. SP-A and SP-D are secreted by alveolar type II cells into the alveolar lumen. The presentation of microorganisms to inflammatory cells is determined in part by SP-A and SP-D, which aggregate and opsonize diverse viral, fungal, and bacterial species. SP-A and SP-D preserve gas exchange by limiting inflammatory cell infiltration and alveolar exudation, both by increasing the rate of clearance of microorganisms from the lung and by directly modulating inflammatory cytokine and oxidant responses. (From McCormack, A.X., Whitsett, J.A., *J Clin. Invest.*, *109*, 707, 2002.)

complex. Accumulating data indicate that the pulmonary collectins affect every phase of the inflammatory response to pulmonary pathogens, exerting direct effects on the growth and viability of microbes, presentation of the microbial antigens to inflammatory cells, and modulation of leukocyte phagocytic, proteolytic, chemotactic, and oxidant responses, as shown schematically in Figure 4.3 (McCormack and Whitsett 2002). Thus, SP-A serves its primary host defense role as a solid-phase, high-density SP-A array that optimizes the surface properties of surfactant and the interception of inhaled microbes at the air–lung interface.

4.3 Connection between Lung Surfactant Layer Fluidity and Human Health

While by no means intending to be a comprehensive review of the role of the PS layer in maintaining lung function and human health, to understand the potential implications of interactions with NPs on lung function, it is necessary to have a broad understanding of the role of lung surfactant fluidity in maintaining a healthy respiratory system. The lung surfactant layer has the dual role of facilitating the expansion of the lungs during inhalation, during which PS flows rapidly into the alveolar air–water interface and thereby stabilizes the alveoli, while also preventing the lungs from collapse when the adsorbed interfacial film is compressed during exhalation (Piknova et al. 2002; Schmidt et al. 2002). By lowering surface tension (γ) to near equilibrium during inspiration, the lung surfactant minimizes the work of breathing. By reducing γ to low values during expiration, surfactant stabilizes the lung at low lung volumes and limits the tendency to develop pulmonary edema (Bakshi et al. 2008).

Alterations in the biochemical composition and biophysical properties of PS are well documented for patients with acute inflammatory lung diseases such as acute respiratory distress syndrome. In these patients, alterations of both the lipid moiety and the surfactant proteins were noted, with a reduced relative content of phosphatidylcholine; an increase in minor phospholipids, including sphingomyelin, phosphatidylinositol, and phosphatidylethanolamine; an increased lavagable protein fraction; and reduced SP-A, SP-B, and SP-C (Schmidt et al. 2002).

Another well-studied lung disease, pneumonia, involves bacterial (endotoxin) induction of changes in PS, either directly on secreted surfactant or indirectly through pulmonary type II epithelial cells (Brogden 1991). The interaction of bacteria or endotoxin with secreted surfactant results in changes in the physical (i.e., density and surface tension) properties of surfactant. In addition, gram-negative bacteria or endotoxin can injure type II epithelial cells, causing them to produce abnormal quantities of surfactant, abnormal concentrations of phospholipids in surfactant, and abnormal compositions (i.e., type and saturation of fatty acids) (Brogden 1991). These changes in surfactant physical properties and/or composition have a deleterious effect on lung function characterized by significant decreases in total lung capacity, static compliance, diffusing capacity, and arterial PO_2, and a significant increase in mean pulmonary arterial pressure, as well as the anatomic changes commonly seen in pneumonia such as pulmonary edema, hemorrhage, and atelectasis (Brogden 1991).

The contribution of airborne pollutants to chronic lung diseases of the airways has been well established; however, their contribution to chronic alveolar diseases is less clear (Anseth et al. 2005). Oxidative inactivation of surfactant by cellular reactive oxygen species, leading to surfactant lipid peroxidation, protein carbonylation, and functional impairment of SP-A, is thought to be the possible mechanism underlying air pollutant–induced surfactant damage. More recently, chronic exposure to airborne pollution has been shown to induce cellular oxidative stress capable of triggering the gelation of lung surfactant, such as that present in patients with genetic abnormalities in lung SP-C associated with familial idiopathic pulmonary fibrosis (Anseth et al. 2005).

Infants born preterm lack the surfactant stores present in the lungs of full-term infants and thus are maladapted to make the transition from a fluid-filled lung to an expanded air-filled lung (Clark 2010). Exogenous surfactant therapy administered at birth or soon after helps the acute transition to air breathing by directly lowering surface tension and by providing a substrate that can be taken up by alveolar epithelial type II cells and facilitate endogenous surfactant production (Clark 2010).

4.4 Effect of NPs on Fluidity of the Lung Surfactant Layer—Role of NP–Protein Adsorption

During interaction with, or passage of NPs through, the alveolar surfactant, the biophysical functioning of the film may be altered (Sachan et al. 2012). Detailed biophysical interaction studies of NPs with the PS film are emerging in the literature, both specifically focused on the lung–air interface (Sachan et al. 2012) and more generally as a paradigm for NP uptake (Wang et al. 2008).

Fan et al. (2011) studied the mechanism of interaction of 90-nm hydroxyapatite NPs (HA-NPs) with a model PS (Infasurf) to understand whether NPs have adverse effects on the biophysical function of PS. Infasurf contains all the hydrophobic components of natural surfactant, including ~90 wt% phospholipids, 5%–8% neutral lipids (mainly cholesterol), and ~2% hydrophobic surfactant proteins (SP-B and SP-C) (Zhang et al. 2011). Using a variety of (bio)physicochemical characterization techniques, including atomic force microscopy (AFM), transmission electron microscopy (TEM), and Langmuir–Blodgett, the molecular mechanism of *in vitro* interaction was assessed. A time-dependent toxicological effect of HA-NPs on the PS was observed, with a significant time-dependent deterioration of the Infasurf layer after exposure to the HA-NPs at 50 μg/ml. The comparison of lateral structure shows variations to both domain formation in surfactant monolayers and organization of surfactant multilayers due to exposure to HA-NPs (Figure 4.4). HA-NPs penetrated the Infasurf films and altered both the compression isotherms and lateral structure. The HA-NPs altered the monolayer-to-multilayer transition of Infasurf. Without the NPs, Infasurf films at the physiologically relevant π (i.e., 50 mN/m) assume a conformation of uniformly distributed fluid phospholipid multilayers with embedded condensed phospholipid domains (as "holes") at the interfacial monolayer. However, after exposure to HA-NPs, the matrix structure disappeared and the multilayers were formed as isolated crystalline folds along the direction of lateral compression, thus causing surfactant inhibition. The inhibition mechanism was found to be due to adsorption of surfactant proteins onto the HA-NPs. Consequently, depletion of proteins from surfactant vesicles caused conversion of original large phospholipid vesicles into much smaller vesicles. These small vesicles, in turn, inhibited the surface activity of surfactant films after adsorption. While this did not translate into observations of cytotoxicity to human bronchial epithelial BEAS-2B cells, it suggests that conventional *in vitro* cytotoxicological testing alone may not be sufficient in evaluating the toxicological effect of inhaled NPs, and that the potential for changes to PS film structure and function, as a result of binding of surfactant proteins, also need to be considered. A point to note here, and for all subsequent studies mentioned, is, in many *in vitro* studies, particles are applied in doses of 1–100 μg per 10^5 cells. Assuming the application of the lowest dose

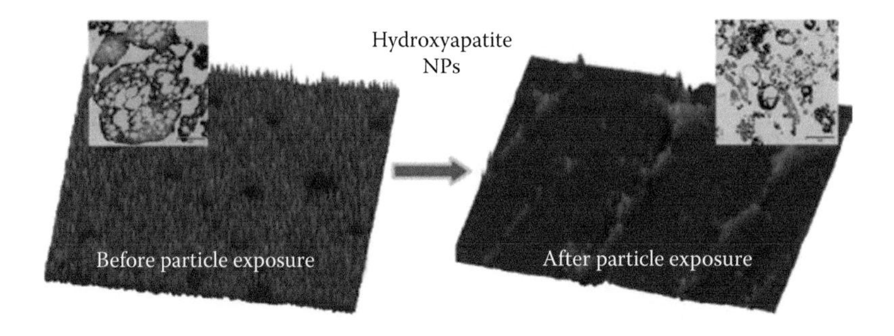

FIGURE 4.4
AFM (and inset TEM) images of pure Infasurf before and after exposure to 50 μg/ml HA-NPs at for 3 h at comparable surface pressures. The scan area of the AFM images was 50 × 50 μm. The comparison of lateral structure shows variations to both domain formation in surfactant monolayers and organization of surfactant multilayers due to exposure to HA-NPs. (From Fan, Q. et al., *ACS Nano 5*, 6410–6416, 2011.)

of 1 μg particles per 10^5 cells and the use of spherical NPs of unit density and a size of 50 nm, 2×10^5 NPs per cell are applied (Schumann et al. 2012). Interestingly, reviews of therapeutic applications of NPs in the lung are generally completely lacking in data regarding the dose of nanomaterials used (Azarmi et al. 2008); therefore, it is difficult to determine whether the NP doses in the studies reported here, and especially where negative impacts on lung fluidity were observed, are realistic exposure doses. However, from aerosol exposure data, at the highest possible NP aerosol number concentration of 1×10^6/ cm^3, an alveolar surface cell will receive, on average, 120 NPs per hour. Even if we consider a maximum factor of 20 for inhomogeneous deposition beyond an otherwise rather homogeneous diffusional deposition in the peripheral lungs, the NP dose to some surface cells may increase by this factor at most (Lynch et al. 2007), suggesting that many of the *in vitro* and model studies discussed here are using greatly exaggerated concentrations.

A study of the interaction of 15-nm gold (Au) NPs with semisynthetic PS (DPPC/palmitoyloleoyl-phosphatidylglycerol [POPG] in the ratio 70:30) in the absence or presence of 1% SP-B demonstrated that low levels of Au NPs (3.7 mol% Au/lipid, 0.98% wt/ wt) impeded the surfactant's ability to reduce surface tension (γ) to low levels during film compression and to respread during film expansion (Bakshi et al. 2008). The surface activity of these surfactant systems was determined via captive bubble tensiometry, showing that DPPC/POPG (70:30) suspensions initially adsorb relatively rapidly, and that inclusion of 1% SP-B augmented adsorption such that the surfactant system reached the equilibrium γ (γ_{eq}) of 23 mN/m at ~15 min. Addition of various amounts of naked Au NP to DPPC/POPG/SP-B mixtures led to a decrease in adsorption. The adsorption in the presence of Au NP was considerably slower and less complete than in the absence of SP-B. TEM analysis demonstrated that the palmitoyl-oleoyl-phosphatidylglycerol appeared to coat the NPs with at least one lipid bilayer but did not affect NP shape or size, whereas the presence of SP-B (1% wt/wt) appeared to induce the formation of elongated strands of interacting threads with the fluid phosphatidylglycerols (PG), as shown in Figure 4.5. By contrast, zwitterionic DPPC-capped Au NP exhibited a different overall organization where the DPPC-capped particles appeared clumped in several layers. Thus, it appeared that Au NP becomes "capped" in the presence of acidic surfactant phospholipids. This interaction was markedly altered in the presence of SP-B, which resulted in highly aggregated Au NP. This latter observation may, in part, explain the highly deleterious effect of the Au NPs on the ability of surfactant phospholipids to adsorb and to reduce surface tension to low values, as determined during surface area expansion/compression cycling. The authors concluded that Au NPs can interact with and sequester PS phospholipids and, if inhaled from the atmosphere, could impede PS function in the lung (Bakshi et al. 2008).

Sachan et al. (2012) investigated the impact of hydrophobic polyorganosiloxane (AmOrSil20, 22 nm in diameter) NPs on the integrity and structural organization of a model PS film using scanning force microscopy (SFM) and electron microscopy to visualize the topology and characterize the localization of NPs within the compressed PS film. The authors showed that the NPs partition in the fluid phase of the compressed film at lower surface pressure, while at higher surface pressure the NPs interact extensively with the surface-associated structures. Major amounts of NPs were retained at the interface and released slowly into the aqueous subphase during repeated compression/expansion cycles. From the fact that the NPs were unable to spontaneously translocate into the subphase from the interfacial PS film and slowly release from the surface layer under repeated compression/expansion cycles, the authors concluded that the presence

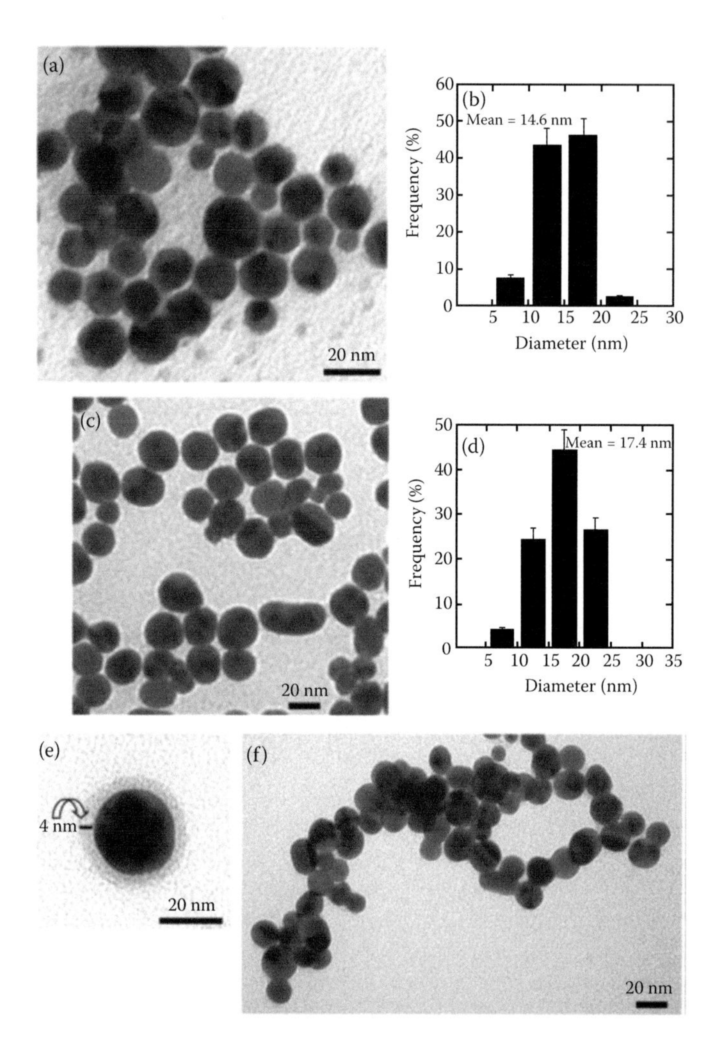

FIGURE 4.5
TEM micrographs of (a) blank (noncapped) Au NPs and (c) POPG-capped Au NPs. The Au NP size distributions for (b) blank Au NPs and (d) POPG-capped Au NPs are depicted. An image consistent with the presence of a POPG bilayer around a single Au NP is shown in (e). Strand formation induced by including 1% wt/wt SP-B with POPG is shown in (f). (From Bakshi, M.S. et al., *Biophys. J.*, *94*, 855, 2008.)

of a surfactant monolayer at the interface offers a high free-energy barrier for such hydrophobic NPs due to their interaction and association with the surfactant components and structures. AFM studies demonstrated that an attractive interaction occurs between the tip and the NP surface, which can result owing to possible NP coverage with lipids, where fatty acid tails are oriented toward the NP surface whereas hydrophilic head groups are exposed to the cantilever tip. They concluded from this that the NPs localized in the compressed PS film are opsonized by the surfactant components. Thus, hydrophobic AmOrSil20 NPs up to ~50 μg/ml do not substantially affect the structural organization and functioning of PS film; however, such NPs do show drastic impacts at higher concentrations, although whether such concentrations represent realistic exposures is a separate question, as alluded to above (Sachan et al. 2012). A related study of the interaction of polyorganosiloxane NPs (AmorSil20, 22 nm in diameter) with lipid monolayers characteristic of alveolar surfactant investigated the formation of domain structures and the changes in the surface pattern induced by NPs using film balance measurements combined with video-enhanced fluorescence microscopy (Harishchandra et al. 2010). The NPs were observed to strongly disturb the domain structure and thus the phase behavior of the fluid and condensed domains and to concentrate at the domain borders. NPs were incorporated into lipid monolayers with a clear preference for defect structures at the fluid–crystalline interface, leading to considerable monolayer expansion and fluidization. NPs remained at the air–water interface, probably by coating themselves with lipids in a self-assembly process, thereby exhibiting hydrophobic surface properties (Harishchandra et al. 2010). Such changes in lipid fluidization could themselves result in altered/enhanced uptake of the NPs (Dawson et al. 2009), and indeed there have already been suggestions that lipid rafts might represent a new route for (viral) entry into cells (Pietiainen et al. 2004), and that NPs are capable of inducing local lipid phase changes depending on their surface charge (Wang et al. 2008), which points to the potential disturbance of PS monolayers by NPs via creation of local raft scenarios. Membranes formed from lipids with phosphocholine head groups were studied by Wang et al. (2008), and anionic NPs were shown to induce local gelation in fluid membranes whereas cationic NPs induced local fluidization of otherwise gelled membranes. A clear dependence on the charge density and charge mobility was observed, with less negatively charged silica NPs resulting in a lesser degree of membrane gelation, while adsorbed DNA, which is also negatively charged, did not induce this effect (Pietiainen et al. 2004; Dawson et al. 2009). Studies such as these remind us that another scientific direction, based on direct physical interactions and modulation of membrane fluidity, could also play a role in NP uptake and potential for toxicity (Dawson et al. 2009).

Multiple other studies assessed similar impacts from a range of NPs, including charged NPs (200-nm carboxyl- or amine-modified polystyrene) aerosolized as droplets (Farnoud and Fiegel 2011); poly(D,L-lactide-*co*-glycolide) NPs with a mean size of 100 nm (Beck-Broichsitter et al. 2011); nanosized and microsized titanium dioxide (TiO_2) particles (Schleh et al. 2009) and nanoclays, such as bentonite NPs (median hydrodynamic diameter, ~153 nm); and others (Kondej and Sosnowski 2012). NP dose, size, and charge dependence effects were all observed, with behavior also depending on the composition of the surfactant film (Schleh et al. 2009; Beck-Broichsitter et al. 2011; Farnoud and Fiegel 2011; Kondej and Sosnowski 2012). An all-atom molecular dynamics simulation of a model PS film interacting with a carbonaceous NP, with a carbon cage structure reminiscent of buckyballs with open ends, demonstrated that the NP affected the structure and packing of the lipids and peptide in the film, and it appears that the NP and peptide repel each other (Choe et al. 2008). The study predicts that the NP can easily penetrate the monolayer; however, further translocation to the water phase is energetically prohibitive.

Interestingly, there were no surfactant proteins used in the two NP–phospholipid fluids interaction studies cited above (Harishchandra et al. 2010; Sachan et al. 2012); thus, the potential mediating or amplification effects of the surfactant proteins could not be assessed. The lipid–peptide components in the surfactant monolayer interact with and complement each other during compression and expansion of the breathing cycle. According to the so-called squeeze-out theory, during compression fluidizing lipids and surfactant proteins are believed to be selectively squeezed out, leaving behind a monolayer, at the interface, enriched in lipids (mainly DPPC) that promotes low surface tension (Watkins 1968; Pastrana-Rios et al. 1994). The excluded substances form the multilamellar structures just beneath the surfactant monolayer and quickly respread on expansion. These topographic structures have been detected and visualized by means of SFM and TEM (Krol et al. 2000a,b), which revealed that SP-B and SP-C are absolutely essential for the formation of the multilayer structures necessary to prevent alveolar collapse as well as adsorption of materials onto the interface, thus facilitating the normal breathing process.

It is clear, however, that for a true understanding of this first critical contact between NPs and the lungs, more complete models of the alveolar lining fluid and the air–liquid interface need to be used to represent the full complexity of the potential interactions. Indeed, Schleh et al. (2013) draw the same conclusion in terms of *in vitro* models for NP toxicity assessment in cellular models, where they suggest that although each inhaled NP, reaching the alveoli, will come into contact with PS, which will probably lead to a surfactant coating, PS components are not commonly integrated in *in vitro* systems. In this commentary, they highlight the efforts by Geiser et al., who have shown that this surfactant coating is able to influence further interaction with cellular systems, and recommend that scientists working with *in vitro* systems and NPs should integrate PS structures into their cellular model to harmonize the *in vitro* systems with the *in vivo* situation.

4.5 NP Uptake via Inhalation—Role of Adsorbed Biomolecules

In the alveolar region of the lung, where 10–100-nm-diameter particles are predicted to deposit efficiently, there is a limited number of cells with which to interact in a healthy organ, namely alveolar macrophages and type I and type II alveolar epithelial cells (Lynch and Elder 2009). Particles that agglomerate and remain in that state in alveolar lining fluid may be taken up by alveolar macrophages and removed via mucociliary clearance. However, this clearance mechanism does not work very efficiently for many NPs (Ahsan et al. 2002), thus promoting their retention in the lungs and possibly leading to interactions with epithelial cells. On the other hand, where NPs bind to surfactant proteins (e.g., SP-A), uptake and hence translocation of the NPs might be facilitated, as type II cells, which cover most of the alveolar region, have SP-A receptors that cause phospholipid uptake (Stevens et al. 1995). NPs could be taken up "accidentally" via this surfactant trafficking route, akin to the well-established route of NP translocation across the blood–brain barrier via adsorption of apolipoprotein A or E (Kreuter 2013). This theory is supported by data from *in vivo* experiments that also indicate NP uptake into epithelia of the respiratory tract via transcytosis and translocation into the lymphatic system or the bloodstream (Oberdörster et al. 2005). Through mechanisms such as endocytosis and passive transcellular or paracellular translocation, NPs can gain access to the interstitial space and blood (Lynch and Elder 2009).

Several studies have now shown that the alveolar epithelium, at least, permits transfer of nanosized particles into the interstitial space. Geiser et al. (2005) reported that a substantial fraction (~20%) of inhaled nanosized TiO_2 particles could be found in alveolar epithelial cells, the interstitium, and blood cells within 1 h of exposure. Other studies have also shown a high degree of interstitialization for nanosized ^{192}Ir (Semmler-Behnke et al. 2007). A recent ultramicroscopic study of uptake and localization of 16-nm Au NPs by rats, exposed via inhalation for 6 h, revealed that alveolar macrophages engulfed inhaled Au NPs very frequently (Takenaka et al. 2012). Internalized Au NPs were found in 94% of macrophages obtained by bronchoalveolar lavage directly (day 0) after a 6-h inhalation period. This high rate of internalization by alveolar macrophages was not expected; therefore, the authors concluded that low lavagable fractions after NP inhalation are not due to the inability of alveolar macrophages in NP recognition and endocytosis, but that other issues have to be considered, including the total NP burden (NP number) in the lung and the rate of elimination of NPs from the alveolar lumen by alveolar macrophages (Takenaka et al. 2012). Although alveolar macrophages efficiently engulf NPs in their immediate vicinity, as shown by Semmler-Behnke et al. (2007), a significant number of NPs deposited far away from the macrophages may escape their endocytotic clearance function, with the result that the interstitium may be the site of uptake/deposition of the NPs.

For nanosized particles delivered via inhalation as singlets, mathematical predictions suggest that they will efficiently deposit via diffusional processes in all regions of the respiratory tract, although the highest fractional deposition for particles of ~10–100 nm occurs in the alveolar region (International Committee on Radiological Protection 1994). Two important anatomical features of this region of the respiratory tract are (i) the large surface area of the alveolar epithelium and (ii) its high degree of vascularization. Deposition also occurs, however, in the tracheobronchial and naso-pharyngeal–laryngeal regions, which contain projections of sensory nerves. Dendrites of the olfactory nerve, for example, project directly into the nasal epithelium (Lynch and Elder 2009).

One important approach to understanding NP uptake following lung exposure is to assess the role of lung surfactant proteins in the uptake. The pulmonary collectins SP-A (630 kDa) and SP-D (520 kDa) are of exceeding interest for bio–nano interactions because they fulfill important immunological functions by acting as opsonins and scavenger molecules (Kishore et al. 2006). Their occurrence at the air–liquid interface of alveoli ideally enables these two proteins to interact with, and bind to, airborne particulate matter deposited into the deep lungs and thus make first contact (Seaton et al. 2010). Furthermore, the fact that SP-A and SP-D can influence the uptake of particulate matter by alveolar macrophages also allocates them a key role in the clearance of inhaled NPs (Wright 2005).

Ruge et al. (2012) compared the effects of SP-A and SP-D on alveolar macrophage clearance of magnetite NPs (mNPs) with different coatings. Both proteins were found to enhance the alveolar macrophage uptake of mNPs compared with pristine NPs; for the hydrophilic starch-coated mNPs, this effect was strongest with SP-D, whereas for the hydrophobic phosphatidylcholine-coated mNPs it was most pronounced with SP-A, as shown in Figure 4.6. Interestingly, in a previous study of cellular binding and uptake of these same NPs by alveolar macrophages, uptake was increased for NPs with adsorbed SP-A, whereas adsorption of bovine serum albumin (BSA), the prevailing protein in plasma, led to a significant decrease of uptake (Ruge et al. 2011). These findings emphasize that the protein coating formed around an NP, and thus the biological behavior, varies with the respective physiological compartment with which an NP first has contact. While this initial corona likely evolves as the NPs are taken up by cells and translocated (Lundqvist et al. 2011), the ultimate fate and behavior are likely linked to the nature of this initial corona.

However, as per comments in Section 4.4 on NP impacts on the phospholipid layer, studies of NP uptake in the presence of surfactant proteins alone are only part of the story. Ruge et al. (2012) also investigated the influence of various surfactant lipids on NP uptake

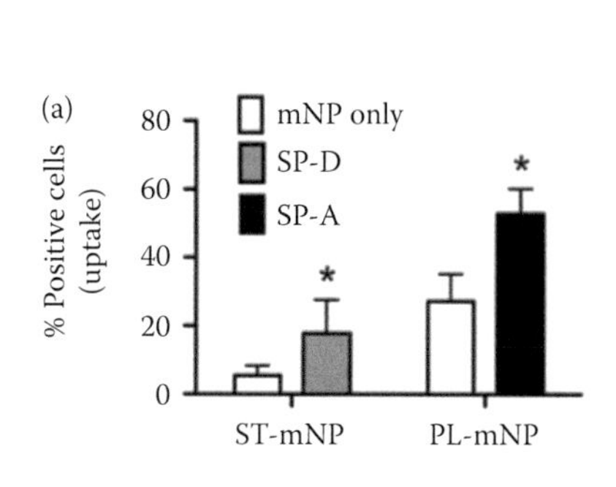

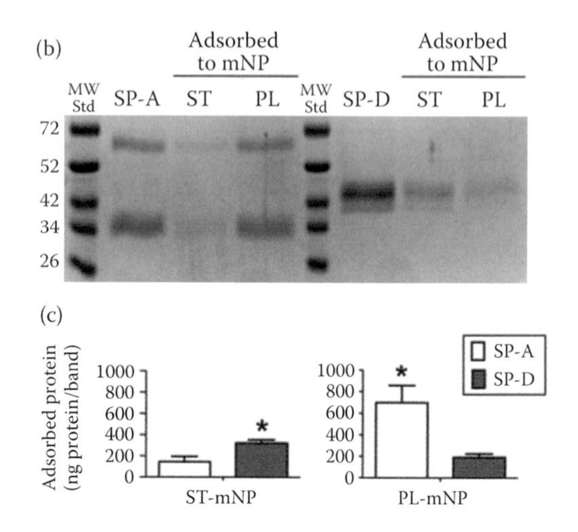

FIGURE 4.6
(a) Dependence of NP uptake by alveolar macrophages on surfactant proteins as studied by confocal microscopy. Representative micrographs are shown for starch-modified magnetite NPs (ST-mNP) or phosphatidylcholine-modified mNPs (PL-mNP) after 90-min exposition to alveolar macrophages (MH-S cells) in the absence (mNP only in buffer) or presence of SP-D (10 μg/ml) or SP-A (10 μg/ml). Particle uptake was determined by visual counting of cells with at least one NP internalized related to total cell count (% positive cells). Data shown as mean ± SD (n = 14 images). Asterisk indicates a significant difference with $p < 0.05$. (b, c) Adsorption of SP-A and SP-D to NPs is affected by particle coating and hydrophobicity. (b) Representative SDS-PAGE gel displaying the adsorbed surfactant proteins (SP-A in the left lanes or SP-D in the right lanes) eluted from starch- (ST) or phosphatidylcholine-modified (PL) mNPs; MW Std, molecular weight standard. (c) Adsorbed amount of protein in nanograms of protein per band.

by alveolar macrophages in the absence and presence of the SP-A or SP-D coronas because lipids are the major surfactant component. Synthetic surfactant lipid and isolated native surfactant preparations significantly modulated the effects exerted by SP-A and SP-D, resulting in comparable levels of macrophage interaction for both starch- and phosphatidylcholine-modified (PL) mNPs. These findings indicate that because of the interplay of both surfactant lipids and proteins, the alveolar macrophage clearance of NPs is essentially the same, regardless of different intrinsic surface properties (Ruge et al. 2012). On the basis of this work, the schematic shown in Figure 4.7 for NP interactions with the lungs was proposed, which reflects the fact that both lipids and surfactant proteins play a role in NP penetration, uptake by alveolar macrophages and subsequent clearance, or translocation of the NPs to the interstitium (Ruge 2012).

Another study by this team demonstrated that there are specific differences in the binding and interaction of metal oxide NPs with SP-A (Handy et al. 2012). Attempts to correlate the adsorption patterns of SP-A with those of commonly used model proteins (e.g., BSA) failed, underscoring the need to apply sufficiently specific and sensitive analytical methods. Additionally, recovery of the SP-A from the NP surface was not always possible because of the very strong binding to particles such as carbon black, despite the harsh conditions used to remove the proteins (boiling in SDS solution) (Handy et al. 2012). Such strong binding of SP-A by inhaled metal oxide NPs could have important toxicological consequences, as potentially SP-A deficiency could occur owing to its accumulation onto particles; however, in the absence of a full understanding of the likely NP exposure amounts and other biomolecules likely to contribute to the corona, this is speculation at present (Handy et al. 2012). Lowered SP-A levels are associated with

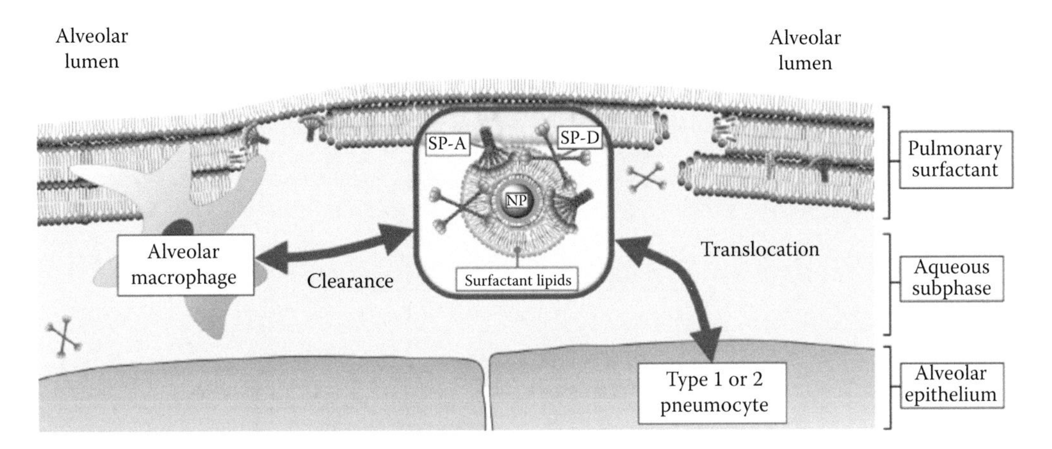

FIGURE 4.7
Pulmonary protein corona influences further biological response such as alveolar macrophage clearance or translocation across the epithelium. (From Ruge, C.A. Bio–nano interactions in the peripheral lungs: Role of pulmonary surfactant components in alveolar macrophage clearance of NPs. Dissertation, Saarbrücken, 2012.)

asthma and allergen-induced bronchial inflammation (Hohlfeld 2002; Hickling et al. 2004). Additionally, in contrast to other proteins such as BSA, SP-A does not seem to significantly deagglomerate large agglomerates of metal oxide NPs, indicating different adsorption mechanisms compared with those active in the well-investigated model protein BSA (Handy et al. 2012).

Single-walled carbon nanotubes (SWCNTs) were found to selectively adsorb two types of the most abundant surfactant phospholipids: phosphatidylcholines and phosphatidylglycerols (Kapralov et al. 2012). Quantitation of adsorbed lipids by liquid chromatography–mass spectrometry (LC–MS), along with the structural assessments of phospholipid binding by AFM and molecular modeling, indicated that the phospholipids (~108 molecules per SWCNT) formed an uninterrupted "coating" whereby the hydrophobic alkyl chains of the phospholipids were adsorbed onto the SWCNT, with the polar head groups pointed away from the SWCNT into the aqueous phase. In addition, the presence of SP-A, SP-B, and SP-D on SWCNTs was determined by LC–MS. Both lipids and SWCNTs were found to bind at a similar binding site involving the trimeric interface of the protein, indicating that SP-D can interact with both lipids and SWCNTs. The interaction of SP-D with DPPC was stabilized via the head group and not via the acyl chains. This indicates that SP-D has the potential to interact with SWCNTs precoated with surfactant phospholipids, whereby the head groups are projected away from the SWCNTs. The exposure of the polar head groups and the masking of the hydrophobic regions of the SWCNT could lead to enhanced recognition of SWCNTs by immune cells expressing specific receptors for lipid-dependent uptake of particulate matter or cellular debris. The authors also demonstrated that the presence of this surfactant coating containing lipids and proteins markedly enhanced the *in vitro* uptake of SWCNTs by macrophages, as shown in Figure 4.8. Taken together, this is the first demonstration of *in vivo* adsorption of surfactant lipids and proteins onto SWCNTs in a physiologically relevant animal model of pulmonary exposure, and the effect of this adsorption on phagocytosis by macrophages. These *in vitro* data suggest that the surfactant coating of SWCNTs may determine the recognition of these nanomaterials *in vivo* and therefore may influence the biodistribution and fate of these nanomaterials, including their biodegradation (Kagan et al. 2010; Shvedova et al. 2012).

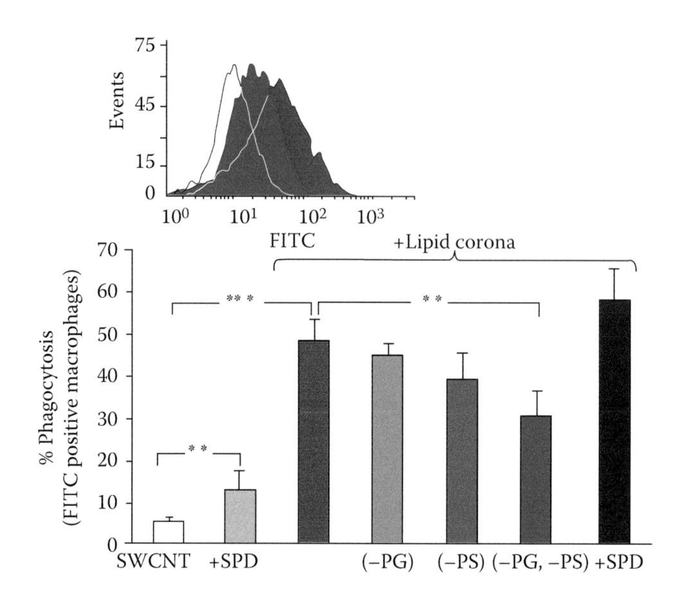

FIGURE 4.8
Effects of surfactant phospholipids and SP-D on the uptake of SWCNTs by RAW264.7 macrophages. RAW264.7 macrophages were cultured overnight, then washed with PBS and incubated with SWCNTs or phospholipid-coated SWCNTs or phospholipid/SP-D-coated SWCNTs for 1 h at 37°C. The cells were then washed, detached, and resuspended in 0.2% trypan blue for flow cytometric analysis. Phagocytosis is reported as the fraction of fluorescein isothiocyanate–positive macrophages. Inset: histogram presentation of flow cytometry data for macrophages incubated with SWCNTs, alone (open trace), phospholipid-coated SWCNTs (red), phospholipid-coated SWCNTs deficient for PG and PS. The data are presented as percent phagocytosis mean values ± SD of three independent experiments; $^{**}p < 0.01$, $^{***}p < 0.001$.

4.6 Is This Role of Surfactant Proteins in Mediating Uptake a Universal Effect?

Given the significant focus on understanding the potential impacts of engineered NPs, it is easy to forget that we have long been exposed to a range of naturally occurring NPs, including, for example, pollen particles. An elegant study by Schleh et al. (2012a) has demonstrated that binding of SP-D mediates the cellular uptake of allergen particles such as subpollen particles (SPPs) within a complex lung cell culture model, and the secretion of cytokines/chemokines. SPPs were found intracellulary in all three of the cell types incorporated into the coculture model: A549 cells (epithelial cells), human monocyte-derived macrophages (MDMs), and monocyte-derived dendritic cells (MDDCs) within the human epithelial airway model. Importantly, SP-D was able to modulate this uptake and the subsequent distribution of the SPPs, as shown in Figure 4.9. This study was able to reliably determine SPP uptake in contrast to binding to the cell membrane, and focused only on the intracellular SPPs. The use of the triple-culture model using human type II pneumocyte-like cells (epithelial cells), human MDMs, and MDDCs allowed the authors to identify modifications to SPP uptake resulting from cell–cell contact or paracrine signaling, since previous studies showed that SP-D was able to modulate the interaction of SPPs with human primary bronchial epithelial cells but not with A549 cells (Schleh et al. 2010), while the A549 cells incorporated in the present epithelial airway model reacted much more sensitively to contact with SP-D and SPP. Contact with other lung cells such as immune cells may modulate the A549 cells to react

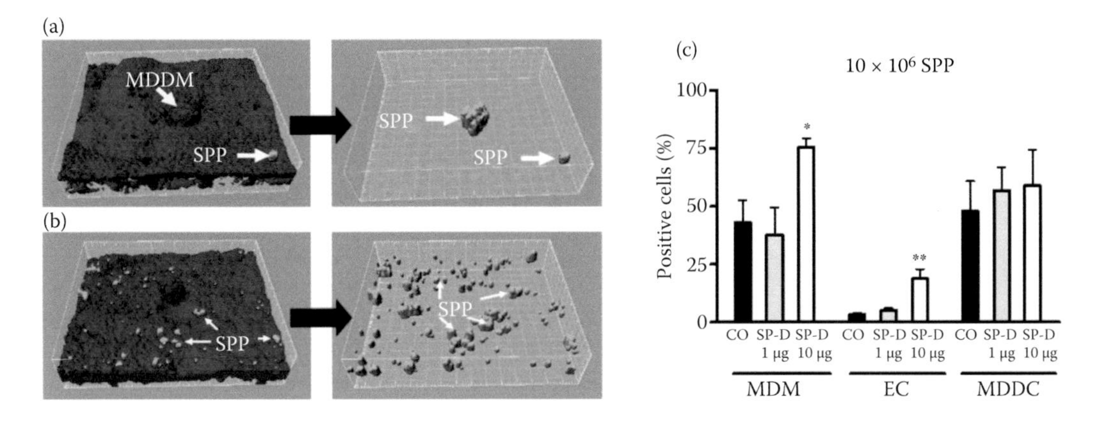

FIGURE 4.9
(a, b) Pictures displaying the strong modulation of SP-D on distribution of SPPs within a human epithelial airway model. Left: Overview of the upper side of the epithelial airway model after incubation with SPPs with or without SP-D. Right: Internalized and attached SPPs after masking of cell borders. Green, SPP (Alexa 488); red, F-actin (rhodamine phalloidin); blue, CD14. Arrows on the left pictures point to SPPs that are attached on the surface of the epithelial cells; arrows on the right pictures point to SPPs inside cells. (c) Influence of SP-D on (A)% cells in an epithelial airway model that have taken up SPPs. CO, control; MDM, monocyte-derived macrophages; EC, epithelial cells; MDDC, monocyte-derived dendritic cells.

more similarly than human primary epithelial cells. Using a complex three-dimensional human epithelial airway model, which simulates the most important barrier functions of the epithelial airway, it has also been shown that SP-D modulated the uptake of SPPs in a cell type–specific way (e.g., increased number of macrophages and epithelial cells that participated in allergen particle uptake) and led to a decreased secretion of pro-inflammatory cytokines (Schleh et al. 2012a). While studies using other natural NPs are limited, it seems likely that this is indeed a universal effect.

4.7 Clear Linkages between Adsorbed Biomolecules and Mediation of NP Toxicity Effects

Kuroda et al. (2006) demonstrated in rats that phospholipid concentrations in lung lavage fluid were increased significantly throughout a 6-month period following crystalline silica exposure and that the increases correlated with the severity of the inflammatory response. However, that was the outcome for a clearly toxic particle, and such particles are likely the exception rather than the rule. Indeed, even with these inherently toxic materials, interaction with biomolecules has been demonstrated to mediate toxicity, at least during the uptake and initial localization stages. For example, *in vitro* studies of uptake of silica NPs in the absence or presence of serum proteins in the cell culture medium resulted in very different uptake amounts and cellular impacts (Lesniak et al. 2012). Silica NPs exposed to cells in the absence of serum had a stronger adhesion to the cell membrane and higher internalization efficiency, compared with their behavior in medium containing serum, when a preformed corona was present on their surface. The different exposure conditions not only affect the uptake levels but also result in differences in the intracellular NP location and impact on cells: cells exposed to silica in serum-free conditions changed their phenotype and assumed a spherical shape, indicative of loss of cell adhesion and cell damage. This also points to the fact that NP behavior is context dependent; thus, *in vitro* studies, especially related to understanding NP impacts on the lung, which is a particularly complex organ, need to consider the total milieu including proteins and other biomolecules that may mediate the NP–cell interactions and thus the NP impact.

An example of a reduction of the toxicity of NPs is given by the study of Gao et al. (2001), who measured apoptosis in rat alveolar macrophage NR8383 cells challenged *in vitro* with respirable quartz or kaolin dust, and with the same dusts pretreated with DPPC, to model conditioning of respired dusts by interaction with a primary phospholipid component of PS. NR8383 cells exposed to native quartz at concentrations from 50 to 400 μg/ml for 6 h showed a dose-dependent increase in apoptosis (measured by several classic end points such as the TUNEL, cell death ELISA, and DNA ladder formation assays), while native kaolin induced significant response only at the higher concentrations and only in the TUNEL and ELISA assays. DPPC pretreatment suppressed quartz activity until 3 days and kaolin activity through 5 days. Cellular release of lactate dehydrogenase, measured in parallel experiments to compare dust apoptotic and necrotic activities, indicated that components of serum as well as surfactant may affect kaolin's *in vitro* expression of those activities (Gao et al. 2001). The fact that the effects were only mitigated for a period of time is consistent with the corona of biomolecules (lipids in this case) being digested by the acidic environment of the lysosomes, which are the final destination of most NPs taken up by cells (Pathak et al. 2008; Sandin et al. 2012).

Another example looked at the effects of diesel particulate material (DPM) on alveolar epithelial cells using A549 cells and assessed the release of interleukin-8 (IL-8) into the conditioned medium (Seagrave et al. 2004). While low concentrations of DPM increased IL-8, higher doses appeared to suppress the response, although this suppression was not related to acute DPM toxicity. The DPM-induced loss was only weakly blocked by a large excess of BSA—a large excess of this nonspecific protein failed to block the binding of IL-8 to the DPM particles. The authors suggest that this observation was particularly important since it suggests the specificity of the IL-8 interaction with DPM particles. While *in vivo* many other proteins are present in the epithelial lining fluid that might compete with IL-8, this specificity is significant. To determine the biological implications of the IL-8 binding, human blood neutrophils were exposed to DPM that had been preincubated with IL-8, then washed to remove free IL-8. The neutrophils changed their shape in a manner suggesting directed movement toward the particles. No morphological change was observed either with carbon black that had been incubated with IL-8 or with DPM alone. These results suggest that DPM not only induces the production of IL-8 by epithelial cells but also binds biologically active chemokine in a particle- and protein-selective manner. DPM-induced inflammatory responses may therefore be more focused or sustained as a result of this binding of inflammatory mediators by DPM (Seagrave et al. 2004). Effects of particle surface conditioning by a primary component of phospholipid PS, diacyl phosphatidylcholine, on *in vitro* expression of genotoxicity by diesel exhaust particles and cytotoxicity by respirable quartz and aluminosilicate kaolin clay particles were reviewed by Wallace et al. (2006). Their conclusion, which has been echoed in several studies since, is that diesel exhaust nanoparticulate material and respirable micrometer-sized mineral dust expression of *in vitro* cytotoxicity or genotoxicity can be strongly affected by particle surface conditioning by a phospholipid component of lung surfactant, modeling an initial *in vivo* phenomenon not usually considered for assays of respirable particle toxicities.

Therefore, particle surface composition and its PS conditioning should be considered in the design and interpretation of *in vitro* cytotoxicity or genotoxicity assays of NP, and for *in vivo* assays that might involve disruption of the pulmonary hypophase in the lung of the animal model, for example, in some localized regions during instillation challenge (Wallace et al. 2006).

There is also an emerging understanding that adsorption of biomolecules can alter the biodistribution of NPs, and thus that different accumulation can result from the same NPs exposed via different routes. A study of the size and charge of Au NPs on their absorption across intestinal barriers and accumulation in rats observed higher accumulation of 18-nm gold particles in the brain compared with even smaller particles (1.4 and 5 nm), which the authors suggested could be related to differences in their protein coatings (Schleh et al. 2012b). They hypothesized that the specific curvature and surface structure of the 18-nm particles alter the structure and function of single adsorbed proteins or selects proteins with an increased epithelial penetration probability compared with the other NPs used. Thereby, a specific increased absorption across intestinal membranes occurs. Importantly, a study from the same laboratory with exactly the same NPs, injected into the tail vein of rats, showed no special modulation of organ accumulations of the 18-nm NPs (Hirn et al. 2011). This supports the interpretation that special intestinal incidents (e.g., protein binding) are responsible for these results. Further (as yet unpublished) data from this group show very clear effects from grafting of specific proteins such as albumin or apolipoprotein E on the organ distribution of NPs, and a clear dependence on the route of exposure (inhalation vs. intravenous). A recent study using a computational multiscale toxicodynamic model, consisting of a collection of coupled toxicodynamic modules developed to describe the dynamics of tissue, with explicit focus on the cells and surfactants that regulate the process of breathing in order to account for processes occurring at multiple biological scales, compared the model predictions with *in vivo* lung function response measurements in mice and analysis of mice lung lavage fluid following exposures to silver and carbon NPs (exact sizes not given) (Mukherjee et al. 2013). The model successfully simulated the time dynamics of the pulmonary alveolar surfactant system in response to particle inhalation, and related surfactant phospholipid levels in the alveolar region to changes in mechanical responses of the lung, with the differences in lung function values between treated and control mice following similar trends as the total phospholipid levels; that is, the differences are minimal at 1 day, increase at 3 days, and decrease again at 7 days (Mukherjee et al. 2013). This represents an important research direction, and an approach to teasing out the impact of NPs, on the basis of their interactions with the components responsible for ensuing lung function, specifically the phospholipids and surfactant proteins, and indeed on the effect of NP physicochemical parameters on these interactions to facilitate the design of safer NPs, especially those for use in medicine.

4.8 Key Directions for Future Research

Having established beyond any doubt that the adsorbed biomolecule corona plays a crucial role in determining the uptake, translocation or clearance, and subsequent impacts of NPs interacting with living systems, there are still a number of important questions that need to be addressed in more specific detail in order to move toward being able to predict from the NP composition the likely interaction partners and their consequences. Some of these questions are highlighted in the following, to stimulate further research into this important topic.

Do lung surfactant proteins reduce or amplify the effects from NPs? Is this dependent on the NP type and/or size? Or the cell type? Or are these effects averaged out when lipids are also included in the biomolecule corona, as was the case in terms of overall particle uptake in surfactant protein and lipid coronas, in studies by Ruge et al.? These findings indicate that because of the interplay of both surfactant lipids and proteins, the alveolar macrophage clearance of NPs is essentially the same, regardless of the different intrinsic surface properties of the tested NPs (Ruge et al. 2012). There is a body of emerging data on this; however, given the enormous diversity of the study conditions (i.e., which surfactants/lipids, which proteins, etc.) and the fact that most studies use overly simplified models of the air–liquid interface, it is difficult at this point to draw any significant conclusions beyond the need to consider all of the relevant components in *in vitro* models and in the design of future experiments to address these questions. Thus, more complete models of the alveolar lining fluid and the air–liquid interface that include surfactant proteins as well as phospholipids need to be used in future studies to represent the full complexity of the potential interactions and facilitate meaningful conclusions regarding the impacts of NPs.

Do the observed effects depend on the protein conformation and form upon binding? For example, are there differences in terms of whether the bound form is the protein monomer subunit, versus oligomer? Are there effects related to unfolding of proteins at specific particle sizes? A recent set of studies have shown that even very small differences in particle size can result in significant differences in protein form/conformation and subsequent NP impacts (Deng et al. 2010, 2012). The number of proteins per NP was shown to vary dramatically for poly(acrylic acid)-coated Au NPs ranging in size from 7 to 22 nm (Deng et al. 2012). Each fibrinogen molecule could accommodate two 7-nm NPs, but only one when the diameter increased to 10 nm. NPs larger than 12 nm bound multiple fibrinogen molecules in a positively cooperative manner. However, in the presence of excess NPs, fibrinogen induced aggregation of the larger NPs (i.e., those that could bind more than one protein molecule). This is consistent with interparticle bridging by the fibrinogen (Deng et al. 2012). This finding is related to the fact that the 5-nm poly(acrylic acid)-coated Au NPs interacted with the Mac-1 receptor of THP-1 cells as a result of changes to the fibrinogen structure due to binding, most likely exposing the γC-terminus and resulting in a significantly increased binding of fibrinogen to Mac-1-receptor-positive THP-1 cells, whereas the 20-nm NPs did not promote cell interaction to the same extent when adjusted for similar protein binding and did not affect the fibrinogen conformation (Deng et al. 2010).

Does the lung surfactant corona stay with the NPs when they reach the circulatory system and translocate to their final destination? Is the final corona a mixture of all proteins encountered along the translocation route? From the data regarding the modulatory effects of the protein corona on biodistribution and impacts cited above, it would seem likely that there is at least some remnant of the initial corona carried with the particles as they translocate. Recovery of NPs following exposure via the lungs and translocation to final organs and assessment of their "final" protein coronas will provide valuable information here. However, the fact that the modulation of impacts seems to be a delay of the effects rather than a complete inhibition (e.g., DPPC pretreatment suppressed quartz activity until 3 days and kaolin activity through 5 days [Gao et al. 2001]), it seems likely that over time the biomolecule component is degraded, and the underlying NPs are reexposed and proceed to induce their original response from the host (e.g., inflammation). An additional area for investigation, particularly for NPs being designed for therapy/nanomedicine, is whether poly(ethylene glycol) coating (commonly used to make surfaces inert to protein binding to reduce opsonization and thereby increase blood circulation time [Jokerst et al. 2011]) attenuate PS adsorption to NPs, and whether this could reduce surfactant-mediated uptake and toxicity.

Are there consequences of lung surfactant proteins being carried to other locations? This is more of a speculative question, however, given how tightly regulated protein signaling pathways are, and the fact that the living world uses a system of compartmentalization through which cells control which reactants colocalize where and when, to build numerous coupled signaling processes that allow for communication and regulation breech of these compartments may impact signalling (Lynch et al. 2006). The fact that NPs can bind and transport proteins that are normally localized to one specific organ to others may have as yet unknown consequences.

4.9 Conclusions

The biomolecule corona that adsorbs to NPs immediately upon contact with the lungs plays a vital role in determining the subsequent fate and behavior of the NPs. The biomolecule corona forms immediately and is very strongly bound, reducing the surface energy of the particles and "pacifying" them in terms of their surface energy. This adsorption of biomolecules can alter both the NPs, in terms of their surface properties (i.e., charge) and their degree of agglomeration and stability, and the adsorbed biomolecules changing their confirmation, orientation, and availability of their binding motifs for interaction with other biomolecules. Thus, the nature of the corona determines the subsequent clearance or translocation mechanism of NPs from the lungs, as well as influences the uptake amount, dosimetry, and biodistribution.

Thus, it is clear that researchers need to consider NPs in a context-dependent manner, and to design assays that are as close as possible to the *in vivo* situation, to be able to draw meaningful conclusions from their data. For example, *in vitro* cell culture models should also include the lung surfactant layer, *in vitro* studies of NP impacts on lung fluid membrane function should include the surfactant proteins as well as the lipids, and indeed corona studies should also include both the lipid and protein components for meaningful correlation with cellular uptake.

While much progress has been made in terms of understanding the interaction of NPs with the lungs, much remains to be done especially in terms of understanding, and in the future being able to predict, the fate and behavior of NPs during and after contact with the lungs. This chapter has highlighted some of the most significant advances to date, and identified some potential research directions to shed further light on how NPs interact with the lungs and the consequences of this for human health in order to ensure the safe and responsible implementation of nanotechnologies.

Acknowledgments

The author acknowledges the stimulating atmosphere and the contributions of her former colleagues at the Centre for BioNano Interactions at University College Dublin. Funding from the EU FP7 NeuroNano project (NMP4-SL-2008-214547) and the QualityNano project (http://www.qualitynano.eu), which is financed by the European Community Research

Infrastructures under the FP7 Capacities Programme (grant no. INFRA-2010-262163), is gratefully acknowledged.

References

Abbott, N.J., Ronnback, L., Hansson, E. 2006. Astrocyte-endothelial interactions at the blood–brain barrier. *Nat. Rev. Neurosci.* *7*, 41–53.

Ahsan, F., Rivas, I.P., Khan, M.A., Torres Suárez, A.I. 2002. Targeting to macrophages: Role of physicochemical properties of particulate carriers—liposomes and microspheres—on the phagocytosis by macrophages. *J. Control. Rel.* *79*, 29–40.

Anseth, J.W., Goffin, A.J., Fuller, G.G., Ghio, A.J., Kao, P.N., Upadhyay, D. 2005. Lung surfactant gelation induced by epithelial cells exposed to air pollution or oxidative stress. *Am. J. Respir. Cell Mol. Biol.* *33*, 161–168.

Azarmi, S., Roa, W.H., Löbenberg, R. 2008. Targeted delivery of nanoparticles for the treatment of lung diseases. *Adv. Drug Deliv. Rev.* *60*, 863–875.

Bakshi, M.S., Zhao, L., Smith, R., Possmayer, F., Petersen, N.O. 2008. Metal nanoparticle pollutants interfere with pulmonary surfactant function *in vitro*. *Biophys. J.* *94*, 855–868.

Beck-Broichsitter, M., Ruppert, C., Schmehl, T., Guenther, A., Betz, T., Bakowsky, U., Seeger, W., Kissel, T., Gessler, T. 2011. Biophysical investigation of pulmonary surfactant surface properties upon contact with polymeric nanoparticles *in vitro*. *Nanomed. Nanotechnol. Biol. Med.* *7*, 341–350.

Borm, P.J.A., Robbins, D., Haubold, S., Kuhlbusch, T., Fissan, H., Donaldson, K., Schins, R.P.F., Stone, V., Kreyling, W., Lademann, J., Krutmann, J., Warheit, D., Oberdorster, E. 2006. The potential risks of nanomaterials: A review carried out for ECETOC. *Part. Fibre Toxicol.* *3*, 11.

Brogden, K.A. 1991. Changes in pulmonary surfactant during bacterial pneumonia. *Antonie Van Leeuwenhoek* *59*, 215–223.

Casals, C. 2001. Role of surfactant protein A (SP-A)/lipid interactions for SP-A functions in the lung. *Pediatr. Pathol. Mol. Med.* *20*, 249–268.

Choe, S., Chang, R., Jeon, J., Violi, A. 2008. Molecular dynamics simulation study of a pulmonary surfactant film interacting with a carbonaceous nanoparticle. *Biophys. J.* *95*, 4102–4114.

Clark, H.W. 2010. Untapped therapeutic potential of surfactant proteins: Is there a case for recombinant SP-D supplementation in neonatal lung disease? *Neonatology* *97*, 380–387.

Crouch, E.C. 1998. Structure, biologic properties, and expression of surfactant protein D (SP-D). *Biochim. Biophys. Acta (BBA) Mol. Basis Dis.* *1408*, 278–289.

Crouch, E., Persson, A., Chang, D., Heusen, J. 1994. Molecular structure of pulmonary surfactant protein D (SP-D). *J. Biol. Chem.* *269*, 17311–17319.

Crouch, E., Wright, J.R. 2001. Surfactant proteins A and D and pulmonary host defense. *Annu. Rev. Physiol.* *63*, 521–554.

Dawson, K.A., Salvati, A., Lynch, I. 2009. Nanotoxicology: Nanoparticles reconstruct lipids. *Nat. Nanotechnol.* 4(2), 84–85.

Deng, Z.J., Liang, M., Toth, I., Monteiro, M.J., Minchin, R.F. 2012. Molecular interaction of poly(acrylic acid) gold nanoparticles with human fibrinogen. *ACS Nano* *6*, 8962–8969.

Deng, Z.J., Mingtao Liang, M., Monteiro, M., Toth, I., Minchin, R.F. 2010. Nanoparticle-induced unfolding of fibrinogen promotes Mac-1 (CD11b/CD18) receptor activation and pro-inflammatory cytokine release. *Nat. Nanotechnol.* *6*, 39–44.

Fan, Q., Wang, Y.E., Zhao, X., Loo, J.S.C., Zuo, Y.Y. 2011. Adverse biophysical effects of hydroxyapatite nanoparticles on natural pulmonary surfactant. *ACS Nano* *5*, 6410–6416.

Farnoud, A.M., Fiegel, J. 2011. Effects of charged polymeric nanoparticles on pulmonary surfactant function. In *Particle Technology Forum*, originally presented on October 17, 2011, 8:55:00–9:20:00.

Gao, N., Keane, M.J., Ong, T., Ye, J., Miller, W.E., Wallace, W.E. 2001. Effects of phospholipid surfactant on apoptosis induction by respirable quartz and kaolin in NR8383 rat pulmonary macrophages. *Toxicol. Appl. Pharmacol. 175*, 217–225.

Geiser, M., Rothen-Rutishauser, B.M., Kapp, N., Schürch, S., Kreyling, W., Schulz, H., Semmler, M., Im Hof, V., Heyder, J., Gehr, P. 2005. Ultrafine particles cross cellular membranes by nonphagocytic mechanisms in lungs and in cultured cells. *Environ. Health Perspect. 113*, 1555–1560.

Haagsman, H.P. 2002. Structural and functional aspects of the collectin SP-A. *Immunobiology 205*, 476–489.

Handy, R.D., Cornelis, G., Fernandes, T., Tsyusko, O., Decho, A., Sabo-Attwood, T., Metcalfe, C., Steevens, J.A., Klaine, S.J., Koelmans, A.A., Horne, N. 2012. Ecotoxicity test methods for engineered nanomaterials: Practical experiences and recommendations from the bench. *Environ. Toxicol. Chem. 31*, 15–31.

Harishchandra, R.K., Saleem, M., Galla, H.J. 2010. Nanoparticle interaction with model lung surfactant monolayers. *J. R. Soc. Interface 7 Suppl 1*, S15–S26.

Hawco, M.W., Coolbear, K.P., Davis, P.J., Keough, K.M. 1981a. Exclusion of fluid lipid during compression of monolayers of mixtures of dipalmitoylphosphatidylcholine with some other phosphatidylcholines. *Biochim. Biophys. Acta 646*, 185–187.

Hawco, M.W., Davis, P.J., Keough, K.M. 1981b. Lipid fluidity in lung surfactant: Monolayers of saturated and unsaturated lecithins. *J. Appl. Physiol. 51*, 509–515.

Hawgood, S., Poulain, F.R. 2001. The pulmonary collectins and surfactant metabolism. *Annu. Rev. Physiol. 63*, 495–519.

Hellstrand, E., Lynch, I., Andersson, A., Drakenberg, T., Dahlbäck, B., Dawson, K.A., Linse, S., Cedervall, T. 2009. Complete high-density lipoproteins in nanoparticle corona. *FEBS J. 276*, 3372–3381.

Hickling, T.P., Clark, H., Malhotra, C.R., Sim, R.B. 2004. Collectins and their role in lung immunity. *J. Leukocyte Biol. 75*, 27–33.

Hirn, S., Semmler-Behnke, M., Schleh, C., Wenk, A., Lipka, J., Schäffler, M., Takenaka, S., Möller, W., Schmid, G., Simon, U., Kreyling, W.G. 2011. Particle size-dependent and surface charge-dependent biodistribution of gold nanoparticles after intravenous administration. *Eur. J. Pharm. Biopharm. 77*, 407–416.

Hohlfeld, J.M. 2002. The role of surfactant in asthma. *Respir. Res. 3*, 4.

International Committee on Radiological Protection. 1994. Human respiratory tract model for radiological protection. A Report of Committee 2 of the ICRP.

Jokerst, J.V., Tatsiana Lobovkina, T., Gambhir, S.S. 2011. Nanoparticle PEGylation for imaging and therapy. *Nanomedicine (Lond.) 6*, 715–728.

Kagan, V.E., Konduru, N.V., Feng, W., Allen, B.L., Conroy, J., Volkov, Y., Vlasova, I.I., Belikova, N.A., Yanamala, N., Kapralov, A., Tyurina, Y.Y., Shi, J., Kisin, E.R., Murray, A.R., Franks, J., Stolz, D., Gou, P., Klein-Seetharaman, J., Fadeel, B., Star, A., Shvedova, A.A. 2010. Carbon nanotubes degraded by neutrophil myeloperoxidase induce less pulmonary inflammation. *Nat Nano 5*(5), 354–359.

Kapralov, A.A., Feng, W.H., Amoscato, A.A., Yanamala, N., Balasubramanian, K., Winnica, D.E., Kisin, E.R., Kotchey, G.P., Gou, P., Sparvero, L.J., Ray, P., Mallampalli, R.K., Klein-Seetharaman, J., Fadeel, B., Star, A., Shvedova, A.A., Kagan, V.E. 2012. Adsorption of surfactant lipids by single-walled carbon nanotubes in mouse lung upon pharyngeal aspiration. *ACS Nano 6*, 4147–4156.

Kishore, U., Greenhough, T.J., Waters, P., Shrive, A.K., Ghai, R., Kamran, M.F., Bernal, A.L., Reid, K.B., Madan, T., Chakraborty, T. 2006. Surfactant proteins SP-A and SP-D: Structure, function and receptors. *Mol. Immunol. 43*, 1293–1315.

Kondej, D., Sosnowski, T.R. 2012. Changes in the activity of the pulmonary surfactant after contact with bentonite nanoclay particles. *Chem. Eng. Trans. 26*, 531–536.

Kreuter, J. 2013. Mechanism of polymeric nanoparticle-based drug transport across the blood–brain barrier (BBB). *J. Microencapsul. 30*, 49–54.

Krol, S., Janshoff, A., Ross, M., Galla, J. 2000a. Structure and function of surfactant protein B and C in lipid monolayers: A scanning force microscopy study. *Phys. Chem. Phys. 2*, 4586–4593.

Krol, S., Ross, M., Sieber, M., Kunneke, S., Galla, H.J., Janshoff, A. 2000b. Formation of three-dimensional protein-lipid aggregates in monolayer films induced by surfactant protein B. *Biophys. J. 79*, 904–918.

Kuroda, K., Morimoto, Y., Ogami, A., Oyabu, T., Nagatomo, H., Hirohashi, M., Yamato, H., Nagafuchi, Y., Tanaka, I. 2006. Phospholipid concentration in lung lavage fluid as biomarker for pulmonary fibrosis. *Inhal. Toxicol. 18*, 389–393.

Lesniak, A., Fenaroli, F., Monopoli, M.P., Aberg, C., Dawson, K.A., Salvati, A. 2012. Effects of the presence or absence of a protein corona on silica nanoparticle uptake and impact on cells. *ACS Nano 6*, 5845–5857.

Lundqvist, M., Stigler, J., Cedervall, T., Berggård, T., Flanagan, M., Lynch, I., Elia, G., Dawson, K.A. 2011. The evolution of the protein corona around nanoparticles: A test study. *ACS Nano 5*, 7503–7509.

Lynch, I., Cedervall, T., Lundqvist, M., Cabaleiro-Lago, C., Linse, S., Dawson, K.A. 2007. The nanoparticle–protein complex as a biological entity; A complex fluids and surface science challenge for the 21st century. *Adv. Colloid Interface Sci. 134–135*, 167–174.

Lynch, I.D., Dawson, K.A., Linse, S. 2006. Detecting cryptic epitopes created by nanoparticles. *Sci. STKE* 2006(327), pe14.

Lynch, I., Elder, A. 2009. Disposition of nanoparticles as a function of their interactions with biomolecules. In *Nanomaterials: Risks and Benefits*, Linkov, I., Steevens, J., Eds. Springer, Dordrecht, The Netherlands.

McCormack, A.X., Whitsett, J.A. 2002. The pulmonary collectins, SP-A and SP-D, orchestrate innate immunity in the lung. *J. Clin. Invest. 109*, 707–712.

Monopoli, M.P., Walczyk, D., Lowry-Campbell, A., Elia, E., Lynch, I., Bombelli, F.B., Dawson, K.A. 2011. Physical–chemical aspects of protein corona: Relevance to *in vitro* and *in vivo* biological impacts of nanoparticles. *J. Am. Chem. Soc. 133*(8), 2525–2534.

Mühlfeld, C., Rothen-Rutishauser, B., Blank, F., Vanhecke, D., Ochs, M., Gehr, P. 2008. Interactions of nanoparticles with pulmonary structures and cellular responses. *Am. J. Physiol. Lung Cell. Mol. Physiol. 294*, L817–L829.

Mukherjee, D., Botelho, D., Gow, A.J., Zhang, J., Georgopoulos, P.G. 2013. Computational multiscale toxicodynamic modeling of silver and carbon nanoparticle effects on mouse lung function. *PLoS One 8*, e80917.

Nel, A., Xia, T., Mädler, L., Li, N. 2006. Toxic potential of materials at the nanolevel. *Science 311*, 622–627.

Oberdörster, G., Oberdörster, E., Oberdörster, J. 2005. Nanotoxicology: An emerging discipline evolving from studies of ultrafine particles. *Environ. Health Perspect. 113*, 823–839.

Pastrana-Rios, B., Flach, C.R., Brauner, J.W., Mautone, A.J., Mendelsohn, R. 1994. A direct test of the 'squeeze-out' hypothesis of lung surfactant function. External reflection FT-IR at the air/water interface. *Biochemistry 33*, 5121–5127.

Pathak, A., Vyas, S.P., Gupta, K.C. 2008. Nano-vectors for efficient liver specific gene transfer. *Int. J. Nanomed. 3*, 31–49.

Pérez-Gil, J. 2008. Structure of pulmonary surfactant membranes and films: The role of proteins and lipid–protein interactions. *Biochim. Biophys. Acta (BBA) Biomembr. 1778*, 1676–1695.

Peters, A., Veronesi, B., Calderon-Garciduenas, L., Gehr, P., Chen, L.C., Geiser, M., Reed, W., Rothen-Rutishauser, B., Schurch, S., Schulz, H. 2006. Translocation and potential neurological effects of fine and ultrafine particles: A critical update. *Part. Fibre Toxicol. 3*, 13.

Pietiainen, V., Marjomaki, V., Upla, P., Pelkmans, L., Helenius, A., Hyypia, T. 2004. Echovirus 1 endocytosis into caveosomes requires lipid rafts, dynamin II, and signaling events. *Mol. Biol. Cell 15*, 4911–4925.

Piknova, B., Schram, V., Hall, S.B. 2002. Pulmonary surfactant: Phase behavior and function. *Curr. Opin. Struct. Biol. 12*, 487–494.

Possmayer, F., Hall, S.B., Haller, T., Petersen, N.O., Zuo, Y.Y., Bernardino de la Serna, J., Postle, A.D., Veldhuizen, R.A.W., Orgeig, S. 2010. Recent advances in alveolar biology: Some new looks at the alveolar interface. *Respir. Physiol. Neurobiol. 173*, S55–S64.

Ruge, C.A. 2012. Bio–nano interactions in the peripheral lungs: Role of pulmonary surfactant components in alveolar macrophage clearance of nanoparticles. Dissertation, Saarbrücken University, Saarbrücken.

Ruge, C.A., Kirch, J., Cañadas, O., Schneider, M., Perez-Gil, J., Schaefer, U.F., Casals, C., Lehr, C.M. 2011. Uptake of nanoparticles by alveolar macrophages is triggered by surfactant protein A. *Nanomedicine 7*, 690–693.

Ruge, C.A., Schaefer, U.F., Herrmann, J., Kirch, J., Cañadas, O., Echaide, M., Pérez-Gil, J., Casals, C., Müller, R., Lehr, C.M. 2012. The interplay of lung surfactant proteins and lipids assimilates the macrophage clearance of nanoparticles. *PLoS One 7*, e40775.

Sachan, A.K., Harishchandra, R.K., Bantz, C., Maskos, M., Reichelt, R., Galla, H.J. 2012. High-resolution investigation of nanoparticle interaction with a model pulmonary surfactant monolayer. *ACS Nano 6*, 1677–1687.

Sandin, P., Fitzpatrick, L.W., Simpson, J.C., Dawson, K.A. 2012. High-speed imaging of Rab family small GTPases reveals rare events in nanoparticle trafficking in living cells. *ACS Nano 6*(2), 1513–1521.

Sano, H., Kuronuma, K., Kudo, K., Mitsuzawa, H., Sato, M., Murakami, S., Kuroki, Y. 2006. Regulation of inflammation and bacterial clearance by lung collectins. *Respirology 11*, S46–S50 Suppl.

Schleh, C., Erpenbeck, V.J., Winkler, C., Lauenstein, H.D., Nassimi, M., Braun, A., Krug, N., Hohlfeld, J.M. 2010. Allergen particle binding by human primary bronchial epithelial cells is modulated by surfactant protein D. *Respir. Res. 11*, 83.

Schleh, C., Hohlfeld, J.M. 2009. Interaction of nanoparticles with the pulmonary surfactant system. *Inhal. Toxicol. 21 Suppl 1*, 97–103.

Schleh, C., Kreyling, W.G., Lehr, C.M. 2013. Pulmonary surfactant is indispensable in order to simulate the *in vivo* situation. *Part. Fibre Toxicol. 10*, 6.

Schleh, C., Mühlfeld, C., Pulskamp, K., Schmiedl, A., Nassimi, M., Lauenstein, H.D., Armin Braun, A., Krug, N., Erpenbeck, V.J., Hohlfeld, J.M. 2009. The effect of titanium dioxide nanoparticles on pulmonary surfactant function and ultrastructure. *Respir. Res. 10*, 90. doi:10.1186/1465-9921-10-90.

Schleh, C., Rothen-Rutishauser, B.M., Blank, F., Lauenstein, H.D., Nassimi, M., Krug, N., Braun, A., Erpenbeck, V.J., Gehr, P., Hohlfeld, J.H. 2012a. Surfactant protein D modulates allergen particle uptake and inflammatory response in a human epithelial airway model. *Respir. Res. 13*, 8.

Schleh, C., Semmler-Behnke, M., Lipka, J., Wenk, A., Hirn, S., Schäffler, M., Schmid, G., Simon, U., Kreyling, W.G. 2012b. Size and surface charge of gold nanoparticles determine absorption across intestinal barriers and accumulation in secondary target organs after oral administration. *Nanotoxicology 6*, 36–46.

Schmidt, R., Meier, U., Markart, P., Grimminger, F., Velcovsky, H.G., Morr, H., Seeger, W., Günther, A. 2002. Altered fatty acid composition of lung surfactant phospholipids in interstitial lung disease. *Am. J. Physiol. Lung Cell. Mol. Physiol. 283*, L1079–L1085.

Schumann, C., Schübbe, S., Cavelius, C., Kraegeloh, A. 2012. A correlative approach at characterizing nanoparticle mobility and interactions after cellular uptake. *J. Biophotonics 5*, 117–127.

Seagrave, J., Knall, C., McDonald, J.D., Mauderly, J.L. 2004. Diesel particulate material binds and concentrates a proinflammatory cytokine that causes neutrophil migration. *Inhal. Toxicol. 16 Suppl 1*, 93–98.

Seaton, B.A., Crouch, E.C., McCormack, F.X., Head, J.F., Hartshorn, K.L., Mendelsohn, R. 2010. Structural determinants of pattern recognition by lung collectins. *Innate Immun. 16*, 143–150.

Semmler-Behnke, M., Takenaka, S., Fertsch, S., Wenk, A., Seitz, J., Mayer, P., Oberdörster, G., Kreyling, W.G. 2007. Efficient elimination of inhaled nanoparticles from the alveolar region: Evidence for interstitial uptake and subsequent reentrainment onto airways epithelium. *Environ. Health Perspect. 115*, 728–733.

Shvedova, A.A., Kapralov, A.A., Feng, W., Kisin, E.R., Murray, A., Mercer, R., St. Croix, C., Lang, M., Watkins, S., Konduru, N., Allen, B.L., Conroy, J., Kotchey, G.P., Mohamed, B.M., Meade, A.D., Volkov, Y., Star, A., Fadeel, B., Kagan, V.E. 2012. Impaired clearance and enhanced pulmonary inflammatory/fibrotic response to carbon nanotubes in myeloperoxidase-deficient mice. *PLoS One 7*(3), e30923. doi:10.1371/journal.pone.0030923.

Stevens, P.A., Wissel, H., Sieger, D., Meienreis-Sudau, V., Rustow, B. 1995. Identification of a new surfactant protein A binding protein at the cell membrane of rat type II pneumocytes. *Biochem. J. 308*, 77–81.

Stone, V., Johnston, H., Clift, M.J. 2007. Air pollution, ultrafine and nanoparticle toxicology: Cellular and molecular interactions. *IEEE Trans. Nanobiosci. 6*, 331–340.

Takenaka, S., Möller, W., Semmler-Behnke, M., Karg, E., Wenk, A., Schmid, O., Stoeger, T., Jennen, L., Aichler, M., Walch, A., Pokhrel, S., Mädler, L., Eickelberg, O., Kreyling, W.G. 2012. Efficient internalization and intracellular translocation of inhaled gold nanoparticles in rat alveolar macrophages. *Nanomedicine (Lond.) 7*, 855–865.

Walczyk, D., Baldelli-Bombelli, F., Campbell, A., Lynch, I., Dawson, K.A. 2010. What the cell "sees" in bionanoscience. *J. Am. Chem. Soc. 132*, 5761–5768.

Wallace, W.E., Keane, M.J., Murray, D.K., Chisholm, W.P., Maynard, A.D., Ong, T.-M. 2006. Phospholipid lung surfactant and nanoparticle surface toxicity: Lessons from diesel soots and silicate dusts. *J. Nanopart. Res. 9*, 23–38.

Wang, B., Zhang, L., Bae, S.C., Granick, S. 2008. Nanoparticle-induced surface reconstruction of phospholipid membranes. *Proc. Natl. Acad. Sci. U.S.A. 105*, 18171–18175.

Watkins, J.C. 1968. The surface properties of pure phospholipids in relation to those of lung extracts. *Biochim. Biophys. Acta 152*, 293–306.

Wright, J.R. 2005. Immunoregulatory functions of surfactant proteins. *Nat. Rev. Immunol. 5*, 58–68.

Zhang, H., Fan, Q., Wang, Y.E., Neal, C.R., Zuo, Y.Y. 2011. Comparative study of clinical pulmonary surfactants using atomic force microscopy. *Biochim. Biophys. Acta 1808*, 1832–1840.

Zuo, Y.Y., Veldhuizen, R.A., Neumann, A.W., Petersen, N.O., Possmayer, F. 2008. Current perspectives in pulmonary surfactant—Inhibition, enhancement and evaluation. *Biochim. Biophys. Acta 1778*, 1947–1977.

5

Interaction with Alveolar Lining Fluid

Vincent Castranova, Dale W. Porter, and Robert R. Mercer

CONTENTS

5.1 Introduction

Nanotechnology is defined as the manipulation of matter on the near-atomic scale to produce novel structures, materials, and devices. Nanoparticles (NPs) are commonly defined as structures having one dimension <100 nm. At the nanoscale, particles exhibit unique physicochemical properties, which may be dramatically different from fine particles of the same composition. Such unique properties are being exploited for a wide range of novel applications and new products. Such applications include medical imaging, targeted drug delivery, bone grafting, antimicrobial products, energy storage devices, conductive fabric, electronics, sensors, structural materials, sporting goods, coatings, paints, cosmetics, and sunscreens. Data indicate that the commercial application of nanotechnology is expanding rapidly and is expected to grow into a trillion dollar industry employing millions of workers worldwide within the next decade (Roco 2004; Lux Research 2007).

Exposure assessment of nanotechnology worksites indicates that aerosolization of NPs is possible during energetic processes. Indeed, NP aerosolization has been reported during transfer, bagging, sonication, weighing, mixing, or vortexing of NPs (Maynard et al. 2004; Han et al. 2008; Methner 2008; Johnson et al. 2010; Lee et al. 2010; Elihn et al. 2011). Therefore, worker inhalation of NPs can be anticipated during production, use, and disposal unless adequate controls and prevention strategies are in place (Maynard and Kuempel 2005).

As NPs are inhaled in the respiratory tract, it is anticipated that upon deposition, NPs would interact with proteins and lipids from fluids lining the airways. Indeed, interaction of NPs with biological fluids has been demonstrated with the adsorption of plasma

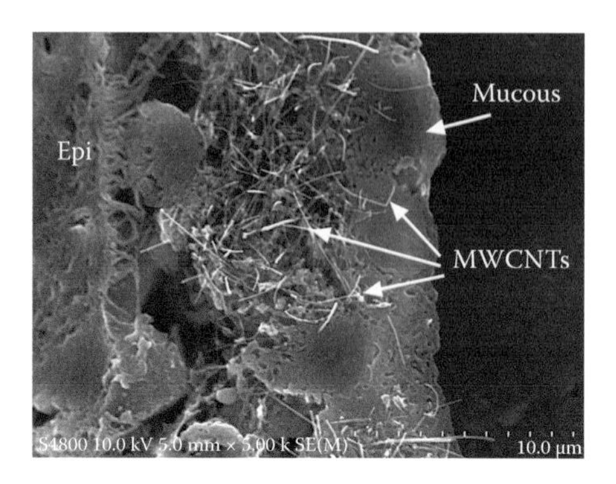

FIGURE 5.1
Field emission scanning electron microscope image of MWCNTs in mucous of mouse bronchus. Representative micrograph of MWCNTs in the airway mucous layer acutely (1 h) after aspiration. The cilia of the ciliated airway epithelial cells (Epi) are visible on the left side of the image with MWCNTs immersed in the mucous layer of the airways. This initial deposition of MWCNTs in the airways has been found to be largely eliminated by the mucocilliary escalator within days after deposition, although the rate of clearance from the airways is neither as rapid nor complete as that typically observed for fine-sized particles.

proteins forming a "corona" around various NPs (Cedervall et al. 2007; Lundqvist et al. 2008; Zhang et al. 2011). In addition, our laboratory has demonstrated that multiwalled carbon nanotubes (MWCNTs) were found in the mucus layer of a bronchiole of an exposed mouse (Figure 5.1). This mucus limits the ability of MWCNTs to enter bronchiolar epithelial cells. In contrast, alveolar surfactant does not appear to be a barrier to the penetration of MWCNTs into alveolar epithelial cells (Mercer et al. 2010). The specific questions to be addressed in this chapter are as follows: (i) Can alveolar surfactant interact with CNTs in the lung? (ii) Can this interaction act to dissociate agglomerated CNT structures in the lung?

5.2 Evidence That Phospholipids and/or Proteins Found in Alveolar Lining Fluid Can Bind to NPs

Evidence exists suggesting that pulmonary surfactant can interact with carbon black or gold NPs *in vitro* (Kendall et al. 2004; Bakshi et al. 2008). Konduru et al. (2009) reported the binding of DSPC (the major component in alveolar surfactant) to single-walled CNTs (SWCNTs) *in vitro*. They reported that sonication of SWCNTs with 2.5 mM disaturated phosphatidylcholine (DSPC) resulted in binding of 462 nmol DSPC per milligram of SWCNTs. Konduru et al. (2009) also reported that phosphatidylserine (PS) can bind to SWCNTs. This coating of PS acts as an "eat me" signal and greatly enhances SWCNT phagocytosis by macrophages *in vitro*. Of interest, binding of phosphatidylcholine (PC) failed to enhance SWCNT uptake by macrophages. Gasser et al. (2010) described the binding of Curosurf 120 (a lipid-based porcine surfactant) to MWCNTs *in vitro*. MWCNTs were suspended in Curosurf 120 (20 mg/ml), sonicated for 15 min, incubated for 24 h, and washed four times with phosphate-buffered saline (PBS). Transmission electron microscopy (TEM) demonstrated lipid binding to the MWCNTs. Thin layer chromatographic analysis indicated that the lipid composition of the bound material was identical to the complete surfactant, suggesting nonspecific binding of lipids to the hydrophobic MWCNTs.

Recently, Salvador-Morales et al. (2007) exposed double-walled CNTs (DWCNTs) to bronchoalveolar lavage (BAL) fluid obtained from patients with alveolar proteinosis (having above normal surfactant protein levels) and quantified protein binding to the nanotubes. The results indicate that low amounts of surfactant proteins A and D (SP-A and SP-D) were able to bind to DWCNTs. Such binding was point specific rather than a coating of the surface. This SP-A and SP-D binding appeared to involve specific functionalized groups (carboxylic, aldehyde, and ketone groups) on the DWCNT surface, since no binding was measurable with pristine DWCNTs.

5.3 Dispersive Effect of Alveolar Lining Fluid on NPs *In Vitro*

NPs agglomerate when suspended in physiological saline. If bioactivity were related to particle number or structure surface area per mass of particles, then exposure of cells or animals (pharyngeal aspiration or intratracheal instillation) to highly agglomerated suspensions of NPs would result in an underestimation of biological potency. Alveolar surfactant, as discussed previously, has been shown to bind to NPs *in vitro*. Since inhaled NPs would be expected to deposit on lung lining fluid, interaction with alveolar surfactant is likely. Therefore, Sager et al. (2007) investigated whether suspension of NPs in diluted lung lining fluid resulted in enhanced particle dispersion. They obtained a diluted preparation of alveolar lining fluid by BAL of naïve rats. Ultrafine carbon black or titanium dioxide NPs were suspended in BAL fluid (5 or 3.5 mg/ml, respectively) and sonicated for 15 min. Suspension of ultrafine carbon black or titanium dioxide in BAL fluid significantly improved the dispersion of these NPs compared with PBS, as judged by light and electron microscopy. DSPC (a major component of alveolar surfactant) alone was not found to be an effective dispersant in this study. Improved dispersion of carbon black NPs increased their inflammatory potential by 8-fold when comparing lavageable neutrophils 1 day after intratracheal instillation of carbon black suspended in BAL fluid versus suspended in PBS (Shvedova et al. 2007). Porter et al. (2008) determined that DSPC plus albumin, at concentrations found in BAL fluid (0.01 mg/ml DSPC and 0.6 mg/ml albumin in PBS), was as effective as diluted alveolar lining fluid in dispersing ultrafine titanium dioxide or carbon black (0.5 mg NPs per milliliter of dispersion medium, sonicated for 30 min). This study also reported that DSPC or albumin alone were not effective in dispersing these NPs. This artificial diluted alveolar lining fluid, designated as dispersion medium by the authors, significantly decreased the mean structure size as determined by dynamic light scattering (Table 5.1). This dispersion medium was also effective in decreasing the agglomeration of MWCNTs in suspension, as shown in Figure 5.2 (1.8 mg MWCNTs per milliliter of dispersion medium, indirect sonication for 5 min, direct sonication for 5 min). Wang et al. (2010a) have confirmed the effectiveness of this dispersion medium in producing a well-dispersed, stable suspension of MWCNTs. Wang et al. (2010b) have employed diluted bovine surfactant (Survanta; 150 μg Survanta/ml PBS) to disperse SWCNTs. Morphometric analysis under a field emission scanning electron microscope indicates that the mean width of SWCNT structures decreased from 12.35 ± 0.76 μm in PBS to 0.38 ± 0.02 μm in diluted Survanta.

TABLE 5.1

Dispersion of NPs in Artificial Diluted Alveolar Lining Fluid

Nanoparticle	Vehicle[a]	Particle Size (nm)[b]
TiO_2	PBS	2849
	BAL fluid	204
	DM	163
Carbon black	PBS	Agglomeration extensive but not quantified
	BAL fluid	131
	DM	93

Source: Modified from D. Porter et al., *Nanotoxicology*, 2, 144, 2008.
Note: Artificial diluted alveolar lining fluid contained 0.01 mg/ml DSPC and 0.6 mg/ml serum albumin in PBS (pH 7.4).
[a] PBS, phosphate-buffered saline (pH 7.4); BAL fluid from naïve rats; DM, dispersion medium, i.e., artificial diluted alveolar lining fluid.
[b] Measured by dynamic light scattering after sonication of NP suspensions for 30 min.

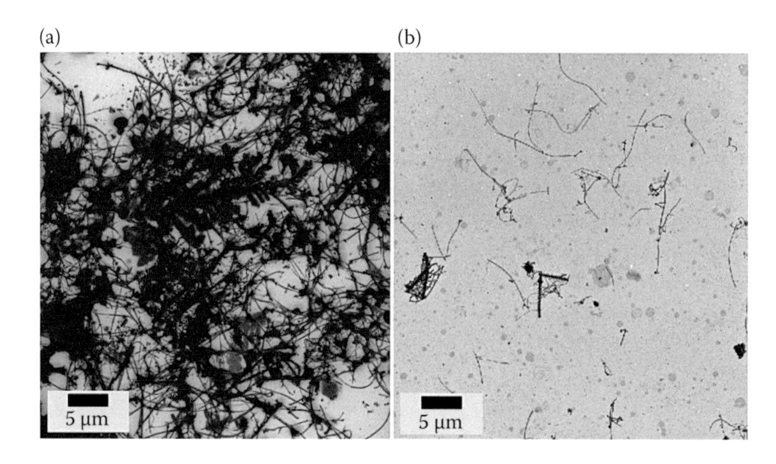

FIGURE 5.2
Transmission electron micrographs demonstrating the effect of dispersion medium on suspension of MWCNTs. (a) Typical agglomeration of MWCNTs in PBS. Such agglomeration is typical of that observed for most NPs as the van der Waals forces of like-to-like attraction are dominant over diffusion/entropic forces at the nanoscale dimensions of these particles. (b) Highly effective dispersion of the MWCNTs when the sample is treated with an artificial diluted lung surfactant (0.01 mg/ml DSPC and 0.6 mg/ml albumin in PBS). (Modified from D. Porter et al., *Nanotoxicology*, 2, 144, 2008.)

5.4 Does Alveolar Lining Fluid Mask the Surface Reactivity of Particles?

From the previous discussion, it is clear that alveolar surfactant can bind to NPs and greatly improve their dispersion in a biologically compatible suspension. Gao et al. (2001) reported that a suspension of fine crystalline silica in the presence of a high concentration of DSPC (5 mg/ml) masked the surface of silica, making it nontoxic *in vitro*. For the dispersion studies described previously, the concentration of DSPC or surfactant was >30 times more dilated than in the Gao et al. (2001) study and, therefore, may not have masked the particle surface. Sager et al. (2007) and Porter et al. (2008) directly evaluated this issue by treating quartz with diluted alveolar lining fluid or dispersion medium (DSPC plus albumin in PBS), respectively. Neither treatment altered the level of silica-induced lung damage or inflammation 1 day after intratracheal instillation in rats or pharyngeal aspiration in mice, respectively. In addition, pulmonary exposure to BAL fluid or dispersion medium did not result in any changes in the levels of inflammatory or injury markers from control. In another study, Cho et al. (2012) evaluated the potential of metal oxide NPs to lyse red blood cells *in vitro*. The potency was related to a positive zeta potential. Coating with BAL fluid decreased both the surface charge on the NPs and their *in vitro* toxicity. Addition of phospholipase A_2 restored the toxicity of the metal oxide NP. Of interest, intratracheal instillation of positively charged metal oxide NPs caused an inflammatory reaction in rats 1 day postexposure. Pretreatment of NPs with BAL fluid did not alter the level of inflammation. The authors proposed that the lipid coating on the NPs was removed by lysosomal enzymes *in vivo* as had been demonstrated previously (Hill et al. 1995). This could explain the lack of masking of quartz bioactivity reported by Sager et al. (2007) and Porter et al. (2008).

5.5 Effect of Improved NP Dispersion on Bioactivity

Shvedova et al. (2008) reported the pulmonary responses to a 4-day inhalation of SWCNTs and compared the bioactivity of inhaled SWCNTs to that for SWCNTs suspended in PBS and administered to mice by pharyngeal aspiration (Shvedova et al. 2005). SWCNT dry aerosols had a count mode aerodynamic diameter of 200 nm, while the SWCNTs in suspension were very agglomerated with many structures >1 μm in diameter. The results indicate that on an equal lung burden basis, lung inflammation, injury, and collagen formation were at least 4-fold greater after inhalation of well-dispersed SWCNTs than aspiration of poorly dispersed SWCNTs (Table 5.2). If steps were taken to achieve a well-dispersed suspension of SWCNTs, the inflammatory and fibrotic potency of SWCNTs increased to a level similar to that seen with inhalation of a dispersed dry aerosol of SWCNTs (Mercer et al. 2008). Indeed, Mercer et al. (2008) demonstrated that SWCNT agglomerates deposit in the terminal bronchioles and proximal alveoli, and induce the rapid formation of granulomatous lesions, while more dispersed SWCNTs deposit in the distal alveoli, migrate into the interstitium of alveoli septa, and induce alveolar interstitial fibrosis (Figure 5.3).

Improved dispersion of NPs by suspension in diluted alveolar lining fluid has been shown to significantly increase bioactivity after pulmonary exposure. Intratracheally instilled

TABLE 5.2

Comparison of Pulmonary Responses to Inhalation versus Aspiration of Raw SWCNTs

	Inhalation (Lung Burden 5 μg/Mouse)	Aspiration (10 μg/Mouse)
PMN (fold increase)[1]	136 ± 20	51 ± 10
BAL protein (% increase from control)[1]	68 ± 3%	35 ± 3%
BAL TGF-β (fold increase)[7]	7.9 ± 7	2.0 ± 0.1
Lung collagen (% increase from control)[7]	127 ± 7%	53 ± 1%

Note: Superscripts indicate days postexposure.

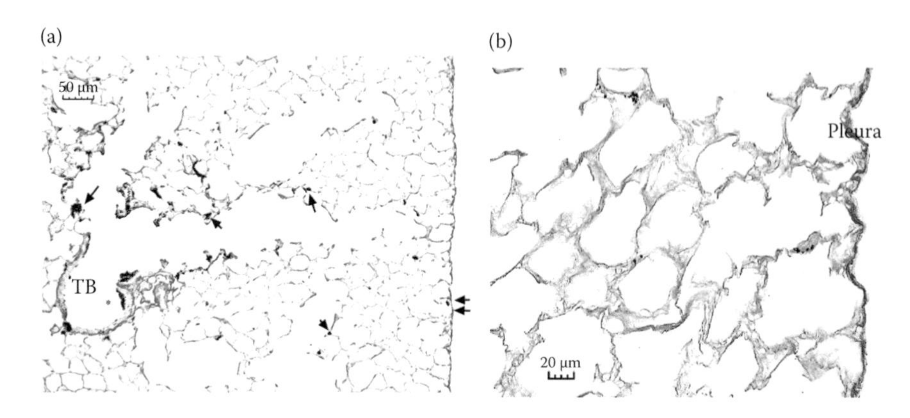

FIGURE 5.3
Distribution of aspirated SWCNTs demonstrated by gold labeling. These light micrographs demonstrate the use of gold labeling to determine the distribution of SWCNTs in the lungs after pharyngeal aspiration. Suspensions of SWCNTs were labeled with 10-nm colloidal gold before aspiration, and then silver-enhanced treatment of the sections was used to make the colloidal gold SWCNTs visible in sections. At 1 day post exposure, large black deposits of silver enhanced, poorly dispersed SWCNTs are concentrated near the proximal alveolar region (a). These are the areas where granulomatous lesions form about the large agglomerates of SWCNTs. As shown in panel b, finer structures of well-dispersed SWCNTs can be seen distributed widely throughout the lungs. These smaller structures were not visible before silver enhancement. The fibrotic response to these widely distributed, well-dispersed SWCNTs was shown to be significantly greater than the agglomerate form. (Figure 5.3 used with permission from R.R. Mercer et al., *Am. J. Physiol. Lung Cell Mol. Physiol.*, 294, L87, 2008.)

ultrafine carbon black NPs caused an 8-fold greater inflammation (measured as neutrophils harvested by BAL of rats 1 day postexposure) when well dispersed in BAL fluid than when poorly dispersed in PBS (Shvedova et al. 2007). The effect of improved dispersion of TiO_2 NPs was also demonstrated by the fact that the inflammatory potency of poorly dispersed TiO_2 NPs suspended in PBS was reported to be similar to fine TiO_2 (Warheit et al. 2006), while the inflammatory potency of well-dispersed TiO_2 NPs suspended in BAL fluid was found to be 40-fold greater than fine TiO_2 (Sager et al. 2008). Additionally, dispersion of SWCNTs in diluted bovine surfactant or MWCNTs in an artificial diluted lung lining fluid induced fibrotic responses in cultured lung fibroblasts (proliferation and collagen production), while agglomerated CNTs suspended in PBS were inactive (Wang et al. 2010a,b).

5.6 Evidence of Interaction of NPs with Alveolar Lining Fluid *In Vivo*

Kapralov et al. (2012) aspirated SWCNTs (80 μg) in mice and recovered the nanotubes by BAL at 2 and 24 h postexposure. SWCNTs were separated from BAL cells by gradient centrifugation, and nonbound material was removed by washing. Lipid constituents bound to the SWCNTs were determined by liquid chromatography mass spectrometry. The composition of lipids bound to SWCNTs deposited in a mouse lung reflected the composition of pulmonary surfactant, that is, phospholipids (mainly DSPC) with apoproteins A, B, and D.

Quantification of the bound lipid indicated that 108 phospholipid molecules were bound to each SWCNT *in vivo*, forming a coating 2.9 nm thick with the hydrophobic alkyl chains of the phospholipids adsorbed to the SWCNTs and the polar head groups exposed to the aqueous phase to enhance dispersion.

An interaction between surfactant and latex particles deposited in the airways of hamsters after inhalation exposure was demonstrated by electron microscopic analysis, where deposited particles appeared to be coated by an osmiophilic film (staining for phospholipid) 15 min after a 30-min inhalation (Gehr et al. 1990; Schurch et al. 1990). Mijailovich et al. (2010) reported similar findings after pulmonary exposure to spores; that is, electron micrographs indicate interaction of alveolar lining fluid with deposited fungal spores. In addition, these surfactant-coated particles appeared to be drawn into the lung lining fluid to allow direct contact between the deposited particles and the epithelial cells. Indeed, Porter et al. (2010) have demonstrated rapid movement of MWCNTs into epithelial cells after aspiration exposure in a mouse model.

5.7 Evidence That NPs Disperse *In Vivo*

The data presented previously indicate that highly diluted alveolar lining fluid (as much as a 50-fold dilution) effectively disperses NPs upon sonication. Jobe and Ikegami (1987) reported that the alveolar surface is lined with fluid containing surfactant with a lipid content of 6.9–10.3 mg lipid per square meter of alveolar surface. Energy for mixing would be the expansion and compression of this surface film during the inhalatory/exhalatory cycle. Levy et al. (2012) reviewed the available literature to determine whether NPs would deagglomerate when deposited on lung lining fluid *in vivo*. They found reports that surfactant can bind to NPs. However, there was a lack of evidence to support NP deagglomeration in the lung. However, data from a recent publication could be interpreted as evidence supporting deagglomeration *in vivo* (Shvedova et al. 2012). In this study, mice were exposed by pharyngeal aspiration to 40 μg of poorly dispersed SWCNTs suspended in PBS. At 1 and 28 days postexposure, mice were killed. The volume and number of large SWCNT agglomerates visualized by light microscopy of lung sections decreased by 80% from day 1 to day 28 postexposure. In preparations of solubilized lungs, the size distribution of SWCNT structures was determined by TEM (Figure 5.4). During the 28-day postexposure period, there was a 37% decline in the number of SWCNT structures >0.75 μm and a corresponding 32% increase in SWCNT structures <0.75 μm. Consistent with the hypothesis that pulmonary surfactant may deagglomerate NPs in the lung are morphometric data from our laboratory (Mercer et al. 2013a). Mice were exposed by inhalation to well-dispersed MWCNTs and killed at various times postexposure. The number of nanotube structures larger than four agglomerated MWCNTs decreased by 13% during 168 days postexposure, while structures containing fewer than four MWCNTs increased by 14% during the 168-day postexposure period (Figure 5.5). In addition, Mercer et al. (2013b) demonstrated a low level of translocated singlet MWCNTs in extrapulmonary tissue of mice 1 day after a 12-day inhalation exposure. The number of MWCNT singlets in these tissues significantly increased from 1 day to 168 days postexposure. These data support some degree of deagglomeration with time in the lung and subsequent translocation of singlet MWCNTs from the lung to systemic organs.

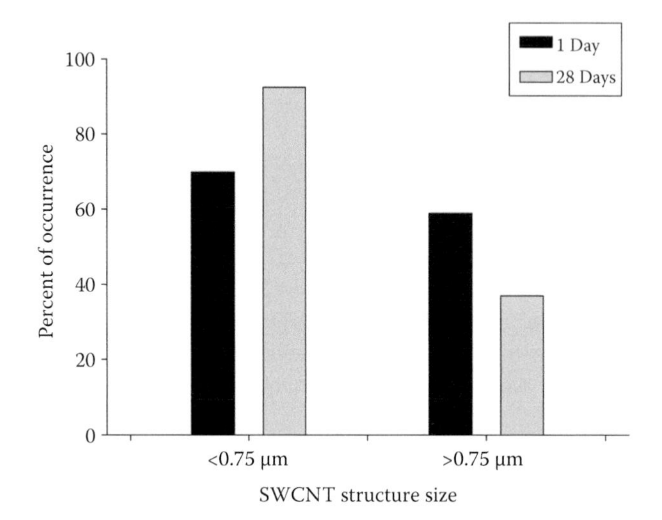

FIGURE 5.4
Frequency of occurrence of SWCNT structure size at 1 and 28 days postexposure. Analysis of the changes in the size distribution of SWCNT structures after aspiration exposure of a poorly dispersed SWCNT preparation. Using TEM analysis of lung digests, the size distribution of SWCNT structures was determined. The results demonstrate a significant decline in the frequency of occurrence of SWCNT structures >0.75 μm and a corresponding increase in SWCNT structures <0.75 μm. (Original figure constructed using data from A.A. Shvedova et al., *PLOS ONE*, 7, e30923, 2012.)

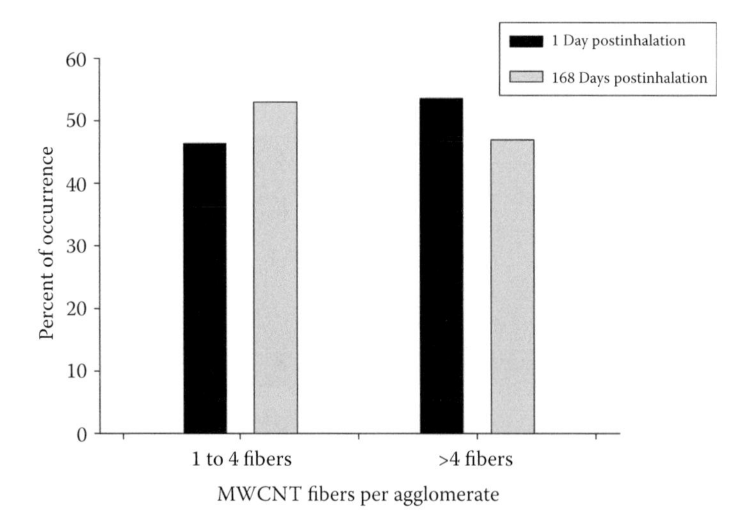

FIGURE 5.5
Changes in MWCNT agglomeration with time in the lungs. Comparison of the number of fibers per agglomerate in the lungs demonstrates that the larger agglomerates are dispersed into smaller structures with time postexposure. As shown in this figure, the number of structures containing one to four fibers increased roughly 14% between 1 and 168 days after inhalation, while the larger agglomerates (more than four fibers) decreased by a comparable 13%. Results based on using combined serial-section and morphometry techniques to determine the frequency of occurrence for MWCNT structures containing one to four fibers and MWCNT structures containing more than four fibers at 1 and 168 days postinhalation exposure.

5.8 Conclusion

There is a growing body of evidence suggesting that alveolar lining fluid can bind to CNTs. Upon sonication, pulmonary surfactant is an effective dispersant for a variety of NPs *in vitro*. Evidence also exists suggesting that the bioactivity of NPs, both *in vitro* and *in vivo*, is significantly enhanced by pulmonary surfactant-dependent dispersion. Although evidence for dispersion of NPs *in vivo* is not extensive, recent data support the possibility of deagglomeration of NPs in the lung. This suggests that NP agglomerates deposited in the lung may represent a slow-release source of smaller, more bioactive structures over a considerable time. Such a possibility would greatly impact risk assessment for NPs.

Disclaimer

The findings and conclusions in this chapter are those of the authors and do not necessarily represent the view of the National Institute for Occupational Safety and Health.

References

Bakshi, M.S., L. Zhao, R. Smith, F. Possmayer, and N.O. Petersen. 2008. Metal nanoparticle pollutants interfere with pulmonary surfactant function *in vitro*. *Biophys. J.* 93: 855–868.

Cedervall, T., I. Lynch, S. Lindman et al. 2007. Understanding the nanoparticle–protein corona using methods to quantify exchange rates and affinities of proteins for nanoparticles. *Proc. Nat. Acad. Sci. U.S.A.* 104: 2050–2055.

Cho, W.-S., R. Duffin, F. Thielbeer et al. 2012. Zeta potential and solubility to toxic ions as mechanisms of lung inflammation caused by metal/metal oxide nanoparticles. *Toxicol. Sci.* 126: 469–477.

Elihn, K., P. Berg, and G. Liden. 2011. Correlation between airborne particle concentration in seven industrial plants and estimated respiratory tract deposition by number, mass and elemental composition. *J. Aerosol Sci.* 42: 127–141.

Gao, N., M. Keane, T. Ong, J. Ye, W. Miller, and W. Wallace. 2001. Effects of phospholipid surfactants on apoptosis induction by respirable quartz and kaolin in NR8383 rat pulmonary macrophages. *Toxicol. Appl. Pharmacol.* 175: 217–225.

Gasser, M., B. Rothen-Rutishauser, H.F. Krug et al. 2010. The adsorption of biomolecules to multiwalled carbon nanotubes is influenced by pulmonary surfactant lipids and surface chemistry. *J. Nanobiotech.* 8: 31.

Gehr, P., S. Schurch, Y. Berthiaume, V. Im Hof, and M. Geiser. 1990. Particle retention in airways by surfactant. *J. Aerosol Mol.* 3: 27–43.

Han, J.H., E.J. Lee, J.H. Lee et al. 2008. Monitoring multiwalled carbon nanotube research facility. *Inhal. Toxicol.* 20: 741–749.

Hill, C.A., W. Wallace, M.E. Keane, and P.S. Mike. 1995. The enzymatic removal of a surfactant coating from quartz and kaolin by P388D1 cells. *Cell. Biol. Toxicol.* 11: 119–128.

Jobe, A., and M. Ikegami. 1987. Surfactant for the treatment of respiratory distress syndrome. *Am. Rev. Respir. Dis.* 136: 1256–1275.

Johnson, D.R., M.M. Methner, A.J. Kennedy, and J.A. Steevens. 2010. Potential for occupational exposure to engineered carbon-based nanomaterial in environmental laboratory studies. *Environ. Health Perspect.* 118: 44–54.
Kapralov, A.A., W.H. Feng, A.A. Amoscato et al. 2012. Absorption of surfactant lipids by single-walled carbon nanotubes in mouse lung upon pharyngeal aspiration. *ACS Nano* 6: 4147–4156.
Kendall, M., L. Brown, and K. Trought. 2004. Molecular adsorption at particle surfaces: A PM toxicity mediation mechanism. *Inhal. Toxicol.* 16(Suppl 1): 99–105.
Konduru, N.V., Y.Y. Tyurina, W. Feng et al. 2009. Phosphatidylserine targets single walled carbon nanotubes to professional phagocytes *in vitro* and *in vivo*. *PLoS One* 4: e4398.
Lee, J.H., S.-B. Lee, G.N. Bae et al. 2010. Exposure assessment of carbon nanotube manufacturing workplaces. *Inhal. Toxicol.* 22: 369–381.
Levy, L., I.S. Chaudhuri, N. Krueger, and R.L. McCunney. 2012. Does carbon black disaggregate in lung fluid? A commentary. *Chem. Res. Toxicol.* 25: 2001–2006.
Lundqvist, M., J. Stigler, G. Elia, I. Lynch, T. Cedervall, and K.A. Dawson. 2008. Nanoparticle size and surface properties determine the protein corona with possible implications for biological impacts. *Proc. Nat. Acad. Sci. U.S.A.* 105: 14265–14270.
Lux Research. 2007. *The Nanotech Report*, 5th ed. New York: Lux Research.
Maynard, A.D., P.A. Baron, M. Foley, A.A. Shvedova, E.R. Kisin, and V. Castranova. 2004. Exposure to carbon nanotube material during the handling of unrefined single walled carbon nanotube material. *J. Toxicol. Environ. Health Part A* 67: 87–107.
Maynard, A.D., and E. Kuempel. 2005. Airborne nanostructural particles and occupational health. *J. Nanopart. Res.* 7: 587–614.
Mercer, R.R., A.F. Hubbs, J.F. Scabilloni et al. 2010. Distribution and persistence of pleural penetrations by multi-walled carbon nanotubes. *Part. Fibre Toxicol.* 7: 28.
Mercer, R.R., J.F. Scabilloni, A.F. Hubbs et al. 2013a. Distribution and fibrotic response following inhalation exposure to multi-walled carbon nanotubes. *Part. Fibre Toxicol.* 10: 33.
Mercer, R.R., J.F. Scabilloni, A.F. Hubbs et al. 2013b. Extrapulmonary transport of MWCNT following inhalation exposure. *Part. Fibre Toxicol.* 10: 38.
Mercer, R.R., J. Scabilloni, L. Wang et al. 2008. Alteration of deposition pattern and pulmonary response as a result of improved dispersion of aspirated single walled carbon nanotubes in a mouse model. *Am. J. Physiol. Lung Cell Mol. Physiol.* 294: L87–L97.
Methner, M.M. 2008. Effectiveness of local exhaust ventilation (LEV) in controlling engineered nanomaterial emissions during reactor clean out operations. *J. Occup. Environ. Hyg.* 5: D63–D69.
Mijailovich, S.M., M. Kojic, and A. Tsuda. 2010. Particle-induced indentation of the alveolar epithelium caused by surface tension forces. *J. Appl. Physiol.* 109: 1179–1194.
Porter, D.W., A. Hubbs, R. Mercer et al. 2010. Mouse pulmonary dose- and time course-responses induced by exposure to multi-walled carbon nanotubes. *Toxicology* 269: 136–147.
Porter, D., K. Sriram, M. Wolfarth et al. 2008. A biocompatible medium for nanoparticle dispersion. *Nanotoxicology* 2: 144–154.
Roco, M.C. 2004. Science and technology integration for increased human potential and societal outcomes. *Ann. N.Y. Acad. Sci.* 1013: 1–6.
Sager, T.M., C. Kommineni, and V. Castranova. 2008. Pulmonary response to intratracheal instillation of ultrafine versus fine titanium dioxide: Role of particle surface area. *Part. Fibre Toxicol.* 5: 17.
Sager, T.M., D.W. Porter, V.A. Robinson, W.G. Lindsley, D. Schwegler-Berry, and V. Castranova. 2007. Improved method to disperse nanoparticles for *in vitro* and *in vivo* investigation of toxicity. *Nanotoxicology* 1: 118–129.
Salvador-Morales, C., P. Townsend, E. Flahauf et al. 2007. Binding of pulmonary surfactant proteins to carbon nanotubes; potential for damage to lung immune defense mechanisms. *Carbon* 45: 607–617.
Schurch, S., P. Gehr, V. Im Hof, M. Geiser, and F. Green. 1990. Surfactant displaces particles toward the epithelium in airways and alveoli. *Respir. Physiol.* 80: 17–32.

Shvedova, A.A., A.A. Kapralov, W.H. Feng et al. 2012. Impaired clearance and enhanced pulmonary inflammatory/fibrotic responses to carbon nanotubes in myeloperoxidase-deficient mice. *PLoS One* 7: e30923.

Shvedova, A.A., E.R. Kisin, R. Mercer et al. 2005. Unusual inflammatory and fibrogenic pulmonary responses to single walled carbon nanotubes in mice. *Am. J. Physiol. Lung Cell Mol. Physiol.* 289: L698–L708.

Shvedova, A.A., E. Kisin, A.R. Murray et al. 2008. Inhalation versus aspiration of single carbon nanotubes in C57BL/6 mice: Inflammation, fibrosis, oxidative stress and mutagenesis. *Am. J. Physiol. Lung Cell Mol. Physiol.* 295: L552–L565.

Shvedova, A.A., T. Sager, A. Murray et al. 2007. Critical issues in the evaluation of possible effects resulting from airborne nanoparticles. In *Nanotechnology: Characterization, Dosing and Health Effects*, eds. N. Monteiro-Riviere and L. Tran, 221–232. Philadelphia, PA: Informa Healthcare.

Wang, L., V. Castranova, A. Mishra et al. 2010b. Dispersion of single-walled carbon nanotubes by a natural lung surfactant for pulmonary *in vitro* and *in vivo* toxicity studies. *Part. Fibre Toxicol.* 7: 31–41.

Wang, X., T. Xia, S.A. Ntim et al. 2010a. Quantitative techniques for assessing and controlling the dispersion and biological effects of multi-walled carbon nanotubes in mammalian tissue culture cells. *ACS Nano* 4: 7241–7252.

Warheit, D., T. Webb, C. Sayes, V. Colvin, and K. Reed. 2006. Pulmonary instillation studies with nanoscale TiO_2 rods and dots in rats: Toxicity is not dependent upon particle size and surface area. *Toxicol. Sci.* 91: 227–236.

Zhang, H., K.E. Burnum, M.L. Luna et al. 2011. Quantitative proteomics analysis of absorbed plasma proteins classifies nanoparticles with different surface properties and sizes. *Proteomics* 11: 4569–4577.

6

Interaction with Lung Macrophages

Barry Weinberger, Patrick J. Sinko, Jeffrey D. Laskin, and Debra L. Laskin

CONTENTS

6.1 Introduction

Administration of therapeutics via inhalation provides an opportunity for direct delivery to the lung epithelium, and nanoparticles (NPs) have emerged as a versatile platform for this. Because of their size (<100 nm), inhaled NP can be effectively deposited in alveoli, where they interact with surfactant proteins and glycoproteins (Muhlfeld et al. 2008). Therapeutic effects at distal sites are dependent on translocation of NP into the underlying interstitium and subsequently into the circulation (Geiser and Kreyling 2010; Kato et al. 2003). However, this process is generally limited and strongly dependent on NP composition, dissolution, surface area, size, and shape (Da Silva et al. 2013; Madl and Pinkerton 2009). Most NP deposited in the lung are removed by macrophages located in the luminal airways and alveoli (Geiser 2010). These cells transport NP from the lung into ciliated airways and the larynx, where they are subject to mucociliary clearance. A less frequent clearance mechanism involves macrophage transport of NP through the interstitium into the tissue lymphatics and thoracic lymph nodes (Geiser 2010). Clearance of NP by lung macrophages poses a significant barrier for both local and systemic delivery of therapeutics. Additionally, the interaction of NP with macrophages can cause cellular activation and/or toxicity, resulting in the release of cytotoxic/pro-inflammatory mediators that can cause collateral tissue injury (Roy et al. 2013a). For the potential therapeutic applications of NP to be realized, it is important to understand how NP can bypass lung macrophage defenses and limit inflammatory responses.

6.2 Lung Macrophages

Macrophages are cellular effectors of the innate immune system. They play an essential role in ridding the body of pathogens, as well as worn-out cells and debris, apoptotic cells, and some tumor cells. They are also among the most active secretory cells in the body, releasing a multitude of mediators that regulate all aspects of host defense, inflammation, wound repair, and homeostasis (Murray and Wynn 2011). Macrophages have also been implicated in tissue injury and disease pathogenesis, as well as in fibrosis and cancer. It is now well established that these diverse activities are mediated by distinct subpopulations of macrophages that develop in response to mediators present in the tissue microenvironment. Two major subpopulations have been characterized, broadly referred to as pro-inflammatory/cytotoxic M1 macrophages and anti-inflammatory/wound repair and immunoregulatory M2 macrophages (Labonte et al. 2014; Laskin et al. 2011; Mantovani et al. 2013) (Table 6.1). The outcome of inflammatory responses to injurious or infectious agents appears to depend on the relative numbers and activities of these two macrophage subpopulations. Thus, when properly controlled, M1 and M2 macrophages work together in a balanced manner to orchestrate inflammatory responses and then to initiate wound repair, leading to restored tissue homeostasis. Conversely, when M1 macrophages persist at injured sites in large numbers, they can overwhelm M2 macrophages and perpetuate tissue injury and disease pathogenesis by releasing excessive quantities of reactive oxygen species (ROS), reactive nitrogen species (RNS), bioactive lipids, proteolytic enzymes, and pro-inflammatory cytokines such as tumor necrosis factor (TNF)α, interleukin-1 (IL-1), and IL-6 (Laskin et al. 2011; Murray and Wynn 2011); on the other hand, disproportionate release of growth factors and mediators involved in extracellular matrix turnover by M2 macrophages can result in aberrant wound healing, tissue fibrosis, and potentially, cancer.

Pulmonary macrophages are among the most prominent macrophage populations in the body. They consist of a heterogeneous group of cells located in large numbers in the alveolar and interstitial spaces, as well as in the airways, pleura, and, in some species, the vasculature (Geiser 2010; Hussell and Bell 2014). Macrophages in the airways and alveolar spaces are major sentinels of the lung, serving as the first line of defense against inhaled xenobiotics. It is these cells that are mainly involved in the response to NP (Geiser 2010). However, the nature of this response depends on the characteristics of the NP, in particular, their size, shape, and surface properties. For example, tungsten carbide–cobalt NP induce the generation of ROS and apoptosis in rat lung macrophages (Zhao et al. 2013), while ZnO NP exert cytotoxicity but do not induce IL-1β (Xia et al. 2013). The initial immune response to NP involves phagocytosis by resident alveolar

TABLE 6.1

Characteristics of M1 and M2 Macrophage Subpopulations

Macrophage Type	Activating Signals	Major Products	Functions	Pathophysiologic Activity
Classically Activated M1 (type I inflammation)	PAMP (e.g., LPS), DAMP (e.g., HMGB1), IFNγ, TNFα	TNFα, IL-1, IL-6, IL-12, IL-15, IL-18, IL-23, ROS, RNS, CCL2, MMP-2, MMP-9	Pro-inflammatory, promote Th1-Thl7 responses; DTH, antimicrobial/ antitumor activity, phagocytosis of apoptotic PMN; antigen presentation	Cytotoxicity, prolong inflammation, promote tissue injury
Alternatively Activated M2				
M2a (type II inflammation)	IL-4/IL-13	IL-10, TGFβ, CTGF, VEGF, PDGF, IL-13, CCL18, galectin-3, MMP12, TIMP1	Mitogenic, promote Th2 responses, promote wound repair, antiparasitic	Suppress antitumor immunity, profibrotic, allergy
M2b (immunoregulatory)	Immune complexes + TLR/IL-1R ligands	IL-10	Antigen presentation, promote Th2 responses, anti-inflammatory	Immune suppression
M2c (deactivated, immunosuppressive)	IL-10, TGFβ, or glucocorticoids	IL-10, TGFβ, pentraxin-3	Anti-inflammatory, promote wound repair, tissue remodeling, angiogenesis	Promote chronic inflammation

Note: CTGF, connective tissue growth factor; DAMP, damage-associated molecular pattern; DTH, delayed type hypersensitivity; LPS, lipopolysaccharide; MMP, matrix metalloproteinases; PAMP, pathogen-associated molecular pattern; PDGF, platelet-derived growth factor; RNS, reactive nitrogen species; ROS, reactive oxygen species; TIMP, tissue inhibitor of metalloproteinase; VEGF, vascular endothelial growth factor.

macrophages; this is associated with an accumulation of inflammatory M1 macrophages in the lung and the release of cytotoxic/pro-inflammatory mediators aimed at ridding the body of the particles and recruiting additional phagocytic cells. Successful sequestering or destruction of NP is followed by M2 macrophage generation of mediators that suppress inflammation and initiate wound repair. However, if macrophages are unable to phagocytize and sequester NP, they respond by initiating a "foreign body reaction." This involves macrophage adhesion to the particle, cellular activation, and continuous production of inflammatory mediators (Anderson et al. 2008). This chronic inflammatory response is associated with macrophage fusion, foreign body giant cell formation, and fibrous capsule formation. It is also frequently associated with granuloma formation and the development of fibrosis. The persistent release of cytotoxic/pro-inflammatory mediators by M1 macrophages can also lead to tissue injury (Figure 6.1).

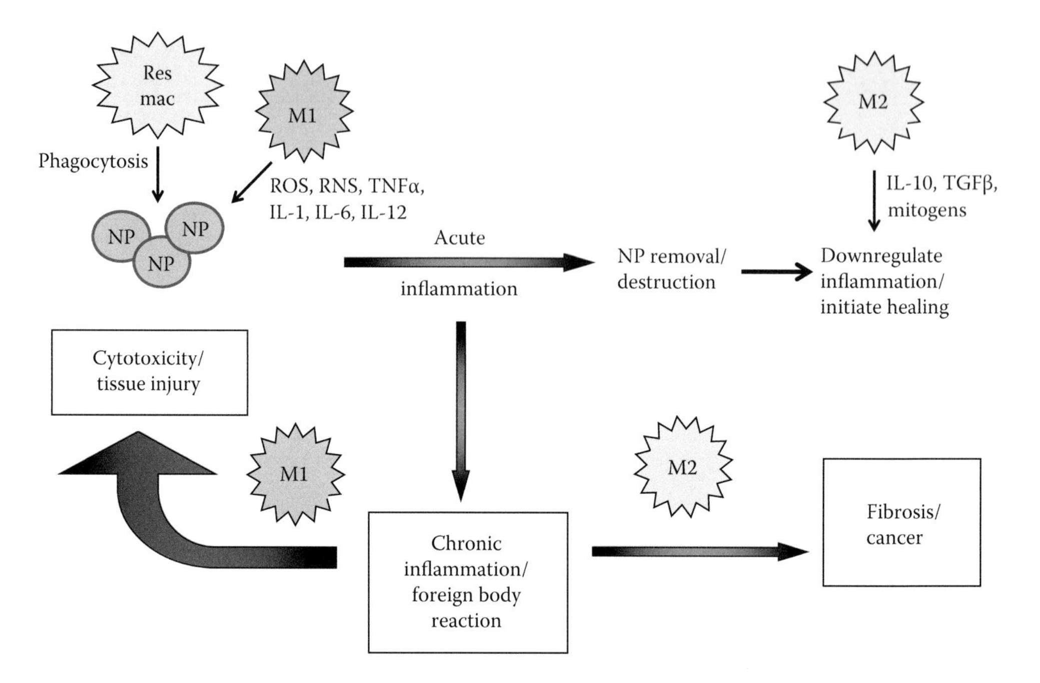

FIGURE 6.1
Lung macrophage responses to NPs. Resident lung macrophages (Res mac), which display an M2-like phenotype, are the first cells to respond to NP, engulfing them mainly via phagocytosis. Inflammatory proteins and chemokines released as a consequence of phagocytosis, or in the presence of injurious NP that cannot be taken up, recruit M1 macrophages that generate pro-inflammatory/cytotoxic mediators (e.g., ROS, RNS, TNFα, IL-1, IL-6, IL-12). Effective removal or destruction of the NP leads to downregulation of acute inflammation and initiation of wound healing by M2 macrophages via the release of IL-10, TGFβ, and various mitogens. The persistence of NP in the lung can lead to a foreign body reaction and chronic inflammation, which can progress to M1 macrophage–mediated cytotoxicity and tissue injury, or to M2 macrophage–mediated fibrosis and potentially cancer. The outcome of the response is dependent on the nature of the NP.

6.3 Uptake of NPs by Macrophages: Opsonization and Phagocytosis

The major pathway utilized by macrophages to engulf solid particles, including NP, is phagocytosis (Zhao et al. 2011). This is an energy-dependent process facilitated by the actin–myosin contractile system of macrophages. The first step in the process is opsonization or coating of the NP, which enhances their visibility to macrophages. Major opsonins in the airways and alveolar spaces include immunoglobulins (IgG or IgM), complement fragments (C3, C4, C5), fibronectin and laminin, as well as pulmonary collectins such as surfactant protein (SP)-A and SP-D (Kuroki et al. 2007; Roy et al. 2013a; Speert 2006). Opsonized NP then attach to macrophages via specific cell surface receptors. A number of alveolar macrophage receptors have been shown to be important in the uptake of NP, including receptors for the Fc component of immunoglobulins, complement (i.e., CR3), and various pulmonary collectins (Laskin et al. 2011). NP may also be taken up by lung macrophages via pattern recognition receptors, including toll-like receptors, as well as scavenger receptors (e.g., mannose receptor, SR-A); interaction of NP with these receptors facilitates opsonin-independent phagocytosis (Chao et al. 2013; Chen et al. 2009; Geiser 2010; Mano et al. 2013). Binding of NP to macrophage receptors triggers an intracellular signaling cascade resulting in actin assembly and the formation of a cell surface extension that encloses and engulfs the particle in an intracellular vesicle. These vesicles or phagosomes mature and fuse with lysosomes, resulting in the formation of phagolysosomes. Depending on the nature of the NP, they are subsequently degraded, and macrophage receptors are recycled back to the cell surface (Kettiger et al. 2013).

6.4 Nonphagocytic Uptake of NPs by Macrophages

NP can also be endocytosed by macrophages via pinocytosis, which involves the formation of large vesicles (0.2–10 μm), resulting in internalization of both fluid and particles (Swanson 2008). The major modes of endocytosis include macropinocytosis, clathrin-mediated endocytosis, and caveolae-mediated endocytosis (Kettiger et al. 2013). Macropinocytosis is a form of bulk uptake of fluid and solid cargo into cytoplasmic vacuoles, called macropinosomes (Falcone et al. 2006). Experimental studies indicate that macropinocytosis contributes to internalization of larger NP, in a nonspecific manner, and often in conjunction with other uptake mechanisms (Buono et al. 2009; Kettiger et al. 2013). Clathrin-mediated endocytosis is considered to be the "classical" and best-characterized route of entry of NP in polarized cells such as endothelial or epithelial cells; it is also involved in macrophage uptake of NP (Geiser 2010). Clathrin-mediated endocytosis is an actin-independent process wherein clathrin-coated pits are formed by cell membrane invaginations; these form coated vesicles (100–120 nm) that fuse with endosomes. Caveolin-mediated endocytosis is the most prominent clathrin-independent uptake mechanism. Caveolae are small, flask-like plasma membrane–associated structures, with a diameter of about 50–80 nm; the main structural component is caveolin-1. Small particles seem to be transported more efficiently into macrophages by caveolin-mediated endocytosis than larger particles. Hence, uptake of 20–40-nm particles is 5–10 times faster than uptake of 100-nm particles (Gratton et al. 2008; Wang et al. 2009). Whereas some NP (e.g., 5-nm silver NP) are taken up by both caveolin- and clathrin-dependent endocytosis, others (e.g., 17–60 nm iron oxide NP) depend mainly on clathrin-mediated endocytosis (Kim and Choi 2012; Yang et al. 2011).

6.5 Factors Controlling Uptake of NPs by Macrophages

Macrophage uptake of NP is related to the physical characteristics of the particles, including size, shape, surface properties, and rigidity.

6.5.1 Size

Generally, the efficiency of pulmonary NP phagocytosis is inversely related to particle size; thus, larger particles are cleared from the lung by pulmonary macrophages more efficiently than NP, and in fact, macrophage phagocytosis is most effective for particles >500 nm (Alexis et al. 2006, 2008; Gehr et al. 2006). For example, polystyrene particles in the range of 250 nm–3 μm are phagocytized more readily than NP <250 nm. Similarly, reduced uptake of

NP relative to microparticles by macrophages has been noted with particles comprising polymers, iron oxide, gold, and iridium (Gonzalez et al. 1996; Raynal et al. 2004; Schafer et al. 1992; Semmler-Behnke et al. 2007; Tabata and Ikada 1988; Takenaka et al. 2006). However, uptake of NP as small as 100 nm in diameter can be augmented by the presence of serum, a process dependent on Fcγ1 receptors (CD64) (Lunov et al. 2011). Although there are reports of even smaller NP being taken up by macrophages, most of these bypass phagocytic uptake by alveolar macrophages (Geiser 2010). Thus, only about 0.1% of 20-nm titanium dioxide (TiO_2) NP are phagocytized by macrophages compared with 100% of micrometer-sized TiO_2 particles (Geiser et al. 2008), and only 1.7% of macrophages contain particles in the first 24 h after aerosol inhalation (Geiser and Kreyling 2010).

6.5.2 Shape

NP can be fabricated in a wide variety of size and shape combinations, including micelles, liposomes, and nanofibers (e.g., nanotubes, rods, and fibers), and the relative width and height dimensions (e.g., aspect ratio) can influence macrophage uptake. Evidence suggests, however, that shape is most directly correlated to macrophage-mediated inflammation and toxicity. Thus, particles with shapes that macrophages are unable to degrade can induce lysosomal disruption and cytotoxicity (Hamilton et al. 2009). For example, whereas TiO_2 by itself is inert, when it is modified into a fiber-like structure with a high aspect ratio (>15 μm length), it induces a robust inflammatory response in alveolar macrophages, characterized by the production of ROS, IL-1, and IL-18, a response initiated by activation of the NALP3 inflammasome (Hamilton et al. 2009). Macrophages also have difficulty taking up particles with very high aspect ratios and high curvature angles (e.g., tubular vs. spherical particles), and their interaction with these particles is associated with increased release of inflammatory mediators and cytotoxicity (Champion and Mitragotri 2006, 2009). For instance, lung macrophages cannot easily incorporate long and rigid fibers into phagosomes; this leads to the release of cytotoxic oxygen radicals and hydrolytic enzymes, which are thought to contribute to chronic lung inflammation (Donaldson et al. 2010). Similarly, the needle-like structure of carbon nanotubules (CNT) can penetrate macrophage membranes, inflicting mechanical damage and cytotoxicity. Conversely, particles with reduced aspect ratios such as ellipsoid particles are more readily ingested by macrophages than spherical particles (Sharma et al. 2010). In studies of inhalation and intratracheal instillation of carbon-based NP, fullerenes (nonlinear) do not induce pulmonary inflammation, while CNT (linear) induce significant inflammatory responses and, with time, granuloma formation and fibrosis (Madl and Pinkerton 2009; Morimoto et al. 2013; Muller et al. 2005; Shvedova et al. 2005, 2008, 2014). In contrast, although aluminum oxide–based "nanowhiskers" (2–4 nm × 2800 nm) cause increases in macrophage accumulation in the lungs of exposed animals, they do not induce pro-inflammatory or cytotoxic responses in these cells (Adamcakova-Dodd et al. 2012).

6.5.3 Surface Properties

Surface properties such as smoothness, charge (zeta potential), and hydrophobicity also influence the uptake of NP by macrophages (Bertrand and Leroux 2012). In biological systems, the surface electrical potential of NP is determined not only by their charge but also by the local environment. The zeta potential is the electrical potential created between the surface of a particle with its associated ions, and its surrounding medium. After ingestion by macrophages, proteolytic enzymes and the acidic pH within phagolysosomes can strip associated proteins from NP, exposing surface charge, and altering their biological activity. In general, positively charged NP are more likely to accumulate in macrophages, when compared with negatively charged or neutral NP owing to steric or electrostatic repulsion (He et al. 2010; Yamaoka et al. 1995). Neutral particles are also opsonized at a much lower rate than charged particles, reducing their uptake by macrophages, and nonionic particles such as polyethylene glycol (PEG) tend to have minimal interactions with macrophages (Chonn et al. 1991; Owens and Peppas 2006; Roser et al. 1998). The strong electrostatic barrier created by charge repulsion also creates a barrier to macrophage-mediated inflammation and cytotoxicity, and it has been suggested that this is more significant than particle size or shape (El Badawy et al. 2011). The preferential interactions of macrophages with negatively charged particles may account for the higher cytotoxicity of anionic cyanoacrylic NP relative to cationic NP (Frohlich 2012; Tomita et al. 2011). Experimental evidence suggests that surface charge–mediated cytotoxicity of NP can result from several mechanisms. For example, many metal and metal oxide NP have highly positive zeta potentials under acidic conditions, which may allow them to damage macrophage phagolysosomal membranes, leading to cytotoxicity and inflammation (Cho et al. 2012b; Donaldson et al. 2013). Kaur and Tikoo (2013) synthesized silver NP with different surface charges by reduction with tannic acid and sodium borohydride, respectively. The tannic acid–reduced NP, with high surface potential, caused significantly increased generation of ROS by macrophages and activation of mitogen-activated protein kinase signaling. NP can also be "functionalized" with antibodies, peptides, or sugars, which can influence their interaction with macrophages (Elsabahy and Wooley 2012). In addition, the chain length of chemical linkers such as PEG, which are used to attach ligands to the surface of NP, may affect macrophage uptake (Cruz et al. 2011). Surface charge may also affect aggregation, which is a key determinant of effective particle size (Kendall and Holgate 2012). Thus, the instability of NP due to aggregation can lead to the generation of larger particles and increased macrophage phagocytosis.

6.5.4 Rigidity

Experimental studies suggest that the flexibility or rigidity of NP also influences their biological activity and interaction with macrophages (Davis 1975; Deshmukh et al. 2012; Fox et al. 2009). The intra- and intermolecular architecture of NP ultimately determines their rigidity. For example, the flexibility of PEGylated materials can be affected by the level of hydration, which is modulated by the amount of PEG and/or the length of PEG chains on the surface. The level of rigidity of NP can also influence their ability to deform to conformations necessary to pass through physiologic pores (Fox et al. 2009). For instance, it is well known that red blood cells, which are highly elastic, traverse the lung capillary bed freely, whereas white blood cells, although only marginally larger in size but much more rigid, traverse the capillary bed much more slowly. However, the biocompatibility of NP is adversely affected when they possess highly rigid structures. Thus, while thin (diameter ~50 nm) MWCNT with a highly crystalline structure can disrupt cell membranes and induce cytotoxicity, thick (diameter ~150 nm) or tangled (diameter ~2–20 nm) MWCNT exhibit significantly lower toxicity (Morimoto et al. 2013). Macrophages use the rigidity of a surface-bound particle to assess how it should be ingested. If the size and shape of NP are equal, macrophages favor phagocytosis of more rigid particles; conversely, less rigid NP are generally taken up by pinocytosis (Beningo and Wang 2002; Geiser 2010).

6.6 Approaches to Reducing the Interaction of NPs with Macrophages

The efficiency of removal of NP by macrophages is determined in large part by the degree of opsonization, which is critical to the process of phagocytic recognition and clearance. Therefore, inhibiting or slowing opsonization of NP is key to the success of drug delivery. Opsonization is promoted by hydrophobicity and by surface charge, with neutral (uncharged) NP exhibiting a much lower opsonization rate than charged particles (Roser et al. 1998). Therefore, one widely used method to slow opsonization is the use of surface-adsorbed or grafted shielding groups that can block the electrostatic and hydrophobic interactions that help opsonins bind to particle surfaces (Owens and Peppas 2006). The most effective strategy is to conjugate PEG and PEG-containing copolymers onto the surface of NP. These polymers are characteristically very flexible and highly hydrophilic, which can help shield hydrophobic or charged particles from opsonins (Owens and Peppas 2006). They are also typically charge neutral, which reduces electrostatic interactions. Thus, PEG substantially reduces nonspecific interactions with proteins through its hydrophilicity and steric repulsion effects, decreasing opsonization and complement activation (Alexis et al. 2008). Decorating NP with PEG polymers has been reported to reduce inflammation and enhance their translocation across the lung epithelium (Ibricevic et al. 2013). Similarly, poly(lactic-co-glycolic) acid (PLGA) NP, which are usually rapidly opsonized and removed by macrophages because of their hydrophobicity, can be protected by albumin, a dysopsonin that reduces the affinity of NP for macrophages by concealing surface targets (Goppert and Muller 2005; Manoocheheri et al. 2013). Surface alterations of CNT by dextran sulfate and of quantum dots by the cross-linker hydrazino nicotinamide have also been shown to be effective in reducing opsonization and macrophage uptake (Jung et al. 2012; Kotagiri et al. 2013).

Surfactant lipids are another potential approach to suppressing the removal of NP by lung macrophages. The alveolar lining fluid consists of an aqueous hypophase and a surface-active lipid–protein mixture known as pulmonary surfactant. Lung surfactant stabilizes alveoli and promotes clearance of inhaled material to maintain the alveoli in a sterile state and dampen inflammation (Clark et al. 2000). Noncellular elements of the air–blood barrier are the first biological materials that make contact with NP after inhalation and deposition in the lungs. The composition of surfactant is about 90% lipid and 10% protein (Goerke 1998). Of the proteins, SP-A and SP-D are collectins that play important roles in lung immunity (Wright 2005). These proteins have been shown to trigger uptake of NP by alveolar macrophages (Ruge et al. 2011, 2012). SP-D-deficient mice exhibit impaired uptake of NP by alveolar macrophages and dendritic cells (Kendall et al. 2013). SP-A and SP-D have differential preferences for hydrophobic and hydrophilic NP, respectively; however, these different affinities are attenuated in the presence of surfactant lipids. Consequently, surfactant lipid supplementation is a potential strategy to suppress collectin activity and optimize the delivery of NP to the lung. Surfactant lipids may also suppress binding of serum albumin to the surface of NP, which promotes uptake by macrophages (Dutta et al. 2007). A commercially available lung surfactant replacement has been shown to down-regulate NP uptake by macrophages *in vitro*, suggesting a potentially feasible approach to decreasing their removal by macrophages (Vranic et al. 2013).

Another strategy to reduce uptake by macrophages is incorporation of NP into microgels for delivery (Wanakule et al. 2012). This is aimed at bypassing the fact that NP (<500 nm) often fail to deposit in the distal lung. Particle diameters between 0.5 and 5 μm are optimal for aerodynamic delivery, but they are readily phagocytized by lung macrophages. To avoid rapid clearance, it is desirable for microparticles to have a diameter >6 μm, and hydrophilic surface chemistry (Ahsan et al. 2002; Champion et al. 2008; El-Sherbiny et al. 2010; Wanakule et al. 2012). To satisfy the need for diameter <5 μm during aerosol delivery, and >6 μm to minimize macrophage uptake, swellable microgel particles have been developed, which increase in size in the lung milieu (Kaggwa et al. 2003). Moreover, NP (<200 nm) have been embedded in these particles, designed to be released after delivery to the distal lung by specific acids present in the tissue. Porous, micron-sized structures assembled from NP for pulmonary delivery have also been described (Tsapis et al. 2002).

6.7 Examples of Macrophage Interactions with NPs

As NP are increasingly being considered for use in drug delivery, their interactions with macrophages are under intense investigation. Examples of some of the responses of macrophages to exposure to NP are described in the following section and summarized in Table 6.2.

6.7.1 Metal Oxide NPs

Superparamagnetic iron oxide NP are used to differentially enhance the contrast of organs on MRI. Iron oxide NP do not elicit direct cytotoxic or pro-inflammatory effects in macrophages, but prime them for increased responsiveness to infectious agents and impair their transition from an M1 to an M2 phenotype (Kodali et al. 2013). Intratracheal instillation of magnetite NP results in macrophage infiltration into the lung, alveolar edema, and granuloma formation resembling a foreign body reaction (Tada et al. 2012). Although not likely to be used for therapeutic delivery of drugs, magnetic NP are being investigated because they have distinct applications in nanomedicine for visualizing or targeting organs or tumors. One candidate is cobalt ferrite NP, which do not appear to exert oxidative toxicity in macrophages (Horev-Azaria et al. 2013).

TiO_2 is an insoluble transitional metal oxide that has numerous industrial uses and potential medical applications. It is ubiquitous in dyes, paints, and consumer products,

TABLE 6.2

Examples of Lung Macrophage Responses to NPs

Nanoparticle	Lung Macrophage Response	References
Iron oxide	Priming for increased responsiveness to inflammatory mediators	Kodali et al. 2013
Magnetite	Inflammation, foreign body reaction	Tada et al. 2012
Titanium oxide	Inflammation; cytotoxicity; release of ROS, IL-1β, TNFα	Kang et al. 2008; Xiong et al. 2013; Yazdi et al. 2010; Oberdorster et al. 1994; Noel et al. 2012
Zinc oxide	Cytotoxicity: release of TNFα, IL-12, macrophage inflammatory protein (MIP)-1α	Raemy et al. 2012; Chang et al. 2013; Adamcakova-Dodd et al. 2014
Carbon nanofibers/ carbon nanotubes	Inflammation; phagocytosis/uptake; foreign body reaction; release of TNFα, IL-1β, IL-6, TGFβ	Warheit et al. 2013; Shvedova et al. 2008; Kobayashi et al. 2011; Mercer et al. 2008
Silver	Release of IL-1, TNFα, IL-6; phagocytosis	Roberts et al. 2013a; Stebounova et al. 2011; Kaewamatawong et al. 2014

including sunscreens and toothpaste. TiO_2 NP are chemically stable and can effectively bind and release drugs in a pH-dependent manner, so they have been investigated as a vehicle for chemotherapeutics (Chen et al. 2011). However, they also exert significant pro-inflammatory and, under some conditions, cytotoxic/pro-oxidant activity. Thus, in macrophages, TiO_2 NP stimulate IL-1β and TNFα release, and respiratory burst activity; this is associated with mitochondrial depolarization and intracellular calcium mobilization (Kang et al. 2008; Xiong et al. 2013; Yazdi et al. 2010). These effects are dependent on the size of the TiO_2 NP, with smaller particles (10–25 nm) exerting stronger inflammatory effects than larger ones (100–250 nm) (Oberdorster et al. 1994). Biologic responses to TiO_2 are also correlated with the size of particle agglomerates in aerosols. Thus, large aggregates (>100 nm) are associated with a 2-fold increased inflammatory response, while smaller aggregates cause increased cytotoxicity and oxidative activity (Noel et al. 2012).

Zinc oxide (ZnO) NP are widely used in topical medications and cosmetics, with airway exposures occurring mainly after incidental inhalation (Vandebriel and De Jong 2012). Responses to ZnO NP are dependent on the mode of exposure, with more rapid responses (i.e., TNFα generation) being observed with aerosols (Raemy et al. 2012). Among metal oxide NP, ZnO is notable for its high solubility (Zhang et al. 2012). Dissolution of ZnO in the lung lining fluid leads to local generation of Zn^{2+}, resulting in significant inflammation and cytotoxicity. Inside macrophage lysosomes, the protein corona surrounding ZnO NP is rapidly digested because of low pH, allowing Zn^{2+} to destabilize the lysosomes (Cho et al. 2012b). High cytosolic levels of Zn^{2+} are found in lung lavage fluid and alveolar macrophages after exposure of rats to ZnO NP (Kao et al. 2012). This is associated with increased lactate dehydrogenase and protein content in lung lavage fluid, markers of lung injury, and granulomatous inflammation in the tissue (Cho et al. 2010, 2012a; Sayes et al. 2007; Warheit et al. 2009). Pro-inflammatory gene expression also increases (Chang et al. 2013). Recent studies have also noted increases in alveolar macrophage production of pro-inflammatory cytokines (IL-12 and MIP-1α) after subacute exposure (2 weeks) of mice to ZnO NP (Adamcakova-Dodd et al. 2014).

6.7.2 Carbon Nanofibers (CNF) and CNT

CNF are cylindrical nanostructures with graphene layers arranged as stacked cones, cups, or plates. CNT are CNF wrapped into cylindrical structures; they are composed of either a single layer (e.g., SWCNT) or multiple layers (e.g., MWCNT) of individual SWCNT stacked within one another. Many practical and industrial uses for these compounds, including vectors for drug delivery, have been explored on the basis of their unusual physical properties. However, aerosol dispersion of CNF, as well as their hydrophobicity, raises concerns regarding their immunotoxic effects. Inhalation studies have shown that CNF are efficiently taken up by lung macrophages (Warheit et al. 2013). Moreover, after 13 weeks of exposure to doses of CNF as low as 2.5 mg/m^3, >90% of alveolar macrophages contain CNF, and inflammatory changes are observed in lung pathology; at 25 mg/m^3, inflammatory markers are also elevated in the lung, indicating dose-dependent activation of lung macrophages. Similarly, inhalation of SWCNT (5 mg/m^3) for 4 days results in macrophage accumulation in the lung and oxidative stress, responses not observed at lower doses (Morimoto et al. 2013; Shvedova et al. 2008). While pulmonary inflammatory responses to inhaled MWCNT were reported to be minimal after exposures up to 14 days (0.3–32 mg/m^3), 3-month exposures produce lung edema and inflammation at doses as low as 0.1–0.4 mg/m^3 (Li et al. 2007; Mitchell et al. 2007; Wang et al. 2014). Of note, multifocal granulomas, typical of a foreign body reaction, and lung fibrosis were also observed after longer exposures, along with increased production of TNFα, IL-1β, IL-6, and TGFβ, consistent with ineffective clearance of particles by macrophages and persistent macrophage activation (Ma-Hock et al. 2009; Pauluhn 2010; Shvedova et al. 2008, 2014; Staal et al. 2014). In cultured macrophages, MWCNT upregulate the expression of cyclooxygenase and inducible nitric oxide synthase, suggesting a potential mechanism mediating inflammatory lung injury (Lee et al. 2012).

Specific physical characteristics of CNT, including tube length and diameter, degree of agglomeration, surface properties, and impurities have been shown to influence macrophage inflammatory responses (Hamilton et al. 2012, 2013a, 2013b; Madl and Pinkerton 2009). Generally, phagocytosis of longer CNT by macrophages is less efficient than shorter particles, leading to greater induction of inflammatory and fibrotic changes in the lung (Hamilton et al. 2013a, 2013b; van Berlo et al. 2014). Agglomeration of particles is highly dependent on the mechanism of dispersion (e.g., droplet vs. dry), with more effective dispersion causing less robust macrophage responses (Morimoto et al. 2013). Differential aggregation states of SWCNT induce distinct biological responses, including cytokine production, granuloma formation, and collagen deposition (Kobayashi et al. 2011; Mercer et al. 2008). The high hydrophobicity of native CNT contributes to their toxicity, resulting in attempts to "functionalize" them for therapeutic uses (i.e., modifying the surface chemistry to increase aqueous solubility). For example, functionalization of SWCNT with $-NH_3$ or MWCNT with –COOH significantly reduces macrophage cytotoxicity and inflammasome activation (Dumortier et al. 2006; Hamilton et al. 2013a, 2013b). Contamination with heavy metals or with amorphous carbon, which can result from the synthetic process, also increases pulmonary inflammation (Hamilton et al. 2012; Lam et al. 2004; Warheit et al. 2004).

6.7.3 Silver NPs

Because of their antimicrobial properties, silver NP have been widely used in consumer products; however, contact via inhalation is primarily due to incidental environmental or occupational exposures (Herzog et al. 2013). Silver NP (25 nm) delivered intranasally are widely distributed in the body, resulting in macrophage-mediated erythrocyte destruction in the spleen, but no inflammatory changes in lung (Genter et al. 2012). Similarly, inhalation of silver NP (30–40 nm) had no effect on alveolar macrophages in rats over the course of several days (Roberts et al. 2013a; Stebounova et al. 2011). In contrast, pro-inflammatory cytokines (IL-1, TNFα, and IL-6) were significantly increased in the lung 28 days after a single intratracheal instillation of silver NP. Particle-laden alveolar macrophages have also been noted in histologic sections of the lung after silver NP exposure, which correlated with lung lesions; however, these effects were transient (Kaewamatawong et al. 2014).

6.7.4 Silica NPs

The effects of exposure to silica NP have been intensely studied because of their presence as an environmental contaminant, and their use in pharmaceuticals, paints, cosmetics, and food. Recent studies have shown that independent of their diameter size, silica particles are efficiently internalized by macrophages via an actin cytoskeleton–dependent pathway, a response associated with lysosomal destabilization, cell death, and IL-1β secretion (Foldbjerg et al. 2013; Kusaka et al. 2014; Sandberg et al. 2012); silica NP also cause severe lung inflammation in mice and pro-inflammatory cytokine release. Engineered cadmium-containing silica NP have also been reported to exert inflammatory effects in rodents after intratracheal instillation, as evidenced by increased production of IL-6 and TGFβ (Coccini et al. 2013). However, carboxy-surface modification of silica NP impairs their agglomeration and ameliorates cytotoxicity and respiratory burst responses in macrophages, when compared with native particles. These differences are likely due to distinct interactions with proteins in the culture medium, resulting in characteristic protein "coronas" that are dependent on surface chemistry and charge (Mortensen et al. 2013).

6.7.5 Engineered NPs Used for Drug Delivery and Diagnosis

Engineered NP are stable solid colloidal particles that are generally <100 nm in diameter. These nanomatrix materials can be modified to allow for different drug release kinetics and/or can be tailored to degrade, on the basis of a particular physical property or trigger. Engineered NP can be prepared from a variety of materials, including proteins, polysaccharides, and polymers. The selection of matrix materials depends on the (i) size of NP required; (ii) inherent properties of the drug (e.g., aqueous solubility and stability); (iii) surface characteristics (e.g., charge and permeability); (iv) degree of biodegradability, biocompatibility, and toxicity; (v) drug release profile desired; and (vi) antigenicity of the final product. Natural and synthetic polymers, proteins, and polysaccharides are examples of NP used for drug delivery. Polymers commonly used to fabricate NP include PEG, PLGA, poly-ε-caprolactone (PCL), poly(cyanoacrylate), poly-D,L-lactide (PLA), polystyrene, and poly(vinylpyrrolidone), and their copolymers. The potential for toxicity of a polymer relates to the specific toxicities of its architectural units (e.g., building blocks, repetitive blocks/subunits, core and shell surfaces, as well as the entire structures). Even if the architectural unit is nontoxic, engineering it into a linear polymer could render it toxic due to shape or size. Biodegradable polymers such as PCL, PGA, PLA, and PLGA are approved for human use (Thomas et al. 2007). However, although encapsulation of drugs into PLGA NP can prevent premature drug degradation, improve drug delivery, and enhance drug efficacy (Cartiera et al. 2009), data on their inflammatory activity and disposition are limited. In RAW 264.7 macrophages, PLGA NP (60–200 nm) are not cytotoxic, but they do induce TNFα release (Xiong et al. 2013). In contrast, PLGA NP have been shown to remain in the lungs of mice for up to 7 days without being cleared or triggering of macrophage inflammatory responses (Roberts et al. 2013b). These findings point to the limitations of *in vitro* studies in transformed macrophages for the assessment of *in vivo* macrophage responses to NP.

6.8 Summary and Conclusions

Macrophages represent the first line of host defense in the lung, and they are the first immune cell type to respond to NP. Depending on the nature of the NP (e.g., size, shape, surface properties, rigidity), this response can involve uptake and removal of NP with no adverse outcome, or macrophage cytotoxicity and the release of inflammatory mediators, chronic inflammation, and with some NP, the development of fibrosis. A number of approaches have been taken to minimize adverse reactions of macrophages with NP and to limit phagocytosis and sequestration. These include reducing NP opsonization, surfactant lipid supplementation, and incorporation of NP into microgels for drug delivery. While these approaches are promising, they require additional investigation and development.

Understanding the factors that contribute to macrophage responses to NP is key to the development of materials that can be used effectively in drug delivery.

Acknowledgments

This work was supported by National Institutes of Health grants AR055073, ES004738, ES005022, AI051214, CA132624, and CA155061.

References

Adamcakova-Dodd, A., Stebounova, L. V., Kim, J. S. et al. 2014. Toxicity assessment of zinc oxide nanoparticles using sub-acute and sub-chronic murine inhalation models. *Part Fibre Toxicol*, 11: 15. Available at http://www.particleandfibretoxicology.com/content/11/1/15.

Adamcakova-Dodd, A., Stebounova, L. V., O'Shaughnessy, P. T. et al. 2012. Murine pulmonary responses after sub-chronic exposure to aluminum oxide-based nanowhiskers. *Part Fibre Toxicol*, 9: 22. Available at http://www.particleandfibretoxicology.com/content/pdf/1743-8977-9-22.pdf.

Ahsan, F., Rivas, I. P., Khan, M. A., and Torres Suarez, A. I. 2002. Targeting to macrophages: Role of physicochemical properties of particulate carriers—Liposomes and microspheres—On the phagocytosis by macrophages. *J Control Release*, 79: 29–40.

Alexis, F., Pridgen, E., Molnar, L. K., and Farokhzad, O. C. 2008. Factors affecting the clearance and biodistribution of polymeric nanoparticles. *Mol Pharmacol*, 5: 505–515.

Alexis, N. E., Lay, J. C., Zeman, K. L. et al. 2006. *In vivo* particle uptake by airway macrophages in healthy volunteers. *Am J Respir Cell Mol Biol*, 34: 305–313.

Anderson, J. M., Rodriguez, A., and Chang, D. T. 2008. Foreign body reaction to biomaterials. *Semin Immunol*, 20: 86–100.

Beningo, K. A., and Wang, Y. L. 2002. Fc-receptor-mediated phagocytosis is regulated by mechanical properties of the target. *J Cell Sci*, 115: 849–856.

Bertrand, N., and Leroux, J. C. 2012. The journey of a drug-carrier in the body: An anatomo-physiological perspective. *J Control Release*, 161: 152–163.

Buono, C., Anzinger, J. J., Amar, M., and Kruth, H. S. 2009. Fluorescent pegylated nanoparticles demonstrate fluid-phase pinocytosis by macrophages in mouse atherosclerotic lesions. *J Clin Invest*, 119: 1373–1381.

Cartiera, M. S., Johnson, K. M., Rajendran, V., Caplan, M. J., and Saltzman, W. M. 2009. The uptake and intracellular fate of PLGA nanoparticles in epithelial cells. *Biomaterials*, 30: 2790–2798.

Champion, J. A., and Mitragotri, S. 2006. Role of target geometry in phagocytosis. *Proc Natl Acad Sci U S A*, 103: 4930–4934.

Champion, J. A., and Mitragotri, S. 2009. Shape induced inhibition of phagocytosis of polymer particles. *Pharm Res*, 26: 244–249.

Champion, J. A., Walker, A., and Mitragotri, S. 2008. Role of particle size in phagocytosis of polymeric microspheres. *Pharm Res*, 25: 1815–1821.

Chang, H., Ho, C. C., Yang, C. S. et al. 2013. Involvement of MyD88 in zinc oxide nanoparticle-induced lung inflammation. *Exp Toxicol Pathol*, 65: 887–896.

Chao, Y., Karmali, P. P., Mukthavaram, R. et al. 2013. Direct recognition of superparamagnetic nanocrystals by macrophage scavenger receptor SR-AI. *ACS Nano*, 7: 4289–4298.

Chen, W., Bian, A., Agarwal, A. et al. 2009. Nanoparticle superstructures made by polymerase chain reaction: Collective interactions of nanoparticles and a new principle for chiral materials. *Nano Lett*, 9: 2153–2159.

Chen, Y., Wan, Y., Wang, Y., Zhang, H., and Jiao, Z. 2011. Anticancer efficacy enhancement and attenuation of side effects of doxorubicin with titanium dioxide nanoparticles. *Int J Nanomed*, 6: 2321–2326.

Cho, W. S., Duffin, R., Poland, C. A. et al. 2010. Metal oxide nanoparticles induce unique inflammatory footprints in the lung: Important implications for nanoparticle testing. *Environ Health Perspect*, 118: 1699–1706.

Cho, W. S., Duffin, R., Poland, C. A. et al. 2012a. Differential pro-inflammatory effects of metal oxide nanoparticles and their soluble ions *in vitro* and *in vivo;* zinc and copper nanoparticles, but not their ions, recruit eosinophils to the lungs. *Nanotoxicology*, 6: 22–35.

Cho, W. S., Duffin, R., Thielbeer, F. et al. 2012b. Zeta potential and solubility to toxic ions as mechanisms of lung inflammation caused by metal/metal oxide nanoparticles. *Toxicol Sci*, 126: 469–477.

Chonn, A., Cullis, P. R., and Devine, D. V. 1991. The role of surface charge in the activation of the classical and alternative pathways of complement by liposomes. *J Immunol*, 146: 4234–4241.

Clark, H. W., Reid, K. B., and Sim, R. B. 2000. Collectins and innate immunity in the lung. *Microbes Infect*, 2: 273–278.

Coccini, T., Barni, S., Vaccarone, R. et al. 2013. Pulmonary toxicity of instilled cadmium-doped silica nanoparticles during acute and subacute stages in rats. *Histol Histopathol*, 28: 195–209.

Cruz, L. J., Tacken, P. J., Fokkink, R., and Figdor, C. G. 2011. The influence of PEG chain length and targeting moiety on antibody-mediated delivery of nanoparticle vaccines to human dendritic cells. *Biomaterials*, 32: 6791–6803.

Da Silva, A. L., Santos, R. S., Xisto, D. G. et al. 2013. Nanoparticle-based therapy for respiratory diseases. *An Acad Bras Cienc*, 85: 137–146.

Davis, M. 1975. Particulate radiopharmaceuticals for pulmonary studies. In *Radiopharmaceuticals*, ed. G. Subramanian, 267–281. New York: Society of Nuclear Medicine.

Deshmukh, M., Kutscher, H. L., Gao, D. et al. 2012. Biodistribution and renal clearance of biocompatible lung targeted poly(ethylene glycol) (PEG) nanogel aggregates. *J Control Release*, 164: 65–73.

Donaldson, K., Murphy, F. A., Duffin, R., and Poland, C. A. 2010. Asbestos, carbon nanotubes and the pleural mesothelium: A review of the hypothesis regarding the role of long fibre retention in the parietal pleura, inflammation and mesothelioma. *Part Fibre Toxicol*, 7: 5. Available at http://www.particleandfibretoxicology.com/content/pdf/1743-8977-7-5.pdf.

Donaldson, K., Schinwald, A., Murphy, F. et al. 2013. The biologically effective dose in inhalation nanotoxicology. *Acc Chem Res*, 46: 723–732.

Dumortier, H., Lacotte, S., Pastorin, G. et al. 2006. Functionalized carbon nanotubes are non-cytotoxic and preserve the functionality of primary immune cells. *Nano Lett*, 6: 1522–1528.

Dutta, D., Sundaram, S. K., Teeguarden, J. G. et al. 2007. Adsorbed proteins influence the biological activity and molecular targeting of nanomaterials. *Toxicol Sci*, 100: 303–315.

El Badawy, A. M., Silva, R. G., Morris, B. et al. 2011. Surface charge-dependent toxicity of silver nanoparticles. *Environ Sci Technol*, 45: 283–287.

Elsabahy, M., and Wooley, K. L. 2012. Design of polymeric nanoparticles for biomedical delivery applications. *Chem Soc Rev*, 41: 2545–2561.

El-Sherbiny, I. M., McGill, S., and Smyth, H. D. 2010. Swellable microparticles as carriers for sustained pulmonary drug delivery. *J Pharm Sci*, 99: 2343–2356.

Falcone, S., Cocucci, E., Podini, P. et al. 2006. Macropinocytosis: Regulated coordination of endocytic and exocytic membrane traffic events. *J Cell Sci*, 119: 4758–4769.

Foldbjerg, R., Wang, J., Beer, C. et al. 2013. Biological effects induced by BSA-stabilized silica nanoparticles in mammalian cell lines. *Chem Biol Interact*, 204: 28–38.

Fox, M. E., Szoka, F. C., and Frechet, J. M. 2009. Soluble polymer carriers for the treatment of cancer: The importance of molecular architecture. *Acc Chem Res*, 42: 1141–1151.

Frohlich, E. 2012. The role of surface charge in cellular uptake and cytotoxicity of medical nanoparticles. *Int J Nanomed*, 7: 5577–5591.

Gehr, P., Blank, F., and Rothen-Rutishauser, B. M. 2006. Fate of inhaled particles after interaction with the lung surface. *Paediatr Respir Rev*, 7 Suppl 1: S73–S75.

Geiser, M. 2010. Update on macrophage clearance of inhaled micro- and nanoparticles. *J Aerosol Med Pulm Drug Deliv*, 23: 207–217.

Geiser, M., Casaulta, M., Kupferschmid, B. et al. 2008. The role of macrophages in the clearance of inhaled ultrafine titanium dioxide particles. *Am J Respir Cell Mol Biol*, 38: 371–376.

Geiser, M., and Kreyling, W. G. 2010. Deposition and biokinetics of inhaled nanoparticles. *Part Fibre Toxicol*, 7: 1–17. Available at http://www.particleandfibretoxicology.com/content/7/1/2.

Genter, M. B., Newman, N. C., Shertzer, H. G., Ali, S. F., and Bolon, B. 2012. Distribution and systemic effects of intranasally administered 25 nm silver nanoparticles in adult mice. *Toxicol Pathol*, 40: 1004–1013.

Goerke, J. 1998. Pulmonary surfactant: Functions and molecular composition. *Biochim Biophys Acta*, 1408: 79–89.

Gonzalez, O., Smith, R. L., and Goodman, S. B. 1996. Effect of size, concentration, surface area, and volume of polymethylmethacrylate particles on human macrophages *in vitro*. *J Biomed Mater Res*, 30: 463–473.

Goppert, T. M., and Muller, R. H. 2005. Adsorption kinetics of plasma proteins on solid lipid nanoparticles for drug targeting. *Int J Pharm*, 302: 172–186.

Gratton, S. E., Ropp, P. A., Pohlhaus, P. D. et al. 2008. The effect of particle design on cellular internalization pathways. *Proc Natl Acad Sci U S A*, 105: 11613–11618.

Hamilton, R. F., Buford, M., Xiang, C., Wu, N., and Holian, A. 2012. NLRP3 inflammasome activation in murine alveolar macrophages and related lung pathology is associated with MWCNT nickel contamination. *Inhal Toxicol*, 24: 995–1008.

Hamilton, R. F., Wu, N., Porter, D. et al. 2009. Particle length-dependent titanium dioxide nanomaterials toxicity and bioactivity. *Part Fibre Toxicol*, 6: 35. Available at http://www.particleandfibretoxicology.com/content/pdf/1743-8977-6-35.pdf.

Hamilton, R. F. Jr., Wu, Z., Mitra, S., Shaw, P. K., and Holian, A. 2013a. Effect of MWCNT size, carboxylation, and purification on *in vitro* and *in vivo* toxicity, inflammation and lung pathology. *Part Fibre Toxicol*, 10: 57. Available at http://www.particleandfibretoxicology.com/content/pdf/1743-8977-10-57.pdf.

Hamilton, R. F., Xiang, C., Li, M. et al. 2013b. Purification and sidewall functionalization of multiwalled carbon nanotubes and resulting bioactivity in two macrophage models. *Inhal Toxicol*, 25: 199–210.

He, C., Hu, Y., Yin, L., Tang, C., and Yin, C. 2010. Effects of particle size and surface charge on cellular uptake and biodistribution of polymeric nanoparticles. *Biomaterials*, 31: 3657–3666.

Herzog, F., Clift, M. J., Piccapietra, F. et al. 2013. Exposure of silver-nanoparticles and silver-ions to lung cells *in vitro* at the air–liquid interface. *Part Fibre Toxicol*, 10: 11. Available at http://www.particleandfibretoxicology.com/content/pdf/1743-8977-10-11.pdf.

Horev-Azaria, L., Baldi, G., Beno, D. et al. 2013. Predictive toxicology of cobalt ferrite nanoparticles: Comparative in-vitro study of different cellular models using methods of knowledge discovery from data. *Part Fibre Toxicol*, 10: 32. Available at http://www.particleandfibretoxicology.com/content/pdf/1743-8977-10-32.pdf.

Hussell, T., and Bell, T. J. 2014. Alveolar macrophages: Plasticity in a tissue-specific context. *Nat Rev Immunol*, 14: 81–93.

Ibricevic, A., Guntsen, S. P., Zhang, K. et al. 2013. PEGylation of cationic, shell-crosslinked-knedel-like nanoparticles modulates inflammation and enhances cellular uptake in the lung. *Nanomedicine*, 9: 912–922.

Jung, K. H., Park, J. W., Paik, J. Y. et al. 2012. Hydrazinonicotinamide prolongs quantum dot circulation and reduces reticuloendothelial system clearance by suppressing opsonization and phagocyte engulfment. *Nanotechnology*, 23: 495102.

Kaewamatawong, T., Banlunara, W., Maneewattanapinyo, P., Thammachareon, C., and Ekgasit, S. 2014. Acute and subacute pulmonary toxicity caused by a single intratracheal instillation of colloidal silver nanoparticles in mice: Pathobiological changes and metallothionein responses. *J Environ Pathol Toxicol Oncol*, 33: 59–68.

Kaggwa, G. B., Carey, M. J., Such, C., and Saunders, B. R. 2003. A new family of water-swellable microgel particles. *J Colloid Interface Sci*, 257: 392–397.

Kang, J. L., Moon, C., Lee, H. S. et al. 2008. Comparison of the biological activity between ultrafine and fine titanium dioxide particles in RAW 264.7 cells associated with oxidative stress. *J Toxicol Environ Health A*, 71: 478–485.

Kao, Y. Y., Chen, Y. C., Cheng, T. J., Chiung, Y. M., and Liu, P. S. 2012. Zinc oxide nanoparticles interfere with zinc ion homeostasis to cause cytotoxicity. *Toxicol Sci*, 125: 462–472.

Kato, T., Yashiro, T., Murata, Y. et al. 2003. Evidence that exogenous substances can be phagocytized by alveolar epithelial cells and transported into blood capillaries. *Cell Tissue Res*, 311: 47–51.

Kaur, J., and Tikoo, K. 2013. Evaluating cell specific cytotoxicity of differentially charged silver nanoparticles. *Food Chem Toxicol*, 51: 1–14.

Kendall, M., Ding, P., Mackay, R. M. et al. 2013. Surfactant protein D (SP-D) alters cellular uptake of particles and nanoparticles. *Nanotoxicology*, 7: 963–973.

Kendall, M., and Holgate, S. 2012. Health impact and toxicological effects of nanomaterials in the lung. *Respirology*, 17: 743–758.

Kettiger, H., Schipanski, A., Wick, P., and Huwyler, J. 2013. Engineered nanomaterial uptake and tissue distribution: From cell to organism. *Int J Nanomed*, 8: 3255–3269.

Kim, S., and Choi, I. H. 2012. Phagocytosis and endocytosis of silver nanoparticles induce interleukin-8 production in human macrophages. *Yonsei Med J*, 53: 654–657.

Kobayashi, N., Naya, M., Mizuno, K. et al. 2011. Pulmonary and systemic responses of highly pure and well-dispersed single-wall carbon nanotubes after intratracheal instillation in rats. *Inhal Toxicol*, 23: 814–828.

Kodali, V., Littke, M. H., Tilton, S. C. et al. 2013. Dysregulation of macrophage activation profiles by engineered nanoparticles. *ACS Nano*, 7: 6997–7010.

Kotagiri, N., Lee, J. S., and Kim, J. W. 2013. Selective pathogen targeting and macrophage evading carbon nanotubes through dextran sulfate coating and PEGylation for photothermal theranostics. *J Biomed Nanotechnol*, 9: 1008–1016.

Kuroki, Y., Takahashi, M., and Nishitani, C. 2007. Pulmonary collectins in innate immunity of the lung. *Cell Microbiol*, 9: 1871–1879.

Kusaka, T., Nakayama, M., Nakamura, K. et al. 2014. Effect of silica particle size on macrophage inflammatory responses. *PLoS One*, 9: e92634. Available at http://www.plosone.org/article/fetchObject.action?uri=info%3Adoi%2F10.1371%2Fjournal.pone.0092634&representation=PDF.

Labonte, A. C., Tosello-Trampont, A. C., and Hahn, Y. S. 2014. The role of macrophage polarization in infectious and inflammatory diseases. *Mol Cells*, 37: 275–285.

Lam, C. W., James, J. T., McCluskey, R., and Hunter, R. L. 2004. Pulmonary toxicity of single-wall carbon nanotubes in mice 7 and 90 days after intratracheal instillation. *Toxicol Sci*, 77: 126–134.

Laskin, D. L., Sunil, V. R., Gardner, C. R., and Laskin, J. D. 2011. Macrophages and tissue injury: Agents of defense or destruction? *Annu Rev Pharmacol Toxicol*, 51: 267–288.

Lee, J. K., Sayers, B. C., Chun, K. S. et al. 2012. Multi-walled carbon nanotubes induce COX-2 and iNOS expression via MAP kinase-dependent and -independent mechanisms in mouse RAW264.7 macrophages. *Part Fibre Toxicol*, 9: 14. Available at http://www.particleandfibre toxicology.com/content/pdf/1743-8977-9-14.pdf.

Li, J. G., Li, W. X., Xu, J. Y. et al. 2007. Comparative study of pathological lesions induced by multi-walled carbon nanotubes in lungs of mice by intratracheal instillation and inhalation. *Environ Toxicol*, 22: 415–421.

Lunov, O., Syrovets, T., Loos, C. et al. 2011. Differential uptake of functionalized polystyrene nanoparticles by human macrophages and a monocytic cell line. *ACS Nano*, 5: 1657–1669.

Madl, A. K., and Pinkerton, K. E. 2009. Health effects of inhaled engineered and incidental nanoparticles. *Crit Rev Toxicol*, 39: 629–658.
Ma-Hock, L., Treumann, S., Strauss, V. et al. 2009. Inhalation toxicity of multiwall carbon nanotubes in rats exposed for 3 months. *Toxicol Sci*, 112: 468–481.
Mano, S. S., Kanehira, K., and Taniguchi, A. 2013. Comparison of cellular uptake and inflammatory response via Toll-like receptor 4 to lipopolysaccharide and titanium dioxide nanoparticles. *Int J Mol Sci*, 14: 13154–13170.
Manoocheheri, S., Darvishi, B., Kamalinia, G. et al. 2013. Surface modification of PLGA nanoparticles via human serum albumin conjugation for controlled delivery of docetaxel. *Daru*, 21: 58.
Mantovani, A., Biswas, S. K., Galdiero, M. R., Sica, A., and Locati, M. 2013. Macrophage plasticity and polarization in tissue repair and remodelling. *J Pathol*, 229: 176–185.
Mercer, R. R., Scabilloni, J., Wang, L. et al. 2008. Alteration of deposition pattern and pulmonary response as a result of improved dispersion of aspirated single-walled carbon nanotubes in a mouse model. *Am J Physiol Lung Cell Mol Physiol*, 294: L87–L97.
Mitchell, L. A., Gao, J., Wal, R. V. et al. 2007. Pulmonary and systemic immune response to inhaled multiwalled carbon nanotubes. *Toxicol Sci*, 100: 203–214.
Morimoto, Y., Horie, M., Kobayashi, N., Shinohara, N., and Shimada, M. 2013. Inhalation toxicity assessment of carbon-based nanoparticles. *Acc Chem Res*, 46: 770–781.
Mortensen, N. P., Hurst, G. B., Wang, W. et al. 2013. Dynamic development of the protein corona on silica nanoparticles: Composition and role in toxicity. *Nanoscale*, 5: 6372–6380.
Muhlfeld, C., Rothen-Rutishauser, B., Blank, F. et al. 2008. Interactions of nanoparticles with pulmonary structures and cellular responses. *Am J Physiol Lung Cell Mol Physiol*, 294: L817–L829.
Muller, J., Huaux, F., Moreau, N. et al. 2005. Respiratory toxicity of multi-wall carbon nanotubes. *Toxicol Appl Pharmacol*, 207: 221–231.
Murray, P. J., and Wynn, T. A. 2011. Protective and pathogenic functions of macrophage subsets. *Nat Rev Immunol*, 11: 723–737.
Noel, A., Maghni, K., Cloutier, Y. et al. 2012. Effects of inhaled nano-TiO_2 aerosols showing two distinct agglomeration states on rat lungs. *Toxicol Lett*, 214: 109–119.
Oberdorster, G., Ferin, J., and Lehnert, B. E. 1994. Correlation between particle size, *in vivo* particle persistence, and lung injury. *Environ Health Perspect*, 102 Suppl 5: 173–179.
Owens, D. E. III, and Peppas, N. A. 2006. Opsonization, biodistribution, and pharmacokinetics of polymeric nanoparticles. *Int J Pharm*, 307: 93–102.
Pauluhn, J. 2010. Subchronic 13-week inhalation exposure of rats to multiwalled carbon nanotubes: Toxic effects are determined by density of agglomerate structures, not fibrillar structures. *Toxicol Sci*, 113: 226–242.
Raemy, D. O., Grass, R. N., Stark, W. J. et al. 2012. Effects of flame made zinc oxide particles in human lung cells—A comparison of aerosol and suspension exposures. *Part Fibre Toxicol*, 9: 33. Available at http://www.particleandfibretoxicology.com/content/pdf/1743-8977-9-33.pdf.
Raynal, I., Prigent, P., Peyramaure, S. et al. 2004. Macrophage endocytosis of superparamagnetic iron oxide nanoparticles: Mechanisms and comparison of ferumoxides and ferumoxtran-10. *Invest Radiol*, 39: 56–63.
Roberts, J. R., McKinney, W., Kan, H. et al. 2013a. Pulmonary and cardiovascular responses of rats to inhalation of silver nanoparticles. *J Toxicol Environ Health A*, 76: 651–668.
Roberts, R. A., Shen, T., Allen, I. C. et al. 2013b. Analysis of the murine immune response to pulmonary delivery of precisely fabricated nano- and microscale particles. *PLoS One*, 8: e62115. Available at http://www.plosone.org/article/fetchObject.action?uri=info%3Adoi%2F10.1371%2Fjournal.pone.0062115&representation=PDF.
Roser, M., Fischer, D., and Kissel, T. 1998. Surface-modified biodegradable albumin nano- and microspheres. II: Effect of surface charges on *in vitro* phagocytosis and biodistribution in rats. *Eur J Pharm Biopharm*, 46: 255–263.
Roy, R., Kumar, S., Tripathi, A., Das, M., and Dwivedi, P. D. 2013a. Interactive threats of nanoparticles to the biological system. *Immunol Lett*, 158: 79–87.

Roy, R., Parashar, V., Chauhan, L. K. et al. 2013b. Mechanism of uptake of ZnO nanoparticles and inflammatory responses in macrophages require PI3K mediated MAPKs signaling. *Toxicol In Vitro*, 28: 457–467.
Ruge, C. A., Kirch, J., Canadas, O. et al. 2011. Uptake of nanoparticles by alveolar macrophages is triggered by surfactant protein A. *Nanomedicine*, 7: 690–693.
Ruge, C. A., Schaefer, U. F., Herrmann, J. et al. 2012. The interplay of lung surfactant proteins and lipids assimilates the macrophage clearance of nanoparticles. *PLoS One*, 7: e40775. Available at http://www.plosone.org/article/fetchObject.action?uri=info%3Adoi%2F10.1371%2Fjournal.pone.0040775&representation=PDF.
Sandberg, W. J., Lag, M., Holme, J. A. et al. 2012. Comparison of non-crystalline silica nanoparticles in IL-1β release from macrophages. *Part Fibre Toxicol*, 9: 32. Available at http://www.particleandfibretoxicology.com/content/pdf/1743-8977-9-32.pdf.
Sayes, C. M., Reed, K. L., and Warheit, D. B. 2007. Assessing toxicity of fine and nanoparticles: Comparing *in vitro* measurements to *in vivo* pulmonary toxicity profiles. *Toxicol Sci*, 97: 163–180.
Schafer, V., von Briesen, H., Andreesen, R. et al. 1992. Phagocytosis of nanoparticles by human immunodeficiency virus (HIV)-infected macrophages: A possibility for antiviral drug targeting. *Pharm Res*, 9: 541–546.
Semmler-Behnke, M., Takenaka, S., Fertsch, S. et al. 2007. Efficient elimination of inhaled nanoparticles from the alveolar region: Evidence for interstitial uptake and subsequent reentrainment onto airways epithelium. *Environ Health Perspect*, 115: 728–733.
Sharma, G., Valenta, D. T., Altman, Y. et al. 2010. Polymer particle shape independently influences binding and internalization by macrophages. *J Control Release*, 147: 408–412.
Shvedova, A. A., Kisin, E. R., Mercer, R. et al. 2005. Unusual inflammatory and fibrogenic pulmonary responses to single-walled carbon nanotubes in mice. *Am J Physiol Lung Cell Mol Physiol*, 289: L698–L708.
Shvedova, A. A., Kisin, E. R., Murray, A. R. et al. 2008. Inhalation vs. aspiration of single-walled carbon nanotubes in C57BL/6 mice: Inflammation, fibrosis, oxidative stress, and mutagenesis. *Am J Physiol Lung Cell Mol Physiol*, 295: L552–L565.
Shvedova, A. A., Yanamala, N., Kisin, E. R. et al. 2014. Long-term effects of carbon containing engineered nanomaterials and asbestos in the lung: One year postexposure comparisons. *Am J Physiol Lung Cell Mol Physiol*, 306: L170–L182.
Speert, D. P. 2006. Bacterial infections of the lung in normal and immunodeficient patients. *Novartis Found Symp*, 279: 42–51.
Staal, Y. C., van Triel, J. J., Maarschalkerweerd, T. V. et al. 2014. Inhaled multiwalled carbon nanotubes modulate the immune response of trimellitic anhydride-induced chemical respiratory allergy in Brown Norway rats. *Toxicol Pathol*, 2014 Apr 3. [Epub ahead of print]. doi: 10.1177/0192623313519874.
Stebounova, L. V., Adamcakova-Dodd, A., Kim, J. S. et al. 2011. Nanosilver induces minimal lung toxicity or inflammation in a subacute murine inhalation model. *Part Fibre Toxicol*, 8: 5. Available at http://www.particleandfibretoxicology.com/content/pdf/1743-8977-8-5.pdf.
Swanson, J. A. 2008. Shaping cups into phagosomes and macropinosomes. *Nat Rev Mol Cell Biol*, 9: 639–649.
Tabata, Y., and Ikada, Y. 1988. Effect of the size and surface charge of polymer microspheres on their phagocytosis by macrophage. *Biomaterials*, 9: 356–362.
Tada, Y., Yano, N., Takahashi, H. et al. 2012. Acute phase pulmonary responses to a single intratracheal spray instillation of magnetite (Fe_3O_4) nanoparticles in Fischer 344 rats. *J Toxicol Pathol*, 25: 233–239.
Takenaka, S., Karg, E., Kreyling, W. G. et al. 2006. Distribution pattern of inhaled ultrafine gold particles in the rat lung. *Inhal Toxicol*, 18: 733–740.
Thomas, V., Dean, D. R., Jose, M. V. et al. 2007. Nanostructured biocomposite scaffolds based on collagen coelectrospun with nanohydroxyapatite. *Biomacromolecules*, 8: 631–637.

Tomita, Y., Rikimaru-Kaneko, A., Hashiguchi, K., and Shirotake, S. 2011. Effect of anionic and cationic *n*-butylcyanoacrylate nanoparticles on NO and cytokine production in Raw264.7 cells. *Immunopharmacol Immunotoxicol*, 33: 730–737.

Tsapis, N., Bennett, D., Jackson, B., Weitz, D. A., and Edwards, D. A. 2002. Trojan particles: Large porous carriers of nanoparticles for drug delivery. *Proc Natl Acad Sci U S A*, 99: 12001–12005.

van Berlo, D., Wilhelmi, V., Boots, A. W. et al. 2014. Apoptotic, inflammatory, and fibrogenic effects of two different types of multi-walled carbon nanotubes in mouse lung. *Arch Toxicol*, 88: 1725–1737.

Vandebriel, R. J., and De Jong, W. H. 2012. A review of mammalian toxicity of ZnO nanoparticles. *Nanotechnol Sci Appl*, 5: 61–71.

Vranic, S., Garcia-Verdugo, I., Darnis, C. et al. 2013. Internalization of SiO_2 nanoparticles by alveolar macrophages and lung epithelial cells and its modulation by the lung surfactant substitute Curosurf. *Environ Sci Pollut Res Int*, 20: 2761–2770.

Wanakule, P., Liu, G. W., Fleury, A. T., and Roy, K. 2012. Nano-inside-micro: Disease-responsive microgels with encapsulated nanoparticles for intracellular drug delivery to the deep lung. *J Control Release*, 162: 429–437.

Wang, X., Shannahan, J. H., and Brown, J. M. 2014. IL-33 modulates chronic airway resistance changes induced by multi-walled carbon nanotubes. *Inhal Toxicol*, 26: 240–249.

Wang, Z., Tiruppathi, C., Minshall, R. D., and Malik, A. B. 2009. Size and dynamics of caveolae studied using nanoparticles in living endothelial cells. *ACS Nano*, 3: 4110–4116.

Warheit, D. B., Laurence, B. R., Reed, K. L. et al. 2004. Comparative pulmonary toxicity assessment of single-wall carbon nanotubes in rats. *Toxicol Sci*, 77: 117–125.

Warheit, D. B., Reed, K. L., and DeLorme, M. P. 2013. Embracing a weight-of-evidence approach for establishing NOAELs for nanoparticle inhalation toxicity studies. *Toxicol Pathol*, 41: 387–394.

Warheit, D. B., Sayes, C. M., and Reed, K. L. 2009. Nanoscale and fine zinc oxide particles: Can *in vitro* assays accurately forecast lung hazards following inhalation exposures? *Environ Sci Technol*, 43: 7939–7945.

Wright, J. R. 2005. Immunoregulatory functions of surfactant proteins. *Nat Rev Immunol*, 5: 58–68.

Xia, T., Hamilton, R. F., Bonner, J. C. et al. 2013. Interlaboratory evaluation of *in vitro* cytotoxicity and inflammatory responses to engineered nanomaterials: The NIEHS Nano GO Consortium. *Environ Health Perspect*, 121: 683–690.

Xiong, S., George, S., Yu, H. et al. 2013. Size influences the cytotoxicity of poly (lactic-co-glycolic acid) (PLGA) and titanium dioxide (TiO_2) nanoparticles. *Arch Toxicol*, 87: 1075–1086.

Yamaoka, T., Kuroda, M., Tabata, Y., and Ikada, Y. 1995. Body distribution of dextran derivatives with electric charges after intravenous administration. *Int J Pharm*, 113: 149–157.

Yang, C. Y., Tai, M. F., Lin, C. P. et al. 2011. Mechanism of cellular uptake and impact of ferucarbotran on macrophage physiology. *PLoS One*, 6: e25524. Available at http://www.plosone.org/article/fetchObject.action?uri=info%3Adoi%2F10.1371%2Fjournal.pone.0025524&representation=PDF.

Yazdi, A. S., Guarda, G., Riteau, N. et al. 2010. Nanoparticles activate the NLR pyrin domain containing 3 (Nlrp3) inflammasome and cause pulmonary inflammation through release of IL-1α and IL-1β. *Proc Natl Acad Sci U S A*, 107: 19449–19454.

Zhang, H., Ji, Z., Xia, T. et al. 2012. Use of metal oxide nanoparticle band gap to develop a predictive paradigm for oxidative stress and acute pulmonary inflammation. *ACS Nano*, 6: 4349–4368.

Zhao, F., Zhao, Y., Liu, Y. et al. 2011. Cellular uptake, intracellular trafficking, and cytotoxicity of nanomaterials. *Small*, 7: 1322–1337.

Zhao, J., Bowman, L., Magaye, R. et al. 2013. Apoptosis induced by tungsten carbide–cobalt nanoparticles in JB6 cells involves ROS generation through both extrinsic and intrinsic apoptosis pathways. *Int J Oncol*, 42: 1349–1359.

Section III

Crossing Epithelial Cells

7

Interactions with Alveolar Epithelium

Farnoosh Fazlollahi, Yong Ho Kim, Arnold Sipos, Zea Borok, Kwang-Jin Kim, and Edward D. Crandall

CONTENTS

7.1 Introduction

Newly developed nanomaterials are being increasingly used in industrial applications and in medicine as cutting-edge tools for therapeutic and diagnostic applications in biomedicine (e.g., highly sensitive and stable probes to detect changes in intracellular molecules) (Cuenca et al. 2006), early detection and diagnosis (Freitas 2005), and targeted delivery of therapeutics to specific cell types (Cuenca et al. 2006; Freitas 2005). Pulmonary delivery of nanomaterials may have certain advantages owing to the large surface area for absorption and the limited proteolytic activity (Sung et al. 2007) of the lungs. Nanoparticles (NPs) can be delivered locally for the treatment of lung diseases, such as cancer, asthma, or cystic fibrosis (Sermet-Gaudelus et al. 2002; Sung et al. 2007). This targeted delivery can potentially result in reducing the overall dose and side effects that result from high levels of systemic exposure. Alternatively, systemic delivery can be achieved by targeting delivery to the alveolar region where the drug can be absorbed through the thin layer of epithelial lining cells into the systemic circulation (Patton et al. 2004; Patton and Byron 2007; Sung et al. 2007). This approach may be desirable to achieve a rapid onset of action, avoidance of first-pass metabolism, and delivery of biotherapeutics that cannot be effective orally (owing to enzymatic degradation and/or poor intestinal membrane permeability) as an alternative to parenteral delivery (Hamman et al. 2005).

Although NP promise to be useful for many biology-related applications, their toxicity in biological tissues and organs is only partially known. Manufactured nanomaterials (or engineered nanomaterials [ENM]) have diameters from 1 nm to several hundred nanometers, with a narrow size distribution. NP are defined as particles with at least one dimension <100 nm. The size of particles is directly linked to their potential for causing adverse health effects. The Environmental Protection Agency (EPA) groups particle pollution into two categories: (i) "inhalable coarse particles" that are >2.5 microns and <10 microns in diameter, and (ii) "fine particles" that are ≤2.5 microns in diameter. The EPA is concerned about particles that are ≤10 microns in diameter because it has been reported that ultrafine ambient particles can pass directly into the circulatory system for systemic dissemination (Brook et al. 2004). The most likely route by which inhaled NP enter the systemic circulation is across alveolar epithelium with its large surface area and thin barrier thickness. The potential mechanisms for the pathophysiologic effects of inhaled NP include lung inflammation and/or direct interactions of NP with end organs. Concerns about the health risks of NP, and interest in their use for therapeutic applications, require increased understanding of interactions of inhaled NP with alveolar epithelium with regard to injury, uptake, and trafficking.

Internalization of NP into cells occurs via endocytic and/or nonendocytic pathways (Dausend et al. 2008; Geiser et al. 2005; Hansen and Nichols 2009; Harush-Frenkel et al. 2007, 2008). It has been reported that the cellular uptake of NP is influenced by composition, size, shape, aspect ratio, surface charge, surface coatings, and state of dispersal (Harush-Frenkel et al. 2007, 2008; Rejman et al. 2004; Yacobi et al. 2008, 2010). Systematic studies of toxicity/trafficking of defined/trackable ENM in the lung, as a portal of entry to the systemic circulation and end organs, will provide insight into the health effects of NP and yield valuable information needed for the safe design of inhaled drug delivery systems. This chapter is focused on interactions of ENM with alveolar epithelial barriers and characteristics that affect such interactions.

7.2 Overview on Health Effects

7.2.1 Health Effects of Ambient Ultrafine Particles (AUFP) and ENM

NP behavior and toxicity in the lungs and translocation from the lungs has been investigated primarily in the environmental health field with a focus on ultrafine particles (<100 nm aerodynamic diameter). Ambient airborne ultrafine particles generated from sources such as automotive combustion engines might contribute to respiratory and cardiovascular disease and mortality (Mossman et al. 2007). The small size of ultrafine particles might be the main factor contributing to their toxicity and adverse effects on human health. Ultrafine particles cross epithelial and endothelial cells by transcytosis into the blood and lymph circulation, resulting in distribution throughout the body to sites such as bone marrow, lymph nodes, spleen, liver, heart, and other organs (Kreyling et al. 2002; Oberdorster et al. 2002), in addition to the central nervous system and ganglia via translocation along axons and dendrites of neurons (Elder et al. 2006). The large surface area per mass of

NP causes them to be more biologically active, thus influencing inflammatory and oxidative stress reactions (Borm and Kreyling 2004; Hoet et al. 2004). However, the quantity of ultrafine particles found in blood and extrapulmonary organs varies widely in different studies, and the relevance of data obtained using rodent models with human health effects remains undetermined. A number of reviews and commentaries regarding applications and potential hazards/toxicity of nanomaterials, especially pertaining to the lung, have appeared recently (Card et al. 2008; Kendall and Holgate 2012; Li et al. 2010; Stella 2011; Warheit et al. 2008; Yacobi et al. 2010). One important note is that these environmental toxicological studies focus on insoluble materials, such as carbon (Gotz et al. 2011; Schinwald et al. 2012) or titanium dioxide (Andersson et al. 2011; Naya et al. 2012), whereas therapeutic nanoparticulate systems are composed of materials that will eventually degrade into biocompatible components. Thus, depending on the speed with which degradation occurs, biodegradable NP may be anticipated to elicit toxicological responses quite different from those of nonbiodegradable materials (Sung et al. 2007).

Since the composition and physicochemical properties of inhaled ambient particles are complex, variable, and poorly defined, mechanistic information on how inhaled ambient particles exert their effects and/or transit from airspaces into the systemic circulation/end organs is not easily obtained. ENM represent a new opportunity for studying material uptake and transport through the lung to elucidate fundamental processes and mechanistic pathways previously not clearly defined. For example, by using trackable and tailored ENM, the specific characteristics of airborne particulate matter that determine exposure-related health outcomes are being identified. Engaging commercial scientific knowledge gathered in the process of developing nanoproducts may support proper risk assessment on new ENM under development (Kendall and Holgate 2012).

With advances in nanotechnology and ENM being incorporated into consumer and industrial products, it is important to understand the impact these products may have on human and environmental health. The "safe by design" approach based on sound scientific evidence is crucial to avoid premature market introduction (Kendall and Holgate 2012). Thus, there are two major considerations from an environmental, health, and safety (EHS) perspective: namely, the interaction of ENM with humans and the possible effects on a wide range of organisms in the environment. Identification of hazardous ENM properties presents a considerable challenge because of the large number of materials being produced and the demonstration that more than a dozen physicochemical properties could potentially contribute to hazardous interactions at the nano–bio interface. This includes properties such as chemical composition, size, shape, aspect ratio, surface charge, hydrophobicity, redox activity, dissolution, crystallinity, surface coatings, and the state of agglomeration or dispersal.

Several studies have used rodents or other animals to demonstrate that inhaled ultrafine particles or ENM enter the systemic circulation and reach end organs at low but finite rates (Heckel et al. 2004; Nel et al. 2006; Oberdorster et al. 2005a,b; Rojas et al. 2011; Sadauskas et al. 2009; Semmler-Behnke et al. 2008; Zhang et al. 2010). A few human studies have been carried out with conflicting findings. One approach is to probe the number of newly emerging ENM and their wide range of properties by using a high-throughput screening platform that utilizes nanomaterial libraries exhibiting a range of compositions and combinatorial properties to study their relationship to specific injury responses. It is also important to find ways of linking this information at the *in vitro* level to biological injury *in vivo* so as to develop a predictive toxicological paradigm that maintains balance between the rate of knowledge generation at the biomolecular and cellular levels versus more relevant but costly experiments in more complex organisms (e.g., mice, rats, or other animal models) (Thomas et al. 2011). We describe in the following the reported effects of various ENM (e.g., carbonaceous NP, quantum dots [QD], silica and metal, and metal oxide NP) on cell viability and related parameters (e.g., oxidative stress and lipid peroxidation) pertaining to mammalian lung cells and tissues. These summaries update a prior review (Yacobi et al. 2011).

7.2.2 Carbonaceous NP-Induced Toxicity in Lung Cells/Tissues

Carbonaceous nanomaterials (including nanodiamonds [ND], graphenes, fullerenes, and carbon nanotubes [CNT] of single- or multiwalled structures) have been utilized in widely varied fields, with their health effects being studied intensely in recent years (Han et al. 2010; Jacobsen et al. 2009; Johnston et al. 2010; Kobayashi et al. 2011; Morimoto et al. 2012; Muller et al. 2008a,b, 2009; Mutlu et al. 2010; Shvedova et al. 2008; Stella 2011; Stoker et al. 2008). For example, fullerenes belong to a family of carbon allotropes (including graphite, [nano]diamond, CNT, and graphene sheet), and their mass ranges from C_{60} through C_{84}. Fullerenes are virtually insoluble in water, although they can be derivatized to become hydrophilic for biomedical application purposes. Fullerenes have useful characteristics, including phototoxicity, oxidant generation, and entrapment of noble gases and metals (Bakry et al. 2007; Bosi et al. 2004; Camps and Hirsch 1997; Deguchi et al. 2007; Partha et al. 2007). There are conflicting reports on fullerene toxicity in various cell types *in vitro* as well as in animal studies (Gharbi et al. 2005; Lewinski et al. 2008; Sayes et al. 2004, 2005, 2007a,b; Scrivens and Tour 1994; Yamago et al. 1995), with some studies showing no fullerene-associated toxicity (Sayes et al. 2007a), while others reported cytotoxicity (which may be associated with oxidative stress and lipid peroxidation, formation of DNA and protein adducts, and necrosis) (Baker et al. 2008; Lewinski et al. 2008; Porter et al. 2006; Sayes et al. 2004, 2005, 2007a,b; Scrivens and Tour 1994; Spohn et al. 2009; Tokuyama et al. 1993; Usenko et al. 2008; Yamawaki and Iwai 2006). When lung resident cells (e.g., monocytes, macrophages, and polymorphonuclear cells [PMN]) of mouse and guinea pigs or human monocytes were exposed for several hours to pristine C_{60} (up to 226 $\mu g/cm^2$), no significant cytotoxicity was found (Lewinski et al. 2008; Porter et al. 2006, 2009). When rats were exposed to nano-C_{60} (colloidal suspension of aggregates of pristine C_{60} in water) and $C_{60}(OH)_{24}$ for 1 day and 3 months, little lung toxicity, except for some inflammation on day 1, was found (Sayes et al. 2007a). Interestingly, nano-C_{60} is reported to be highly cytotoxic (~3–4 orders of magnitude) in human dermal fibroblasts, lung epithelial cells, and normal human astrocytes, while exposure of the same cell types to a highly water-soluble fullerol of $C_{60}(OH)_{24}$ does not exhibit much toxicity (Isakovic et al. 2006; Lewinski et al. 2008; Sayes et al. 2004). Only marginal influences of C_{60} on rat lung tissue have been documented (Baker et al. 2008). Interestingly, nano-C_{60} may be useful in protection against liver toxicity caused by carbon tetrachloride owing, in part, to its free radical–scavenging properties (Gharbi et al. 2005).

ND have been utilized in various fields, including biosensors (e.g., protein immobilization), drug delivery, and imaging of cells and tissues; however, specific information on their kinetics/dynamics in biological cells/tissues and toxicity is only slowly emerging. Many reviews have appeared recently, describing the applications of ND in biomedicine (Barnard 2009; Barnard et al. 2005; Bianco et al. 2008; Holt 2007; Lam and Ho 2009; Mochalin et al. 2012; Xing and Dai 2009). A recent study on ND-induced toxicity in rodent lungs indicates that intratracheal instillation of ND into mouse lungs leads to biodistribution of ND to the spleen, liver, bone, and heart, and that ND could induce dose-dependent toxicity to the lung, liver, kidney, and blood (Zhang et al. 2010). The specific trafficking mechanisms for ND across lung alveolar epithelium and ND-induced injury, if any, to alveolar epithelium are unknown.

The CNT family of carbonaceous NP is composed of multiwalled CNT (MWCNT) and single-walled CNT (SWCNT). Exposure to SWCNT or MWCNT was reported to lead to inflammation (via PMN) with granulomatous lesions in the lung (Kobayashi et al. 2011; Morimoto et al. 2012; Shvedova et al. 2005). It appears that inflammatory activity (assessed as increased number of PMN in bronchoalveolar lavage fluid) is commensurate with an increased surface area of CNT, although impurities (including toxic metals such as Co, Fe, Ni, and Mo, all of which have documented toxic effects) associated with CNT may be the cause for CNT-induced pulmonary toxicity (Donaldson et al. 2006; Helland et al. 2007; Johnston et al. 2010; Morimoto et al. 2012; Muller et al. 2008b). Amorphous carbons associated with SWCNT are thought to cause multifocal granulomas in the lung, since exposure of the lungs to purer SWCNT does not result in granulomas. van der Waals forces between CNT in air or in aqueous solutions cause them to form large aggregates, which can be >100 μm in diameter. In this regard, the aggregation state of carbonaceous NP turns out to be an important factor for causing toxicity in the lung, as exposure to agglomerated CNT results in granulomatous inflammation, but well-dispersed CNT lead to thickening of the alveolar wall with fewer granulomatous lesions in the lung. Several studies investigated the role of the CNT aspect ratio (i.e., length divided by diameter) in lung toxicity (Han et al. 2010; Shvedova et al. 2008; Warheit et al. 2004). Recently, lung toxicity in rodents was reported to be largely determined by the aggregation state of CNT; however, the aspect ratio of CNT appears to contribute much less to toxicity (Mutlu et al. 2010). It is noteworthy that SWCNT, on an equal mass basis, exhibit much more toxicity in the lungs than carbon black or quartz (Hurt et al. 2006; Jia et al. 2005). Of particular note is that when SWCNT are inhaled into mouse lungs along with pathogenic bacteria, clearance of bacteria from the lung significantly slows down, suggesting that the presence of SWCNT in the lung contributes to impairment of normal lung clearance of bacteria and/or decreases lung resistance to pathogenic organisms (Shvedova et al. 2008). Interestingly, inhalation of SWCNT dispersed in surfactant into mouse lungs led to much less lung toxicity than that caused by pristine SWCNT (Mutlu et al. 2010). These reports clearly indicate that more in-depth studies of physicochemical properties of CNT are required to better understand lung toxicology associated with this class of ENM.

MWCNT appear to manifest lung-specific and/or systemic effects in a manner dependent on their delivery methods of instillation versus inhalation into the lung (Cesta et al. 2010; Mitchell et al. 2007; Porter et al. 2010; Ryman-Rasmussen et al. 2009a,b). For example, MWCNT (0.3–5 mg/m^3), when inhaled by C57BL/6 adult (10–12 weeks old) male mice, did not result in significant lung inflammation or tissue damage, with no changes in blood cell counts in bronchoalveolar lavage fluid (Mitchell et al. 2007). After 14 days of MWCNT exposure, systemic immunosuppression was evident, including decreased natural killer cell function without changed expression of cytokine genes in lung tissue, suggesting that prolonged effects of MWCNT inhalation into the lung cause systemic immunosuppression (Mitchell et al. 2007). In another report, however, MWCNT inhaled into mice lungs were reported to cause pleural inflammation (with no overt changes in lung pathology) (Ryman-Rasmussen et al. 2009a). After a single inhalation exposure of 30 mg/m^3 for 6 h, embedded MWCNT were evident in the subpleural wall and within subpleural macrophages, along with an increased presence of mononuclear cell aggregates on the pleural surface after 1 day. Subpleural fibrosis increased after 2 and 6 weeks following MWCNT inhalation. No major effects were seen in mice that inhaled carbon black NPs or a lower dose of MWCNT at 1 mg/m^3 (Ryman-Rasmussen et al. 2009a). Instilled MWCNT, on the other hand, cause primarily lung pathologic effects. For example, when MWCNT were delivered into the lungs of 7-week-old male C57BL/6J mice by instillation (by pharyngeal aspiration) at 10–80 mcg total mass and studied at 1, 7, 28, and 56 days postexposure, pulmonary inflammation and damage (e.g., blood cell counts in bronchoalveolar lavage fluid) was dose dependent and peaked at 7 days postexposure, returning to normal levels by 56 days postexposure, except for the 40 mcg MWCNT dose that led to still greater damage than vehicle control. MWCNT instillation caused rapid development of pulmonary fibrosis by 7 days postexposure, suggesting that MWCNT instilled into mouse lungs rapidly produces adverse lung health effects (Porter et al. 2010). These reports strongly suggest that nanomaterial effects on lungs (and end organs) may be dependent on the method of delivery to the lung, playing an important role in ensuing changes in lung/systemic (patho)physiology.

Graphenes, representing a single atomic carbon layer, have received intense attention in recent years for applications in biomedicine as sensors and building blocks for tissue engineering (Artiles et al. 2011; Feng and Liu 2011; Kuila et al. 2011; Liu et al. 2012; Pumera 2012; Sanchez et al. 2012). Intratracheal instillation of graphene oxide (GO) into mouse lungs has been reported to lead to severe and persistent lung injury, where the rate of mitochondrial respiration and the generation of reactive oxygen species (ROS) in GO-containing lung cells all increased, and subsequently, inflammatory and apoptotic pathways were activated. On the other hand, instillation of pristine graphenes or nonoxidized graphenes dispersed in a block copolymer, Pluronic, led to minimal injury to lung cells, suggesting that the state of covalent oxidation in graphenes may be an important factor for pulmonary toxicity (Duch et al. 2011). Whether graphenes can enter alveolar epithelial cells (AEC) and, if so, by what mechanisms they cross the distal air–blood barrier of the mammalian lung, largely remain to be determined.

7.2.3 QD-Induced Toxicity in Lung Cells/Tissues

Semiconductor nanocrystals (e.g., QD) have been utilized widely for bioimaging (both *in vitro* and *in vivo*) (Qian et al. 2007; Schipper et al. 2007) and other nanomedicine applications (including drug delivery and phototherapeutics) (Chakravarthy et al. 2011; Morosini et al. 2011; Walther et al. 2008; Wang et al. 2010), due in part to their stable and much more sensitive optical properties compared with conventional fluorochromes. It is generally believed that the core materials of QD (e.g., cadmium) might cause toxicity to cells and tissues, which is a concern for utilizing QD as a stable biomarker (Bottrill and Green 2011; Geys et al. 2008). Other parameters associated with QD, including water dispersibility, biocompatibility, chemical stability, and robust optical properties, are investigated and being improved over the years with some success (Ghasemi et al. 2009). For example, chemical coatings (e.g., surface layering of QD with inert/nontoxic substances) have been reported to yield relatively nontoxic QD probes by preventing QD degradation and/or leaching of the toxic core materials and improving biocompatibility (Daou et al. 2009). In this regard, use of functionalizable silica shells (Erogbogbo et al. 2010), poly(ethylene glycol) (PEG) attachment to QD surface (Mohs et al. 2009), or adsorption of bovine serum albumin to QD surfaces (Chu et al. 2010) has been reported to decrease QD toxicity.

Toxicity associated with QD exposure of cells/tissues is reported to be related to oxidant stress (Li et al. 2009; Lovric et al. 2005a,b). On the other hand, intact QD were demonstrated to cause cell injury directly (Cho et al. 2007; Choi et al. 2007; Male et al. 2008; Stern et al. 2008). Some efforts to decrease QD toxicity have been reported, in that intravenous injection of QD (composed of a high-bandgap ZnTe shell with CdSe core [i.e., CdSe/ZnTe

QD]) into mice resulted in clearance of QD from blood (into urine) after 2–4 weeks, with no evidence for major organ damage (Law et al. 2009). As seen above, the important determinants of cytotoxicity afforded by QD include composition of shell and core of QD. Studies of QD trafficking in mammalian lungs *in vivo* indicate that acute pulmonary inflammation/edema and hepatic necrosis developed in hyperlipidemic ApoE$^{-/-}$ mice at 24 h after intratracheal instillation of CdTe QD (with either aminated or carboxylated surface) (Jacobsen et al. 2009). When PEG-coated CdSeTe/ZnS QD were administered intravenously in ICR (CD-1) mice, QD were found in the spleen, liver, and kidneys for at least 28 days after the injection, with notable renal toxicity (e.g., disruption of mitochondrial function) (Lin et al. 2008). Pulmonary vascular thrombosis, with distribution of QD mainly in the lung, liver, and blood, was noted following intravenous injection of CdSe/ZnS QD into Balb/c mice (Geys et al. 2008). Moreover, negatively charged QD were found in the same study to lead to much more coagulation than positively charged QD, with decreased thrombotic effects in animals (Geys et al. 2008). As seen above, it appears that the toxicity afforded by QD can be dictated by leaching of toxic core material(s), coating material(s), and surface charge conferred on QD.

7.2.4 Silica NP-Induced Toxicity in Lung Cells/Tissues

Another widely utilized class of ENM is amorphous silica (SiO_2) NP (SNP) as bioimaging materials and drug delivery vehicles (Anglin et al. 2008; Tasciotti et al. 2011). Amorphous SNP toxicity has been thought to be much less than crystalline SNP (i.e., quartz NP), although the physicochemical structure of amorphous SNP may also influence their cellular (toxic) responses (Gazzano et al. 2012). Inhalation of crystalline SNP has been documented to lead to silicosis, while intratracheal instillation of amorphous SNP (Carbosil, ~7 nm) into the rat lung was shown to significantly increase solute permeability of the lung air–blood barrier and inflammation/edema with a rat alveolar epithelial type II (AT2) cell marker being found in lung lavage fluid (Evans et al. 2006). SNP appears to confer toxicity to lung cells/tissues (especially alveolar epithelial type I [AT1] cells), which may take place via increased oxidant stress and lipid peroxidation/cell membrane leakage (and necrosis).

7.2.5 Metal Nanoparticle (MNP) or Metal Oxide Nanoparticle (MONP) Toxicity in Lung Cells/Tissues

MNP are utilized widely in various industrial applications (Cohen et al. 2007; Doria et al. 2012; Dreaden and El-Sayed 2012; Lok et al. 2007). Of particular importance, several MNP and MONP have been used as surrogates for AUFP. It has been well established that several MNP (e.g., composed of Fe, Zn, V, and Ni) and MONP (e.g., composed of Fe_2O_3, ZnO, V_2O_5, and NiO) are toxic and cause lung inflammation and injury (Ban et al. 2012; Horie et al. 2012; Kim et al. 2010). Of these, inhaled Au NP in rat lungs were found to be taken up into endosomes of alveolar macrophages, AT1 cells, and AT2 cells, leading to Au NP found in the systemic circulation following pulmonary instillation (Takenaka et al. 2006). Intratracheally instilled Ag NP appear to be rapidly cleared from lung airspaces into the systemic circulation and distributed into the brain, spleen, liver, kidney, and heart (Takenaka et al. 2001). Of note, agglomerated Ag NP were found in macrophages, suggesting that nonagglomerated Ag NP may have evaded engulfment and continued their journey into the circulation from the lung airspaces. Of the many MONP studied, ZnO NP, generated in high production volume in industrial sites (for smelting, welding, galvanizing, and brass plating) (Barceloux 1999; Lam et al. 1985, 1988) and already part of consumer products, may pose significant health risks. For example, ZnO NP causes acute metal fume fever (accompanied by chills, nausea, fatigue, fever, headache, and muscle aches) when inhaled at high concentrations. Exposure of guinea pig lungs to freshly generated ZnO NP led to significant injury to the lungs with increased wet-to-dry weight ratios and cellular infiltrates in alveoli (Lam et al. 1988). Similarly, exposure of a human AEC line (A549 cells) to ZnO NP was reported to decrease cell viability and increase DNA damage (Karlsson et al. 2008). Recently, we reported that exposure of primary rat AEC monolayers (RAECM) to ZnO NP results in severe damage to cell membranes, increase in cellular ROS, and decrease in mitochondrial activity in a dose- and time-dependent manner (largely irreversible), which appear to be, at least in part, mediated by free Zn^{2+} released from ZnO NP (Kim et al. 2010). Currently, we have limited understanding of how MNP/MONP enter pneumocytes, cross the lung air–blood barrier, and/or their fate in the cells. Many studies suggest that the determining factors for the cellular fate/transcellular trafficking properties of these NP are strongly governed by physicochemical characteristics (e.g., size, surface charge, shape [e.g., sphere vs. rod], and composition) of these NP (Goodman et al. 2004; McIntosh et al. 2001; Niemeyer and Mirkin 2004; Wang et al. 2008).

7.3 Mechanisms Underlying ENM-Induced Toxicity and Injury to Cells

From many studies reported on the (potential) toxicity of ENM on various cells and tissues, several possible mechanisms underlying (cyto)toxicity have emerged, including (i) dissolution or release of ions from ENM, (ii) disruption of cell membrane integrity by ENM, (iii) binding of ENM to protein or DNA and subsequent damage to these molecules, and (iv) increased levels of cellular ROS and (consequent) apoptosis/necrosis. The current paradigm for ENM-induced (cyto)toxicity is that ENM physicochemical properties (such as size [i.e., surface area], shape [including structural distortion and surface striation], capping [or encapsulating] agent, surface charge, and purity) all contribute to the toxicity of ENM (Hamoir et al. 2003; Kim et al. 2010; Nemmar et al. 2002a,b, 2003, 2004, 2005, 2006, 2007; Nemmar and Inuwa 2008; Ryman-Rasmussen et al. 2007; Yacobi et al. 2008). Epidemiological studies have indicated that individuals with preexisting pulmonary disease are more susceptible to the adverse health effects of inhaled particulates (Brook et al. 2002, 2004; Cesta et al. 2010; Pope and Dockery 2006; Ryman-Rasmussen et al. 2009b). Experimental studies strengthened this suggestion, showing that lung epithelial cells or macrophages "primed" with either tumor necrosis factor-α or lipopolysaccharide (LPS) had an increased inflammatory response on exposure to particulates (Imrich et al. 1999, 2007; Stringer and Kobzik 1998). It has been reported that carbon black particles (14 nm) aggravated bacterial endotoxin-induced lung inflammation (Inoue et al. 2006a,b, 2007). Moreover, pretreatment of rat lungs with LPS was seen to increase the extent of extrapulmonary translocation of radiolabeled polystyrene particles into the blood (Chen et al. 2006).

Currently, we have limited understanding of the mechanisms underlying specific ENM-induced (cyto)toxicity. For example, various ENM (e.g., C_{60}, carbon black, GO, SWCNT/MWCNT, Ag NP, TiO_2 NP, Fe_2O_3 NP, Al_2O_3 NP, ZrO_2 NP, Si_3N_4 NP, and MnO_2 NP) have been reported to cause *in vitro* cytotoxicity (Bottini et al. 2007; Gurr et al. 2005; Hussain et al. 2005; Sayes et al. 2005; Soto et al. 2007). Similarly, instillation of SWCNT (which contain variable levels of residual metals) intratracheally into mice lungs led to the dose-dependent development of epithelioid granulomas, whereas instillation into rat lungs of the SWCNT resulted in a transient inflammatory response with multifocal granulomas in a dose-independent manner (Lam et al. 2004). Interestingly, inflammatory and granulomatous responses appear to be confined near dense SWCNT aggregates localized in lung airspaces. In contrast, well-dispersed SWCNT found in regions away from where aggregated SWCNT were localized may contribute to alveolar wall thickening and diffuse interstitial fibrosis (Shvedova et al. 2005). When ultrafine carbon black or fine crystalline silica dust were utilized at an equal mass-based dose, inflammatory responses (and associated injury) appear to be less, without apparent alveolar wall thickening or granulomas. As seen above, there are numerous difficulties inherent in animal studies, including species differences, uneven distribution, and ENM/AUFP agglomeration in airspaces of the lung following instillation/inhalation, purity (and/or physicochemical characteristics) of ENM, and formidable laboratory-to-laboratory variations inherent within experimental design/techniques/execution.

The most likely route for the entry of inhaled nanomaterials into the systemic circulation is across the alveolar epithelium, with its very large surface area and thin barrier thickness (Furuyama et al. 2009; Nemmar et al. 2001, 2002a; Oberdorster 2001; Yang et al. 2008). Further knowledge about the mechanisms by which ENM injure and/or are translocated into/across the alveolar epithelium is thus of considerable importance for understanding the interactions of nanomaterials with the lung (and effects on other end organs). Nanomaterials may exert greater toxicity owing to higher specific surface area when compared with micrometer-sized materials of the same composition. For example, very low solubility and higher toxicity levels were reported for nanosized carbon black and TiO_2 when compared with larger particles of the same composition (Donaldson et al. 2005). In some cases, specific mineral components may be the predominating toxic component in nanomaterials (especially AUFP). For example, fine-sized crystalline silica (e.g., quartz dust) is much more reactive and causes more severe inflammation when inhaled into the lungs than TiO_2 (Oberdorster et al. 1994). Similarly, different degrees of lung inflammation and injury were reported for nanosized particles (composed of NiO, Co_3O_4, TiO_2, and carbon black) instilled into rat lungs (Dick et al. 2003), where the severity of lung injury appears to correlate with the particles' abilities to generate free radicals and cause oxidant injury. It appears that size (and thus total surface area), composition, and surface reactivity all are important determinants of the toxicity of nanomaterials.

7.4 Mechanisms Underlying ENM Trafficking

Since the physicochemical properties of inhaled ambient particles are poorly defined, mechanisms underlying trafficking of inhaled AUFP from lung airspaces into the systemic circulation (and subsequently to extrapulmonary end organs) are not easily obtainable, necessitating the use of surrogate labeled ENM for such studies. In general, macromolecules (and ENM) may gain access into the cell interior via either endocytic or nonendocytic pathways. These substances, internalized into "epithelial cells," may traverse the cell (i.e., transcytosis) and be delivered to the contralateral side of the epithelial barrier. Many reports showed that ENM inhaled into the mammalian lungs appear in the lymphatics and/or systemic circulation (Conner and Schmid 2003; Ferin et al. 1992; Geiser et al. 2005). We describe in the following the major endocytic and nonendocytic processes that may play roles in ENM trafficking in (lung) cells and tissues.

7.4.1 Endocytic Processes

The pathways of endocytosis are either clathrin dependent or clathrin independent. While clathrin-dependent pathways are dynamin dependent, clathrin-independent pathways are either dynamin dependent (i.e., caveolin-mediated endocytosis) or dynamin independent (i.e., phagocytosis and macropinocytosis) (Wieffer et al. 2009). It is reasonable to speculate that smaller NP may be internalized into cells via nonspecific fluid-phase endocytosis (e.g., micropinocytosis taking place via clathrin- and caveolin-independent pathways). Specialized endocytosis, requiring receptors congregated in caveolae or clathrin-coated pits, may also allow sampling of extracellular fluid in the vicinity, and thus NP contained therein, to be internalized along with ligand–receptor complex(es).

The mechanisms by which proteins involved in clathrin-mediated endocytosis recruit cargo into developing clathrin-coated pits to consequently form clathrin-coated vesicles are becoming better understood (Brodsky et al. 2001; da Costa et al. 2003; McMahon and Boucrot 2011; Meier and Greber 2004). The clathrin-mediated pathway is versatile, as many different cargoes (e.g., transferrin and polymeric immunoglobulin A) and extracellular particles (e.g., Ad5 viruses) can be packaged for transport using a range of accessory adaptor proteins (McMahon and Boucrot 2011; Meier et al. 2002; Meier and Greber 2004). As many proteins involved in clathrin-mediated endocytosis have been recognized and characterized, investigators can make use of specific small-molecule inhibitors and protein complementation at different places along the pathway to interfere with this process in a defined manner (Doherty and McMahon 2009). Molecular strategies for inhibition of clathrin-mediated endocytosis include overexpression of the clathrin hub domain (Brodsky et al. 2001; Meier et al. 2002), as well as siRNA and antisense approaches to inhibit clathrin heavy chain expression (Hinrichsen et al. 2003). Clathrin-coated pits have an observed upper limit of about 200 nm external diameter; however, the size of these pits varies between different cell types within the same species and depends on the size of its cargo(es) (McMahon and Boucrot 2011). Dynamin is a large GTPase that forms a helical polymer around the constricted neck of invaginated vesicles and, upon GTP hydrolysis, mediates the fission of these vesicles from the plasma membrane. This step irreversibly releases the clathrin-coated vesicles into the interior of the cell (Doherty and McMahon 2009).

The term lipid raft generally defines an assembly of specific lipids, usually glycosphingolipids and cholesterol, into a more ordered domain within the lipid bilayer of cell plasma membranes (Pelkmans 2005). Lipid rafts are involved in many different endocytic pathways, including caveolin-mediated endocytosis, CLathrin-Independent Carriers/GPI-Enriched Endocytic Compartments (CLIC/GEEC) endocytosis, Arf6-mediated endocytosis, flotillin-mediated endocytosis, and macropinocytosis (Ivanov 2008; Doherty and McMahon 2009). One of these pathways, caveolin-mediated endocytosis, involves the formation of flask-shaped invaginations (~60 nm) that are lined by the structural protein caveolin and enriched in cholesterol and sphingolipids (Cohen et al. 2004; Parton and Richards 2003). The lung is richly endowed with caveolae, especially within endothelial and AT1 cells (Cohen et al. 2004; Parton and Richards 2003). The most effective approach to disruption of caveolar function is to treat cells with sterol-binding agents that sequester cholesterol (Neufeld et al. 1996; Orlandi and Fishman 1998; Rothberg et al. 1992). A more specific strategy to inhibit caveolin-mediated endocytosis is utilization of a dominant-negative caveolin mutant (Pelkmans and Helenius 2002; Roy et al. 1999). Dynamin, which plays a crucial role in internalization of clathrin-coated vesicles, is essential for the fission of caveolin-coated vesicles for subsequent internalization (Doherty and McMahon 2009).

Some clathrin-independent and caveolin-independent pathways also are in operation within lipid raft domains of cell plasma membranes (Doherty and McMahon 2009; Nichols and Lippincott-Schwartz 2001). These pathways are termed receptor-independent fluid-phase endocytosis and can occur either through micropinocytosis by invagination of the plasma membrane, which is then pinched off, resulting in small vesicles (<100 nm diameter) in the cytoplasm, or by macropinocytosis forming large intracellular vacuoles (up to 5.0 μm in diameter) called macropinosomes (Amyere et al. 2002). Macropinocytosis usually occurs in highly ruffled regions of the plasma membrane by actin polymerization. Ruffled extensions of the plasma membrane form around regions of extracellular fluid, with subsequent internalization of large amounts of plasma membrane and uptake of a large volume of fluids (Doherty and McMahon 2009).

Phagocytosis is similar to macropinocytosis in its reliance on the actin cytoskeleton and use of specific signaling effectors (Dehio et al. 1998; Meier and Greber 2004). Like macropinocytosis, phagocytosis is sensitive to inhibitors of phosphatidylinositol (or phosphoinositide) 3 kinase; however, there are no specific inhibitors for phagocytosis alone, making distinction of involved pathways difficult. These processes involve the uptake of larger membrane areas by clathrin- or caveolin-mediated endocytosis (Doherty and McMahon 2009). Phagocytosis takes place in many cell types (including pneumocytes; Corrin 1970); however, it is most important in specialized cells such as dendritic cells and macrophages, where it plays an important role in cellular immunity (Doherty and McMahon 2009).

7.4.2 Nonendocytic Mechanisms

There are also nonendocytic mechanisms by which ENM may enter cells and/or traffic across the barrier composed of cells joined together (i.e., epithelium or endothelium). The nonendocytic mechanisms for ENM trafficking across epithelial barriers include restricted passive diffusion of ENM via paracellular pathways (tight junctions), translocation of ENM through "holes" (or defects) created by ENM in the cell plasma membrane, or apparent "diffusion" of ENM across the cell plasma membrane (without creating defects/holes). Translocation of ENM via the tight junctional route across undamaged epithelial barriers may be limited to smaller ENM whose diameter is <5 nm or so, since it has been shown that hydrophilic solutes, whose Stokes–Einstein radii range larger than ~6 nm, may be effectively excluded from passive diffusion via paracellular pathways (Bindslev and Wright 1976; Kim and Crandall 1983; Matsukawa et al. 1997; Wright and Pietras 1974) and that paracellular pathways of several epithelial barriers can be modeled to have equivalent water-filled pores whose radii range from 5 to 6 nm (Matsukawa et al. 1997; Wright and Pietras 1974). However, if ENM interactions with an epithelial barrier lead to "opening" the pathway between epithelial cells, it is expected that passive diffusion of ENM via altered tight junctions could take place (Shimada et al. 2006).

Creation of holes or defects by ENM in the artificial lipid bilayer and/or cell plasma membrane has been documented (Hong et al. 2006; Mecke et al. 2005). Noncytotoxic concentrations of polycationic NP were reported to create nanoscale holes (measurable as 30–2000 pA currents across cell membranes) in cell plasma membranes, consequently allowing leakage of a cytosolic enzyme (i.e., lactate dehydrogenase [LDH]) in human embryonic kidney 293A and human epidermoid carcinoma KB cells (Chen et al. 2009). Technically, the formation of nanoscale holes can be confirmed by monitoring leakage of small ions in a patch-clamp setting or assessment of fluorescein isothiocyanate or propidium iodide dye leakage into the cytosol over time (Hong et al. 2006; Leroueil et al. 2007, 2008; Yacobi et al. 2010). ENM (especially those that are hydrophobic) may, with or without disturbance in the cell plasma membrane, diffuse across lipid bilayers of cell plasma membranes, although this potential trafficking mechanism is currently under intense investigation but poorly understood.

ENM internalization into cells may be enhanced due to pro-inflammatory stimulation and/or impairment of proliferative activity as a result of direct and/or indirect interactions of cells with ENM. In other words, increased translocation of ENM into cells or across epithelial or endothelial barriers may result from ENM toxicity leading to oxidant stress, opening of tight junctions, inflammation, and/or (partial) destruction of epithelial and endothelial cells. In line with this supposition, ultrafine particles appearing in extrapulmonary organs following intratracheal instillation or inhalation in rodents may be related to deranged air–blood barrier and/or inflamed lung resident/recruited blood-borne cells.

7.5 Nanomaterial Interactions with Primary Rat/Mouse AEC Monolayers

Studies of *in vivo* and/or *in situ* lungs have yielded important findings on AUFP/ENM-induced health effects, albeit extracting specific mechanistic information from such studies is difficult. We have been utilizing a much simpler *in vitro* model of primary mammalian AEC monolayers cultivated on permeable supports (Cheek et al. 1989a,b) to investigate vectorial transport of ions, small solutes, and various macromolecules across the *in vitro* rat and mouse alveolar epithelial barriers (RAECM and MAECM, respectively). RAECM and MAECM exhibit a spontaneous potential difference (PD) of >10 mV (apical-side negative), resistance (Rt) of >2000 $\Omega{\cdot}cm^2$, and short-circuit current (Isc) of ~5 $\mu A/cm^2$ (Cheek et al. 1989b; Demaio et al. 2009). We have utilized RAECM and MAECM to investigate the mechanisms for alveolar epithelial transcytosis and metabolism of various peptides, proteins, and other macromolecules (e.g., dextrans) (Kim and Crandall 1996; Kim and Malik 2003; Kim et al. 2003, 2004; Widera et al. 2003a,b).

Exposure of RAECM to ultrafine ambient particle suspensions (UAPS) (containing mixtures of organic and elemental carbons, nitrate, sulfate, and trace amounts of metals) resulted in significantly decreased Rt (which did not recover to control level) (Yacobi et al. 2007). UAPS-induced decreases in Rt were dose and time dependent, with an apparent half-maximal concentration (EC_{50}) of ~9 μg/mL. Interestingly, paracellular diffusion of mannitol or inulin across UAPS-exposed RAECM was not significantly different from that across control RAECM, suggesting that decreases in Rt may reflect UAPS-induced damage to cell plasma membranes (but not at the tight junctional complex). Exposure of RAECM to SWCNT led to a rapid and significant decrease in Rt, which recovered toward control level after 24 h of exposure. When CdSe/ZnS (core/shell) QD (with surface coat of either chitosan [positively charged] or alginate [negatively charged]) were added to apical fluid of RAECM, the Rt of positively charged QD-exposed RAECM significantly decreased (transiently), with recovery toward control after 24 h of exposure, while negatively charged QD had a lesser effect. Unlike other ENM or AUFP/UAPS, which all exhibit decreased Rt upon exposure of RAECM, neither positively charged (amidinated) nor negatively charged (carboxylated) polystyrene NP (PNP) elicited significant changes in Rt. LDH release from UAPS- or SWCNT-exposed RAECM did not increase, suggesting relatively intact cell plasma membranes following incubation with either of these two NP. These data suggest that apical exposure of RAECM to some ENM (or AUFP) may result in alterations in alveolar epithelial barrier properties (Yacobi et al. 2007).

As noted previously, it is generally accepted that ENM interactions, and their fate(s) in cells/tissues (including AEC), are predominantly governed by the physicochemical properties (e.g., composition, shape, size, and surface charge/substances) of ENM/AUFP. ENM-induced alveolar epithelial injury is expected to be associated with alterations in cell plasma membrane integrity, cell–cell junctions, and/or mitochondrial function (e.g., generation of ROS). We have shown that ZnO NP exposure results in severe disruption of both transcellular and paracellular (tight junctional) pathways of RAECM. Cell plasma membrane damage by ZnO NP appears to be associated with elevation of intracellular ROS, since increases in ROS were abrogated in RAECM pretreated with *N*-acetylcysteine (Kim et al. 2010). These findings suggest that interactions of ZnO NP and/or excess free Zn^{2+} (released from ZnO NP) with RAECM lead to increased intracellular ROS, cell plasma membrane damage, and cell injury/death (Kim et al. 2010).

Alveolar epithelial translocation of PNP is highly dependent on the surface charge of PNP. The transalveolar epithelial trafficking rate of amidinated PNP is ~20–40 times greater than that of carboxylated PNP of comparable size. Larger (i.e., 100 or 120 nm) PNP are translocated across RAECM at three to four times slower rates than smaller (20 nm) PNP of similar surface charge. Confocal microscopy of RAECM exposed to PNP for 24 h revealed intracellular localization of PNP without evidence for PNP association at nuclei or tight junctions (Yacobi et al. 2008). Transcellular PNP trafficking across RAECM appears to utilize cellular energy–requiring process(es) (Yacobi et al. 2008).

ENM trafficking across an artificial lipid bilayer on filter (ALBF) model (Yacobi et al. 2010), whose lipid compositions are similar to native rat AT2 cells (Kikkawa et al. 1975), showed that exposure of the ALBF to PNP yielded 20–40 times greater translocation rates of amidinated PNP than those of carboxylated PNP, similar to those observed in RAECM (although the absolute fluxes of PNP across the ALBF were two to three times less than those across RAECM) (Yacobi et al. 2010). The apparent permeability (Papp) of either Na ion (using ^{22}Na) or ^{14}C-mannitol was not changed in the presence or absence of PNP, suggesting that PNP translocation across the ALBF (or RAECM) does not take place via "holes" created in the cell plasma membranes (Yacobi et al. 2010).

Disruption of tight junctional pathways with EGTA, which reduces the Rt of RAECM to zero, did not affect PNP trafficking rates, suggesting that PNP do not traverse tight junctional paracellular pathways (Yacobi et al. 2010). Major endocytosis mechanisms (including phagocytosis, macropinocytosis, clathrin-dependent endocytosis, or caveolin-dependent endocytosis) do not appear to mediate the transalveolar epithelial trafficking of PNP (Yacobi et al. 2010). PNP trafficking rates assessed in the presence of inhibitors of phagocytosis, macropinocytosis, and other well-defined endocytosis (e.g., clathrin- or caveolin-dependent processes) did not decrease. Colocalization of PNP with caveolin-1, clathrin heavy chain, early endosome antigen-1, cholera toxin B (a ligand known to be internalized via caveolin-mediated endocytosis), or wheat germ agglutinin (a ligand known to be endocytosed via fluid-phase adsorptive endocytosis) was not observed using confocal microscopy (Yacobi et al. 2010). These findings strongly suggest that translocation of PNP across RAECM takes place primarily transcellularly (but not via classic endocytic pathways), but may occur via a "diffusional" pathway through the lipid bilayer of cell plasma membranes (Yacobi et al. 2010).

To evaluate if interactions (including trafficking characteristics) of NP with epithelial cells are cell type specific, the characteristics of PNP translocation across Madin–Darby canine kidney cell II monolayers (MDCK-II) were investigated for comparison with those across RAECM (Fazlollahi et al. 2011). Trafficking rates of amidine-modified PNP were 500 times faster than those of 20 and 100 nm carboxylate-modified PNP. Exposure of MDCK-II monolayers to methyl-β-cyclodextrin (an inhibitor of caveolin-mediated endocytosis) did not decrease PNP flux, whereas dansylcadaverine (an inhibitor of clathrin-mediated endocytosis)- and dynasore (an inhibitor of dynamin)-treated MDCK-II monolayers exhibited ~80% decreases in PNP flux. Intracellular colocalization of PNP with clathrin heavy chain was observed by confocal laser scanning microscopy. These findings suggest that translocation of PNP across MDCK-II (i) occurs via clathrin-mediated endocytosis and (ii) is dependent on the physicochemical properties of PNP. We recently reported that similar endocytic mechanisms exist in MAECM for positively (but not negatively) charged PNP (Fazlollahi et al. 2013), suggesting that uptake/trafficking of NPs into/across epithelia is dependent both on the properties of the NPs and the specific epithelial cell type.

7.6 Summary

Inhaled ENM appear to be subject to translocation across the lung air–blood barrier, as evidenced by the biodistribution of ENM in extrapulmonary end organs in rodent and human studies. Of the ENM studied to date, CNT, ZnO NP, and QD cause pulmonary toxicity manifested by inflammatory responses and/or oxidant stress. It remains unresolved if the toxicity of inhaled ENM is due to pulmonary inflammation leading to systemic manifestations alone, or if ENM distribution to end organs is required for systemic manifestations by direct effects. ENM may translocate across alveolar epithelium via either nonendocytic or endocytic pathways. ENM trafficking appears to be governed by ENM physicochemical characteristics in a species-specific manner. As human beings are exposed to ENM (and AUFP), incidentally or accidentally, and are thus prone to subsequent health effects, better understanding of ENM/AUFP interactions with and trafficking across lung alveolar epithelium remains important. Rapid progress in biomedical applications of therapeutic/diagnostic ENM, including drug delivery (to lung parenchymal cells and/or the systemic circulation) via the lung, molecular imaging, biomarkers, biosensors, and tissue engineering, combined with potential toxicity, makes understanding ENM interactions with alveolar epithelium crucial to the safe and effective use of these products in humans.

Acknowledgments

This work was supported, in part, by research grants from the National Institutes of Health (ES017034, ES018782, HL038578, HL038621, HL062569, HL064365, HL089445, HL 108634, and HL112638), Hastings Foundation, and Whittier Foundation. Zea Borok is Ralph Edgington Chair in Medicine. Edward D. Crandall is Hastings Professor and Kenneth T. Norris Jr. Chair of Medicine.

References

Amyere, M., M. Mettlen, P. Van Der Smissen, A. Platek, B. Payrastre, A. Veithen, and P. J. Courtoy. 2002. Origin, originality, functions, subversions and molecular signalling of macropinocytosis. *Int J Med Microbiol* 291 (6–7):487–494.

Andersson, P. O., C. Lejon, B. Ekstrand-Hammarstrom, C. Akfur, L. Ahlinder, A. Bucht, and L. Osterlund. 2011. Polymorph- and size-dependent uptake and toxicity of TiO_2 nanoparticles in living lung epithelial cells. *Small* 7 (4):514–523.

Anglin, E. J., L. Cheng, W. R. Freeman, and M. J. Sailor. 2008. Porous silicon in drug delivery devices and materials. *Adv Drug Deliv Rev* 60 (11):1266–1277.

Artiles, M. S., C. S. Rout, and T. S. Fisher. 2011. Graphene-based hybrid materials and devices for biosensing. *Adv Drug Deliv Rev* 63 (14–15):1352–1360.

Baker, G. L., A. Gupta, M. L. Clark, B. R. Valenzuela, L. M. Staska, S. J. Harbo, J. T. Pierce, and J. A. Dill. 2008. Inhalation toxicity and lung toxicokinetics of C_{60} fullerene nanoparticles and microparticles. *Toxicol Sci* 101 (1):122–131.

Bakry, R., R. M. Vallant, M. Najam-ul-Haq, M. Rainer, Z. Szabo, C. W. Huck, and G. K. Bonn. 2007. Medicinal applications of fullerenes. *Int J Nanomed* 2 (4):639–649.

Ban, M., I. Langonne, N. Huguet, and M. Goutet. 2012. Effect of submicron and nano-iron oxide particles on pulmonary immunity in mice. *Toxicol Lett* 210 (3):267–275.

Barceloux, D. G. 1999. Zinc. *J Toxicol Clin Toxicol* 37 (2):279–292.

Barnard, A. S. 2009. Diamond standard in diagnostics: Nanodiamond biolabels make their mark. *Analyst* 134 (9):1751–1764.

Barnard, A. S., S. P. Russo, and I. K. Snook. 2005. Simulation and bonding of dopants in nanocrystalline diamond. *J Nanosci Nanotechnol* 5 (9):1395–1407.

Bianco, A., K. Kostarelos, and M. Prato. 2008. Opportunities and challenges of carbon-based nanomaterials for cancer therapy. *Expert Opin Drug Deliv* 5 (3):331–342.

Bindslev, N., and E. M. Wright. 1976. Effect of temperature on nonelectrolyte permeation across the toad urinary bladder. *J Membr Biol* 29 (3):265–288.

Borm, P. J., and W. Kreyling. 2004. Toxicological hazards of inhaled nanoparticles—Potential implications for drug delivery. *J Nanosci Nanotechnol* 4 (5):521–531.

Bosi, S., L. Feruglio, T. Da Ros, G. Spalluto, B. Gregoretti, M. Terdoslavich, G. Decorti, S. Passamonti, S. Moro, and M. Prato. 2004. Hemolytic effects of water-soluble fullerene derivatives. *J Med Chem* 47 (27):6711–6715.

Bottini, M., F. D'Annibale, A. Magrini, F. Cerignoli, Y. Arimura, M. I. Dawson, E. Bergamaschi, N. Rosato, A. Bergamaschi, and T. Mustelin. 2007. Quantum dot-doped silica nanoparticles as probes for targeting of T-lymphocytes. *Int J Nanomed* 2 (2):227–233.

Bottrill, M., and M. Green. 2011. Some aspects of quantum dot toxicity. *Chem Commun (Camb)* 47 (25):7039–7050.

Brodsky, F. M., C. Y. Chen, C. Knuehl, M. C. Towler, and D. E. Wakeham. 2001. Biological basket weaving: Formation and function of clathrin-coated vesicles. *Annu Rev Cell Dev Biol* 17:517–568.

Brook, R. D., J. R. Brook, B. Urch, R. Vincent, S. Rajagopalan, and F. Silverman. 2002. Inhalation of fine particulate air pollution and ozone causes acute arterial vasoconstriction in healthy adults. *Circulation* 105 (13):1534–1536.

Brook, R. D., B. Franklin, W. Cascio, Y. Hong, G. Howard, M. Lipsett, R. Luepker, M. Mittleman, J. Samet, S. C. Smith Jr., and I. Tager. 2004. Air pollution and cardiovascular disease: A statement for healthcare professionals from the Expert Panel on Population and Prevention Science of the American Heart Association. *Circulation* 109 (21):2655–2671.

Camps, X., and A. Hirsch. 1997. Efficient cyclopropanation of C_{60} from malonates. *JCS Perkin Trans* 1:1595–1596.

Card, J. W., D. C. Zeldin, J. C. Bonner, and E. R. Nestmann. 2008. Pulmonary applications and toxicity of engineered nanoparticles. *Am J Physiol Lung Cell Mol Physiol* 295 (3):L400–L411.

Cesta, M. F., J. P. Ryman-Rasmussen, D. G. Wallace, T. Masinde, G. Hurlburt, A. J. Taylor, and J. C. Bonner. 2010. Bacterial lipopolysaccharide enhances PDGF signaling and pulmonary fibrosis in rats exposed to carbon nanotubes. *Am J Respir Cell Mol Biol* 43 (2):142–151.

Chakravarthy, K. V., B. A. Davidson, J. D. Helinski, H. Ding, W. C. Law, K. T. Yong, P. N. Prasad, and P. R. Knight. 2011. Doxorubicin-conjugated quantum dots to target alveolar macrophages and inflammation. *Nanomedicine* 7 (1):88–96.

Cheek, J. M., M. J. Evans, and E. D. Crandall. 1989a. Type I cell-like morphology in tight alveolar epithelial monolayers. *Exp Cell Res* 184 (2):375–387.

Cheek, J. M., K. J. Kim, and E. D. Crandall. 1989b. Tight monolayers of rat alveolar epithelial cells: Bioelectric properties and active sodium transport. *Am J Physiol* 256 (3 Pt 1):C688–C693.

Chen, J., J. A. Hessler, K. Putchakayala, B. K. Panama, D. P. Khan, S. Hong, D. G. Mullen, S. C. Dimaggio, A. Som, G. N. Tew, A. N. Lopatin, J. R. Baker, M. M. Holl, and B. G. Orr. 2009. Cationic nanoparticles induce nanoscale disruption in living cell plasma membranes. *J Phys Chem B* 113 (32):11179–11185.

Chen, J., M. Tan, A. Nemmar, W. Song, M. Dong, G. Zhang, and Y. Li. 2006. Quantification of extrapulmonary translocation of intratracheal-instilled particles *in vivo* in rats: Effect of lipopolysaccharide. *Toxicology* 222 (3):195–201.

Cho, S. J., D. Maysinger, M. Jain, B. Roder, S. Hackbarth, and F. M. Winnik. 2007. Long-term exposure to CdTe quantum dots causes functional impairments in live cells. *Langmuir* 23 (4):1974–1980.

Choi, H. S., W. Liu, P. Misra, E. Tanaka, J. P. Zimmer, B. Itty Ipe, M. G. Bawendi, and J. V. Frangioni. 2007. Renal clearance of quantum dots. *Nat Biotechnol* 25 (10):1165–1170.

Chu, M., F. Wu, Q. Zhang, T. Liu, Y. Yu, A. Ji, K. Xu, Z. Feng, and J. Zhu. 2010. A novel method for preparing quantum dot nanospheres with narrow size distribution. *Nanoscale* 2 (4):542–547.

Cohen, A. W., R. Hnasko, W. Schubert, and M. P. Lisanti. 2004. Role of caveolae and caveolins in health and disease. *Physiol Rev* 84 (4):1341–1379.

Cohen, M. S., J. M. Stern, A. J. Vanni, R. S. Kelley, E. Baumgart, D. Field, J. A. Libertino, and I. C. Summerhayes. 2007. *In vitro* analysis of a nanocrystalline silver-coated surgical mesh. *Surg Infect (Larchmt)* 8 (3):397–403.

Conner, S. D., and S. L. Schmid. 2003. Regulated portals of entry into the cell. *Nature* 422 (6927):37–44.

Corrin, B. 1970. Phagocytic potential of pulmonary alveolar epithelium with particular reference to surfactant metabolism. *Thorax* 25 (1):110–115.

Cuenca, A. G., H. Jiang, S. N. Hochwald, M. Delano, W. G. Cance, and S. R. Grobmyer. 2006. Emerging implications of nanotechnology on cancer diagnostics and therapeutics. *Cancer* 107 (3):459–466.

da Costa, S. R., C. T. Okamoto, and S. F. Hamm-Alvarez. 2003. Actin microfilaments et al.—The many components, effectors and regulators of epithelial cell endocytosis. *Adv Drug Deliv Rev* 55 (11):1359–1383.

Daou, T. J., L. Li, P. Reiss, V. Josserand, and I. Texier. 2009. Effect of poly(ethylene glycol) length on the *in vivo* behavior of coated quantum dots. *Langmuir* 25 (5):3040–3044.

Dausend, J., A. Musyanovych, M. Dass, P. Walther, H. Schrezenmeier, K. Landfester, and V. Mailander. 2008. Uptake mechanism of oppositely charged fluorescent nanoparticles in HeLa cells. *Macromol Biosci* 8 (12):1135–1143.

Deguchi, S., T. Yamazaki, S. A. Mukai, R. Usami, and K. Horikoshi. 2007. Stabilization of C_{60} nanoparticles by protein adsorption and its implications for toxicity studies. *Chem Res Toxicol* 20 (6):854–858.

Dehio, C., E. Freissler, C. Lanz, O. G. Gomez-Duarte, G. David, and T. F. Meyer. 1998. Ligation of cell surface heparan sulfate proteoglycans by antibody-coated beads stimulates phagocytic uptake into epithelial cells: A model for cellular invasion by *Neisseria gonorrhoeae*. *Exp Cell Res* 242 (2):528–539.

Demaio, L., W. Tseng, Z. Balverde, J. R. Alvarez, K. J. Kim, D. G. Kelley, R. M. Senior, E. D. Crandall, and Z. Borok. 2009. Characterization of mouse alveolar epithelial cell monolayers. *Am J Physiol Lung Cell Mol Physiol* 296 (6):L1051–L1058.

Dick, C. A., D. M. Brown, K. Donaldson, and V. Stone. 2003. The role of free radicals in the toxic and inflammatory effects of four different ultrafine particle types. *Inhal Toxicol* 15 (1):39–52.

Doherty, G. J., and H. T. McMahon. 2009. Mechanisms of endocytosis. *Annu Rev Biochem* 78: 857–902.

Donaldson, K., R. Aitken, L. Tran, V. Stone, R. Duffin, G. Forrest, and A. Alexander. 2006. Carbon nanotubes: A review of their properties in relation to pulmonary toxicology and workplace safety. *Toxicol Sci* 92 (1):5–22.

Donaldson, K., L. Tran, L. A. Jimenez, R. Duffin, D. E. Newby, N. Mills, W. MacNee, and V. Stone. 2005. Combustion-derived nanoparticles: A review of their toxicology following inhalation exposure. *Part Fibre Toxicol* 2:10.

Doria, G., J. Conde, B. Veigas, L. Giestas, C. Almeida, M. Assuncao, J. Rosa, and P. V. Baptista. 2012. Noble metal nanoparticles for biosensing applications. *Sensors (Basel)* 12 (2):1657–1687.

Dreaden, E. C., and M. A. El-Sayed. 2012. Detecting and destroying cancer cells in more than one way with noble metals and different confinement properties on the nanoscale. *Acc Chem Res* 45 (11):1854–1865.

Duch, M. C., G. R. Budinger, Y. T. Liang, S. Soberanes, D. Urich, S. E. Chiarella, L. A. Campochiaro, A. Gonzalez, N. S. Chandel, M. C. Hersam, and G. M. Mutlu. 2011. Minimizing oxidation and stable nanoscale dispersion improves the biocompatibility of graphene in the lung. *Nano Lett* 11 (12):5201–5207.

Elder, A., R. Gelein, V. Silva, T. Feikert, L. Opanashuk, J. Carter, R. Potter, A. Maynard, Y. Ito, J. Finkelstein, and G. Oberdorster. 2006. Translocation of inhaled ultrafine manganese oxide particles to the central nervous system. *Environ Health Perspect* 114 (8):1172–1178.

Erogbogbo, F., K. T. Yong, R. Hu, W. C. Law, H. Ding, C. W. Chang, P. N. Prasad, and M. T. Swihart. 2010. Biocompatible magnetofluorescent probes: Luminescent silicon quantum dots coupled with superparamagnetic iron(III) oxide. *ACS Nano* 4 (9):5131–5138.

Evans, S. A., A. Al-Mosawi, R. A. Adams, and K. A. Berube. 2006. Inflammation, edema, and peripheral blood changes in lung-compromised rats after instillation with combustion-derived and manufactured nanoparticles. *Exp Lung Res* 32 (8):363–378.

Fazlollahi, F., S. Angelow, N. R. Yacobi, R. Marchelletta, A. S. Yu, S. F. Hamm-Alvarez, Z. Borok, K. J. Kim, and E. D. Crandall. 2011. Polystyrene nanoparticle trafficking across MDCK-II. *Nanomedicine* 7 (5):588–594.

Fazlollahi, F., Y. H. Kim, A. Sipos, S. F. Hamm-Alvarez, Z. Borok, K. J. Kim, and E. D. Crandall. 2013. Nanoparticle translocation across mouse alveolar epithelial cell monolayers: Species-specific mechanisms. *Nanomedicine* 9 (6):786–794.

Feng, L., and Z. Liu. 2011. Graphene in biomedicine: Opportunities and challenges. *Nanomedicine (Lond)* 6 (2):317–324.

Ferin, J., G. Oberdorster, and D. P. Penney. 1992. Pulmonary retention of ultrafine and fine particles in rats. *Am J Respir Cell Mol Biol* 6 (5):535–542.

Freitas, R. A. Jr. 2005. Nanotechnology, nanomedicine and nanosurgery. *Int J Surg* 3 (4):243–246.

Furuyama, A., S. Kanno, T. Kobayashi, and S. Hirano. 2009. Extrapulmonary translocation of intratracheally instilled fine and ultrafine particles via direct and alveolar macrophage-associated routes. *Arch Toxicol* 83 (5):429–437.

Gazzano, E., M. Ghiazza, M. Polimeni, V. Bolis, I. Fenoglio, A. Attanasio, G. Mazzucco, B. Fubini, and D. Ghigo. 2012. Physicochemical determinants in the cellular responses to nanostructured amorphous silicas. *Toxicol Sci* 128 (1):158–170.

Geiser, M., B. Rothen-Rutishauser, N. Kapp, S. Schurch, W. Kreyling, H. Schulz, M. Semmler, V. Im Hof, J. Heyder, and P. Gehr. 2005. Ultrafine particles cross cellular membranes by nonphagocytic mechanisms in lungs and in cultured cells. *Environ Health Perspect* 113 (11):1555–1560.

Geys, J., A. Nemmar, E. Verbeken, E. Smolders, M. Ratoi, M. F. Hoylaerts, B. Nemery, and P. H. Hoet. 2008. Acute toxicity and prothrombotic effects of quantum dots: Impact of surface charge. *Environ Health Perspect* 116 (12):1607–1613.

Gharbi, N., M. Pressac, M. Hadchouel, H. Szwarc, S. M. Wilson, and F. Moussa. 2005. C_{60} fullerene is a powerful antioxidant *in vivo* with no acute or subacute toxicity. *Nano Lett* 5:2578–2585.

Ghasemi, Y., P. Peymani, and S. Afifi. 2009. Quantum dot: Magic nanoparticle for imaging, detection and targeting. *Acta Biomed* 80 (2):156–165.

Goodman, C. M., C. D. McCusker, T. Yilmaz, and V. M. Rotello. 2004. Toxicity of gold nanoparticles functionalized with cationic and anionic side chains. *Bioconjug Chem* 15 (4):897–900.

Gotz, A. A., A. Vidal-Puig, H. G. Rodel, M. H. de Angelis, and T. Stoeger. 2011. Carbon-nanoparticle-triggered acute lung inflammation and its resolution are not altered in PPARgamma-defective (P465L) mice. *Part Fibre Toxicol* 8:28.

Gurr, J. R., A. S. Wang, C. H. Chen, and K. Y. Jan. 2005. Ultrafine titanium dioxide particles in the absence of photoactivation can induce oxidative damage to human bronchial epithelial cells. *Toxicology* 213 (1–2):66–73.

Hamman, J. H., G. M. Enslin, and A. F. Kotze. 2005. Oral delivery of peptide drugs: Barriers and developments. *BioDrugs* 19 (3):165–177.

Hamoir, J., A. Nemmar, D. Halloy, D. Wirth, G. Vincke, A. Vanderplasschen, B. Nemery, and P. Gustin. 2003. Effect of polystyrene particles on lung microvascular permeability in isolated perfused rabbit lungs: Role of size and surface properties. *Toxicol Appl Pharmacol* 190 (3):278–285.

Han, S. G., R. Andrews, and C. G. Gairola. 2010. Acute pulmonary response of mice to multi-wall carbon nanotubes. *Inhal Toxicol* 22 (4):340–347.

Hansen, C. G., and B. J. Nichols. 2009. Molecular mechanisms of clathrin-independent endocytosis. *J Cell Sci* 122 (Pt 11):1713–1721.

Harush-Frenkel, O., N. Debotton, S. Benita, and Y. Altschuler. 2007. Targeting of nanoparticles to the clathrin-mediated endocytic pathway. *Biochem Biophys Res Commun* 353 (1):26–32.

Harush-Frenkel, O., E. Rozentur, S. Benita, and Y. Altschuler. 2008. Surface charge of nanoparticles determines their endocytic and transcytotic pathway in polarized MDCK cells. *Biomacromolecules* 9 (2):435–443.

Heckel, K., R. Kiefmann, M. Dorger, M. Stoeckelhuber, and A. E. Goetz. 2004. Colloidal gold particles as a new *in vivo* marker of early acute lung injury. *Am J Physiol Lung Cell Mol Physiol* 287 (4):L867–L878.

Helland, A., P. Wick, A. Koehler, K. Schmid, and C. Som. 2007. Reviewing the environmental and human health knowledge base of carbon nanotubes. *Environ Health Perspect* 115 (8): 1125–1131.

Hinrichsen, L., J. Harborth, L. Andrees, K. Weber, and E. J. Ungewickell. 2003. Effect of clathrin heavy chain- and alpha-adaptin-specific small inhibitory RNAs on endocytic accessory proteins and receptor trafficking in HeLa cells. *J Biol Chem* 278 (46):45160–45170.

Hoet, P. H., I. Bruske-Hohlfeld, and O. V. Salata. 2004. Nanoparticles—Known and unknown health risks. *J Nanobiotechnol* 2 (1):12.

Holt, K. B. 2007. Diamond at the nanoscale: Applications of diamond nanoparticles from cellular biomarkers to quantum computing. *Philos Trans A Math Phys Eng Sci* 365 (1861):2845–2861.

Hong, S., P. R. Leroueil, E. K. Janus, J. L. Peters, M. M. Kober, M. T. Islam, B. G. Orr, J. R. Baker Jr., and M. M. Banaszak Holl. 2006. Interaction of polycationic polymers with supported lipid bilayers and cells: Nanoscale hole formation and enhanced membrane permeability. *Bioconjug Chem* 17 (3):728–734.

Horie, M., H. Fukui, S. Endoh, J. Maru, A. Miyauchi, M. Shichiri, K. Fujita, E. Niki, Y. Hagihara, Y. Yoshida, Y. Morimoto, and H. Iwahashi. 2012. Comparison of acute oxidative stress on rat lung induced by nano and fine-scale, soluble and insoluble metal oxide particles: NiO and TiO(2). *Inhal Toxicol* 24 (7):391–400.

Hurt, R. H., M. Monthioux, and A. Kane. 2006. Toxicology of carbon nanomaterials: Status, trends, and perspectives on the special issue. *Carbon* 44:1028–1033.

Hussain, S. M., K. L. Hess, J. M. Gearhart, K. T. Geiss, and J. J. Schlager. 2005. *In vitro* toxicity of nanoparticles in BRL 3A rat liver cells. *Toxicol In Vitro* 19 (7):975–983.

Imrich, A., Y. Y. Ning, H. Koziel, B. Coull, and L. Kobzik. 1999. Lipopolysaccharide priming amplifies lung macrophage tumor necrosis factor production in response to air particles. *Toxicol Appl Pharmacol* 159 (2):117–124.

Imrich, A., Y. Ning, J. Lawrence, B. Coull, E. Gitin, M. Knutson, and L. Kobzik. 2007. Alveolar macrophage cytokine response to air pollution particles: Oxidant mechanisms. *Toxicol Appl Pharmacol* 218 (3):256–264.

Inoue, K., H. Takano, R. Yanagisawa, S. Hirano, T. Kobayashi, Y. Fujitani, A. Shimada, and T. Yoshikawa. 2007. Effects of inhaled nanoparticles on acute lung injury induced by lipopolysaccharide in mice. *Toxicology* 238 (2–3):99–110.

Inoue, K., H. Takano, R. Yanagisawa, S. Hirano, M. Sakurai, A. Shimada, and T. Yoshikawa. 2006a. Effects of airway exposure to nanoparticles on lung inflammation induced by bacterial endotoxin in mice. *Environ Health Perspect* 114 (9):1325–1330.

Inoue, K., H. Takano, R. Yanagisawa, T. Ichinose, M. Sakurai, and T. Yoshikawa. 2006b. Effects of nanoparticles on cytokine expression in murine lung in the absence or presence of allergen. *Arch Toxicol* 80 (9):614–619.

Isakovic, A., Z. Markovic, B. Todorovic-Markovic, N. Nikolic, S. Vranjes-Djuric, M. Mirkovic, M. Dramicanin, L. Harhaji, N. Raicevic, Z. Nikolic, and V. Trajkovic. 2006. Distinct cytotoxic mechanisms of pristine versus hydroxylated fullerene. *Toxicol Sci* 91 (1):173–183.

Ivanov, A. I. 2008. Pharmacological inhibition of endocytic pathways: Is it specific enough to be useful? *Methods Mol Biol* 440:15–33.

Jacobsen, N. R., P. Moller, K. A. Jensen, U. Vogel, O. Ladefoged, S. Loft, and H. Wallin. 2009. Lung inflammation and genotoxicity following pulmonary exposure to nanoparticles in $ApoE^{-/-}$ mice. *Part Fibre Toxicol* 6:2.

Jia, G., H. Wang, L. Yan, X. Wang, R. Pei, T. Yan, Y. Zhao, and X. Guo. 2005. Cytotoxicity of carbon nanomaterials: Single-wall nanotube, multi-wall nanotube, and fullerene. *Environ Sci Technol* 39 (5):1378–1383.

Johnston, H. J., G. R. Hutchison, F. M. Christensen, S. Peters, S. Hankin, K. Aschberger, and V. Stone. 2010. A critical review of the biological mechanisms underlying the *in vivo* and *in vitro* toxicity of carbon nanotubes: The contribution of physico–chemical characteristics. *Nanotoxicology* 4 (2):207–246.

Karlsson, H. L., P. Cronholm, J. Gustafsson, and L. Moller. 2008. Copper oxide nanoparticles are highly toxic: A comparison between metal oxide nanoparticles and carbon nanotubes. *Chem Res Toxicol* 21:1726–1732.

Kendall, M., and S. Holgate. 2012. Health impact and toxicological effects of nanomaterials in the lung. *Respirology* 17 (5):743–758.

Kikkawa, Y., K. Yoneda, F. Smith, B. Packard, and K. Suzuki. 1975. The type II epithelial cells of the lung. II. Chemical composition and phospholipid synthesis. *Lab Invest* 32 (3):295–302.

Kim, K. J., and E. D. Crandall. 1983. Heteropore populations of bullfrog alveolar epithelium. *J Appl Physiol* 54 (1):140–146.

Kim, K. J., and E. D. Crandall. 1996. Models for investigation of peptide and protein transport across cultured mammalian respiratory epithelial barriers. *Pharm Biotechnol* 8:325–346.

Kim, K. J., T. E. Fandy, V. H. Lee, D. K. Ann, Z. Borok, and E. D. Crandall. 2004. Net absorption of IgG via FcRn-mediated transcytosis across rat alveolar epithelial cell monolayers. *Am J Physiol Lung Cell Mol Physiol* 287 (3):L616–L622.

Kim, K. J., and A. B. Malik. 2003. Protein transport across the lung epithelial barriers. *Am J Physiol Lung Cell Mol Physiol* 284 (2):L247–L259.

Kim, K. J., Y. Matsukawa, H. Yamahara, V. K. Kalra, V. H. Lee, and E. D. Crandall. 2003. Absorption of intact albumin across rat alveolar epithelial cell monolayers. *Am J Physiol Lung Cell Mol Physiol* 284 (3):L458–L465.

Kim, Y. H., F. Fazlollahi, I. M. Kennedy, N. R. Yacobi, S. F. Hamm-Alvarez, Z. Borok, K. J. Kim, and E. D. Crandall. 2010. Alveolar epithelial cell injury due to zinc oxide nanoparticle exposure. *Am J Respir Crit Care Med* 182 (11):1398–1409.

Kobayashi, N., M. Naya, K. Mizuno, K. Yamamoto, M. Ema, and J. Nakanishi. 2011. Pulmonary and systemic responses of highly pure and well-dispersed single-wall carbon nanotubes after intratracheal instillation in rats. *Inhal Toxicol* 23 (13):814–828.

Kreyling, W. G., M. Semmler, F. Erbe, P. Mayer, S. Takenaka, H. Schulz, G. Oberdorster, and A. Ziesenis. 2002. Translocation of ultrafine insoluble iridium particles from lung epithelium to extrapulmonary organs is size dependent but very low. *J Toxicol Environ Health A* 65 (20):1513–1530.

Kuila, T., S. Bose, P. Khanra, A. K. Mishra, N. H. Kim, and J. H. Lee. 2011. Recent advances in graphene-based biosensors. *Biosens Bioelectron* 26 (12):4637–4648.

Lam, C. W., J. T. James, R. McCluskey, and R. L. Hunter. 2004. Pulmonary toxicity of single-wall carbon nanotubes in mice 7 and 90 days after intratracheal instillation. *Toxicol Sci* 77 (1):126–134.

Lam, H. F., L. C. Chen, D. Ainsworth, S. Peoples, and M. O. Amdur. 1988. Pulmonary function of guinea pigs exposed to freshly generated ultrafine zinc oxide with and without spike concentrations. *Am Ind Hyg Assoc J* 49 (7):333–341.

Lam, H. F., M. W. Conner, A. E. Rogers, S. Fitzgerald, and M. O. Amdur. 1985. Functional and morphologic changes in the lungs of guinea pigs exposed to freshly generated ultrafine zinc oxide. *Toxicol Appl Pharmacol* 78 (1):29–38.

Lam, R., and D. Ho. 2009. Nanodiamonds as vehicles for systemic and localized drug delivery. *Expert Opin Drug Deliv* 6 (9):883–895.

Law, W. C., K. T. Yong, I. Roy, H. Ding, R. Hu, W. Zhao, and P. N. Prasad. 2009. Aqueous-phase synthesis of highly luminescent CdTe/ZnTe core/shell quantum dots optimized for targeted bioimaging. *Small* 5 (11):1302–1310.

Leroueil, P. R., S. A. Berry, K. Duthie, G. Han, V. M. Rotello, D. Q. McNerny, J. R. Baker Jr., B. G. Orr, and M. M. Holl. 2008. Wide varieties of cationic nanoparticles induce defects in supported lipid bilayers. *Nano Lett* 8 (2):420–424.

Leroueil, P. R., S. Hong, A. Mecke, J. R. Baker Jr., B. G. Orr, and M. M. Banaszak Holl. 2007. Nanoparticle interaction with biological membranes: Does nanotechnology present a Janus face? *Acc Chem Res* 40 (5):335–342.

Lewinski, N., V. Colvin, and R. Drezek. 2008. Cytotoxicity of nanoparticles. *Small* 4 (1):26–49.

Li, J. J., S. Muralikrishnan, C. T. Ng, L. Y. Yung, and B. H. Bay. 2010. Nanoparticle-induced pulmonary toxicity. *Exp Biol Med (Maywood)* 235 (9):1025–1033.

Li, K. G., J. T. Chen, S. S. Bai, X. Wen, S. Y. Song, Q. Yu, J. Li, and Y. Q. Wang. 2009. Intracellular oxidative stress and cadmium ions release induce cytotoxicity of unmodified cadmium sulfide quantum dots. *Toxicol In Vitro* 23 (6):1007–1013.

Lin, P., J. W. Chen, L. W. Chang, J. P. Wu, L. Redding, H. Chang, T. K. Yeh, C. S. Yang, M. H. Tsai, H. J. Wang, Y. C. Kuo, and R. S. Yang. 2008. Computational and ultrastructural toxicology of a nanoparticle, Quantum Dot 705, in mice. *Environ Sci Technol* 42 (16):6264–6270.

Liu, Y., X. Dong, and P. Chen. 2012. Biological and chemical sensors based on graphene materials. *Chem Soc Rev* 41 (6):2283–2307.

Lok, C. N., C. M. Ho, R. Chen, Q. Y. He, W. Y. Yu, H. Sun, P. K. Tam, J. F. Chiu, and C. M. Che. 2007. Silver nanoparticles: Partial oxidation and antibacterial activities. *J Biol Inorg Chem* 12 (4): 527–534.

Lovric, J., H. S. Bazzi, Y. Cuie, G. R. Fortin, F. M. Winnik, and D. Maysinger. 2005a. Differences in subcellular distribution and toxicity of green and red emitting CdTe quantum dots. *J Mol Med* 83 (5):377–385.

Lovric, J., S. J. Cho, F. M. Winnik, and D. Maysinger. 2005b. Unmodified cadmium telluride quantum dots induce reactive oxygen species formation leading to multiple organelle damage and cell death. *Chem Biol* 12 (11):1227–1234.

Male, K. B., B. Lachance, S. Hrapovic, G. Sunahara, and J. H. Luong. 2008. Assessment of cytotoxicity of quantum dots and gold nanoparticles using cell-based impedance spectroscopy. *Anal Chem* 80 (14):5487–5493.

Matsukawa, Y., V. H. Lee, E. D. Crandall, and K. J. Kim. 1997. Size-dependent dextran transport across rat alveolar epithelial cell monolayers. *J Pharm Sci* 86 (3):305–309.

McIntosh, C. M., E. A. Esposito III, A. K. Boal, J. M. Simard, C. T. Martin, and V. M. Rotello. 2001. Inhibition of DNA transcription using cationic mixed monolayer protected gold clusters. *J Am Chem Soc* 123 (31):7626–7629.

McMahon, H. T., and E. Boucrot. 2011. Molecular mechanism and physiological functions of clathrin-mediated endocytosis. *Nat Rev Mol Cell Biol* 12 (8):517–533.

Mecke, A., I. J. Majoros, A. K. Patri, J. R. Baker Jr., M. M. Holl, and B. G. Orr. 2005. Lipid bilayer disruption by polycationic polymers: The roles of size and chemical functional group. *Langmuir* 21 (23):10348–10354.

Meier, O., K. Boucke, S. V. Hammer, S. Keller, R. P. Stidwill, S. Hemmi, and U. F. Greber. 2002. Adenovirus triggers macropinocytosis and endosomal leakage together with its clathrin-mediated uptake. *J Cell Biol* 158 (6):1119–1131.

Meier, O., and U. F. Greber. 2004. Adenovirus endocytosis. *J Gene Med* 6 Suppl 1:S152–S163.

Mitchell, L. A., J. Gao, R. V. Wal, A. Gigliotti, S. W. Burchiel, and J. D. McDonald. 2007. Pulmonary and systemic immune response to inhaled multiwalled carbon nanotubes. *Toxicol Sci* 100 (1): 203–214.

Mochalin, V. N., O. Shenderova, D. Ho, and Y. Gogotsi. 2012. The properties and applications of nanodiamonds. *Nat Nanotechnol* 7 (1):11–23.

Mohs, A. M., H. Duan, B. A. Kairdolf, A. M. Smith, and S. Nie. 2009. Proton-resistant quantum dots: Stability in gastrointestinal fluids and implications for oral delivery of nanoparticle agents. *Nano Res* 2 (6):500–508.

Morimoto, Y., M. Horie, N. Kobayashi, N. Shinohara, and M. Shimada. 2012. Inhalation toxicity assessment of carbon-based nanoparticles. *Acc Chem Res* 46 (3):770–781.

Morosini, V., T. Bastogne, C. Frochot, R. Schneider, A. Francois, F. Guillemin, and M. Barberi-Heyob. 2011. Quantum dot-folic acid conjugates as potential photosensitizers in photodynamic therapy of cancer. *Photochem Photobiol Sci* 10 (5):842–851.

Mossman, B. T., P. J. Borm, V. Castranova, D. L. Costa, K. Donaldson, and S. R. Kleeberger. 2007. Mechanisms of action of inhaled fibers, particles and nanoparticles in lung and cardiovascular diseases. *Part Fibre Toxicol* 4:4.

Muller, J., I. Decordier, P. H. Hoet, N. Lombaert, L. Thomassen, F. Huaux, D. Lison, and M. Kirsch-Volders. 2008a. Clastogenic and aneugenic effects of multi-wall carbon nanotubes in epithelial cells. *Carcinogenesis* 29 (2):427–433.

Muller, J., M. Delos, N. Panin, V. Rabolli, F. Huaux, and D. Lison. 2009. Absence of carcinogenic response to multiwall carbon nanotubes in a 2-year bioassay in the peritoneal cavity of the rat. *Toxicol Sci* 110 (2):442–448.

Muller, J., F. Huaux, A. Fonseca, J. B. Nagy, N. Moreau, M. Delos, E. Raymundo-Pinero, F. Beguin, M. Kirsch-Volders, I. Fenoglio, B. Fubini, and D. Lison. 2008b. Structural defects play a major role in the acute lung toxicity of multiwall carbon nanotubes: Toxicological aspects. *Chem Res Toxicol* 21 (9):1698–1705.

Mutlu, G. M., G. R. Budinger, A. A. Green, D. Urich, S. Soberanes, S. E. Chiarella, G. F. Alheid, D. R. McCrimmon, I. Szleifer, and M. C. Hersam. 2010. Biocompatible nanoscale dispersion of single-walled carbon nanotubes minimizes *in vivo* pulmonary toxicity. *Nano Lett* 10 (5): 1664–1670.

Naya, M., N. Kobayashi, M. Ema, S. Kasamoto, M. Fukumuro, S. Takami, M. Nakajima, M. Hayashi, and J. Nakanishi. 2012. *In vivo* genotoxicity study of titanium dioxide nanoparticles using comet assay following intratracheal instillation in rats. *Regul Toxicol Pharmacol* 62 (1):1–6.

Nel, A., T. Xia, L. Madler, and N. Li. 2006. Toxic potential of materials at the nanolevel. *Science* 311 (5761):622–627.

Nemmar, A., S. Al-Maskari, B. H. Ali, and I. S. Al-Amri. 2007. Cardiovascular and lung inflammatory effects induced by systemically administered diesel exhaust particles in rats. *Am J Physiol Lung Cell Mol Physiol* 292 (3):L664–L670.

Nemmar, A., J. Hamoir, B. Nemery, and P. Gustin. 2005. Evaluation of particle translocation across the alveolo-capillary barrier in isolated perfused rabbit lung model. *Toxicology* 208 (1):105–113.

Nemmar, A., P. H. Hoet, B. Vanquickenborne, D. Dinsdale, M. Thomeer, M. F. Hoylaerts, H. Vanbilloen, L. Mortelmans, and B. Nemery. 2002a. Passage of inhaled particles into the blood circulation in humans. *Circulation* 105 (4):411–414.

Nemmar, A., M. F. Hoylaerts, P. H. Hoet, D. Dinsdale, T. Smith, H. Xu, J. Vermylen, and B. Nemery. 2002b. Ultrafine particles affect experimental thrombosis in an *in vivo* hamster model. *Am J Respir Crit Care Med* 166 (7):998–1004.

Nemmar, A., M. F. Hoylaerts, P. H. Hoet, and B. Nemery. 2004. Possible mechanisms of the cardiovascular effects of inhaled particles: Systemic translocation and prothrombotic effects. *Toxicol Lett* 149 (1–3):243–253.

Nemmar, A., M. F. Hoylaerts, and B. Nemery. 2006. Effects of particulate air pollution on hemostasis. *Clin Occup Environ Med* 5 (4):865–881.

Nemmar, A., and I. M. Inuwa. 2008. Diesel exhaust particles in blood trigger systemic and pulmonary morphological alterations. *Toxicol Lett* 176 (1):20–30.

Nemmar, A., B. Nemery, P. H. Hoet, J. Vermylen, and M. F. Hoylaerts. 2003. Pulmonary inflammation and thrombogenicity caused by diesel particles in hamsters: Role of histamine. *Am J Respir Crit Care Med* 168 (11):1366–1372.

Nemmar, A., H. Vanbilloen, M. F. Hoylaerts, P. H. Hoet, A. Verbruggen, and B. Nemery. 2001. Passage of intratracheally instilled ultrafine particles from the lung into the systemic circulation in hamster. *Am J Respir Crit Care Med* 164 (9):1665–1668.

Neufeld, E. B., A. M. Cooney, J. Pitha, E. A. Dawidowicz, N. K. Dwyer, P. G. Pentchev, and E. J. Blanchette-Mackie. 1996. Intracellular trafficking of cholesterol monitored with a cyclodextrin. *J Biol Chem* 271 (35):21604–21613.

Nichols, B. J., and J. Lippincott-Schwartz. 2001. Endocytosis without clathrin coats. *Trends Cell Biol* 11 (10):406–412.

Niemeyer, C. M., and C. A. Mirkin. 2004. *Nanobiotechnology*, 1st ed. Weinheim: Wiley-VCH.

Oberdorster, G. 2001. Pulmonary effects of inhaled ultrafine particles. *Int Arch Occup Environ Health* 74 (1):1–8.

Oberdorster, G., J. Ferin, and B. E. Lehnert. 1994. Correlation between particle size, *in vivo* particle persistence, and lung injury. *Environ Health Perspect* 102 Suppl 5:173–179.

Oberdorster, G., A. Maynard, K. Donaldson, V. Castranova, J. Fitzpatrick, K. Ausman, J. Carter, B. Karn, W. Kreyling, D. Lai, S. Olin, N. Monteiro-Riviere, D. Warheit, and H. Yang. 2005a. Principles for characterizing the potential human health effects from exposure to nanomaterials: Elements of a screening strategy. *Part Fibre Toxicol* 2:8.

Oberdorster, G., E. Oberdorster, and J. Oberdorster. 2005b. Nanotoxicology: An emerging discipline evolving from studies of ultrafine particles. *Environ Health Perspect* 113 (7):823–839.

Oberdorster, G., Z. Sharp, V. Atudorei, A. Elder, R. Gelein, A. Lunts, W. Kreyling, and C. Cox. 2002. Extrapulmonary translocation of ultrafine carbon particles following whole-body inhalation exposure of rats. *J Toxicol Environ Health A* 65 (20):1531–1543.

Orlandi, P. A., and P. H. Fishman. 1998. Filipin-dependent inhibition of cholera toxin: Evidence for toxin internalization and activation through caveolae-like domains. *J Cell Biol* 141 (4):905–915.

Partha, R., M. Lackey, A. Hirsch, S. W. Casscells, and J. L. Conyers. 2007. Self assembly of amphiphilic C_{60} fullerene derivatives into nanoscale supramolecular structures. *J Nanobiotechnology* 5:6.

Parton, R. G., and A. A. Richards. 2003. Lipid rafts and caveolae as portals for endocytosis: New insights and common mechanisms. *Traffic* 4 (11):724–738.

Patton, J. S., and P. R. Byron. 2007. Inhaling medicines: Delivering drugs to the body through the lungs. *Nat Rev Drug Discov* 6 (1):67–74.

Patton, J. S., C. S. Fishburn, and J. G. Weers. 2004. The lungs as a portal of entry for systemic drug delivery. *Proc Am Thorac Soc* 1 (4):338–344.

Pelkmans, L. 2005. Secrets of caveolae- and lipid raft-mediated endocytosis revealed by mammalian viruses. *Biochim Biophys Acta* 1746 (3):295–304.

Pelkmans, L., and A. Helenius. 2002. Endocytosis via caveolae. *Traffic* 3 (5):311–320.

Pope, C. A. III, and D. W. Dockery. 2006. Health effects of fine particulate air pollution: Lines that connect. *J Air Waste Manag Assoc* 56:709–742.

Porter, A. E., M. Gass, J. S. Bendall, K. Muller, A. Goode, J. N. Skepper, P. A. Midgley, and M. Welland. 2009. Uptake of noncytotoxic acid-treated single-walled carbon nanotubes into the cytoplasm of human macrophage cells. *ACS Nano* 3 (6):1485–1492.

Porter, A. E., K. Muller, J. Skepper, P. Midgley, and M. Welland. 2006. Uptake of C_{60} by human monocyte macrophages, its localization and implications for toxicity: Studied by high resolution electron microscopy and electron tomography. *Acta Biomater* 2 (4):409–419.

Porter, D. W., A. F. Hubbs, R. R. Mercer, N. Wu, M. G. Wolfarth, K. Sriram, S. Leonard, L. Battelli, D. Schwegler-Berry, S. Friend, M. Andrew, B. T. Chen, S. Tsuruoka, M. Endo, and V. Castranova. 2010. Mouse pulmonary dose- and time course-responses induced by exposure to multi-walled carbon nanotubes. *Toxicology* 269 (2–3):136–147.

Pumera, M. 2012. Graphene, carbon nanotubes and nanoparticles in cell metabolism. *Curr Drug Metab* 13 (3):251–256.

Qian, J., K. T. Yong, I. Roy, T. Y. Ohulchanskyy, E. J. Bergey, H. H. Lee, K. M. Tramposch, S. He, A. Maitra, and P. N. Prasad. 2007. Imaging pancreatic cancer using surface-functionalized quantum dots. *J Phys Chem B* 111 (25):6969–6972.

Rejman, J., V. Oberle, I. S. Zuhorn, and D. Hoekstra. 2004. Size-dependent internalization of particles via the pathways of clathrin- and caveolae-mediated endocytosis. *Biochem J* 377 (Pt 1):159–169.

Rojas, S., J. D. Gispert, R. Martin, S. Abad, C. Menchon, D. Pareto, V. M. Victor, M. Alvaro, H. Garcia, and J. R. Herance. 2011. Biodistribution of amino-functionalized diamond nanoparticles. *In vivo* studies based on 18F radionuclide emission. *ACS Nano* 5 (7):5552–5559.

Rothberg, K. G., J. E. Heuser, W. C. Donzell, Y. S. Ying, J. R. Glenney, and R. G. Anderson. 1992. Caveolin, a protein component of caveolae membrane coats. *Cell* 68 (4):673–682.

Roy, S., R. Luetterforst, A. Harding, A. Apolloni, M. Etheridge, E. Stang, B. Rolls, J. F. Hancock, and R. G. Parton. 1999. Dominant-negative caveolin inhibits H-Ras function by disrupting cholesterol-rich plasma membrane domains. *Nat Cell Biol* 1 (2):98–105.

Ryman-Rasmussen, J. P., M. F. Cesta, A. R. Brody, J. K. Shipley-Phillips, J. I. Everitt, E. W. Tewksbury, O. R. Moss, B. A. Wong, D. E. Dodd, M. E. Andersen, and J. C. Bonner. 2009a. Inhaled carbon nanotubes reach the subpleural tissue in mice. *Nat Nanotechnol* 4 (11):747–751.

Ryman-Rasmussen, J. P., J. E. Riviere, and N. A. Monteiro-Riviere. 2007. Surface coatings determine cytotoxicity and irritation potential of quantum dot nanoparticles in epidermal keratinocytes. *J Invest Dermatol* 127 (1):143–153.

Ryman-Rasmussen, J. P., E. W. Tewksbury, O. R. Moss, M. F. Cesta, B. A. Wong, and J. C. Bonner. 2009b. Inhaled multiwalled carbon nanotubes potentiate airway fibrosis in murine allergic asthma. *Am J Respir Cell Mol Biol* 40 (3):349–358.

Sadauskas, E., N. R. Jacobsen, G. Danscher, M. Stoltenberg, U. Vogel, A. Larsen, W. Kreyling, and H. Wallin. 2009. Biodistribution of gold nanoparticles in mouse lung following intratracheal instillation. *Chem Cent J* 3:16.

Sanchez, V. C., A. Jachak, R. H. Hurt, and A. B. Kane. 2012. Biological interactions of graphene-family nanomaterials: An interdisciplinary review. *Chem Res Toxicol* 25 (1):15–34.

Sayes, C. M., J. D. Fortner, W. Guo, D. Lyon, A. M. Boyd, K. D. Ausman, Y. J. Tao, B. Sitharaman, L. J. Wilson, J. B. Hughes, J. L. West, and V. L. Colvin. 2004. The differential cytotoxicity of water-soluble fullerenes. *Nano Lett* 4:1881–1887.

Sayes, C. M., A. M. Gobin, K. D. Ausman, J. Mendez, J. L. West, and V. L. Colvin. 2005. Nano-C_{60} cytotoxicity is due to lipid peroxidation. *Biomaterials* 26 (36):7587–7595.

Sayes, C. M., A. A. Marchione, K. L. Reed, and D. B. Warheit. 2007a. Comparative pulmonary toxicity assessments of C_{60} water suspensions in rats: Few differences in fullerene toxicity *in vivo* in contrast to *in vitro* profiles. *Nano Lett* 7 (8):2399–2406.

Sayes, C. M., K. L. Reed, and D. B. Warheit. 2007b. Assessing toxicity of fine and nanoparticles: Comparing *in vitro* measurements to *in vivo* pulmonary toxicity profiles. *Toxicol Sci* 97 (1): 163–180.

Schinwald, A., F. A. Murphy, A. Jones, W. MacNee, and K. Donaldson. 2012. Graphene-based nanoplatelets: A new risk to the respiratory system as a consequence of their unusual aerodynamic properties. *ACS Nano* 6 (1):736–746.

Schipper, M. L., Z. Cheng, S. W. Lee, L. A. Bentolila, G. Iyer, J. Rao, X. Chen, A. M. Wu, S. Weiss, and S. S. Gambhir. 2007. microPET-based biodistribution of quantum dots in living mice. *J Nucl Med* 48 (9):1511–1518.

Scrivens, W., and J. M. Tour. 1994. Synthesis of ^{14}C-labeled C_{60}, its suspension in water, and its uptake by human keratinocytes. *J Am Chem Soc* 116:4517–4518.

Semmler-Behnke, M., W. G. Kreyling, J. Lipka, S. Fertsch, A. Wenk, S. Takenaka, G. Schmid, and W. Brandau. 2008. Biodistribution of 1.4- and 18-nm gold particles in rats. *Small* 4 (12):2108–2111.

Sermet-Gaudelus, I., Y. Le Cocguic, A. Ferroni, M. Clairicia, J. Barthe, J. P. Delaunay, V. Brousse, and G. Lenoir. 2002. Nebulized antibiotics in cystic fibrosis. *Paediatr Drugs* 4 (7):455–467.

Shimada, A., N. Kawamura, M. Okajima, T. Kaewamatawong, H. Inoue, and T. Morita. 2006. Translocation pathway of the intratracheally instilled ultrafine particles from the lung into the blood circulation in the mouse. *Toxicol Pathol* 34:949–957.

Shvedova, A. A., J. P. Fabisiak, E. R. Kisin, A. R. Murray, J. R. Roberts, Y. Y. Tyurina, J. M. Antonini, W. H. Feng, C. Kommineni, J. Reynolds, A. Barchowsky, V. Castranova, and V. E. Kagan. 2008. Sequential exposure to carbon nanotubes and bacteria enhances pulmonary inflammation and infectivity. *Am J Respir Cell Mol Biol* 38 (5):579–590.

Shvedova, A. A., E. R. Kisin, R. Mercer, A. R. Murray, V. J. Johnson, A. I. Potapovich, Y. Y. Tyurina, O. Gorelik, S. Arepalli, D. Schwegler-Berry, A. F. Hubbs, J. Antonini, D. E. Evans, B. K. Ku, D. Ramsey, A. Maynard, V. E. Kagan, V. Castranova, and P. Baron. 2005. Unusual inflammatory and fibrogenic pulmonary responses to single-walled carbon nanotubes in mice. *Am J Physiol Lung Cell Mol Physiol* 289 (5):L698–L708.

Soto, K., K. M. Garza, and L. E. Murr. 2007. Cytotoxic effects of aggregated nanomaterials. *Acta Biomater* 3 (3):351–358.

Spohn, P., C. Hirsch, F. Hasler, A. Bruinink, H. F. Krug, and P. Wick. 2009. C_{60} fullerene: A powerful antioxidant or a damaging agent? The importance of an in-depth material characterization prior to toxicity assays. *Environ Pollut* 157 (4):1134–1139.

Stella, G. M. 2011. Carbon nanotubes and pleural damage: Perspectives of nanosafety in the light of asbestos experience. *Biointerphases* 6 (2):P1–P17.

Stern, S. T., B. S. Zolnik, C. B. McLeland, J. Clogston, J. Zheng, and S. E. McNeil. 2008. Induction of autophagy in porcine kidney cells by quantum dots: A common cellular response to nanomaterials? *Toxicol Sci* 106 (1):140–152.

Stoker, E., F. Purser, S. Kwon, Y. B. Park, and J. S. Lee. 2008. Alternative estimation of human exposure of single-walled carbon nanotubes using three-dimensional tissue-engineered human lung. *Int J Toxicol* 27 (6):441–448.

Stringer, B., and L. Kobzik. 1998. Environmental particulate-mediated cytokine production in lung epithelial cells (A549): Role of preexisting inflammation and oxidant stress. *J Toxicol Environ Health A* 55 (1):31–44.

Sung, J. C., B. L. Pulliam, and D. A. Edwards. 2007. Nanoparticles for drug delivery to the lungs. *Trends Biotechnol* 25 (12):563–570.

Takenaka, S., E. Karg, W. G. Kreyling, B. Lentner, W. Moller, M. Behnke-Semmler, L. Jennen, A. Walch, B. Michalke, P. Schramel, J. Heyder, and H. Schulz. 2006. Distribution pattern of inhaled ultrafine gold particles in the rat lung. *Inhal Toxicol* 18 (10):733–740.

Takenaka, S., E. Karg, C. Roth, H. Schulz, A. Ziesenis, U. Heinzmann, P. Schramel, and J. Heyder. 2001. Pulmonary and systemic distribution of inhaled ultrafine silver particles in rats. *Environ Health Perspect* 109 Suppl 4:547–551.

Tasciotti, E., B. Godin, J. O. Martinez, C. Chiappini, R. Bhavane, X. Liu, and M. Ferrari. 2011. Near-infrared imaging method for the *in vivo* assessment of the biodistribution of nanoporous silicon particles. *Mol Imaging* 10 (1):56–68.

Thomas, C. R., S. George, A. M. Horst, Z. Ji, R. J. Miller, J. R. Peralta-Videa, T. Xia, S. Pokhrel, L. Madler, J. L. Gardea-Torresdey, P. A. Holden, A. A. Keller, H. S. Lenihan, A. E. Nel, and J. I. Zink. 2011. Nanomaterials in the environment: From materials to high-throughput screening to organisms. *ACS Nano* 5 (1):13–20.

Tokuyama, H., S. Yamago, E. Nakamura, T. Shiraki, and Y. Sugiura. 1993. Photoinduced biochemical activity of fullerene carboxylic acid. *J Am Chem Soc* 115:7918–7919.

Usenko, C. Y., S. L. Harper, and R. L. Tanguay. 2008. Fullerene C_{60} exposure elicits an oxidative stress response in embryonic zebrafish. *Toxicol Appl Pharmacol* 229 (1):44–55.

Walther, C., K. Meyer, R. Rennert, and I. Neundorf. 2008. Quantum dot-carrier peptide conjugates suitable for imaging and delivery applications. *Bioconjug Chem* 19 (12):2346–2356.

Wang, L., W. Yang, P. Read, J. Larner, and K. Sheng. 2010. Tumor cell apoptosis induced by nanoparticle conjugate in combination with radiation therapy. *Nanotechnology* 21 (47):475103.

Wang, Y., P. H. Camargo, S. E. Skrabalak, H. Gu, and Y. Xia. 2008. A facile, water-based synthesis of highly branched nanostructures of silver. *Langmuir* 24 (20):12042–12046.

Warheit, D. B., B. R. Laurence, K. L. Reed, D. H. Roach, G. A. Reynolds, and T. R. Webb. 2004. Comparative pulmonary toxicity assessment of single-wall carbon nanotubes in rats. *Toxicol Sci* 77 (1):117–125.

Warheit, D. B., C. M. Sayes, K. L. Reed, and K. A. Swain. 2008. Health effects related to nanoparticle exposures: Environmental, health and safety considerations for assessing hazards and risks. *Pharmacol Ther* 120 (1):35–42.

Widera, A., K. Beloussow, K. J. Kim, E. D. Crandall, and W. C. Shen. 2003a. Phenotype-dependent synthesis of transferrin receptor in rat alveolar epithelial cell monolayers. *Cell Tissue Res* 312 (3):313–318.

Widera, A., K. J. Kim, E. D. Crandall, and W. C. Shen. 2003b. Transcytosis of GCSF-transferrin across rat alveolar epithelial cell monolayers. *Pharm Res* 20:1231–1238.

Wieffer, M., T. Maritzen, and V. Haucke. 2009. SnapShot: Endocytic trafficking. *Cell* 137 (2):382 e381–e383.

Wright, E. M., and R. J. Pietras. 1974. Routes of nonelectrolyte permeation across epithelial membranes. *J Membr Biol* 17 (3):293–312.

Xing, Y., and L. Dai. 2009. Nanodiamonds for nanomedicine. *Nanomedicine (Lond)* 4 (2):207–218.

Yacobi, N. R., L. Demaio, J. Xie, S. F. Hamm-Alvarez, Z. Borok, K. J. Kim, and E. D. Crandall. 2008. Polystyrene nanoparticle trafficking across alveolar epithelium. *Nanomedicine* 4 (2):139–145.

Yacobi, N. R., F. Fazlollahi, Y. H. Kim, A. Sipos, Z. Borok, K. J. Kim, and E. D. Crandall. 2011. Nanomaterial interactions with and trafficking across the lung alveolar epithelial barrier: Implications for health effects of air-pollution particles. *Air Qual Atmos Health* 4:65–78.

Yacobi, N. R., N. Malmstadt, F. Fazlollahi, L. DeMaio, R. Marchelletta, S. F. Hamm-Alvarez, Z. Borok, K. J. Kim, and E. D. Crandall. 2010. Mechanisms of alveolar epithelial translocation of a defined population of nanoparticles. *Am J Respir Cell Mol Biol* 42 (5):604–614.

Yacobi, N. R., H. C. Phuleria, L. Demaio, C. H. Liang, C. A. Peng, C. Sioutas, Z. Borok, K. J. Kim, and E. D. Crandall. 2007. Nanoparticle effects on rat alveolar epithelial cell monolayer barrier properties. *Toxicol In Vitro* 21 (8):1373–1381.

Yamago, S., H. Tokuyama, E. Nakamura, K. Kikuchi, S. Kananishi, K. Sueki, H. Nakahara, S. Enomoto, and F. Ambe. 1995. *In vivo* biological behavior of a water-miscible fullerene: ^{14}C labeling, absorption, distribution, excretion and acute toxicity. *Chem Biol* 2 (6):385–389.

Yamawaki, H., and N. Iwai. 2006. Cytotoxicity of water-soluble fullerene in vascular endothelial cells. *Am J Physiol Cell Physiol* 290 (6):C1495–C1502.

Yang, W., J. I. Peters, and R. O. Williams III. 2008. Inhaled nanoparticles—A current review. *Int J Pharm* 356 (1–2):239–247.

Zhang, X., J. Yin, C. Kang, J. Li, Y. Zhu, W. Li, Q. Huang, and Z. Zhu. 2010. Biodistribution and toxicity of nanodiamonds in mice after intratracheal instillation. *Toxicol Lett* 198 (2):237–243.

8

Mechanistic Aspects of Cellular Uptake

Lennart Treuel, Xiue Jiang, and Gerd Ulrich Nienhaus

CONTENTS

8.1 Introduction

Nanoparticles (NPs) can efficiently intrude living cells by utilizing the cellular endocytosis machinery. Whereas only specialized cells such as macrophages are capable of phagocytosis, a form of endocytosis of larger particles, almost all cells can incorporate NPs by pinocytosis. Four different basic pinocytic mechanisms are currently distinguished: macropinocytosis, clathrin-mediated endocytosis, caveolae-mediated endocytosis, and mechanisms independent of clathrin and caveolin (Conner and Schmid 2003; Jiang et al. 2011). The physicochemical properties of NPs, including size (Chithrani et al. 2006; Rejman et al. 2004), shape (Chithrani et al. 2006), surface charge (Arbab et al. 2003; Labhasetwar et al. 1998; Sun et al. 2005), and surface chemistry (Holzapfel et al. 2006; Labhasetwar et al. 1998; Nativo et al. 2008), have been identified as strongly modulating cellular uptake efficiency.

Upon incorporation by an organism, NPs typically interact with extracellular biomolecules dissolved in body fluids, including proteins, sugars, and lipids, before attaching to cell membranes. These interactions frequently result in a layer of proteins adsorbed onto the NP surfaces, the so-called protein corona (Cedervall et al. 2007a,b; de Paoli Lacerda et al. 2010; Gebauer et al. 2012; Lundqvist et al. 2008; Lynch 2007; Röcker et al. 2009; Shang et al. 2007, 2012; Treuel and Nienhaus 2012; Treuel et al. 2010). Consequently, the cell surface receptors activating the endocytosis machinery actually encounter NPs enshrouded with biomolecules rather than bare particles. This is an important issue for NP design and engineering efforts that aim to intentionally enhance or suppress cellular uptake.

A variety of studies have reported the effects of protein corona formation on cellular response to NP exposure. For example, uptake of carboxyl-functionalized NPs by HeLa cells was shown to be strongly suppressed in the presence of blood plasma proteins in comparison with bare NPs (Jiang et al. 2010b). Immunoglobulin binding caused NP opsonization, thereby promoting receptor-mediated phagocytosis (Owens and Peppas 2006). Suppressing protein absorption onto NPs by coating them with polyethylene glycol decreased NP uptake by macrophages (Kah et al. 2009) and resulted in longer circulation times and altered biodistribution upon injection into mice (Owens and Peppas 2006). It was also pointed out that adsorbed proteins may be transported across membranes alongside the NPs, bringing them into compartments that they would not normally reach (Klein 2007). Understanding the formation and persistence of the protein corona on NPs is therefore critically important for the elucidation, interpretation, and assessment of cellular NP uptake.

In addition to active cellular uptake, NPs may also enter cells via other mechanisms, for example, passive bilayer penetration (Rothen-Rutishauser et al. 2006; Wang et al. 2012). The ability of NPs to adhere to and penetrate cell membranes was shown to depend on their physical properties, including size, surface composition, and surface charge (Dif et al. 2008; Laurencin et al. 2010; Leroueil et al. 2007; Roiter et al. 2008; Verma et al. 2008; Zhang and Yang 2011). Small, positively charged NPs were observed to pass through cell membranes, leading to membrane rupture and noticeable cytotoxicity (Cho et al. 2009; Kostarelos et al. 2007; Leroueil et al. 2008; Yu et al. 2007). Even particles >500 nm in diameter were reported to penetrate cell membranes by inducing strong local membrane deformations (Zhao et al. 2011). Membrane disruption can be reduced or even entirely avoided by a suitable design of the surface structure and charge density (Lin et al. 2010; Verma et al. 2008). We note, however, that the surface properties of NPs in the organism may be significantly modified by prior biomolecule adsorption.

Besides the ability of organisms and cells to incorporate NPs, yet another issue crucial to the assessment of biological consequences of NP exposure is their degradation in the biological environment. This may involve removal of the protein corona and the original surface functionalization, which typically encloses the core particle. Depending on the nature of the core material, exposure of the NP core may subsequently cause its degradation and eventually lead to its complete dissolution, and molecules or metal ions thereby released into the biological environment may elicit toxic effects. The adverse effects of metal ions (e.g., heavy metal ions and silver ions) have been well documented (Ratte 1999; Silver 1996; Stohs and Bagchi 1995). Degradation of NPs will generally lead to a combination of ionic/molecular toxicity and toxicity aspects related to the particulate nature of the material. Recent studies of silver NPs showed that formation of a protein corona was strongly affected by the presence of a polymeric surface coating around these NPs (Treuel et al. 2010), which reduced serum albumin affinity to the NP surfaces by around one order of magnitude. It should be noted in this context that the severe cytotoxicity of silver NPs is largely caused by the release of silver ions rather than by particulate properties (Greulich et al. 2009; Kittler et al. 2010a,b).

In this chapter, we will discuss a number of experiments performed with the aim to better understand the mechanisms by which NPs interact with cellular membranes. A detailed understanding of these interactions at the molecular level will, on the one hand, greatly assist the engineering of NPs that do not penetrate cells so as to avoid adverse effects, which is highly relevant in the context of NP-based contrast agents that are widely used in medical diagnosis. On the other hand, this knowledge is also important for developing NPs designed for selective uptake by specific cells, which is of utmost relevance for targeted drug delivery.

8.2 Effects of Protein Corona on Uptake Efficiency

To elucidate how the presence of a protein adsorption layer around NPs affects their cellular uptake, Jiang and coworkers (2010b) used quantitative confocal microscopy to compare NP uptake by live HeLa cells in the presence or absence of human transferrin (TF) and human serum albumin (HSA). These studies were conducted using small (diameter ~10 nm) iron–platinum (FePt) NPs coated with a carboxyl-functionalized polymer and fluorescently labeled by incorporation of DY-636 dye molecules within the polymer shell. Complete adsorption layers of TF and HSA were found for protein concentrations above ~100 μM in phosphate-buffered saline (PBS), as determined by fluorescence correlation spectroscopy. HeLa cells were exposed to NPs suspended in PBS with or without 100 μM TF or HSA, and NP uptake was studied by spinning disk confocal fluorescence microscopy. NPs were observed to accumulate on the cell membranes within minutes after exposure, and uptake required ~1 h to saturate. In comparison with bare NPs, significantly reduced amounts of NPs were incorporated by the cells if the NPs were coated with TF or HSA. These experiments were complemented by controls to test the ability of HeLa cells to internalize fluorescently labeled TF and HSA without NPs. Strong uptake of TF, which is well known to be internalized by its cognate receptor, was observed, whereas HSA was barely endocytosed by HeLa cells under otherwise identical conditions.

The pronounced but similar uptake suppression of NPs coated with TF and HSA was in marked contrast to the behavior of the uptake of the individual proteins. The reduced uptake of TF-coated NPs may be explained by assuming that the cellular endocytosis machinery is occupied with internalization of the freely dissolved (100 μM) protein, which was present in 10^5-fold molar excess over the 1 nM NPs. Alternatively, TF may adsorb to the NPs in a specific manner, such that the receptor-binding site of the TF molecule is concealed, thereby blocking the specific receptor interaction that triggers uptake. For HSA, it appears that the protein corona may act as a protective layer, shielding the carboxyl-functionalized NP surface from direct interactions with receptors on the cell membrane.

This work presents a vivid example of how protein adsorption onto NPs markedly changes their uptake behavior, underscoring that it is crucially important to understand NP–protein interactions at the molecular level.

8.3 Analysis of Endocytosis Pathways Involved in NP Uptake

As discussed in Section 8.1, several different pinocytosis pathways exist by which NPs can actively be incorporated by the cell. To explore which pathways are involved, inhibitory drugs are available that can be used to specifically interfere with individual endocytic pathways. Using this strategy, Jiang and coworkers (2011) investigated the uptake of fluorescent polystyrene NPs by mesenchymal stem cells (MSCs). Uptake was observed and quantified by confocal fluorescence microscopy. Experiments were carried out in PBS to avoid complications arising from protein adsorption to the NPs. Two types of fluorescent anionic polystyrene (PS) NPs with essentially identical sizes (~100 nm) and ζ-potentials were compared: carboxyl-functionalized PS NPs (CPS) and plain PS NPs; both were coated in addition with anionic detergent for colloidal stabilization.

Unlike smaller NPs of 10 nm diameter, these larger particles were not observed to adsorb onto the cell membrane (Figure 8.1a), indicating that they were immediately internalized upon making contact with the membrane. CPS NPs were internalized more rapidly and accumulated within cells to a much greater extent than plain PS NPs. To investigate the cellular uptake mechanisms for anionic plain PS and carboxyl-functionalized CPS NPs, cells were incubated with NPs in the presence of inhibitory drugs.

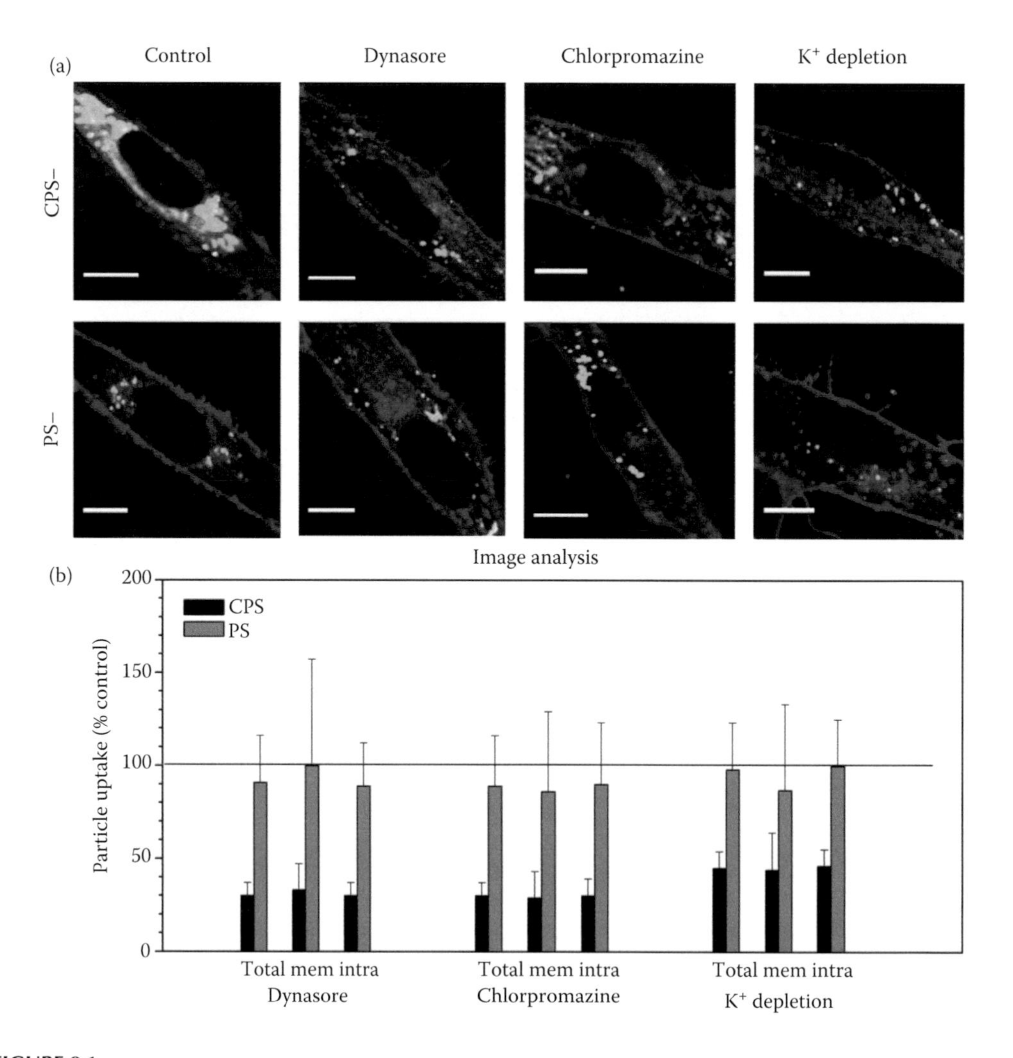

FIGURE 8.1
Inhibitor effects on the uptake of CPS and plain PS NPs by MSCs. (a) Confocal fluorescence images of MSCs (red membrane stain) after 1-h incubation with 75 mg mL^{-1} NPs (green), without inhibitor (control) and in the presence of dynasore (80 mM), chlorpromazine (50 mM), or potassium depletion, respectively. (b) Quantified NP uptake (total, membrane-associated, and intracellular fractions) in the presence of dynasore, chlorpromazine, and potassium depletion (normalized to untreated control), determined by quantitative image analysis. (From X. Jiang et al., *Nanoscale*, 3, 2028, 2011. Reproduced by permission of The Royal Society of Chemistry.)

Pretreatment of cells with the inhibitor dynasore, which suppresses endocytic processes involving dynamin, a large multidomain protein involved in clathrin- and caveolin-mediated endocytosis (Macia et al. 2006), resulted in a ~70% reduction in the uptake of CPS NPs relative to untreated controls, but had no effect on the uptake of PS NPs (Figure 8.1a). These observations were corroborated by quantitative image analysis (Figure 8.1b), which clearly demonstrated that accumulation of NPs both in the intracellular space and close to the plasma membrane were affected to a similar degree. These results suggest that the carboxylic acid groups on the CPS NPs give rise to a strong preference for a dynamin-dependent endocytosis pathway.

Effects similar to those observed with dynasore were also seen with chlorpromazine, which interferes with clathrin-mediated endocytosis by disrupting the assembly of the clathrin lattice forming the endocytic pit at the plasma membrane (Wang et al. 1993). Quantitative analysis (Figure 8.1) showed that the uptake of CPS NPs was likewise suppressed to about 30%, whereas, again, little effect was seen for PS NPs.

In addition, uptake experiments were carried out with potassium-depleted cells as an alternative strategy for inhibiting clathrin-mediated endocytosis (Larkin et al. 1983), and uptake of CPS NPs was found to be reduced to about 40%, whereas the uptake of PS NPs was again unaffected (Figure 8.1).

Taken together, these inhibition studies demonstrate that uptake of CPS NPs proceeds predominantly via dynamin- and clathrin-dependent pathways, whereas uptake of PS NPs apparently utilize dynamin- and clathrin-independent mechanisms. The observation that markedly different mechanisms are involved in the endocytosis of two types of NPs with identical properties except for the surface functionalities underscores the importance of this property in determining cellular uptake pathways. Active incorporation of NPs proceeds via specific interactions between NP surfaces and cell surface receptors, which subsequently may activate various signaling pathways. Moreover, incorporated NPs may further interact with intracellular epitopes. Therefore, surface functionality, by mediating or facilitating these interactions, may affect biological interactions more generally and at many levels.

8.4 Dose Dependence of NP Uptake

In our studies of NP uptake by cultured cells, we observed that NPs with diameters of <10 nm coated the cell membrane, whereas large NPs (100 nm) never accumulated there. A small NP may not be able to trigger endocytosis by itself because it interacts with an insufficient number of receptors. In this case, more than one NP would be necessary to activate pit formation, and a nonlinear dependence of the uptake yield on the overall NP concentration will result. This effect was investigated by monitoring the incorporation of 8-nm-diameter D-penicillamine-coated quantum dots (DPA-QDs) by live human cancer (HeLa) cells. NP uptake was assessed quantitatively using spinning disk and 4Pi confocal microscopies (Jiang et al. 2010a). Following exposure of HeLa cells with PBS solution containing 10 nM DPA-QDs, cells were imaged for typically 1 h. Within 1 min, QDs were observed at the cell membrane and subsequently accumulated there. Over time, more and more QDs were internalized by the cells, forming large clusters in the perinuclear region within 1 h. A significant fraction of endocytosed QDs was localized in lysosomes, and some of these QD clusters were actively transported to the cell periphery and exocytosed.

Confocal images taken after 1-h incubation at different QD concentrations (10, 3, and 1 nM) (Figure 8.2a through c) revealed the dependence of DPA-QD accumulation, both at the plasma membrane and within the cell, on NP concentration. At a concentration of 10 nM, roughly equal amounts of QDs were found associated with cell membranes and inside cells after 1 h. With decreasing QD concentration, the membrane-associated fraction decreased in a linear fashion, whereas the intracellular fraction decreased much more strongly (Figure 8.2d). Kinetic analysis of these processes (Figure 8.2e) supported these findings (Jiang et al. 2010a). At a 1 nM NP concentration, only very few bright spots were visible inside the cells. These findings support the notion that, for very small NPs, a critical threshold density of QDs on the cell membrane has to be exceeded to trigger the internalization process.

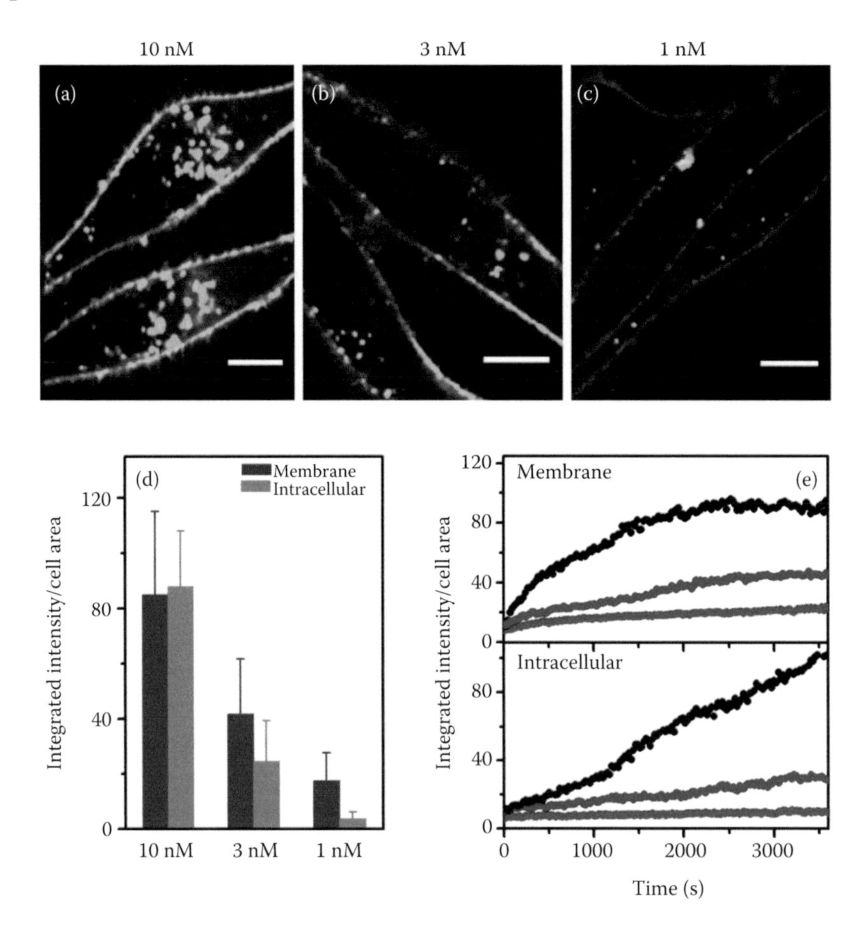

FIGURE 8.2
Dose-dependent uptake of DPA-QDs by HeLa cells. (a through c) Confocal fluorescence images of HeLa cells (scale bar, 10 μm) after 1-h exposure to PBS solutions containing DPA-QDs at concentrations of (a) 10 nM, (b) 3 nM, and (c) 1 nM. (d) Dose dependence of DPA-QD uptake by HeLa cells (after 1-h incubation), determined by quantitative analysis of membrane-associated (dark gray bar) and intracellular (light gray bar) fluorescence. (e) Kinetics of DPA-QD association with HeLa cells (upper panel: membrane-associated, lower panel: intracellular) at particle concentrations of 10 nM (black), 3 nM (blue), and 1 nM (red). (Reprinted with permission from Jiang et al. 2010a, 6787. Copyright 2010 American Chemical Society.)

8.5 Modeling Forces Acting on Membrane-Bound NPs

Carboxydextran-coated superparamagnetic iron oxide NPs of 60 nm (SPIO) and 20 nm (USPIO) diameters are widely used as contrast agents in magnetic resonance imaging. Lunov et al. (2011) analyzed the uptake of these NPs by human macrophages and found that endocytosis occurred via a clathrin-mediated, scavenger receptor A–dependent mechanism. They measured NP uptake as a function of time and presented a mathematical model that allows several mechanistic parameters to be estimated. This model describes internalization by initial binding of NPs to receptors and subsequent wrapping a membrane patch around them for internalization. The total uptake time was assumed to be dominated by the wrapping time, τ_w. Earlier models (Gao et al. 2005) had suggested that NP diffusion through the membrane controls the wrapping time.

By using quantitative confocal fluorescence microscopy, the time dependence of NP uptake, $N(t)$, was observed to be exponential

$$N(t) = N_S(1 - \exp[-t/T]), \tag{8.1}$$

where N_S is the number of NPs at saturation (~10^7), and T is the characteristic time (~1 h). At short times after exposing the cells to the NPs, the uptake rates, $dN(0)/dt$, for SPIO and USPIO were ~25,000 and ~2500 s^{-1}, respectively. The overall uptake rate per cell, dN/dt, can be recast as the rate per individual clathrin-coated pit-forming event, dn/dt

$$\frac{dn}{dt} = \frac{a^2}{L^2}\frac{dN}{dt}, \tag{8.2}$$

by introducing two parameters: the lateral dimension of the macrophage, L (~20 μm), and the characteristic footprint of an individual endocytic pit, a. However, neither dn/dt nor a are directly accessible from kinetic experiments on entire cells. The rate dn/dt, however, is the inverse of the uptake time, which may be approximated by the wrapping time, τ_w. A simple force model was introduced to calculate the wrapping time (Figure 8.3a)

$$F(x) = F_{el}(x) + 6\pi\eta R\frac{dx}{dt}. \tag{8.3}$$

Here, the total force that the cytoskeletal actin structure has to exert along the direction perpendicular to the membrane, $F(x)$, consists of the elastic membrane deformation force, $F_{el}(x)$, and the viscous drag force, which, in turn, depends on the cytoplasmic viscosity, η, the particle radius, R, and the velocity, dx/dt. The attachment of the NP during incorporation must be ensured by receptor interactions (f_{rec} in Figure 8.3a). From these calculations, τ_w was obtained as a function of $F(x)$, as shown in Figure 8.3b. For the same actin force, the model reveals that SPIO and USPIO wrapping times differ by a factor of ~10. Thus, the rate of individual uptake events, dn/dt, can be estimated as 10–100 s^{-1}, yielding reasonable values of 0.3–4 μm for the characteristic length scale a associated with clathrin pits. Given that the overall number of receptors per cell is 2×10^4–4×10^4 (Pitas et al. 1992), one can estimate that ~2–20 receptors are involved in NP binding during an individual endocytosis event.

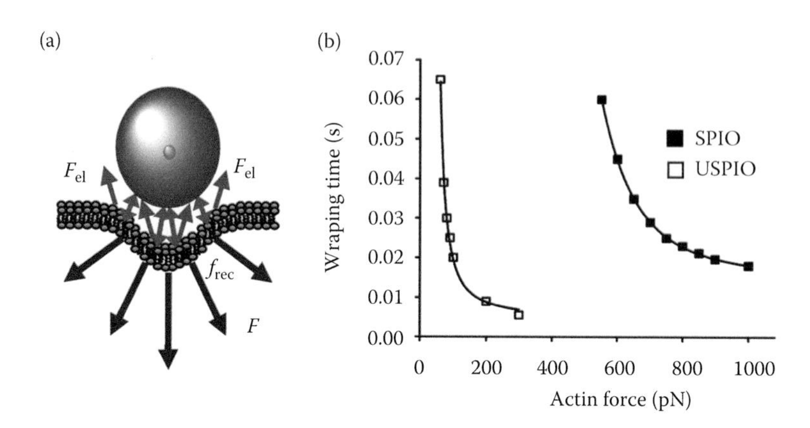

FIGURE 8.3
Forces acting on receptor-associated NPs. (a) *F*: total force from the cytoskeleton, F_{el}: elastic forces of the deformed membrane, f_{rec}: receptor interactions ensuring NP attachment during incorporation. (b) Calculated wrapping time of NPs as a function of actin force. (Reprinted from *Biomaterials*, 32, O. Lunov, V. Zablotskii, T. Syrovets, C. Röcker, K. Tron, G. U. Nienhaus, and T. Simmet, Modeling receptor-mediated endocytosis of polymer-functionalized iron oxide NPs by human macrophages, 547–555, Copyright 2011, with permission from Elsevier.)

Overall, the analysis of the experimental data obtained by confocal microscopy using the model presented by Lunov et al. (2011) produced reasonable parameters and emphasized the value of even quite simplistic physical models for furthering our understanding of complex biological processes.

8.6 Passive Uptake of NPs

In addition to active cellular uptake, NPs have also been reported to translocate passively through cellular membranes. Because red blood cells (RBCs) are highly specialized and lack a nucleus, most organelles, and endocytic machinery (Underhill and Ozinsky 2002), they have frequently been used as model systems to study the passive transport of NPs across cellular membranes (Geiser et al. 2005; Mühlfeld et al. 2008; Rothen-Rutishauser et al. 2006; Wang et al. 2012).

Recently, Wang et al. (2012) studied the interactions between DPA-coated QDs and RBCs. RBCs were incubated with 10 nM DPA-QDs in PBS for different periods, pelleted by centrifugation, and imaged by confocal fluorescence microscopy. Merged bright-field and confocal fluorescence images of cells exposed to DPA-QDs for 1, 4, and 8 h, and of control (unexposed) cells are shown in Figure 8.4a. These images clearly reveal that DPA-QDs adhere to RBC membranes, and that the number of fluorescence spots, either close to the cell membranes or within cells, increases with exposure time.

The internalization kinetics was calculated from the integrated fluorescence intensity of the internal spots (red circles), normalized to the cellular area in the observation plane (Figure 8.4b). A half-life of 1.7 h was determined for the internalized fraction. The images showed that the adsorbed DPA-QDs did not induce a strong local membrane deformation and that penetration of DPA-QDs into RBCs apparently did not disturb the integrity of the

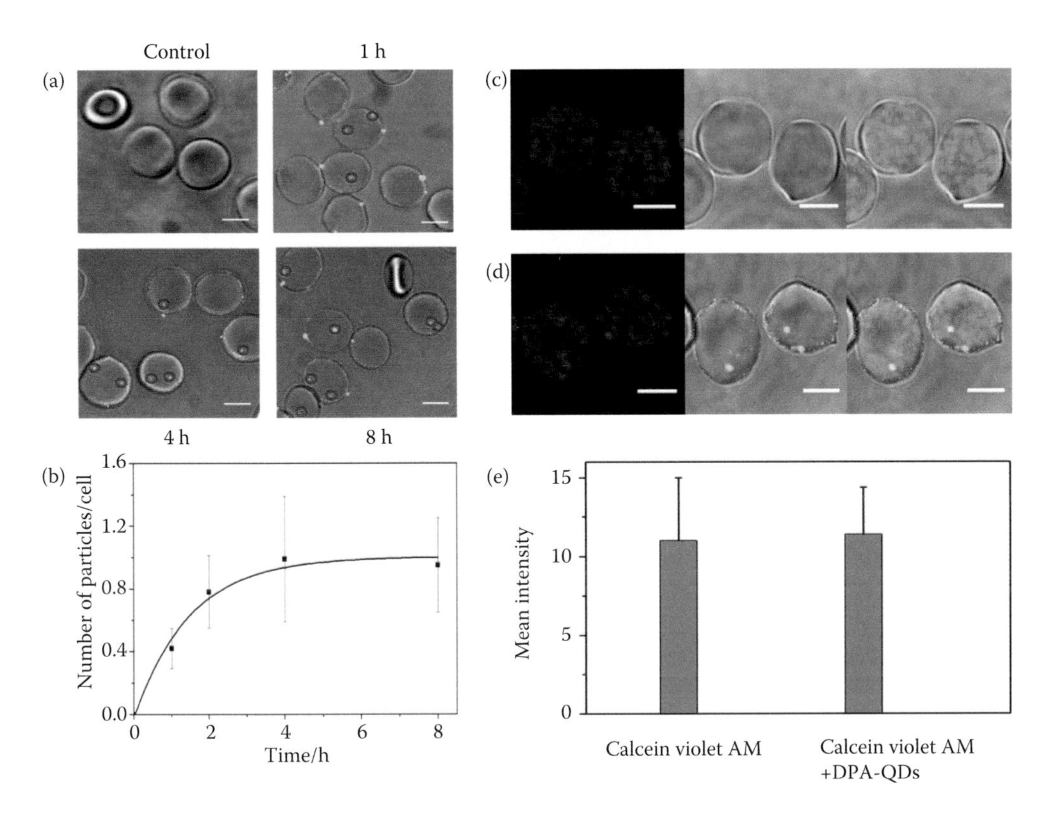

FIGURE 8.4
(a) Overlaid bright-field and confocal fluorescence images (scale bar, 5 μm) of DPA-QD internalization by RBCs after incubating with PBS (control) and 10 nM DPA-QDs for 1, 4, and 8 h. (b) Normalized fluorescence intensity of intracellular bright spots plotted as a function of time. (c, d) Images from fluorescence microscopy experiments (scale bar, 5 μm) probing the integrity of the RBC plasma membrane during DPA-QD uptake. RBCs were incubated with (c) calcein violet AM and subsequently (d) with 10 nM DPA-QDs for 6 h. (e) Mean fluorescence intensities of calcein violet AM–labeled cells. (Reprinted with permission from Wang et al. 2012, 1251. Copyright 2012 American Chemical Society.)

membrane or induce formation of pores. Of course, these statements only hold for spatial scales that can be resolved by optical microscopy.

The integrity of the RBC membrane during DPA-QD internalization was therefore further examined by quantifying the escape of a tracer dye from the cytosol. RBCs were preincubated with calcein violet AM, a cell-membrane-permeant dye that becomes impermeant after hydrolysis by intracellular esterases. Subsequently, the cells were incubated with DPA-QDs for 6 h. Intracellular calcein fluorescence was quantified by analysis of fluorescence microscopy images acquired at the end of the incubation (Figure 8.4c, d). These data clearly showed that internalization of DPA-QDs was not associated with any detectable loss of intracellular fluorescence (leakage of dye), suggesting that the RBC membranes remained intact during NP penetration of the bilayer.

These data were further complemented by electrochemical studies of the interaction between DPA-QDs and a planar model membrane. Briefly, vesicles with the lipid content of the outer or inner leaflets of the RBC lipid bilayer were fused onto a gold electrode by the interaction between the vesicles and the hydrophobic surface of a self-assembled monolayer of 1-dodecanethiol preadsorbed onto the electrode. The outer lipid leaflet consisted mainly of phosphatidylcholine and sphingomyelin, and the inner one mainly of phosphatidylethanolamine and phosphatidylserine; cholesterol was evenly distributed.

With these bilayers prepared on a gold electrode, cyclic voltammograms of 5 mM $[Fe(CN)_6]^{3-}$ in a solution of 0.1 M KCl were acquired before and after treating the membrane with DPA-QD solution for several hours. A leakage current due to the presence of NPs was not detectable, confirming the notion that the presence of DPA-QDs does not cause formation of pores in the outer and inner lipid layers of the model membrane, which would allow ions to penetrate the bilayer and diffuse to the gold electrode.

Surface-enhanced infrared absorption spectroscopy (SEIRAS) was also carried out using the same model-membrane preparation. SEIRAS allows infrared spectra to be acquired from extremely thin molecular layers, by utilizing the strong enhancement of the infrared absorption of molecules in near-field distance to nanostructured metal films (Jiang et al. 2008; Osawa 2001). The analyses of the frequencies of the stretching modes of the CH_2 groups in the lipid tails revealed that the bilayer structure is softened in the presence of DPA-QDs interacting with either side of the model membrane. Its higher dynamics may facilitate penetration of DPA-QDs into the lipid bilayer without pore formation.

As was shown earlier (Lin et al. 2010; Verma et al. 2008), NP-induced hole formation in a membrane can be avoided by suitably modulating the NP surface charge density or structure. Cationic Au NPs with 50% charge density (relative to hydrophobic ligands) effectively penetrated the lipid membrane without forming holes, while significant membrane disruption occurred at higher charge densities. Certain cell-penetrating peptides are also known to translocate across membranes without lipid bilayer disruption (Herbig et al. 2006; Patel et al. 2007; Thorén et al. 2003). The DPA ligand bound to the QDs used in the study by Wang et al. (2012) is a small, zwitterionic amino acid and, at neutral pH, the charges on the amino and carboxylic acid groups are balanced. This zwitterionic functionalization may be responsible for the ability of these NPs to penetrate lipid membranes without compromising bilayer integrity.

8.7 Conclusions and Outlook

The results that we have discussed in this chapter have a number of implications for NP exposure to cells and entire organisms, including humans. It has become very clear in recent years that a protein corona forms around NPs upon their exposure to biofluids, and that this layer can markedly modify cellular uptake (Jiang et al. 2010b; Treuel and Nienhaus 2012). The current state of knowledge is, however, still insufficient, especially with respect to the detailed physicochemical processes occurring on a molecular level. Little is known thus far about the effect of the adsorption process on the structure of the proteins. On some NP surfaces, proteins may maintain their native conformations, whereas on other NP surfaces they may undergo severe denaturation. Besides such structural details, the kinetics of formation, stability, and aging effects are still only poorly understood. Moreover, the development of the protein corona under complex biological conditions, with proteins exchanging with a multitude of competing proteins, needs to be addressed. The consequences of the protein corona and, hence, its effects on the biological behavior of NPs, are still elusive and need further attention (Casals et al. 2010; Chithrani and Chan 2007; Ehrenberg et al. 2009; Oberdörster 2010; Treuel and Nienhaus 2012).

It is now well established that NP uptake involves well-known cellular pinocytotic mechanisms (Conner and Schmid 2003; Jiang et al. 2011; Rothen-Rutishauser et al. 2006). Factors such as NP size (Chithrani et al. 2006; Rejman et al. 2004), shape (Chithrani et al. 2006), surface charge (Arbab et al. 2003; Labhasetwar et al. 1998; Sun et al. 2005), and surface chemistry (Holzapfel et al. 2006; Labhasetwar et al. 1998; Nativo et al. 2008) have been clearly identified as influential parameters governing the uptake routes and efficiencies.

As we have discussed in this chapter, for smaller NPs, a critical threshold density on the cell membrane must be exceeded to trigger the internalization process, as inferred from the nonlinear dependence of the uptake on NP concentration. Frequently, *in vitro* experiments are carried out with higher NP concentrations than those expected for environmental exposure of cells. It is therefore quite possible that such threshold densities may not be reached in real environmental scenarios. However, these effects may bear relevance in biomedical applications. Especially in targeted drug delivery, it is of utmost importance to know the uptake yields at particular NP concentrations as well as the specific pathways that are utilized by the NPs.

Fairly simple physical models have been developed and tested that can reveal the key parameters of endocytosis. Their ability to provide meaningful results that correlate well with experimental data obtained, e.g., by confocal microscopy, underscores their value for advancing our understanding of complex biological processes. Quantitative modeling of NP uptake, however, is still at an early stage and needs further development.

The ability of NPs to adhere to and penetrate cell membranes by processes that do not involve any cellular uptake machinery is well documented. Such processes depend on the NP physical properties, including size, surface composition, and surface charge (Dif et al. 2008; Laurencin et al. 2010; Leroueil et al. 2007; Roiter et al. 2008; Verma et al. 2008; Zhang and Yang 2011). While some NP types were shown to rupture the plasma membrane, eliciting noticeable cytotoxic effects (Cho et al. 2009; Kostarelos et al. 2007; Leroueil et al. 2008; Yu et al. 2007), other studies demonstrated that NPs may enter cells without causing membrane leakage (Wang et al. 2012). Overall, the way in which NP surface properties govern their behavior in passive uptake has not been characterized. We stress that passive uptake routes may well play an important role during long-term exposure to low NP concentrations. Whenever threshold densities of NPs on cellular membranes cannot be reached, passive uptake of NPs could become a significant contribution to their overall internalization.

In conclusion, the general mechanisms of NP uptake by cells have been identified and the first correlations between uptake behavior and NP properties have been demonstrated; yet, many details remain unknown. The importance of specific interactions between NPs and endocytosis receptors is emerging. The role of physical and chemical parameters of NPs, as well as the potentially decisive influence of the protein corona on this behavior, needs further attention. In addition, the possibility that different cell types have different specific responses to NP exposures must be addressed.

Acknowledgments

LT and GUN acknowledge financial support by the Deutsche Forschungsgemeinschaft (DFG) through the Center for Functional Nanostructures (CFN) and Schwerpunktprogramm (SPP) 1313. XJ acknowledges grant support by the Youth Foundation of China (21105097), President Funds of the Chinese Academy of Sciences.

References

Arbab, A. S., L. A. Bashaw, B. R. Miller, E. K. Jordan, B. K. Lewis, H. Kalish, and J. A. Frank. 2003. Characterization of biophysical and metabolic properties of cells labeled with superparamagnetic iron oxide nanoparticles and transfection agent for cellular MR imaging. *Radiology* 229:838–846.

Casals, E., T. Pfaller, A. Duschl, G. J. Oostingh, and V. Puntes. 2010. Time evolution of the nanoparticle protein corona. *ACS Nano* 4:3623–3632.

Cedervall, T., I. Lynch, M. Foy, T. Berggård, S. C. Donnelly, G. Cagney, S. Linse, and K. A. Dawson. 2007a. Detailed identification of plasma proteins adsorbed on copolymer nanoparticles. *Angew. Chem. Int. Ed.* 46:5754–5756.

Cedervall, T., I. Lynch, S. Lindman, T. Berggård, E. Thulin, H. Nilsson, K. A. Dawson, and S. Linse. 2007b. Understanding the nanoparticle–protein corona using methods to quantify exchange rates and affinities of proteins for nanoparticles. *Proc. Natl. Acad. Sci. U.S.A.* 104:2050–2055.

Chithrani, B. D., and W. C. W. Chan. 2007. Elucidating the mechanism of cellular uptake and removal of protein-coated gold nanoparticles of different sizes and shapes. *Nano Lett.* 7:1542–1550.

Chithrani, B. D., A. A. Ghazani, and W. C. W. Chan. 2006. Determining the size and shape dependence of gold nanoparticle uptake into mammalian cells. *Nano Lett.* 6:662–668.

Cho, E. C., J. Xie, P. A. Wurm, and Y. Xia. 2009. Understanding the role of surface charges in cellular adsorption versus internalization by selectively removing gold nanoparticles on the cell surface with a I_2/KI etchant. *Nano Lett.* 9:1080–1084.

Conner, S., and S. L. Schmid. 2003. Regulated portals of entry into the cell. *Nature* 422:37–44.

de Paoli Lacerda, S. H., J. J. Park, C. Meuse, D. Pristinski, M. L. Becker, A. Karim, and J. Douglas. 2010. Interaction of gold nanoparticles with common human blood proteins. *ACS Nano* 4:365–379.

Dif, A., E. Henry, F. Artzner, M. Baudy-Floc'h, M. Schmutz, M. Dahan, and V. Marchi-Artzner. 2008. Interaction between water-soluble peptidic CdSe/ZnS nanocrystals and membranes: Formation of hybrid vesicles and condensed lamellar phases. *J. Am. Chem. Soc.* 130: 8289–8296.

Ehrenberg, M. S., A. E. Friedman, J. N. Finkelstein, G. Oberdörster, and J. L. McGrath. 2009. The influence of protein adsorption on nanoparticle association with cultured endothelial cells. *Biomaterials* 30:603–610.

Gao, H., W. Shi, and L. B. Freund. 2005. Mechanics of receptor-mediated endocytosis. *Proc. Natl. Acad. Sci. U.S.A.* 102:9469–9474.

Gebauer, J. S., M. Malissek, S. Simon, S. K. Knauer, M. Maskos, R. H. Stauber, W. Peukert, and L. Treuel. 2012. Impact of the nanoparticle–protein corona on colloidal stability and protein structure. *Langmuir* 28:9181–9906.

Geiser, M., B. Rothen-Rutishauser, N. Kapp, S. Schürch, W. Kreyling, H. Schulz, M. Semmler, V. Im Hof, J. Heyder, and P. Gehr. 2005. Ultrafine particles cross cellular membranes by nonphagocytic mechanisms in lungs and in cultured cells. *Environ. Health Perspect.* 113:1555–1560.

Greulich, C., S. Kittler, M. Epple, G. Muhr, and M. Köller. 2009. Studies on the biocompatibility and the interaction of silver nanoparticles with human mesenchymal stem cells (hMSCs). *Langenbecks Arch. Surg.* 394:495–502.

Herbig, M. E., F. Assi, M. Textor, and H. P. Merkle. 2006. The cell penetrating peptides pVEC and W2-pVEC induce transformation of gel phase domains in phospholipid bilayers without affecting their integrity. *Biochemistry* 45:3598–3609.

Holzapfel, V., M. Lorenz, C. K. Weiss, H. Schrezenmeier, K. Landfester, and V. Mailänder. 2006. Synthesis and biomedical applications of functionalized fluorescent and magnetic dual reporter nanoparticles as obtained in the miniemulsion process. *J. Phys.: Condens. Matter.* 18:S2581–S2594.

Jiang, X., A. Musyanovych, C. Röcker, K. Landfester, V. Mailänder, and G. U. Nienhaus. 2011. Specific effects of surface carboxyl groups on anionic polystyrene particles in their interactions with mesenchymal stem cells. *Nanoscale* 3:2028–2035.

Jiang, X., C. Röcker, M. Hafner, S. Brandholt, R. M. Dörlich, and G. U. Nienhaus. 2010a. Endo- and exocytosis of zwitterionic quantum dot nanoparticles by live HeLa cells. *ACS Nano* 4:6787–6797.

Jiang, X., S. Weise, M. Hafner, C. Röcker, F. Zhang, W. J. Parak, and G. U. Nienhaus. 2010b. Quantitative analysis of the protein corona on FePt nanoparticles formed by transferrin binding. *J. R. Soc. Interface* 7:S5–S13.

Jiang, X., E. Zaitseva, M. Schmidt, F. Siebert, M. Engelhard, R. Schlesinger, K. Ataka, R. Vogel, and J. Heberle. 2008. Resolving voltage-dependent structural changes of a membrane photoreceptor by surface-enhanced IR difference spectroscopy. *Proc. Natl. Acad. Sci. U.S.A.* 105:12113–12117.

Kah, J. C., K. Y. Wong, K. G. Neoh, J. H. Song, J. W. Fu, S. Mhaisalkar, M. Olivo, and C. J. Sheppard. 2009. Critical parameters in the pegylation of gold nanoshells for biomedical applications: An *in vitro* macrophage study. *J. Drug Target.* 17:181–193.

Kittler, S., C. Greulich, J. Diendorf, M. Köller, and M. Epple. 2010a. The toxicity of silver nanoparticles increases during storage due to slow dissolution under release of silver ions. *Chem. Mater.* 22:4548–4554.

Kittler, S., C. Greulich, J. S. Gebauer, J. Diendorf, L. Treuel, L. Ruiz, J. M. Gonzalez-Calbet, M. Vallet-Regi, R. Zellner, M. Köller, and M. Epple. 2010b. The influence of proteins on the dispersability and cell-biological activity of silver nanoparticles. *J. Mater. Chem.* 20:512–518.

Klein, J. 2007. Probing the interactions of proteins and nanoparticles. *Proc. Natl. Acad. Sci. U.S.A.* 104:2029–2030.

Kostarelos, K., L. Lacerda, G. Pastorin, W. Wu, S. Wieckowski, J. Luangsivilay, S. Godefroy, D. Pantarotto, J.-P. Briand, S. Muller, M. Prato, and A. Bianco. 2007. Cellular uptake of functionalized carbon nanotubes is independent of functional group and cell type. *Nat. Nanotechnol.* 2:108–113.

Labhasetwar, V., C. Song, W. Humphrey, R. Shebuski, and R. J. Levy. 1998. Arterial uptake of biodegradable nanoparticles: Effect of surface modifications. *J. Pharmaceut. Sci.* 87:1229–1234.

Larkin, J., M. Brown, J. Goldstein, and R. Anderson. 1983. Depletion of intracellular potassium arrests coated pit formation and receptor-mediated endocytosis in fibroblasts. *Cell* 33:273–285.

Laurencin, M., T. Georgelin, B. Malezieux, J.-M. Siaugue, and C. Ménager. 2010. Interactions between giant unilamellar vesicles and charged core–shell magnetic nanoparticles. *Langmuir* 26:16025–16030.

Leroueil, P. R., S. A. Berry, K. Duthie, G. Han, V. M. Rotello, D. Q. McNerny, J. R. Baker, B. G. Orr, and M. M. Banaszak Holl. 2008. Wide varieties of cationic nanoparticles induce defects in supported lipid bilayers. *Nano Lett.* 8:420–424.

Leroueil, P. R., S. Hong, A. Mecke, J. R. Baker, B. G. Orr, and M. M. Banaszak Holl. 2007. Nanoparticle interaction with biological membranes: Does nanotechnology present a Janus face? *Acc. Chem. Res.* 40:335–342.

Lin, J., H. Zhang, Z. Chen, and Y. Zheng. 2010. Penetration of lipid membranes by gold nanoparticles: Insights into cellular uptake, cytotoxicity, and their relationship. *ACS Nano* 4:5421–5429.

Lundqvist, M., J. Stigler, G. Elia, I. Lynch, T. Cedervall, and K. A. Dawson. 2008. Nanoparticle size and surface properties determine the protein corona with possible implications for biological impacts. *Proc. Natl. Acad. Sci. U.S.A.* 105:14265–14270.

Lunov, O., V. Zablotskii, T. Syrovets, C. Röcker, K. Tron, G. U. Nienhaus, and T. Simmet. 2011. Modeling receptor-mediated endocytosis of polymer-functionalized iron oxide nanoparticles by human macrophages. *Biomaterials* 32:547–555.

Lynch, I. 2007. Are there generic mechanisms governing interactions between nanoparticles and cells? Random epitope mapping for the outer layer of the protein–material interface. *Physica A* 373:511–520.

Macia, E., M. Ehrlich, R. Massol, E. Boucrot, C. Brunner, and T. Kirchhausen. 2006. Dynasore, a cell-permeable inhibitor of dynamin. *Dev. Cell* 10:839–850.

Mühlfeld, C., P. Gehr, and B. Rothen-Rutishauser. 2008. Translocation and cellular entering mechanisms of nanoparticles in the respiratory tract. *Swiss Med. Wkly.* 138:387–391.

Nativo, P., I. A. Prior, and M. Brust. 2008. Uptake and intracellular fate of surface-modified gold nanoparticles. *ACS Nano* 2:1639–1644.

Oberdörster, G. 2010. Safety assessment for nanotechnology and nanomedicine: Concepts of nanotoxicology. *J. Intern. Med.* 267:89–105.

Osawa, M. 2001. Surface-enhanced infrared absorption. *Top. Appl. Phys.* 81:163–187.

Owens, D. E., and N. A. Peppas. 2006. Opsonization, biodistribution, and pharmacokinetics of polymeric nanoparticles. *Int. J. Pharm.* 307:93–102.

Patel, L., J. Zaro, and W.-C. Shen. 2007. Cell penetrating peptides: Intracellular pathways and pharmaceutical perspectives. *Pharmaceut. Res.* 24:1977–1992.

Pitas, R. E., A. Friera, J. McGuire, and S. Dejager. 1992. Further characterization of the acetyl LDL (scavenger) receptor expressed by rabbit smooth muscle cells and fibroblasts. *Arterioscler. Thromb. Vasc. Biol.* 12:1235–1244.

Ratte, H. T. 1999. Bioaccumulation and toxicity of silver compounds: A review. *Environ. Toxicol. Chem.* 18:89–108.

Rejman, J., V. Oberle, I. Zuhorn, and D. Hoekstra. 2004. Size-dependent internalization of particles via the pathways of clathrin-and caveolae-mediated endocytosis. *Biochem. J.* 377:159–169.

Röcker, C., M. Pötzl, F. Zhang, W. J. Parak, and G. U. Nienhaus. 2009. A quantitative fluorescence study of protein monolayer formation on colloidal nanoparticles. *Nat. Nanotechnol.* 4:577–580.

Roiter, Y., M. Ornatska, A. R. Rammohan, J. Balakrishnan, D. R. Heine, and S. Minko. 2008. Interaction of nanoparticles with lipid membrane. *Nano Lett.* 8:941–944.

Rothen-Rutishauser, B. M., S. Schürch, B. Haenni, N. Kapp, and P. Gehr. 2006. Interaction of fine particles and nanoparticles with red blood cells visualized with advanced microscopic techniques. *Environ. Sci. Technol.* 40:4353–4359.

Shang, L., S. Brandholt, F. Stockmar, V. Trouillet, M. Bruns, and G. U. Nienhaus. 2012. Effect of protein adsorption on the fluorescence of ultrasmall gold nanoclusters. *Small* 8:661–665.

Shang, L., Y. Wang, J. Jiang, and S. Dong. 2007. pH-dependent protein conformational changes in albumin:gold nanoparticle bioconjugates: A spectroscopic study. *Langmuir* 23:2714–2721.

Silver, S. 1996. Bacterial resistances to toxic metal ions—A review. *Gene* 179:9–19.

Stohs, S. J., and D. Bagchi. 1995. Oxidative mechanisms in the toxicity of metal ions. *Free Radic. Biol. Med.* 18:321–336.

Sun, X. K., R. Rossin, J. L. Turner, M. L. Becker, M. J. Joralemon, M. J. Welch, and K. J. Wooley. 2005. An assessment of the effects of shell cross-linked nanoparticle size, core composition, and surface PEGylation on *in vivo* biodistribution. *Biomacromolecules* 6:2541–2554.

Thorén, P. E. G., D. Persson, P. Isakson, M. Goksör, A. Önfelt, and B. Nordén. 2003. Uptake of analogs of penetratin, Tat(48–60) and oligoarginine in live cells. *Biochem. Biophys. Res. Commun.* 307:100–107.

Treuel, L., M. Malissek, J. S. Gebauer, and R. Zellner. 2010. The influence of surface composition of nanoparticles on their interactions with serum albumin. *Chem. Phys. Chem.* 11:3093–3099.

Treuel, L., and G. U. Nienhaus. 2012. Toward a molecular understanding of nanoparticle–protein interactions. *Biophys. Rev.* 4:137–147.

Underhill, D. M., and A. Ozinsky. 2002. Phagocytosis of microbes: Complexity in action. *Annu. Rev. Immunol.* 20:825–852.

Verma, A., O. Uzun, Y. Hu, Y. Hu, H.-S. Han, N. Watson, S. Chen, D. J. Irvine, and F. Stellacci. 2008. Surface-structure-regulated cell-membrane penetration by monolayer-protected nanoparticles. *Nat. Mater.* 7:588–595.

Wang, L., K. Rothberg, and R. Anderson. 1993. Mis-assembly of clathrin lattices on endosomes reveals a regulatory switch for coated pit formation. *J. Cell. Biol.* 123:1107–1117.

Wang, T., J. Bai, X. Jiang, and G. U. Nienhaus. 2012. Cellular uptake of nanoparticles by membrane penetration: A study combining confocal microscopy with FTIR spectroelectrochemistry. *ACS Nano* 6:1251–1259.

Yu, J., S. A. Patel, and R. M. Dickson. 2007. *In vitro* and intracellular production of peptide-encapsulated fluorescent silver nanoclusters. *Angew. Chem. Int. Ed.* 46:2028–2030.

Zhang, X., and S. Yang. 2011. Nonspecific adsorption of charged quantum dots on supported zwitterionic lipid bilayers: Real-time monitoring by quartz crystal microbalance with dissipation. *Langmuir* 27:2528–2535.

Zhao, Y., X. Sun, G. Zhang, B. G. Trewyn, I. I. Slowing, and V. S. Y. Lin. 2011. Interaction of mesoporous silica nanoparticles with human red blood cell membranes: Size and surface effects. *ACS Nano* 5:1366–1375.

9

Cellular Uptake and Intracellular Trafficking

Barbara Rothen-Rutishauser, Dagmar A. Kuhn, Dimitri Vanhecke, Fabian Herzog, Alke Petri-Fink, and Martin J.D. Clift

CONTENTS

9.1 Introduction

The detailed mechanisms whereby nanoparticles (NPs) may interact with target cells and how this can influence cell reactions are still poorly understood. The principal interaction of NPs with a biological environment occurs at the cellular level through their interaction with structural and functional cell compartments (e.g., nucleus and organelles) (Muhlfeld et al. 2008b). For clarity within this chapter, NPs are defined as a nano-object (a material with one, two, or three external dimensions in the nanoscale, 1–100 nm) with all three external dimensions in the nanoscale (ISO Report 2008). The aim of this chapter is to discuss, on the basis of existing literature, the potential uptake mechanism of NPs into cells as well as their possible interaction with intracellular compartments.

With every breath we take, not only oxygen is inhaled but millions of (nano)particles also enter the respiratory system. The deposition of these (nano)particles in the lung is size dependent (Heyder et al. 1986; Patton and Byron 2007). Specifically, NPs can be released into the environment from combustion-derived processes, or, in an occupational setting via the use of NP-containing consumer products, such as aerosol sprays. Besides the geometry of the airways and their breathing pattern(s), the specific (nano)particle size is important for studying the deposition and clearance of (nano)particles in the respiratory tract. Following inhalation, airborne particles deposit in the different regions of the respiratory tract in a size-dependent manner (Heyder et al. 1986; Patton and Byron 2007). Larger particles (1–10 μm) preferentially deposit in larger conducting airways (i.e., trachea and bronchi), whereas smaller particles (i.e., NPs, <100 nm) localize to more peripheral lung regions (i.e., alveoli) (Oberdorster et al. 2005). Once deposited, particles interact with various structural barriers (Rothen-Rutishauser et al. 2009), such as the pulmonary surfactant, and are displaced via wetting forces into the aqueous hypophase (Gehr et al. 1990). As a result of this displacement, NPs come into contact with >40 different cell types in the respiratory tract. These cell types comprise epithelial and endothelial cells, fibroblasts, nerve cells, lymphoid cells, and immune cells such as macrophages and dendritic cells (Ochs and Weibel 2008). The epithelium is *the* primary cellular barrier endowed with tight junctions (Rothen-Rutishauser et al. 2012; Schneeberger and Lynch 1984) whose effectiveness is greatly enhanced by the aqueous surface lining layer and the mucociliary escalator covering the conducting airways (Kilburn 1968). A second tier of structural barriers further comprises cells from both the innate and the adaptive immune system, which are positioned throughout the respiratory tree. These cells are located above and below the respiratory epithelium, and their activities are tightly regulated to protect the integrity of the airways and the vital gas-exchange region (Wikstrom and Stumbles 2007). The innate response is largely governed by macrophages (Lohmann-Matthes et al. 1994; Peters-Golden 2004), whereas respiratory tract dendritic cells are responsible for the adaptive immune response (Holt and Stumbles 2000; Vermaelen and Pauwels 2005).

It has been proposed that the permeability of the lung tissue barrier to NPs is controlled at the epithelial and endothelial levels (Meiring et al. 2005). If NPs pass through the epithelial barrier, they may then further progress through the basement membrane and subsequent subepithelial connective tissue layer, eventually interacting with endothelial cells that line the capillaries. Since endothelial cells play an important role in inflammatory processes (Michiels 2003), the interaction of NPs with these cells can induce heightened pro-inflammatory stimuli through altering endothelial cell function and viability.

As previously highlighted, NPs have the capacity to enter various cells, which then might lead to an altered (toxic) cellular response. The aim of this chapter is to provide a comprehensive overview concerning the entry mechanisms of NPs into single cells and their resultant intracellular trafficking, to provide the reader with a basic understanding of NP–cell interactions.

9.2 Eukaryotic Cell

All eukaryotic cells contain functionally distinct and complex compartments enclosed within internal membranes, which have become highly specialized (Alberts et al. 1998). Membranes surround the nucleus ("karyon" in Greek) and organelles such as mitochondria, endoplasmic reticulum, Golgi apparatus, peroxisomes, lysosomes, and endosomes (Figure 9.1). There is a continuous exchange between internal membrane-bounded

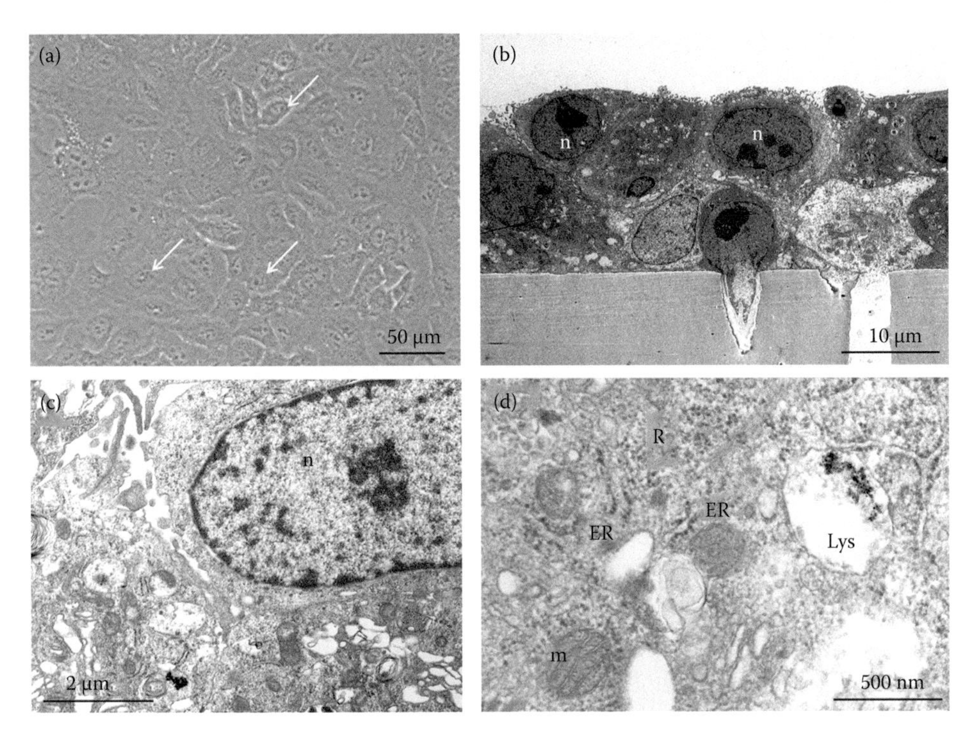

FIGURE 9.1
Light microscopy and conventional TEM pictures of A549 epithelial cells. (a) Phase contrast pictures of the confluent monolayer; arrows point to the cell nuclei. (b through d) Different intracellular structures and compartments can be identified at different magnifications. n, nucleus; m, mitochondria; ER, endoplasmic reticulum; R, ribosomes; Lys, lysosomes.

compartments and the outside of the cell by endocytosis and exocytosis. In addition, all eukaryotic cells have an internal cytoskeleton composed of microtubules, microfilaments, and intermediate filaments, which play an important role in defining the eukaryotic cell organization and shape (Alberts et al. 1998). Within this chapter, only the structure and function of animal cells will be focused on.

9.3 Membrane Structures

The nucleus and other intracellular organelles are enclosed in the animal cell by a lipid bilayer containing distinct proteins (Warren and Wickner 1996). All cell membranes have a common structure: they consist of a very thin film of lipid and protein molecules, held together mainly by noncovalent interactions (Kendall 2007; Singer and Nicolson 1972). The lipid molecules are arranged as a continuous double layer about 4–5 nm thick in cell membranes. This lipid bilayer serves as a relatively impermeable barrier to the passage of most water-soluble molecules. The transmembrane protein molecules mediate specific functions as transporting proteins serving as pumps or channels across the bilayer or catalyzing membrane-associated reactions. Some proteins serve as structural links that connect the cytoskeleton through the lipid bilayer to the extracellular matrix or an adjacent cell by integrins and cadherins, while others serve as receptors to detect and transduce chemical signals into the cell environment (Eisenberg et al. 1984). It is important to note that the membrane is not a rigid structure but rather a dynamic system (for a review, see Ritchie and Spector 2007). This has been described as the fluid mosaic model by Singer and Nicolson in 1972, who defines biological membranes as a two-dimensional (2D) liquid in which lipid and protein molecules can diffuse. Because of this fluidity of the phospholipid bilayer, the components of the membranes can be arranged in a non-homogeneous distribution, and areas can differ in its lipid composition (for a review see, Sonnino and Prinetti 2013). Especially the clustering of sphingolipids and cholesterol into such rafts has been associated with specific cellular and/or biological functions (Simons and Ikonen 1997).

9.4 Cellular Compartments

The different organelles that can be found in eukaryotic (animal) cells have been described extensively by others (Alberts et al. 1998; The University of Arizona 1997) and are only briefly summarized, as follows:

- The cell nucleus contains the genetic material deoxyribonucleic acid (DNA) and a nucleolus within a double membrane called the nuclear envelope. The nucleolus produces ribonucleic acid (RNA).
- Mitochondria are the power plants of the cell, producing energy by combining oxygen with food molecules to make adenosine triphosphate (ATP). They have an outer membrane and an inner membrane that is folded into cristae to increase the surface area.
- Ribosomes make proteins that are either used within or transported out of the cell.
- The rough endoplasmic reticulum is in structural continuity with the outer membrane of the nuclear envelope, and it specializes in the synthesis and transport of lipids and membrane proteins. Its outer face is covered with ribosomes that are engaged in protein synthesis.
- The smooth endoplasmic reticulum lacks attached ribosomes and is involved in lipid metabolism.
- The Golgi apparatus is the processing plant of the cell. It is a system of stacked, membrane-bound, and flattened sacs involved in processing proteins and lipids into different forms and transports them to the appropriate organelles within the cell.
- Endosomes are membrane-bound compartments (i.e., vesicles) inside eukaryotic cells. There is an intracellular trafficking from primary endocytotic vesicles toward late endosomes and finally lysosomes.
- Lysosomes contain hydrolytic enzymes involved in intracellular digestions.
- Peroxisomes contain oxidative enzymes that break down hydrogen peroxide, which is a by-product of cell metabolism.
- Vacuoles are essentially storage units.

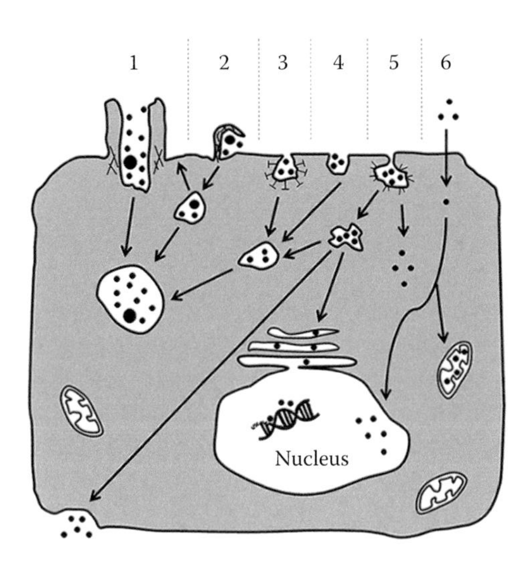

FIGURE 9.2
Possible mechanisms for NPs to be taken up by cells and their subsequent intracellular trafficking. NPs may be actively incorporated via phagocytosis (1), macropinocytosis (2), clathrin-dependent endocytosis (3), clathrin- and caveolae-independent endocytosis (4), or caveolae-mediated endocytosis (5). Particles that were internalized via active uptake are commonly transported via vesicular structures that then fuse to phagolysosomes or endosomes (1 through 5). Sometimes, they might be exocytosed upon macropinocytosis (2). Alternatively, they may also be carried to the cytosol or be transported via caveosomes to the endoplasmic reticulum or cross the cell as part of transcytotic processes (5). Besides active transport, NPs may also enter the cell passively via diffusion through the plasma membrane (6). From the cytoplasm, they may then gain access to subcellular compartments such as the nucleus and mitochondria (6). However, further research is needed to clarify if a particular entering mechanism and a certain intracellular localization may elicit specific cellular responses. (Figure and legend have been modified from Brandenberger C et al., *Small*, 6, 1669, 2010; Muhlfeld C et al., *Swiss Med. Wkly.*, 138, 387, 2008; Rothen-Rutishauser B et al., Nanoparticle–cell membrane interactions. In *Particle–Lung Interactions*, (eds) Gehr P, Mühlfeld Ch, Rothen-Rutishauser B, and Blank F (New York: Informa Healthcare), pp. 226–242, 2009; Unfried K et al., *Nanotoxicology*, 1, 1, 2007. Reprinted from Jud C et al., *Swiss Med. Wkly.*, 143, w13758, 2013. With permission.)

9.5 Exocytosis and Endocytosis of Macromolecules and Particles

Most cells can secrete and ingest macromolecules and particles by exocytosis and endocytosis (Alberts et al. 1998). The plasma membrane of the cells is a dynamic structure and segregates the chemically distinct intracellular milieu (the cytoplasm) from the extracellular environment by coordinating the entry and exit of small and large molecules. Figure 9.2 summarizes different possible cellular entering and intracellular trafficking mechanisms of NPs. It is important to note that viruses can take advantage of endocytotic pathways to enter host cells (for a review, see Barrow et al. 2013), and that this knowledge can be used for the development of new NP platforms for drug or vaccine delivery (Plummer and Manchester 2013).

9.6 Endocytosis

While essentially small molecules are able to traverse the plasma membrane through the action of integral membrane protein pumps or channels, macromolecules must be carried into the cells in membrane-bound vesicles derived from the invagination and pinching-off of pieces of the plasma membrane to form endocytic vesicles (Alberts et al. 1998). This process is termed endocytosis, and two types of endocytosis are distinguished on the basis of the size of endocytic vesicles formed: pinocytosis ("cellular drinking") involves the ingestion of fluid and molecules via small vesicles (<0.15 μm in diameter), whereas phagocytosis involves the ingestion of large particles such as microorganisms and cell debris, resulting in the formation of large vesicles called phagosomes (generally >0.25 μm in diameter).

The term pinocytosis describes macropinocytosis, clathrin-mediated endocytosis, caveolin-mediated endocytosis, and clathrin- and caveolin-independent endocytosis (Conner and Schmid 2003). Phagocytosis and macropinocytosis are both dependent on actin (Aderem and Underhill 1999). Phagocytosis is carried out by professional phagocytes (i.e., monocytes/macrophages, neutrophils, dendritic cells), which form intracellular phagosomes. Macromolecule or particle internalization is initiated by the interaction of specific receptors on the surface of the phagocyte with ligands on the surface of the molecule. This leads to the polymerization of actin at the site of ingestion, and after internalization the phagosome matures by a series of fusion and fission events with components of the endocytic pathway, culminating in the formation of the mature phagolysosome. Since endosome–lysosome trafficking occurs primarily in association with microtubules, phagosome maturation requires the coordinated interaction of the actin- and tubulin-based cytoskeletons (Conner and Schmid 2003).

Macropinocytosis triggers actin formation, and the macropinosomes form large intracellular vesicles. However, instead of invaginating a ligand-coated particle, they collapse onto and fuse with the plasma membrane to generate large endocytic vesicles called macropinosomes, which sample large volumes of extracellular milieu. Activation of antigen-presenting dendritic cells triggers extensive and prolonged macropinocytotic activity, enabling these cellular sentinels to sample large volumes of the extracellular milieu and to fulfill their role in immune surveillance (Mellman and Steinman 2001). However, little is known about the nature of the entire uptake process (Conner and Schmid 2003).

Caveolin-mediated endocytosis is mostly used for the transport of serum proteins, and the uptake is triggered by the cargo molecules themselves. Caveolae are static structures that are flask-shaped invaginations of the plasma membrane of a diameter of 50–100 nm observed in several types, including capillary endothelium, type I alveolar epithelial cells, smooth muscle cells, and fibroblasts, and are slow in uptake (Conner and Schmid 2003). The protein that gives shape and structure is called caveolin-1, which is a dimeric protein and binds cholesterol onto the cellular surface for intracellular trafficking (lipid homeostasis) (Rothberg et al. 1992). This mechanism, however, is poorly understood. It is generally referred to as cholesterol-rich microdomains, called rafts, with a diameter of 40–50 nm. Their unique lipid composition provides a physical basis for specific sorting of membrane proteins, glycoproteins, and/or glycolipids based on their transmembrane regions (Anderson and Jacobson 2002; Edidin 2001). These small rafts can presumably be captured by and internalized within any endocytic vesicle.

Clathrin-mediated endocytosis is very well studied and it is, like most pinocytic pathways, a form of receptor-mediated endocytosis. It constitutively occurs in all mammalian cells and carries out the continuous uptake of essential nutrients such as the cholesterol-laden low-density lipoprotein particles that bind to the low-density lipoprotein receptor and iron-laden transferrin that binds to transferrin receptors (Brodsky et al. 2001; Schmid 1997). Clathrin-mediated endocytosis involves the concentration of high-affinity transmembrane receptors and their bound ligands into "coated pits" on the plasma membrane, which are formed by the assembly of cytosolic proteins, the main assembly protein being clathrin. Coated pits invaginate and pinch-off to form endocytic vesicles, termed clathrin-coated vesicles, which are encapsulated by a polygonal clathrin coat and carry concentrated receptor–ligand complexes into the cell (Conner and Schmid 2003). Clathrin is a three-legged structure, called triskelion, formed by three clathrin heavy chains, each with a tightly associated clathrin light chain (Brodsky et al. 2001; Kirchhausen 2000).

9.7 Exocytosis

Two pathways of exocytosis have been described, one constitutive and one that is regulated (Alberts et al. 1998; Burgess and Kelly 1987). Constitutive exocytosis can be performed by all cells. Proteins are secreted by being packaged into transport vesicles in the Golgi apparatus and then carried to and incorporated into the plasma membrane. The regulated pathway is found in cells that are specialized for secreting their products—such as hormones, neurotransmitters, or digestive enzymes. This process is performed rapidly, and on demand, which is often triggered by an influx of Ca^{2+}.

9.8 Methods to Study NP Uptake into Cells

To understand how NPs interact with cellular systems, it is important to detect and to localize them within single cells (Muhlfeld et al. 2007; Rothen-Rutishauser et al. 2011). Once intracellular NPs are identified, their distribution in different cellular compartments, such as endosomes, lysosomes, mitochondria, Golgi/endoplasmic reticulum, the nucleus, or cytosol, may also provide some indication about their potential impact on cellular homeostasis.

9.9 Microscopy Methods

During the last years, many new NPs labeled with fluorophores or with fluorescence specificity have been introduced (Lin et al. 2009), such as semiconductor nanocrystals (i.e., quantum dots) that are robust and bright light emitters (Alivisatos et al. 2005). Another very promising technique is the design of the so-called core–shell NPs with fluorophores embedded in their shell (Chastellain et al. 2004; Rodriguez-Lorenzo et al. 2014). Such fluorescently labeled NPs can also be visualized by laser scanning microscopy (LSM) combined with digital image restoration to analyze their intracellular localization. However, it is important to understand that the resolution in LSM is limited to 200 nm lateral and 500–700 nm in the axial dimensions.

Additional modules can enhance selected features of the LSM data. Computational algorithms (deconvolution) may additionally improve the resolution by 2- to 3-fold (van der Voort and Strasters 1995). Spinning disk modules drastically improve the time dimension and can capture the dynamics of very fast events. 4PI confocal fluorescence microscopy techniques improve the z-resolution, and 2PI microscopy allows imaging at a much higher depth, suitable for imaging living tissue or organs (Jiang et al. 2010). Thus, considering the possibility of visualizing the uptake of fluorescently labeled NPs into living cells and of performing colocalization studies with fluorescently labeled organelles in 3D objects (fixed or living cells) (Lehmann et al. 2010), LSM provides an excellent tool to gain new insights into NP–cell interactions and NP uptake mechanisms (Figure 9.3).

Owing to the small size of NPs, their intracellular identification can be challenging and time consuming. For the detection of both fluorescent and/or electron-dense NPs, light and transmission electron microscopy (TEM) methods can be applied (Muhlfeld et al. 2007). TEM, with a resolution range from angstrom to nanometers, is the method of choice for detection of (electron-dense) NPs. However, if cellular structures are visualized, samples must be fixed and dehydrated; therefore, TEM cannot offer information on time resolution and dynamics. In addition, stereological approaches are required to understand the 3D distribution of NPs within a defined reference volume (Brandenberger et al. 2010c).

The presence of other cellular and noncellular nanosized structures in TEM cell samples such as ribosomes or glycogen, which may resemble NPs in size, morphology, and electron density, can bias the precise intracellular identification of NPs. Hence, elemental analysis, which can confirm the presence of NPs inside the cell, is recommended (Brandenberger et al. 2010b; Muhlfeld et al. 2007).

Conventional TEM, however, provides only 2D information of very thin specimen slices. The development of electron tomography made it possible to obtain spatial information about the 3D structure of an object by combining tilted views (Muhlfeld et al. 2007). This technique greatly helps with the characterization of NPs and the discrimination between genuine particles and agglomerated NPs (Figure 9.4).

To conclude this section, the authors emphasize using a combination of different techniques for one and the same NP to obtain a more complete picture about NP localization inside cells.

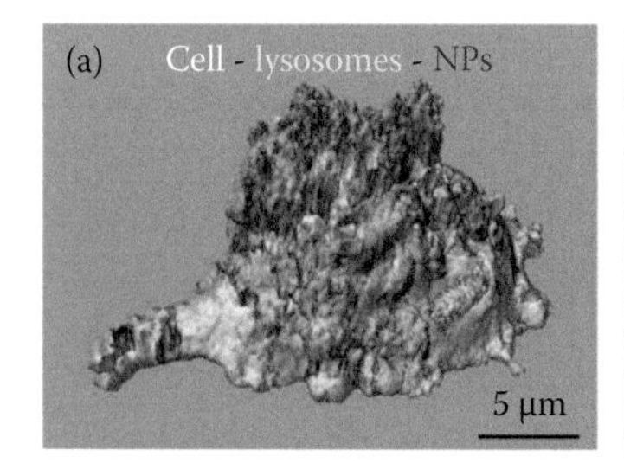

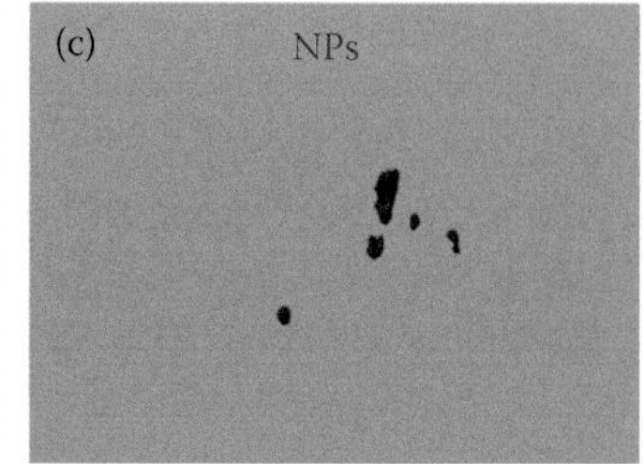

FIGURE 9.3
Visualization of intracellular NPs by LSM. The internalization of fluorescent-magnetic hybrid NPs in a dendritic cell in culture (light grey and transparent) is shown in a 3D reconstruction (a). In (b), the localization of the particles with lysosomes is shown. The NPs alone are shown in (c).

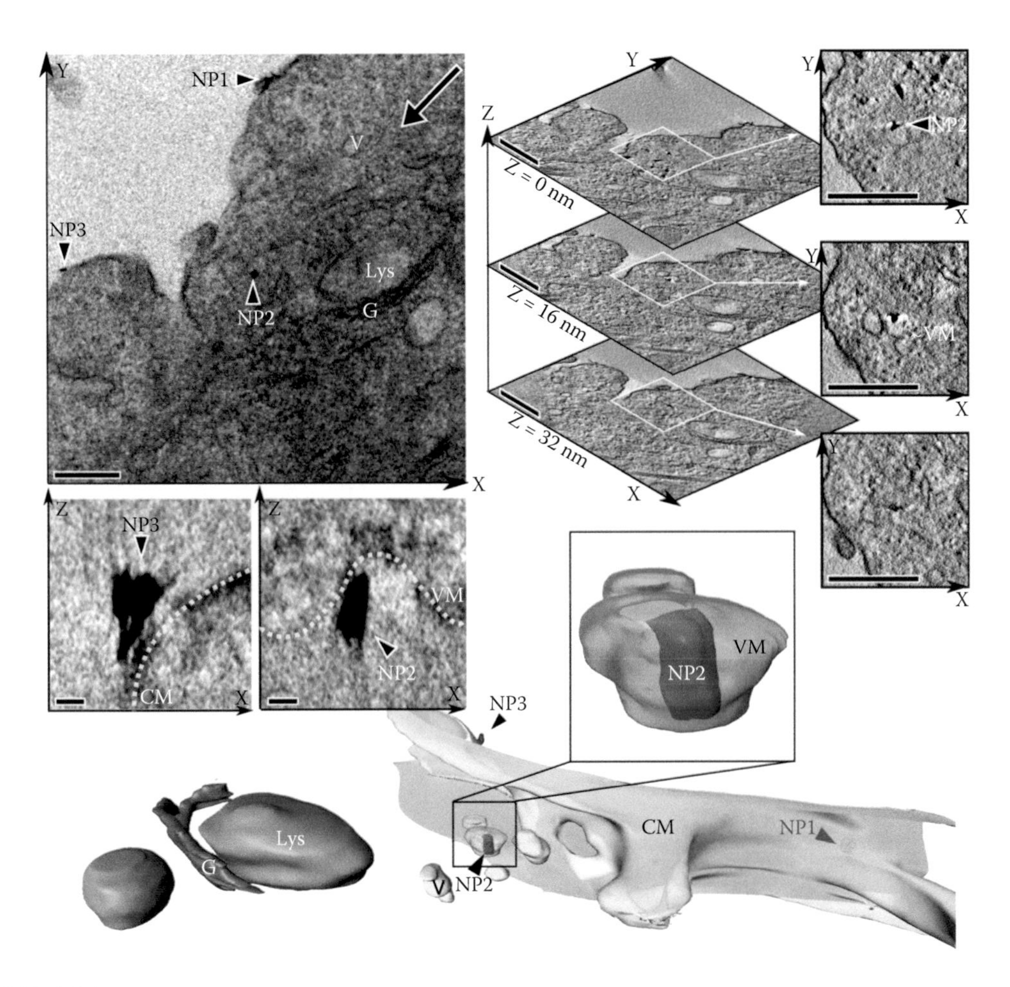

FIGURE 9.4
Electron tomography of three gold NP aggregates (NP1–3) in and near an alveolar macrophage. The 350-nm-thick section was reconstructed in 3D at a lateral pixel size of 2.3 nm and an axial pixel size of 4.0 nm. Top left: common cellular features can be recognized despite the relative thick section (350 nm). The arrow shows the angle at which the rendered scene was captured (see below). Top right: The reconstructed 3D data allow virtual browsing through the 3D stack along the Z-axis (16-nm steps). Each of the three computational slices contains about 4-nm axial information (estimated thickness of a slice, equivalent to the pixel size in Z). Browsing through a higher magnification of NP2 (see insets on the right) clearly shows that the aggregate is completely surrounded by a vesicular membrane, most probably an endosome. Middle: NP3 (left) and NP2 (right) as seen from the XZ plane. NP3 is not intracellular but closely associated with the cell membrane (dotted line). The XZ view of NP2 confirms that the aggregate is completely taken up by a vesicle. Both XZ images were digitally magnified slightly beyond the resolution limit. Bottom: 3D rendered scene of the tomographic data showing the relations between the cellular features. Note that NP1 is extracellular and can be seen through the semitransparent cell membrane. Amira software was used for the rendering. Bars: 100 nm, except the XZ planes: 10 nm. NP1–3, gold NP aggregate 1–3; Lys, lysosome; V, vesicle, probably an endosome; G, Golgi apparatus; CM, cell membrane; VM, vesicular membrane.

9.10 Inhibitors of Endocytosis

To test for the most important cellular endocytotic uptake mechanism of NPs, specific pharmacological inhibitors of different pathways can be used (Ivanov 2008). It is important to highlight that the use of inhibitors has to be optimized for each cell type and for each NP since an inhibitor might show a high specificity in one experiment but cause side effects in another experimental setup (Rothen-Rutishauser et al. 2014). The use of positive controls to show that an inhibitor only affects one endocytotic pathway without interfering with other uptake mechanism(s) is mandatory (Liu et al. 2007). There are many different inhibitors described, and we will focus only on the most commonly used drugs to study NP uptake.

Cytochalasin D can depolymerize actin filaments (Axline and Reaven 1974; Cooper 1987) and can therefore be used to study the actin-dependent uptake mechanism. Larger-sized particles of any material can be used to run the experiment under controlled conditions.

The most commonly used inhibitor for clathrin-mediated endocytosis is chlorpromazine, which induces a loss of clathrin and AP2 adaptor complex from the surface of the cell (Ivanov 2008). Furthermore, chlorpromazine is specific in inhibiting the uptake of the serum protein transferrin (Perumal et al. 2008); therefore, fluorescently labeled transferrin can be used for investigating clathrin-mediated endocytosis.

Caveolae and lipid raft internalizations were shown to be inhibited by nystatin, filipin, and methyl-β-cyclodextrin by distortion of the structure and function of the membrane (Ivanov 2008). These inhibitors build aggregates, which accumulate cholesterol and separate it from the membrane structures. Cholera toxin subunit b (Puri et al. 2001) and serum albumin (Ikehata et al. 2008) have been shown to enter cells by caveolin-mediated endocytosis; therefore, both proteins can be used to control the inhibitors in any experimental setting.

9.11 Cellular Uptake of NPs

Different uptake mechanisms of NPs into cells and intracellular trafficking described thus far have been discussed in detail in various reviews (Brandenberger et al. 2010c; Muhlfeld et al. 2008a; Rothen-Rutishauser et al. 2009; Unfried et al. 2007). Depending on the size, shape, material, and surface coating of the material and the cell type, possible uptake can occur by phagocytosis, macropinocytosis, clathrin- and caveolae-mediated endocytosis, clathrin- and caveolae-independent endocytosis, and NP diffusion/transport across the cell plasma membrane.

9.12 Limitations/Challenges

It is well known that the physicochemical properties of NPs (e.g., material, size [distribution], morphology, surface charge, surface functionalization, and surface roughness/stiffness) can influence how cells internalize NPs (Harush-Frenkel et al. 2008; Mahmoudi et al. 2011; Stark 2011; Verma and Stellacci 2010). When NPs enter a biological interface, the particles can be coated with proteins that are present within a biological fluid such as blood serum (when NPs are translocated or injected directly into the bloodstream) (Lynch et al. 2007) or surfactant proteins (when NPs are inhaled) (Schleh et al. 2013). Ruge and colleagues (2011) have recently demonstrated that surfactant protein A can trigger the uptake of magnetite NPs by alveolar macrophages. In addition, it has been suggested that the amount and presentation of the proteins on the NP surface rather than the particles themselves are responsible for any possible cellular effect (Lynch et al. 2007). This has also been demonstrated in a study where surfactant coating of multiwalled carbon nanotubes and not their functionalization is the key factor in determining their ability to induce a cell response (Gasser et al. 2012).

It has been shown in several studies that particle mass transport processes such as diffusion and sedimentation are the rate-limiting factors for cellular NP uptake (Limbach et al. 2005; Raemy et al. 2010); therefore, the protein–corona interactions can possibly change the NP properties (Lynch et al. 2007), leading to aggregation and the higher sedimentation velocity of aggregates in *in vitro* experimental conditions (i.e., in cell culture medium), which can result in a misrepresentative higher NP uptake (Teeguarden et al. 2007).

Not only the surface of the NPs is important, but also, just recently, a new perspective has been described, namely the "cell vision," which describes the phenomenon that the same NPs can trigger different responses and mechanisms in different cell types (Mahmoudi et al. 2012). It has been shown that different cells *in vitro* take up different numbers when using exactly the same NP type, that is, for instance, gold–NPs with a core diameter of 1.4 nm (Coulter et al. 2012). A recent study also reports that uptake of NPs and translocation across alveolar epithelial cell monolayers is species dependent (Fazlollahi et al. 2013). However, direct comparison of the various data is difficult owing to the number of variables within the experiments such as cell type, NP exposure route and time, concentration, and NP size. Further research is needed to reveal more details about this complex field, and within this book this would be out of context.

9.13 NP Uptake Mechanisms

Since it is somehow difficult to obtain complete and detailed information from published studies regarding the physicochemical properties of NPs in different suspension media, the culture conditions, and the details of NP exposure and analysis, the data in this subchapter must be interpreted with caution since the same experimental conditions might have not always been used.

It has been shown that in contrast to fine polystyrene particles (1 μm), the internalization of NPs (78 nm) after cytochalasin D treatment (inhibition of phagocytosis) could not be blocked completely in cultured macrophages, indicating that cellular uptake of NPs does not occur only via any actin-based mechanism (Geiser et al. 2005). The same particles have been used in another study with red blood cells since they do not have any phagocytic receptors on their surface nor the necessary intracellular structures for phagocytosis or endocytosis. Polystyrene particles of 1 μm size were found attached to the surface of the red blood cells, whereas 78 nm NPs could be detected inside the cells; thus, they must have entered the cells by nonphagocytic or nonendocytic mechanisms (Rothen-Rutishauser et al. 2006). Nonphagocytotic uptake was also demonstrated for virosomes with different surface charges and a diameter of 120–180 nm (Hofer et al. 2009).

Surface charge–dependent uptake was described in Madin–Darby canine kidney type II cells. Both cationic and anionic NPs were taken up by clathrin-mediated endocytosis, and a small fraction was observed to be taken up by macropinocytosis. It was shown that cationic NPs were found to be promising drug carriers because of their better penetration through cells and their attraction to the negatively charged plasma membrane of the cells

(Harush-Frenkel et al. 2008); however, another hypothesis is that the cationic NPs aggregated more in the culture medium than the others, and therefore more NPs reached the cell surface.

When testing A549 lung epithelial cells and the uptake of dendrimers (nanostructured polymeric biomaterials), it was again shown that the surface charge of dendrimers was crucial for the different uptake pathways. Anionic dendrimers were found to commonly be internalized via caveolin-mediated endocytosis, while the neutral and cationic dendrimers were taken up by clathrin- and caveolin-independent endocytosis (Perumal et al. 2008).

Particles smaller than 200 nm were found to be predominantly taken up by clathrin-mediated endocytosis in the murine melanoma cell line B16-F10, while larger particles (500 nm) were internalized by caveolin-mediated endocytosis (Rejman et al. 2004). Clathrin-mediated uptake was also observed for fluorescein isothiocyanate chitosan NPs (200 nm) into A549 cell (Huang et al. 2002).

As previously mentioned, the uptake mechanism for one and the same NP into different cell types can also vary. For instance, the particle uptake of fetal bovine serum–treated titanium dioxide NPs into A549 cells and H1299, two human lung cell lines, is different, and it has been shown that in A549 cells, uptake of titanium dioxide NPs involves a clathrin-dependent pathway, whereas in H1299 a caveolae- and clathrin-independent pathway was observed (Tedja et al. 2012). By using specific inhibitors, it could also be shown that D-penicillamine-coated quantum dots of 4 nm radius were predominantly internalized by clathrin-mediated endocytosis and to a smaller extent by macropinocytosis (Jiang et al. 2010). A quantitative analysis by stereology on TEM demonstrated that clathrin- as well as caveolin-mediated endocytosis are the main uptake mechanism for PEGylated gold–NPs (15 nm) into A549 cells, whereas citrate-stabilized gold–NPs were mainly taken up by macropinocytosis (Brandenberger et al. 2010c), which might again been explained by the different agglomeration behaviors of the two different surfaces.

Despite the experimental differences, all of the previously presented endocytic pathways have at least one aspect in common: the internalized particle is ultimately located in an intracellular vesicle. However, there are studies that have reported the intracellular localization of NPs of different materials to not be membrane bound, thus indicating alternative pathways for particles to enter cells (Geiser et al. 2005; Lesniak et al. 2005; Mu et al. 2012; Rothen-Rutishauser et al. 2006). Besides other possible mechanisms, passive diffusion through membrane pores and passive uptake by van der Waals or steric interactions (subsumed as adhesive interactions) (Rimai et al. 2000) are suggested by the authors of these studies. However, it remains to be determined which chemical and physical properties of the cellular membrane and particles are responsible for the translocation of NPs into cells, the nucleus, and organelles both *in vitro* and *in vivo*.

9.14 Intracellular Trafficking of NPs

For the design of new NPs as potential pharmaceutical applications (i.e., vehicles) for targeted drug delivery, or to understand their toxic potential of any NPs, it is imperative to understand not only the cellular entry mechanisms but also the intracellular localization and trafficking. As already stated earlier, the method of choice for the detection of NPs depends on the characteristics of the NPs (fluorescence, size, and structure or electron density) and on the cellular structure of interest. In brief, fluorescent microscopy or LSM allows real-time analysis or colocalization studies with fluorescently labeled structures (Lehmann et al. 2010; Rothen-Rutishauser et al. 2011). For electron-dense NPs, TEM enables the highest resolution to be achieved as well as allows the quantification of NPs within subcellular structures (Muhlfeld et al. 2007).

In an initial project, we have used iron–platinum NPs with a polymer shell and that were fluorescently labeled (Lehmann et al. 2010). A gold core coated with the identical polymer shell (gold NPs) was also included. Utilizing laser scanning and electron microscopy techniques, we observed that the iron–platinum NPs penetrated all three cell types investigated, but to a higher extent in monocyte-derived macrophages and dendritic cells than epithelial cells. Colocalization was detected by LSM in lysosomes, but not for mitochondria or the cell nuclei. By comparing different NP types with the mere polymer shell, we observed that NPs, combined with these shells, were responsible for triggering pro-inflammatory effects, but not the shell alone. On the basis of these results, it was concluded that the uptake and pro-inflammatory response generated by NPs depended on incubation time, cell type, and cell culture conditions.

In many studies in which NP uptake has been studied by TEM, the particles have been found as aggregates in vesicles (Limbach et al. 2005; Stearns et al. 2001; Takenaka et al. 2006, 2012). However, in all these studies, only conventional TEM was used to detect the particles. It is possible that single NPs or small aggregates of a few particles could not be identified. Sophisticated microscopic methods are needed to detect and localize different NP types inside cells or tissues (Brandenberger et al. 2010b; Muhlfeld et al. 2007). One example was the application of electron energy loss spectroscopy to identify titanium dioxide– or silver-enhanced gold NPs inside cells where the particles were identified as membrane-bound larger aggregates and as free smaller aggregates, or individual particles in the cytoplasm (Rothen-Rutishauser et al. 2007a).

NP penetration *in vitro* into cells does not exclusively occur by any of the expected endocytic processes, but also by other yet unknown mechanisms that had been called adhesive interactions, as, for instance, electrostatic, van der Waals, or steric interactions (Geiser et al. 2005; Rimai et al. 2000; Rothen-Rutishauser 2007b). Particles within cells have been found not to be membrane bound (Geiser et al. 2005; Kapp et al. 2004; Rothen-Rutishauser et al. 2007a). Hence, they have direct access to cytoplasmic proteins, important biochemical molecules in organelles such as the respiratory chain in the mitochondria and the DNA in the nucleus, which may greatly enhance their hazardous potential. It is generally accepted that besides the cell membrane, the mitochondria and cell nucleus are the major cell compartments relevant for possible NP-induced toxicity (Unfried et al. 2007). One study previously showed that carbon-based NPs, such as the Buckminster fullerene (C_{60}), can enter macrophages and are distributed within the cytoplasm, lysosomes and the nucleus (Porter et al. 2007). Other *in vitro* experiments revealed penetration of NPs into mitochondria of macrophages and epithelial cells, associated with oxidative stress and mitochondrial damage (Li et al. 2003). This penetration of NPs led to a loss of cristae in the mitochondria. The inner mitochondrial membrane, the structure carrying the respiratory chain, which converts molecular oxygen into energy stored as ATP, was destroyed by NPs. Regardless of the mechanisms causing this mitochondrial damage, there is evidence that it is the organic substances attached to the particle that are responsible for this (Xia et al. 2004).

Penetration of particles into the nucleus has been shown in a number of *in vivo* studies for inhaled titanium dioxide NPs in rats (Geiser et al. 2005; Gurr et al. 2005) and *in vitro* studies in different cell types for nanowires (Liu et al. 2003), gold NPs (Rothen-Rutishauser et al. 2007a), polystyrene NPs (Rothen-Rutishauser et al. 2009), and C_{60} (Porter et al. 2007).

Tsoli and colleagues (2005) have shown that Au_{55} clusters interact with DNA of different cell lines in a way that may be the reason for the strong toxicity of these tiny 1.4-nm particles. For quantum dots, it was shown that the smaller they are, the greater the toxicity, and this was suggested to be because the smaller quantum dots could penetrate the nuclear envelope (Lovric et al. 2005).

To avoid aggregation of NPs, which might provide a false-positive result, different approaches have been used recently. One is the *in vivo* inhalation of low doses of 5-nm gold NPs as small nanosized and not agglomerated particles (Takenaka et al. 2012) and the other one is the air–liquid exposure of nebulized single gold NPs at different (non-overload) doses over the surface of 3D lung cell cultures (Brandenberger et al. 2010a,c). Analysis in both studies was performed by TEM combined with stereological approaches, and showed that most of the gold NPs were localized inside vesicles and agglomeration of NPs within vesicles increased with time in both studies. In the latter *in vitro* study, other cellular compartments were also analyzed (Brandenberger et al. 2010c) and no particles were found in the nucleus, mitochondria, or Golgi/ER. In addition, it could be shown that the cytosol is not the preferred compartment for gold NPs with different surface coatings; however, significantly more PEGylated gold NPs than citrate-stabilized NPs were present there.

9.15 NP-Induced Cell Responses

Once inside the cells, NPs may cause several biological responses including the generation of reactive oxygen species (Gonzalez-Flecha 2004), the enhanced expression of proinflammatory cytokines (Muller et al. 2005), and DNA strand breaks (Vinzents et al. 2005). However, the precise mechanism of possible NP toxicology is still not fully understood (Clift et al. 2011). Currently, the hypothesis that NPs induce adverse cellular effects via oxidative means (oxidative stress paradigm) (Donaldson et al. 2003) is used as a basis for many NP-based investigations. Recently, additional paradigms have been suggested for nanomaterials, such as the fiber paradigm (Donaldson and Tran 2004) and the theory of genotoxicity (Schins and Knaapen 2007). In addition, some important aspects about the use of reliable methods and realistic test conditions to study possible risks of NPs have recently been reviewed in several publications. For instance, the in-depth characterization of the nanomaterial, the use of suspension versus air–liquid exposures, the use of realistic doses, and the validity of the selected test methods have been highlighted (Krug and Wick 2011; Paur et al. 2011).

9.16 Conclusions

There is convincing evidence that NPs are taken up by any cell, with most studies suggesting an endocytotic uptake route, which might determine their intracellular trafficking and fate. NPs, however, are not limited to such active uptake processes, as they have also been seen to be located free within the cytosol, thus indicating passive entry into the cell. Independent of the uptake mechanism, the interactions of NPs with living matter and quantification of NP uptake by cells has to be performed under controlled experimental conditions and involve cutting-edge microscopy techniques combined with stereological analysis. By using these new and promising approaches, further research must be performed to investigate and specify which NP properties influence their uptake and intracellular trafficking. If we can provide a substantial understanding of NPs at this biointerface, we will have the possibility to conceive and develop new NPs to realize the advantageous applications posed by nanotechnology (e.g., biotechnology and medicine).

Acknowledgments

Part of the work outlined above was supported by the Swiss National Science Foundation (PP00P2_123373, 320030_138365/1), the National Research Program 64, the German Research Foundation (SPP1313), the Federal Office of Public Health (Switzerland), the Swiss Nanoscience Institute (SNI) within the National Center of Research (NCCR) in Nanoscale Science, and the Adolphe Merkle Foundation.

References

Aderem A, and Underhill D M 1999 Mechanisms of phagocytosis in macrophages; *Annu. Rev. Immunol.* 17 593–623.

Alberts B, Bray D, Johnson A, Lewis J, Raff M, Roberts K, and Walter P 1998 *Essential Cell Biology. An Introduction to the Molecular Biology of the Cell* (Garland Publishing, Inc., New York).

Alivisatos A P, Gu W, and Larabell C 2005 Quantum dots as cellular probes; *Annu. Rev. Biomed. Eng.* 7 55–76.

Anderson R G, and Jacobson K 2002 A role for lipid shells in targeting proteins to caveolae, rafts, and other lipid domains; *Science (New York, N. Y.)* 296 1821–1825.

Axline S G, and Reaven E P 1974 Inhibition of phagocytosis and plasma membrane mobility of the cultivated macrophage by cytochalasin B. Role of subplasmalemmal microfilaments; *J. Cell Biol.* 62 647–659.

Barrow E, Nicola A V, and Liu J 2013 Multiscale perspectives of virus entry via endocytosis; *Virol. J.* 10 177.

Brandenberger C, Clift M J, Vanhecke D, Muhlfeld C, Stone V, Gehr P, and Rothen-Rutishauser B 2010b Intracellular imaging of nanoparticles: Is it an elemental mistake to believe what you see?; *Part. Fibre Toxicol.* 7 15.

Brandenberger C, Muhlfeld C, Ali Z, Lenz A G, Schmid O, Parak W J, Gehr P, and Rothen-Rutishauser B 2010c Quantitative evaluation of cellular uptake and trafficking of plain and polyethylene glycol-coated gold nanoparticles; *Small* 6 1669–1678.

Brandenberger C, Rothen-Rutishauser B, Mühlfeld C, Schmid O, Ferron G A, Maier K L, Gehr P, and Lenz A G 2010a Effects and uptake of gold nanoparticles deposited at the air–liquid interface of a human epithelial airway model; *Toxicol. Appl. Pharmacol.* 242 56–65.

Brodsky F M, Chen C Y, Knuehl C, Towler M C, and Wakeham D E 2001 Biological basket weaving: Formation and function of clathrin-coated vesicles; *Annu. Rev. Cell Dev. Biol.* 17 517–568.

Burgess T L, and Kelly R B 1987 Constitutive and regulated secretion of proteins; *Annu. Rev. Cell Biol.* 3 243–293.

Chastellain M, Petri A, and Hofmann H 2004 Particle size investigations of a multistep synthesis of PVA coated superparamagnetic nanoparticles; *J. Colloid Interface Sci.* 278 353–360.
Clift M J, Gehr P, and Rothen-Rutishauser B 2011 Nanotoxicology: A perspective and discussion of whether or not *in vitro* testing is a valid alternative; *Arch. Toxicol.* 85 723–731.
Conner S D, and Schmid S L 2003 Regulated portals of entry into the cell; *Nature* 422 37–44.
Cooper J A 1987 Effects of cytochalasin and phalloidin on actin; *J. Cell Biol.* 105 1473–1478.
Coulter J A, Jain S, Butterworth K T, Taggart L E, Dickson G R, McMahon S J, Hyland W B, Muir M F, Trainor C, Hounsell A R, O'Sullivan J M, Schettino G, Currell F J, Hirst D G, and Prise K M 2012 Cell type-dependent uptake, localization, and cytotoxicity of 1.9 nm gold nanoparticles; *Int. J. Nanomed.* 7 2673–2685.
Donaldson K, Stone V, Borm P J, Jimenez L A, Gilmour P S, Schins R P, Knaapen A M, Rahman I, Faux S P, Brown D M, and MacNee W 2003 Oxidative stress and calcium signaling in the adverse effects of environmental particles (PM10); *Free Radic. Biol. Med.* 34 1369–1382.
Donaldson K, and Tran C L 2004 An introduction to the short-term toxicology of respirable industrial fibres; *Mutat. Res.* 553 5–9.
Edidin M 2001 Membrane cholesterol, protein phosphorylation, and lipid rafts; *Sci. STKE* 2001 E1.
Eisenberg D, Schwarz E, Komaromy M, and Wall R 1984 Analysis of membrane and surface protein sequences with the hydrophobic moment plot; *J. Mol. Biol.* 179 125–142.
Fazlollahi F, Kim Y H, Sipos A, Hamm-Alvarez S F, Borok Z, Kim K J, and Crandall E D 2013 Nanoparticle translocation across mouse alveolar epithelial cell monolayers: Species-specific mechanisms; *Nanomedicine* 9 786–794.
Gasser M, Wick P, Clift M J, Blank F, Diener L, Yan B, Gehr P, Krug H F, and Rothen-Rutishauser B 2012 Pulmonary surfactant coating of multi-walled carbon nanotubes (MWCNTs) influences their oxidative and pro-inflammatory potential *in vitro*; *Part. Fibre Toxicol.* 9 17.
Gehr P, Schürch S, Berthiaume Y, Im Hof V, and Geiser M 1990 Particle retention in airways by surfactant; *J. Aerosol Med.* 3 27–43.
Geiser M, Rothen-Rutishauser B, Kapp N, Schurch S, Kreyling W, Schulz H, Semmler M, Im Hof V, Heyder J, and Gehr P 2005 Ultrafine particles cross cellular membranes by nonphagocytic mechanisms in lungs and in cultured cells; *Environ. Health Perspect.* 113 1555–1560.
Gonzalez-Flecha B 2004 Oxidant mechanisms in response to ambient air particles; *Mol. Aspects Med.* 25 169–182.
Gurr J R, Wang A S, Chen C H, and Jan K Y 2005 Ultrafine titanium dioxide particles in the absence of photoactivation can induce oxidative damage to human bronchial epithelial cells; *Toxicology* 213 66–73.
Harush-Frenkel O, Rozentur E, Benita S, and Altschuler Y 2008 Surface charge of nanoparticles determines their endocytic and transcytotic pathway in polarized MDCK cells; *Biomacromolecules* 9 435–443.
Heyder J, Gebhart J, Rudolf G, Schiller C F, and Stahlhofen W 1986 Deposition of particles in the human respiratory tract in the size range 0.005–15 μm; *J. Aerosol Sci.* 17 811–825.
Hofer U, Lehmann A D, Waelti E, Amacker M, Gehr P, and Rothen-Rutishauser B 2009 Virosomes can enter cells by non-phagocytic mechanisms; *J. Liposome Res.* 19 301–309.
Holt P G, and Stumbles P A 2000 Characterization of dendritic cell populations in the respiratory tract; *J. Aerosol Med.* 13 361–367.
Huang M, Ma Z, Khor E, and Lim L Y 2002 Uptake of FITC–chitosan nanoparticles by A549 cells; *Pharm. Res.* 19 1488–1494.
Ikehata M, Yumoto R, Nakamura K, Nagai J, and Takano M 2008 Comparison of albumin uptake in rat alveolar type II and type I-like epithelial cells in primary culture; *Pharm. Res.* 25 913–922.
ISO Report 2008 ISO/TS 27687.
Ivanov A I 2008 *Exocytosis and Endocytosis* (Humana Press, Totowa, NJ).
Jiang X, Rocker C, Hafner M, Brandholt S, Dorlich R M, and Nienhaus G U 2010 Endo- and exocytosis of zwitterionic quantum dot nanoparticles by live HeLa cells; *ACS Nano* 4 6787–6797.
Jud C, Clift M J, Petri-Fink A, and Rothen-Rutishauser B 2013 Nanomaterials and the human lung: What is known and what must be deciphered to realise their potential advantages?; *Swiss Med. Wkly.* 143 w13758.

Kapp N, Kreyling W, Schulz H, Im Hof V, Gehr P, Semmler M, and Geiser M 2004 Electron energy loss spectroscopy for analysis of inhaled ultrafine particles in rat lungs; *Microsc. Res. Tech.* 63 298–305.

Kendall M 2007 Fine airborne urban particles (PM2.5) sequester lung surfactant and amino acids from human lung lavage; *Am. J. Physiol. Lung Cell. Mol. Physiol.* 293 L1053–L1058.

Kilburn K H 1968 A hypothesis for pulmonary clearance and its implications; *Am. Rev. Respir. Dis.* 98 449–463.

Kirchhausen T 2000 Clathrin; *Annu. Rev. Biochem.* 69 699–727.

Krug H F, and Wick P 2011 Nanotoxicology: An interdisciplinary challenge; *Angew. Chem. Int. Ed. Engl.* 50 1260–1278.

Lehmann A D, Parak W J, Zhang F, Zulqurnain A, Röcker C, Nienhaus G U, Gehr P, and Rothen-Rutishauser B 2010 Fluorescent-magnetic hybrid nanoparticles induce a dose-dependent increase of the pro-inflammatory response in lung cells *in vitro* correlated with intracellular localization; *Small* 6 753–762.

Lesniak W, Bielinska A U, Sun K, Janczak K W, Shi X, Baker J R Jr., and Balogh L P 2005 Silver/dendrimer nanocomposites as biomarkers: Fabrication, characterization, *in vitro* toxicity, and intracellular detection; *Nano Lett.* 5 2123–2130.

Li N, Sioutas C, Cho A, Schmitz D, Misra C, Sempf J, Wang M, Oberley T, Froines J, and Nel A 2003 Ultrafine particulate pollutants induce oxidative stress and mitochondrial damage; *Environ. Health Perspect.* 111 455–460.

Limbach L K, Li Y, Grass R N, Brunner T J, Hintermann M A, Muller M, Gunther D, and Stark W J 2005 Oxide nanoparticle uptake in human lung fibroblasts: Effects of particle size, agglomeration, and diffusion at low concentrations; *Environ. Sci. Technol.* 39 9370–9376.

Lin C A, Yang T Y, Lee C H, Huang S H, Sperling R A, Zanella M, Li J K, Shen J L, Wang H H, Yeh H I, Parak W J, and Chang W H 2009 Synthesis, characterization, and bioconjugation of fluorescent gold nanoclusters toward biological labeling applications; *ACS Nano* 3 395–401.

Liu Y, Meyer-Zaika W, Franzka S, Schmid G, Tsoli M, and Kuhn H 2003 Gold-cluster degradation by the transition of B-DNA into A-DNA and the formation of nanowires; *Angew. Chem. Int. Ed. Engl.* 42 2853–2857.

Liu Y, Steiniger S C, Kim Y, Kaufmann G F, Felding-Habermann B, and Janda K D 2007 Mechanistic studies of a peptidic GRP78 ligand for cancer cell-specific drug delivery; *Mol. Pharm.* 4 435–447.

Lohmann-Matthes M L, Steinmuller C, and Franke-Ullmann G 1994 Pulmonary macrophages; *Eur. Respir. J.* 7 1678–1689.

Lovric J, Bazzi H S, Cuie Y, Fortin G R, Winnik F M, and Maysinger D 2005 Differences in subcellular distribution and toxicity of green and red emitting CdTe quantum dots; *J. Mol. Med. (Berl.)* 83 377–385.

Lynch I, Cedervall T, Lundqvist M, Cabaleiro-Lago C, Linse S, and Dawson K A 2007 The nanoparticle–protein complex as a biological entity; a complex fluids and surface science challenge for the 21st century; *Adv. Colloid Interface Sci.* 134–135 167–174.

Mahmoudi M, Lynch I, Ejtehadi M R, Monopoli M P, Bombelli F B, and Laurent S 2011 Protein–nanoparticle interactions: Opportunities and challenges; *Chem. Rev.* 111 5610–5637.

Mahmoudi M, Saeedi-Eslami S N, Shokrgozar M A, Azadmanesh K, Hassanlou M, Kalhor H R, Burtea C, Rothen-Rutishauser B, Laurent S, Sheibani S, and Vali H 2012 Cell "vision": Complementary factor of protein corona in nanotoxicology; *Nanoscale* 4 5461–5468.

Meiring J J, Borm P J, Bagate K, Semmler M, Seitz J, Takenaka S, and Kreyling W G 2005 The influence of hydrogen peroxide and histamine on lung permeability and translocation of iridium nanoparticles in the isolated perfused rat lung; *Part. Fibre Toxicol.* 2 3.

Mellman I, and Steinman R M 2001 Dendritic cells: Specialized and regulated antigen processing machines; *Cell* 106 255–258.

Michiels C 2003 Endothelial cell functions; *J. Cell Physiol.* 196 430–443.

Mu Q, Hondow N S, Ski L, Brown A P, Jeuken L J, and Routledge M N 2012 Mechanism of cellular uptake of genotoxic silica nanoparticles; *Part. Fibre Toxicol.* 9 29.

Muhlfeld C, Gehr P, and Rothen-Rutishauser B 2008a Translocation and cellular entering mechanisms of nanoparticles in the respiratory tract; *Swiss Med. Wkly.* 138 387–391.

Muhlfeld C, Rothen-Rutishauser B, Blank F, Vanhecke D, Ochs M, and Gehr P 2008b Interactions of nanoparticles with pulmonary structures and cellular responses; *Am. J. Physiol. Lung Cell. Mol. Physiol.* 294 L817–L829.

Muhlfeld C, Rothen-Rutishauser B, Vanhecke D, Blank F, Gehr P, and Ochs M 2007 Visualization and quantitative analysis of nanoparticles in the respiratory tract by transmission electron microscopy; *Part. Fibre Toxicol.* 4 11.

Muller J, Huaux F, Moreau N, Misson P, Heilier J F, Delos M, Arras M, Fonseca A, Nagy J B, and Lison D 2005 Respiratory toxicity of multi-wall carbon nanotubes; *Toxicol. Appl. Pharmacol.* 207 221–231.

Oberdorster G, Oberdorster E, and Oberdorster J 2005 Nanotoxicology: An emerging discipline evolving from studies of ultrafine particles; *Environ. Health Perspect.* 113 823–839.

Ochs M, and Weibel E 2008 Functional design of the human lung for gas exchange; in *Fishman's Pulmonary Diseases and Disorders* (eds) Fishman A P, Elias J A, Fishman J A, Grippi M A, Senior R M, and Pack A (McGraw Hill, New York).

Patton J S, and Byron P R 2007 Inhaling medicines: Delivering drugs to the body through the lungs; *Nat. Rev. Drug Discov.* 6 67–74.

Paur H R, Cassee F R, Teeguarden J G, Fissan H, Diabate S, Aufderheide M, Kreyling W, Hänninen O, Kasper G, Riediker M, Rothen-Rutishauser B, and Schmid O 2011 In-vitro cell exposure studies for the assessment of nanoparticle toxicity in the lung—A dialog between aerosol science and biology; *J. Aerosol Sci.* 42 668–692.

Perumal O P, Inapagolla R, Kannan S, and Kannan R M 2008 The effect of surface functionality on cellular trafficking of dendrimers; *Biomaterials* 29 3469–3476.

Peters-Golden M 2004 The alveolar macrophage: The forgotten cell in asthma; *Am. J. Respir. Cell Mol. Biol.* 31 3–7.

Plummer E M, and Manchester M 2013 Endocytic uptake pathways utilized by CPMV nanoparticles; *Mol. Pharm.* 10 26–32.

Porter A E, Gass M, Muller K, Skepper J N, Midgley P, and Welland M 2007 Visualizing the uptake of C60 to the cytoplasm and nucleus of human monocyte-derived macrophage cells using energy-filtered transmission electron microscopy and electron tomography; *Environ. Sci. Technol.* 41 3012–3017.

Puri V, Watanabe R, Singh R D, Dominguez M, Brown J C, Wheatley C L, Marks D L, and Pagano R E 2001 Clathrin-dependent and -independent internalization of plasma membrane sphingolipids initiates two Golgi targeting pathways; *J. Cell Biol.* 154 535–547.

Raemy D O, Limbach L K, Rothen-Rutishauser B, Grass R N, Gehr P, Birbaum K, Brandenberger C, Gunther D, and Stark W J 2010 Cerium oxide nanoparticle uptake kinetics from the gas-phase into lung cells *in vitro* is transport limited; *Eur. J. Pharm. Biopharm.* 77 368–375.

Rejman J, Oberle V, Zuhorn I S, and Hoekstra D 2004 Size-dependent internalization of particles via the pathways of clathrin- and caveolae-mediated endocytosis; *Biochem. J.* 377 159–169.

Rimai D S, Quesnel D J, and Busnaia A A 2000 The adhesion of dry particles in the nanometer to micrometer size range; *Colloids Surf. A. Physicochem. Eng. Aspects* 165 3–10.

Ritchie K, and Spector J 2007 Single molecule studies of molecular diffusion in cellular membranes: Determining membrane structure; *Biopolymers* 87 95–101.

Rodriguez-Lorenzo L, Fytianos K, Blank F, von Garnier Ch, Rothen-Rutishauser B, and Petri-Fink A 2014 Fluorescence-encoded gold nanoparticles: Library design and modulation of cellular uptake into dendritic cells; *Small* 10 1341–50.

Rothberg K G, Heuser J E, Donzell W C, Ying Y S, Glenney J R, and Anderson R G 1992 Caveolin, a protein component of caveolae membrane coats; *Cell* 68 673–682.

Rothen-Rutishauser B, Blank F, Mühlfeld C, and Gehr P 2009 Nanoparticle–cell membrane interactions. In *Particle–Lung Interactions* (eds) Gehr P, Mühlfeld C, Rothen-Rutishauser B, and Blank F (Informa Healthcare, New York), pp. 226–242.

Rothen-Rutishauser B, Blank F, Petri-Fink A, Clift M J D, Geiser T, and von G C 2011 Intracellular localisation of fluorescently labelled nanoparticles; *G. I. T. Imaging Microsc.* 1 30–33.

Rothen-Rutishauser B, Clift M J D, Jud C, Fink A, and Wick P 2012 Human epithelial cells *in vitro*—Are they an advantageous tool to help understand the nanomaterial–biological barrier interaction?; *Euro Nanotox Letters* 1 1–20.

Rothen-Rutishauser B, Kuhn D A, Ali Z, Gasser M, Amin F, Parak W J, Vanhecke D, Fink A, Gehr P, and Brandenberger C 2014 Quantification of gold nanoparticle cell uptake under controlled biological conditions and adequate resolution; *Nanomedicine (Lond.)* 9 607–621.

Rothen-Rutishauser B, Muhlfeld C, Blank F, Musso C, and Gehr P 2007a Translocation of particles and inflammatory responses after exposure to fine particles and nanoparticles in an epithelial airway model; *Part. Fibre Toxicol.* 4 9.

Rothen-Rutishauser B, Schurch S, and Gehr P 2007b Interaction of particles with membranes; in *The Toxicology of Particles* (eds) Donaldson K, and Borm P (Taylor & Francis Group, LLC, Boca Raton, FL) pp. 139–160.

Rothen-Rutishauser B M, Schurch S, Haenni B, Kapp N, and Gehr P 2006 Interaction of fine particles and nanoparticles with red blood cells visualized with advanced microscopic techniques; *Environ. Sci. Technol.* 40 4353–4359.

Ruge C A, Kirch J, Canadas O, Schneider M, Perez-Gil J, Schaefer U F, Casals C, and Lehr C M 2011 Uptake of nanoparticles by alveolar macrophages is triggered by surfactant protein A; *Nanomedicine* 7 690–693.

Schins R P, and Knaapen A M 2007 Genotoxicity of poorly soluble particles; *Inhal. Toxicol.* 19 Suppl 1 189–198.

Schleh C, Kreyling W G, and Lehr C M 2013 Pulmonary surfactant is indispensable in order to simulate the *in vivo* situation; *Part. Fibre Toxicol.* 10 6.

Schmid S L 1997 Clathrin-coated vesicle formation and protein sorting: An integrated process; *Annu. Rev. Biochem.* 66 511–548.

Schneeberger E E, and Lynch R D 1984 Tight junctions. Their structure, composition, and function; *Circ. Res.* 55 723–733.

Simons K, and Ikonen E 1997 Functional rafts in cell membranes; *Nature* 387 569–572.

Singer S J, and Nicolson G L 1972 The fluid mosaic model of the structure of cell membranes; *Science* 175 720–731.

Sonnino S, and Prinetti A 2013 Membrane domains and the "lipid raft" concept; *Curr. Med. Chem.* 20 4–21.

Stark W J 2011 Nanoparticles in biological systems; *Angew. Chem. Int. Ed. Engl.* 50 1242–1258.

Stearns R C, Paulauskis J D, and Godleski J J 2001 Endocytosis of ultrafine particles by A549 cells; *Am. J. Respir. Cell Mol. Biol.* 24 108–115.

Takenaka S, Karg E, Kreyling W G, Lentner B, Moller W, Behnke-Semmler M, Jennen L, Walch A, Michalke B, Schramel P, Heyder J, and Schulz H 2006 Distribution pattern of inhaled ultrafine gold particles in the rat lung; *Inhal. Toxicol.* 18 733–740.

Takenaka S, Moller W, Semmler-Behnke M, Karg E, Wenk A, Schmid O, Stoeger T, Jennen L, Aichler M, Walch A, Pokhrel S, Madler L, Eickelberg O, and Kreyling W G 2012 Efficient internalization and intracellular translocation of inhaled gold nanoparticles in rat alveolar macrophages; *Nanomedicine (Lond.)* 7 855–865.

Tedja R, Lim M, Amal R, and Marquis C 2012 Effects of serum adsorption on cellular uptake profile and consequent impact of titanium dioxide nanoparticles on human lung cell lines; *ACS Nano* 6 4083–4093.

Teeguarden J G, Hinderliter P M, Orr G, Thrall B D, and Pounds J G 2007 Particokinetics *in vitro*: Dosimetry considerations for *in vitro* nanoparticle toxicity assessments; *Toxicol. Sci.* 95 300–312.

The University of Arizona, Department of Biochemistry and Molecular Biophysics 1997 *Prokaryotes, Eukaryotes, & Viruses Tutorial.* http://www.biology.arizona.edu/cell_bio/tutorials/pev/main.html.

Tsoli M, Kuhn H, Brandau W, Esche H, and Schmid G 2005 Cellular uptake and toxicity of Au55 clusters; *Small* 1 841–844.

Unfried K, Albrecht C, Klotz L O, von Mikecz A, Grether-Beck S, and Schins R P 2007 Cellular responses to nanoparticles: Target structures and mechanisms; *Nanotoxicology* 1 1–20.

van der Voort H T M, and Strasters K C 1995 Restoration of confocal images for quantitative image analysis.; *J. Microsc.* 178 165–181.

Verma A, and Stellacci F 2010 Effect of surface properties on nanoparticle–cell interactions; *Small* 6 12–21.

Vermaelen K, and Pauwels R 2005 Pulmonary dendritic cells; *Am. J. Respir. Crit. Care Med.* 172 530–551.

Vinzents P S, Moller P, Sorensen M, Knudsen L E, Hertel O, Jensen F P, Schibye B, and Loft S 2005 Personal exposure to ultrafine particles and oxidative DNA damage; *Environ. Health Perspect.* 113 1485–1490.

Warren G, and Wickner W 1996 Organelle inheritance; *Cell* 84 395–400.

Wikstrom M E, and Stumbles P A 2007 Mouse respiratory tract dendritic cell subsets and the immunological fate of inhaled antigens; *Immunol. Cell Biol.* 85 182–188.

Xia T, Korge P, Weiss J N, Li N, Venkatesen M I, Sioutas C, and Nel A 2004 Quinones and aromatic chemical compounds in particulate matter induce mitochondrial dysfunction: Implications for ultrafine particle toxicity; *Environ. Health Perspect.* 112 1347–1358.

Section IV

Translocation

10

Translocation across the Air–Blood Tissue Barrier

Fabian Blank, Christophe von Garnier, Peter Gehr, and Barbara Rothen-Rutishauser

CONTENTS

10.1 Introduction

Through its enormous surface, the lung provides a vast interface for exposure with inhaled particulates (Gehr et al. 1978). On the one hand, the interaction of nanosized particles with the human body, in particular with the respiratory tract as their main portal of entry, has become a fundamental topic in the field of environmental/epidemiologic monitoring where inhalational exposure to ambient, combustion-derived particles is being increasingly recognized as an important cause of excess cardiovascular morbidity and mortality in areas with air pollution (Mills et al. 2009; Oberdorster 2001; Wichmann et al. 2000). On the other hand, in the area of innovative biomedical applications, the potential benefits of manufactured nanoparticles (NPs) are increasingly being recognized. In particular, nanosized carriers have been proposed as promising novel diagnostic, therapeutic, and vaccination approaches for a variety of human diseases (Choi and Frangioni 2010; Foged et al. 2005; Nembrini et al. 2011; Zrazhevskiy et al. 2010). In particular, recent advances in both drug formulation and inhalation device design are creating new opportunities for inhaled drug and vaccine delivery as an alternative to oral or parenteral delivery methods (Tonnis et al. 2012). For example, in contrast to subcutaneous or intramuscular injection, pulmonary drug or vaccine delivery does not face drawbacks such as limited acceptance (needle phobia) and transmission of diseases by needle stick injury, and does not require trained health-care workers (Giudice and Campbell 2006; Kersten and Hirschberg 2004). Furthermore, pulmonary vaccine delivery is expected to be more efficient: the respiratory tract is continuously exposed to large amounts of inhaled and deposited airborne antigen, ranging from innocuous substances (e.g., house dust mite allergen) to potentially detrimental pathogens (e.g., bacteria, virus), which require an immune barrier consisting of several populations of immune cells (macrophages, dendritic cells [DCs], B cells, monocytes) in close proximity to the epithelium for protection (von Garnier et al. 2007). Those key players of pulmonary immunity provide promising targets for modulation by vaccines. The major advantages of aerosol delivery over other routes of administration are instant access, the high ratio of the drug deposited within the lungs noninvasively (Gautam et al. 2003); the large absorptive surface area (>140 m^2); and the thin alveolar epithelium (0.1–0.2 μm) (Huang and Wang 2006; Niven 1995), rapid onset of drug absorption, comparatively lower enzymatic activity, and avoidance of the hepatic first-pass effect (Misra 2010). For assessing the potential risks and benefits of inhaled (nano)particles to human health, and to optimize the application of drugs or vaccines via the pulmonary route, two primary steps within the characterization of NPs are dosimetry and biokinetics (Kreyling et al. 2012). Dosimetry provides the essential base for the toxicological evaluation of any kind of NP (Kreyling et al. 2012). Furthermore, evaluation of the right dosimetry is indispensable in the development and characterization of biomedical NPs for diagnostic or therapeutic use. Biokinetics studies should be based on a broad range of administered particle doses. Special attention needs to be directed toward a too high particle burden applied, which can cause significant biological and also toxic responses (e.g., overload of alveolar macrophages leading to neutrophilia in the alveolar lumen, or epithelial damage) and which may affect the biokinetics fate of the NPs (Kreyling et al. 2012). It is therefore essential that in a biokinetics study, a baseline situation showing physiological (healthy) condition is available for comparison. Hence, the amount of administered particles in biokinetics studies should be as low as possible to carefully analyze the translocation to target organs (Kreyling et al. 2012). However, since the focus of this chapter is on the translocation across the air–blood tissue barrier (i.e., in the alveolar region) of NPs, the emphasis of the present text will be put on the biokinetics toward systemic circulation rather than on NP dose metrics. The biokinetics of NPs depend, on the one hand, on particle characteristics such as size, shape, protein binding, agglomeration, hydrophobicity, and surface charge. On the other hand, biokinetics depend on the type of application to the human body and consequently on the interaction (i.e., translocation) of the NPs with the barriers present at the site of application. The crucial feature in the characterization of an NP entering the human respiratory tract is its potential to be interacting with the various parenchymal lung cells and in translocation. NPs may either get trapped within the diversity of biological barriers protecting the respiratory tract from harmful inhaled pathogen, or they may overcome existing barriers and enter the blood circulation. Entering the systemic circulation is a very crucial step in determining the fate of inhaled NPs because it allows particles direct translocation to secondary organs such as the liver, kidney, heart, and brain (Semmler-Behnke et al. 2007). This significant step in NP translocation will be discussed in the present chapter, which will be subdivided into topics covering (i) deposition and persistence in the area close to the blood circulation (i.e., the alveoli), (ii) translocation to the respiratory capillaries, and (iii) persistence in the blood circulation.

10.2 Deposition and Persistence in the Alveolar Region

The retention time of NPs in the lung depends on deposition site and on the interaction of the particles with the inner lung surface. In general, it is short for particles deposited in the conducting airways due to the efficient mucociliary escalator and cough clearance (time span up to 24 h); however, the duration of clearance increases with airway generation number as a consequence of the increased pathway length and decreased mucus transport velocity in distant airways (Kreyling et al. 2012). However, in the alveolar region, particle clearance takes longer because of the absence of a rapid clearance mechanism such as the mucociliary escalator.

After inhalation and deposition of particles in the gas-exchange region, they are immediately retained on the lung epithelium by wetting forces caused by the surfactant. In this process, particles are displaced into the liquid phase below the surfactant film and toward the epithelium. This process occurs regardless of particle shape, surface topography, and surface free energy (Geiser et al. 2003; Schurch et al. 1990). This process has only been demonstrated for microparticles (1–6 μm) thus far; however, the same is expected for NPs since this process becomes even more efficient with decreasing particle size (Kreyling et al. 2012). In the alveolar region, there are three major pathways of particle clearance: (i) uptake by alveolar macrophages and migration from the alveolar region to the ciliated airways for mucociliary transport to the larynx, with subsequent swallowing into the gastrointestinal tract; (ii) uptake by DCs with subsequent transport to the lung-draining lymph nodes (Desch et al. 2011; Jakubzick et al. 2006, 2008; Thornton et al. 2012), or free drainage of particles to the lung-draining lymph nodes (Choi et al. 2010); and (iii) translocation across the epithelium into the blood circulation and subsequent accumulation in secondary organs. While only the first pathway leads to an effective clearance of particulates from the respiratory tract, the latter two are routes for further potential interaction with secondary compartments of the human body: particles entering the pulmonary lymphatic system may affect downstream immune responses, which are triggered in the lung-draining lymph nodes, while entering of the bloodstream may lead to access to and interaction with secondary organs. However, there are also studies reporting persistence of particles in the alveolar space, that is, particles that are accessible to bronchoalveolar lavage (BAL) after exposure. A recent study monitoring deposition and persistence of freshly generated iridium NP aerosols of 20 or 80 nm median diameter has demonstrated that about 80% of the 20-nm NPs and 70% of the 80-nm NPs were no longer accessible to BAL after 24 h and must, therefore, have been translocated in the pulmonary tissue (Semmler et al. 2004; Semmler-Behnke et al. 2007). These fractions, which were calculated relative to the contemporary lung burden, remained unchanged during the next 6 months. Similar results were also found after 24 h when inhaling freshly generated 25-nm elemental carbon and 20-nm TiO_2 NP aerosols in rats (Kreyling et al. 2009). However, similar studies using microparticles (fluorescent 2 μm and radiolabeled 3.5 and 10 μm polystyrene particles) inhaled in rodents demonstrated a longer persistence of particles in the alveolar region compared with NPs: >80% of the lung-retained fraction was collected by exhaustive BAL throughout the entire retention time of 6 months (Lehnert et al. 1989; Oberdorster et al. 1994). Similar results were found in a study with 1-μm particles inhaled in rats using oral–pharyngeal aspiration, where >95% of particles were recovered from the lung after exposure compared with 70%–80% of 20-nm and 89%–95% of 100-nm spheres (Sarlo et al. 2009). Note that the burden of microparticles as a fraction of the initial dose is declining rather rapidly due to continuous alveolar macrophage–mediated clearance toward the ciliated airways and larynx. One percent to 3% of particles are initially cleared per day, which declines exponentially with time such that after about 200 days, >90% of microparticles are cleared (Kreyling 1990; Kreyling et al. 2012). A similar efficiency of clearance during a 6-month period was also found with NPs that were initially not accessible to BAL but shown to be cleared by macrophage-mediated transport (BAL) later on (Semmler et al. 2004; Semmler-Behnke et al. 2007). The most likely mechanism of the observed NP reappearance on the epithelial surface is presumed to be via macrophage-mediated uptake and migration in the interstitium. Therefore, there is evidence for a long-term retention of NPs within the interstitium, unlike microparticles that are contained in alveolar space within alveolar macrophages. However, much of the data on retention of NPs in the alveolar space (like the results shown above) is collected from studies using rodents. Interestingly, a study comparing long-term relocation and accumulation of various dusts (diesel soot, coal dust, and a combination of both) in monkeys and rats has shown that there seems to be no significant difference in the retention of NPs between rodent lungs and simian lungs (Nikula et al. 1997), which, together with models using dogs and according to their specific biokinetics, may serve as more appropriate models for NP biokinetics in humans (Kreyling 1990). In contrast, studies focusing on

(a)

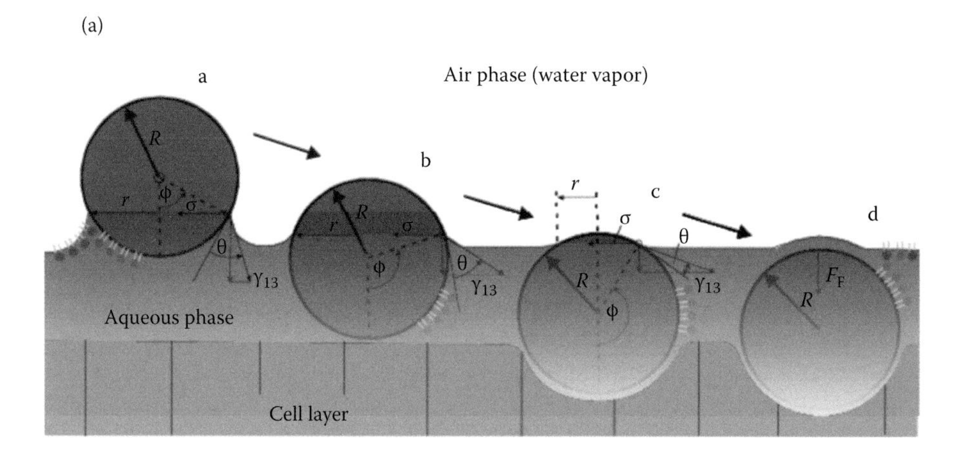

(b)

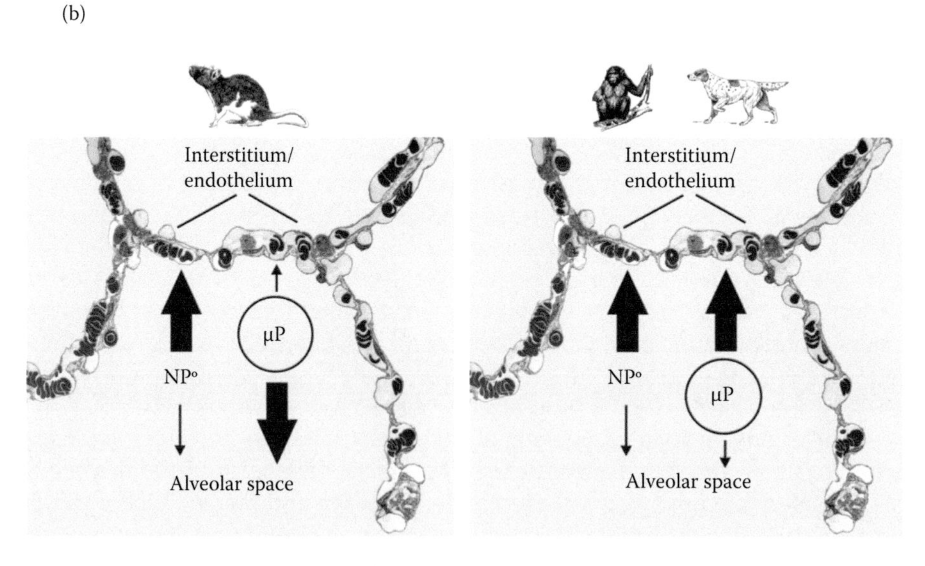

FIGURE 10.1
Particle deposition and translocation in the alveolar region. (a) Particles deposited on the liquid lining layer covering the alveolar epithelium are rapidly displaced into the aqueous phase and toward the epithelial cells owing to wetting forces caused by the surfactant film at the air–liquid interface. Displacement has been experimentally demonstrated for microsized particles only but is expected to affect NPs in the same way. (b) NPs rapidly translocate inside the interstitium after alveolar deposition. This has been shown with rodent (left) and canine and simian models (right). However, microsized particles have been shown to persist longer in the alveolar space in rodent models, while they were also found translocating inside the interstitium in canine and simian models. (Panel b adapted from Kreyling, W.G. et al. *Acc Chem Res*, 46, 714, 2012.)

lung retention and clearance of insoluble, radiolabeled microparticles in dogs have shown that particles did not remain on the epithelium as in rodents but had been relocated into the interstitium with a half-life of 150 days (Ferron et al. 1990; Heyder et al. 2009; Kreyling et al. 1999). In addition, in a similar study using fluorescently labeled polystyrene particles to study the morphological location of microparticles in lung tissues using fluorescence microscopy, an interstitial retention of up to 70% of all lung-retained particles was shown after 6 months (Kreyling et al. 2001). Finally, macrophage-mediated clearance kinetics of microparticles from the human, simian, and canine lungs have been shown to be 10 times slower than clearance in rodents (Kreyling et al. 2012). Deposition and retention of NPs in the alveolar space may be summarized as follows (Figure 10.1).

Pulmonary surfactant allows a rapid displacement of inhaled particles into the liquid phase and toward the alveolar epithelium with subsequent particle–cell interaction (e.g., epithelial cells, alveolar macrophages, DCs) and/or translocation.

NPs rapidly translocate to the interstitium after deposition in the alveolar region. However, depending on the model used, biokinetics of microparticles differ from NPs: while microparticles remain on the alveolar epithelium after deposition in rodents, they translocate in the interstitium of human, simian, and canine lungs.

In both microparticles and NPs, long-term macrophage-mediated clearance seems not to depend on the particle retention site (i.e., on the alveolar epithelium vs. interstitium) of all species studied.

There is a species-specific dynamics in macrophage-mediated particle clearance from the alveolar space, which is 10-fold lower in human, simian, and canine lungs compared with rodent lungs.

10.3 Translocation across the Air–Blood Tissue Barrier

The translocation of inhaled and deposited insoluble NPs into the blood circulation is very controversial and also very much dependent on the species and NPs used for the studies, but also on the experimental approach such as inhalation or installation (Kreyling et al. 2012; Muhlfeld et al. 2008).

One of the first studies focusing on this topic was done in 2002 by Nemmar and colleagues and has shown that particulate antigens <0.1 μm are able to cross the air–blood barrier of the lung, and thus can enter blood circulation (Nemmar et al. 2002). This is, however, the only study that describes a rapid and significant translocation of inhaled carbonaceous NPs to the systemic circulation and other extra pulmonary organs, whereas most other studies only detected a low degree of translocation for iridium (Kreyling et al. 2002) or carbonaceous NPs (Mills et al. 2006; Wiebert et al. 2006). The studies of Nemmar et al. (2002) and Mills et al. (2006) had a very similar design, and Mills et al. have provided a convincing discussion that the strong translocation observed by Nemmar et al. was mainly related to the translocation of soluble pertechnetate that was cleaved from the carbonaceous particles.

A more recent study performed by intratracheal instillation of engineered polyethylene glycol (PEG)-modified gold NPs in rats confirmed the minor translocation into the circulation and accumulation in secondary organs (Lipka et al. 2010). Another inhalation study performed with cerium oxide NPs also showed that the amount of material is highest in the lung tissue, and only small fraction of cerium oxide is systemically available

(Geraets et al. 2012). It is therefore currently accepted that the degree to which inhaled NPs translocate to the circulation is rather small; however, the cumulative effects of this translocation are lacking thus far, and chronic exposure studies need to be done in the future.

After antigens have passed through the epithelial barrier, they may pass through the basement membrane and subsequently through the subepithelial connective tissue layer and eventually come into contact with endothelial cells lining the capillaries. Since endothelial cells play an important role in inflammation processes (Michiels 2003), particles might affect endothelial cell function and viability inducing pro-inflammatory stimuli. It has been proposed that the permeability of the lung tissue barrier to NPs is controlled at the epithelial and endothelial levels (Meiring et al. 2005). Another important issue in the analysis of translocation processes is whether the transport is random or controlled. By applying a newly developed stereological method to compare NP distribution, it could be shown that the distribution of inhaled titanium dioxide NPs in the rat lung is not random, thus pointing at a regulated translocation process rather than a diffusive particle movement (Muhlfeld et al. 2007). However, how and why this process is regulated is currently unknown.

10.4 Persistence in Blood Circulation

Inhaled NPs access the blood circulation by crossing the air–blood barrier (as discussed in Section 10.3) of the respiratory tract. NPs may leave the circulation either independently (and translocate to secondary organs or tissue) through pores and fenestrae present throughout the vascular endothelium and/or they may be cleared from the circulation mainly by the mononuclear phagocyte system (MPS) present in the reticular connective tissue of the lymph nodes, spleen, and liver. There is the possibility of reentering the circulation after translocation to and accumulation in secondary organs. Finally, there are also indications that NPs may enter and accumulate inside cells of the blood circulation. The persistence of NPs in the blood circulation depends on a number of different factors, which affect the leaving and the (re-)entering of the circulation. These factors, the serum protein corona and opsonization being the most important, will be discussed in this section.

Upon entering the bloodstream, interaction with any protein of the complete plasma proteome (expected to contain around 3700 proteins; Lynch et al. 2007) may be possible. About 50 of these plasma proteins have positively been identified in association with NPs (Aggarwal et al. 2009). It is very important to keep in mind the interaction of NPs with blood constituents when comparing results from *in vitro* experiments to the *in vivo* situation. Some of the proteins with potential to interact with NPs belong to the opsonin family (e.g., immunoglobulin G) and lead to recognition by the scavenger receptor on the macrophage cell surface followed by internalization (Patel 1992), largely affecting toxicokinetic parameters such as the rate of clearance from blood, volume of distribution, organ distribution, and route and rate of clearance from the organism (Goppert and Muller 2005). Opsonization often causes a high loss of NPs from the circulation (Opanasopit et al. 2002), as well as fast transfer to and concentration in organs of the MPS such as in the liver and spleen (Aggarwal et al. 2009).

The process of opsonization is the major factor inducing the removal of NPs from the blood circulation by the MPS. Therefore, the surface characteristics of NPs that favor or disfavor opsonization profoundly influence their toxicokinetics (Landsiedel et al. 2012).

Moreover, in contrast to the effect of opsonins, dysopsonins (e.g., albumin apolipoprotein A-I, A-IV, C-III, and H) reduce the affinity of NPs to the MPS (Mehnert and Mader 2001; Muller and Keck 2004; Goppert and Muller 2005). For example, it was shown with metal oxides incubated with dysopsonins (BSA, IgA), resulting in decreased resorption (Patil et al. 2007) and longer persistence in the blood circulation (Joshi and Muller 2009).

A number of studies have shown that the surface charge of NPs is crucially affecting the binding of opsonins to and therefore also the recognition of NPs by the MPS. The half-life of NPs in the blood circulation greatly depends on their uptake by the MPS (Landsiedel et al. 2012). A study using negatively charged liposomes (ζ potential ~−40 mV) with a diameter of ~200 nm demonstrated an increased uptake of liposomes by the MPS in the mouse liver compared with neutral liposomes of the same size. Furthermore, when the surface charge was shielded by the conjugation with PEG, the ζ potential was reduced to ~−15 mV, leading to a reduced rate of uptake by the liver and to a prolonged circulation in the blood (Levchenko et al. 2002). Positively charged NPs often form large aggregates upon entering the blood circulation due to interactions with negatively charged serum proteins. After dissociation, they may redistribute to the liver (Zhang et al. 2005). Aggregates may even cause transient embolism in the lung capillaries (Zhang et al. 2005). Therefore, positively charged NPs often show a rapid blood clearance phase with a large quantity accumulating in the liver and the lung. Neutral NPs were shown to have a decreased rate of uptake by the MPS and hence a prolonged half-life in the blood. Therefore, they are available for increased uptake by organs depending on additional factors such as the tissue/blood partition coefficient, the ratio of the NP concentration in the tissue related to that in the venous blood leaving the organ (Landsiedel et al. 2012).

As already mentioned, a hydrophilic and neutral surface of NP disfavors the binding of opsonins (Landsiedel et al. 2012). Consequently, a commonly used strategy to minimize opsonization is to conjugate the relatively inert hydrophilic and neutral polymer PEG onto the surface of NPs to provide, in addition, steric hindrance to protein binding. It has been shown for many types (including liposomes and polymer-based NPs) of NPs that PEGylation reduces the uptake by the MPS system and therefore prolongs the persistence of NPs in the blood circulation (Li and Huang 2008). In particular, studies have shown that particles coated with varying molecular masses or changing surface densities of PEG show different results in opsonization and biodistribution: protein adsorption to poly(lactic acid) NPs decreased with increasing molecular mass of PEG and also decreased with increasing PEG-coating density. Consequently, the PEG coatings decreased uptake by MPS proportional to the surface density of PEG (Gref et al. 2000). However, while PEGylation can help in avoiding rapid recognition by the MPS, complete avoidance is rarely achieved as NPs may still get recognized and taken up by the MPS, as it has been demonstrated in a recent study (Paciotti et al. 2004). The reason for this is currently unknown and has not been thoroughly investigated. Finally, instead of PEG, various other polymers have been used to avoid immune recognition and/or influence the pattern of proteins that bind to the surface of NPs (Aggarwal et al. 2009).

Besides interaction with proteins of the blood plasma, NPs may also directly interact with cells of the systemic circulation. Of particular interest in this regard are the red blood cells (RBCs) because they lack receptors and an actomyosin system for active uptake (i.e., endocytosis, phagocytosis) (Rothen-Rutishauser et al. 2006). To study the potential interaction of NPs with RBCs, commercially available polystyrene (labeled with yellow green), gold, and TiO_2 particles of different sizes and with different surface modifications were used for *in vitro* exposure of human RBCs for 4–24 h. Interactions of cells with fluorescently labeled particles were analyzed using laser scanning microscopy, and exposures with TiO_2 and gold NPs were investigated using transmission electron microscopy and electron filtering transmission electron microscopy. Results showed that particles with a size of 200 nm and smaller were able to enter the cells, while larger particles remained outside the cells (Figure 10.2). Furthermore, internalized particles were not found inside membrane-bound vesicles and therefore may have direct access to intracellular organelles, proteins, and DNA, which might have toxicological consequences (Hagens et al. 2007). Consequently, in contrast to surface modification and surface charge, particle size alone was affecting the interaction of particles with RBCs: only NPs were able to cross the cell membrane of RBCs by processes other than phagocytosis or endocytosis (Rothen-Rutishauser et al. 2006). Whether the entering of RBCs by NPs is affecting the persistence of particles in the systemic circulation and their biodistribution is currently unknown.

Although most NPs are cleared from the systemic circulation via uptake by the MPS as already mentioned, NPs may also leave blood vessels and independently enter and accumulate in secondary organs. In this regard, the pore size of endothelial walls in various organs represents a barrier for nanomaterials and at the same time allows selective accumulation (Landsiedel et al. 2012). While small molecular drugs are able to diffuse through the capillary wall into the tissue, NPs may pass through the gaps between the endothelial cells. Endothelia of the blood vessels are either continuous or fenestrated or discontinuous. Kidneys, digestive mucosa, and glands have fenestrated endothelia with fenestrae of about 60 nm. The liver, spleen, and bone marrow have discontinuous endothelia with pores of 50–100 nm (Li et al. 2010). Consequently, the liver, spleen, bone marrow, and also many tumors possess a leaky endothelial wall, which allows for a large translocation of nanomaterials and therefore represent locations for NP accumulation in the body. In addition, a very interesting study demonstrated that the proteomics of the endothelium is unique in every tissue. Consequently, specific targeting of nanomaterials to individual organs and tissues is, in principle, possible by targeting individual microvessels (Simonson and Schnitzer 2007).

Finally, it has also been shown that, after leaving the systemic circulation, a small fraction of NPs may be subject to retranslocation from tissue and organs and therefore may reenter the circulation, resulting in further accumulation in various organs and tissues (Kreyling et al. 2012). For this experiment, freshly generated aerosols of 20-nm iridium, elemental carbon, and TiO_2 NPs were inhaled for 1–2 h in healthy adult rats, and particle

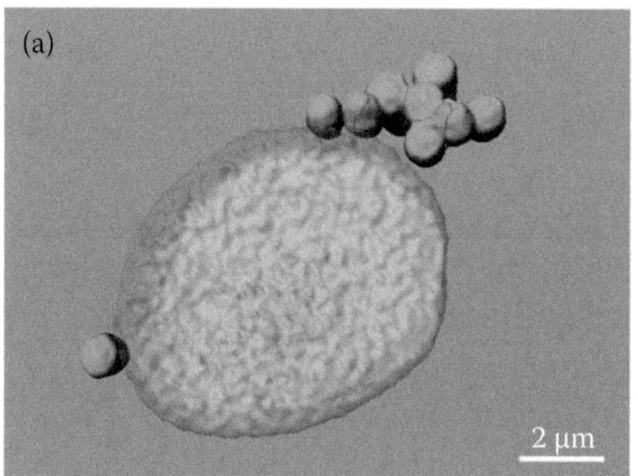

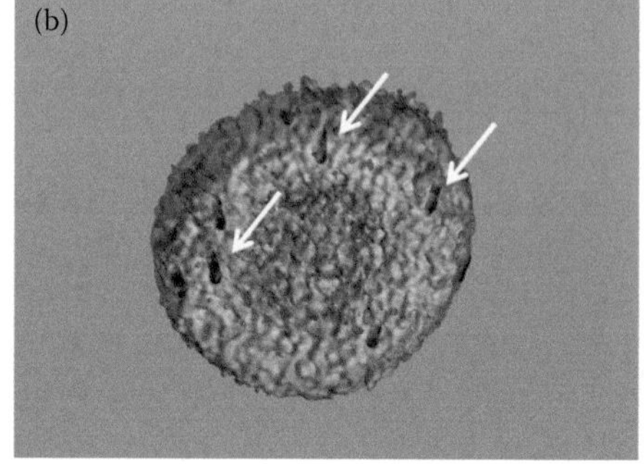

FIGURE 10.2
Interaction of micro- and NPs with RBCs. Laser scanning micrographs of fluorescent polystyrene particles and RBCs: (a) 1-µm particles were attached to the cells and were not observed inside cells, (b) whereas numerous NPs were found inside the cells (white arrows). The images represent three-dimensional reconstructions, and the RBCs are visualized in a transparent mode.

distribution after inhalation was monitored. Interestingly, low but detectable NP concentrations of each NP were found in the blood circulation during several weeks after exposure, suggesting not completely stationary conditions of translocated and accumulated NPs in the organism (Kreyling et al. 2012).

The factors affecting the persistence of NPs in the systemic circulation may be summarized as follows: after entering the blood circulation, NPs may be coated by proteins of the blood plasma and opsonized, which renders them more or less available for the MPS as the major clearance mechanism of NPs from the circulation. Particle surface charge is the main attribute affecting binding of opsonins to NPs, with negatively charged particles showing the most efficient binding. In contrast, the most important attribute affecting the entering of particles in RBCs is particle size. While microparticles are not able to be taken up actively, NPs of different materials and surface characteristics may enter the cells by mechanisms other than phagocytosis or endocytosis. The pores and fenestrae of the vascular endothelium play a crucial role in NP translocation from the bloodstream to secondary organs, and NPs may also reenter systemic circulation from secondary organs or tissues to accumulate in another part of the body or to be cleared from the circulation.

10.5 Conclusion

Intense research in the fields of biotechnology and nanotechnology has led to the design of novel drug delivery devices. The applicability of such carrier systems depends on their physicochemical characteristics, payload, and ability to interact with target cells through a specific route of administration. Pulmonary delivery of drugs or vaccines using nanocarriers poses a promising new strategy, which is more efficient, less invasive, and therefore more patient-friendly. Nanocarriers represent powerful tools for pulmonary drug delivery because interactions of NPs with the alveolar tissue show rapid translocation to the interstitial area followed by access to the systemic circulation. In addition, persistence of NPs in the blood may be controlled by modifying particle characteristics such as surface charge, particle size, or using specific molecules for recognition by the immune system (Rodriguez et al. 2013). Features beneficial for pulmonary drug delivery may also apply to delivery of vaccines: rapid translocation to the pulmonary interstitium enables nanocarriers to interact quickly with important key players of adaptive immunity such as DCs, resulting in the modulation of downstream immune responses (Blank et al. 2013). Despite a considerable number of nanocarriers already being used in clinical applications, our knowledge regarding toxicological and immune responses related to such particulates remains incomplete, and detailed mechanisms of action still require clarification, in particular in the field of pulmonary delivery.

References

Aggarwal, P., Hall, J.B., McLeland, C.B., Dobrovolskaia, M.A., and McNeil, S.E. 2009. Nanoparticle interaction with plasma proteins as it relates to particle biodistribution, biocompatibility and therapeutic efficacy. *Adv Drug Deliv Rev 61*, 428–437.

Blank, F., Stumbles, P.A., Seydoux, E., Holt, P.G., Fink, A., Rothen-Rutishauser, B., Strickland, D.H., and von Garnier, C. 2013. Size-dependent uptake of particles by pulmonary APC populations and trafficking to regional lymph nodes. *Am J Respir Cell Mol Biol 49*, 67–77.

Choi, H.S., Ashitate, Y., Lee, J.H., Kim, S.H., Matsui, A., Insin, N., Bawendi, M.G., Semmler-Behnke, M., Frangioni, J.V., and Tsuda, A. 2010. Rapid translocation of nanoparticles from the lung airspaces to the body. *Nat Biotechnol 28*, 1300–1303.

Choi, H.S., and Frangioni, J.V. 2010. Nanoparticles for biomedical imaging: Fundamentals of clinical translation. *Mol Imaging 9*, 291–310.

Desch, A.N., Randolph, G.J., Murphy, K., Gautier, E.L., Kedl, R.M., Lahoud, M.H., Caminschi, I., Shortman, K., Henson, P.M., and Jakubzick, C.V. 2011. CD103(+) pulmonary dendritic cells preferentially acquire and present apoptotic cell-associated antigen. *J Exp Med 208*, 1789–1797.

Ferron, G.A., Kreyling, W.G., Furst, G., Neuner, M., Schumann, G., and Heyder, J. 1990. Long-term exposure of dogs to a sulfite aerosol. 3. Effect of lung clearance. *J Aerosol Sci 21*, S479–S482.

Foged, C., Brodin, B., Frokjaer, S., and Sundblad, A. 2005. Particle size and surface charge affect particle uptake by human dendritic cells in an *in vitro* model. *Int J Pharm 298*, 315–322.

Gautam, A., Waldrep, J.C., and Densmore, C.L. 2003. Aerosol gene therapy. *Mol Biotechnol 23*, 51–60.

Gehr, P., Bachofen, M., and Weibel, E.R. 1978. The normal human lung: Ultrastructure and morphometric estimation of diffusion capacity. *Respir Physiol 32*, 121–140.

Geiser, M., Matter, M., Maye, I., Im Hof, V., Gehr, P., and Schurch, S. 2003. Influence of airspace geometry and surfactant on the retention of man-made vitreous fibers (MMVF 10a). *Environ Health Perspect 111*, 895–901.

Geraets, L., Oomen, A.G., Schroeter, J.D., Coleman, V.A., and Cassee, F.R. 2012. Tissue distribution of inhaled micro- and nano-sized cerium oxide particles in rats: Results from a 28-day exposure study. *Toxicol Sci 127*, 463–473.

Giudice, E.L., and Campbell, J.D. 2006. Needle-free vaccine delivery. *Adv Drug Deliv Rev 58*, 68–89.

Goppert, T.M., and Muller, R.H. 2005. Adsorption kinetics of plasma proteins on solid lipid nanoparticles for drug targeting. *Int J Pharm 302*, 172–186.

Gref, R., Luck, M., Quellec, P., Marchand, M., Dellacherie, E., Harnisch, S., Blunk, T., and Muller, R.H. 2000. 'Stealth' corona-core nanoparticles surface modified by polyethylene glycol (PEG): Influences of the corona (PEG chain length and surface density) and of the core composition on phagocytic uptake and plasma protein adsorption. *Colloid Surf B 18*, 301–313.

Hagens, W.I., Oomen, A.G., de Jong, W.H., Cassee, F.R., and Sips, A.J.A.M. 2007. What do we (need to) know about the kinetic properties of nanoparticles in the body? *Regul Toxicol Pharm 49*, 217–229.

Heyder, J., Beck-Speier, I., Ferron, G.A., Josten, M., Karg, E., Kreyling, W.G., Lenz, A.G., Maier, K.L., Reitmeier, P., Ruprecht, L. et al. 2009. Long-term responses of canine lungs to acidic particles. *Inhal Toxicol 21*, 920–932.

Huang, Y.Y., and Wang, C.H. 2006. Pulmonary delivery of insulin by liposomal carriers. *J Control Release 113*, 9–14.

Jakubzick, C., Helft, J., Kaplan, T.J., and Randolph, G.J. 2008. Optimization of methods to study pulmonary dendritic cell migration reveals distinct capacities of DC subsets to acquire soluble versus particulate antigen. *J Immunol Methods 337*, 121–131.

Jakubzick, C., Tacke, F., Llodra, J., van Rooijen, N., and Randolph, G.J. 2006. Modulation of dendritic cell trafficking to and from the airways. *J Immunol 176*, 3578–3584.

Joshi, M.D., and Muller, R.H. 2009. Lipid nanoparticles for parenteral delivery of actives. *Eur J Pharm Biopharm 71*, 161–172.

Kersten, G., and Hirschberg, H. 2004. Antigen delivery systems. *Expert Rev Vaccines 3*, 453–462.

Kreyling, W.G. 1990. Interspecies comparison of lung clearance of insoluble particles. *J Aerosol Med 3*, S93–S110.

Kreyling, W.G., Dirscherl, P., Ferron, G.A., Heilmann, P., Josten, M., Miaskowski, U., Neuner, M., Reitmeir, P., Ruprecht, L., Schumann, G. et al. 1999. Health effects of sulfur-related environmental air pollution. III. Nonspecific respiratory defense capacities. *Inhal Toxicol 11*, 391–422.

Kreyling, W.G., Schumann, G., and Ziesenis, A. 2001. Particles are predominantly transported from the canine epithelium towards the interstitial spaces and not to larynx! Analogy to human lungs? *Am J Respir Crit Care Med 163*, A166.

Kreyling, W.G., Semmler, M., Erbe, F., Mayer, P., Takenaka, S., Schulz, H., Oberdorster, G., and Ziesenis, A. 2002. Translocation of ultrafine insoluble iridium particles from lung epithelium to extrapulmonary organs is size dependent but very low. *J Toxicol Environ Health A 65*, 1513–1530.

Kreyling, W.G., Semmler-Behnke, M., Seitz, J., Scymczak, W., Wenk, A., Mayer, P., Takenaka, S., and Oberdorster, G. 2009. Size dependence of the translocation of inhaled iridium and carbon nanoparticle aggregates from the lung of rats to the blood and secondary target organs. *Inhal Toxicol 21 Suppl 1*, 55–60.

Kreyling, W.G., Semmler-Behnke, M., Takenaka, S., and Moller, W. 2012. Differences in the biokinetics of inhaled nano- versus micrometer-sized particles. *Acc Chem Res 46*, 714–722.

Landsiedel, R., Fabian, E., Ma-Hock, L., van Ravenzwaay, B., Wohlleben, W., Wiench, K., and Oesch, F. 2012. Toxico-/biokinetics of nanomaterials. *Arch Toxicol 86*, 1021–1060.

Lehnert, B.E., Valdez, Y.E., and Tietjen, G.L. 1989. Alveolar macrophage particle relationships during lung clearance. *Am J Respir Cell Mol Biol 1*, 145–154.

Levchenko, T.S., Rammohan, R., Lukyanov, A.N., Whiteman, K.R., and Torchilin, V.P. 2002. Liposome clearance in mice: The effect of a separate and combined presence of surface charge and polymer coating. *Int J Pharm 240*, 95–102.

Li, M., Al-Jamal, K.T., Kostarelos, K., and Reineke, J. 2010. Physiologically based pharmacokinetic modeling of nanoparticles. *ACS Nano 4*, 6303–6317.

Li, S.D., and Huang, L. 2008. Pharmacokinetics and biodistribution of nanoparticles. *Mol Pharmaceut 5*, 496–504.

Lipka, J., Semmler-Behnke, M., Sperling, R.A., Wenk, A., Takenaka, S., Schleh, C., Kissel, T., Parak, W.J., and Kreyline, W.G. 2010. Biodistribution of PEG-modified gold nanoparticles following intratracheal instillation and intravenous injection. *Biomaterials 31*, 6574–6581.

Lynch, I., Cedervall, T., Lundqvist, M., Cabaleiro-Lago, C., Linse, S., and Dawson, K.A. 2007. The nanoparticle-protein complex as a biological entity; A complex fluids and surface science challenge for the 21st century. *Adv Colloid Interface Sci 134–135*, 167–174.

Mehnert, W., and Mader, K. 2001. Solid lipid nanoparticles—Production, characterization and applications. *Adv Drug Deliv Rev 47*, 165–196.

Meiring, J.J., Borm, P.J., Bagate, K., Semmler, M., Seitz, J., Takenaka, S., and Kreyling, W.G. 2005. The influence of hydrogen peroxide and histamine on lung permeability and translocation of iridium nanoparticles in the isolated perfused rat lung. *Part Fibre Toxicol 2*, 3.

Michiels, C. 2003. Endothelial cell functions. *J Cell Physiol 196*, 430–443.

Mills, N.L., Amin, N., Robinson, S.D., Anand, A., Davies, J., Patel, D., de la Fuente, J.M., Cassee, F.R., Boon, N.A., MacNee, W. et al. 2006. Do inhaled carbon nanoparticles translocate directly into the circulation in humans? *Am J Respir Crit Care Med 173*, 426–431.

Mills, N.L., Donaldson, K., Hadoke, P.W., Boon, N.A., MacNee, W., Cassee, F.R., Sandstrom, T., Blomberg, A., and Newby, D.E. 2009. Adverse cardiovascular effects of air pollution. *Nat Clin Pract Cardiovasc Med 6*, 36–44.

Misra A. (ed.) 2010. *Challenges in Delivery of Therapeutic Genomics and Proteomics*. 1st edition. Elsevier, London, UK.

Muhlfeld, C., Mayhew, T.M., Gehr, P., and Rothen-Rutishauser, B. 2007. A novel quantitative method for analyzing the distributions of nanoparticles between different tissue and intracellular compartments. *J Aerosol Med 20*, 395–407.

Muhlfeld, C., Rothen-Rutishauser, B., Blank, F., Vanhecke, D., Ochs, M., and Gehr, P. 2008. Interactions of nanoparticles with pulmonary structures and cellular responses. *Am J Physiol Lung Cell Mol Physiol 294*, L817–L829.

Muller, R.H., and Keck, C.M. 2004. Challenges and solutions for the delivery of biotech drugs—A review of drug nanocrystal technology and lipid nanoparticles. *J Biotechnol 113*, 151–170.

Nembrini, C., Stano, A., Dane, K.Y., Ballester, M., van der Vlies, A.J., Marsland, B.J., Swartz, M.A., and Hubbell, J.A. 2011. Nanoparticle conjugation of antigen enhances cytotoxic T-cell responses in pulmonary vaccination. *Proc Natl Acad Sci U S A 108*, E989–E997.

Nemmar, A., Hoet, P.H.M., Vanquickenborne, B., Dinsdale, D., Thomeer, M., Hoylaerts, M.F., Vanbilloen, H., Mortelmans, L., and Nemery, B. 2002. Passage of inhaled particles into the blood circulation in humans. *Circulation 105*, 411–414.

Nikula, K.J., Avila, K.J., Griffith, W.C., and Mauderly, J.L. 1997. Sites of particle retention and lung tissue responses to chronically inhaled diesel exhaust and coal dust in rats and cynomolgus monkeys. *Environ Health Perspect 105*, 1231–1234.

Niven, R.W. 1995. Delivery of biotherapeutics by inhalation aerosol. *Crit Rev Ther Drug 12*, 151–231.

Oberdorster, G. 2001. Pulmonary effects of inhaled ultrafine particles. *Int Arch Occup Environ Health 74*, 1–8.

Oberdorster, G., Ferin, J., and Lehnert, B.E. 1994. Correlation between particle-size, *in-vivo* particle persistence, and lung injury. *Environ Health Perspect 102*, 173–179.

Opanasopit, P., Nishikawa, M., and Hashida, M. 2002. Factors affecting drug and gene delivery: Effects of interaction with blood components. *Crit Rev Ther Drug Carrier Syst 19*, 191–233.

Paciotti, G.F., Myer, L., Weinreich, D., Goia, D., Pavel, N., McLaughlin, R.E., and Tamarkin, L. 2004. Colloidal gold: A novel nanoparticle vector for tumor directed drug delivery. *Drug Deliv 11*, 169–183.

Patel, H.M. 1992. Serum opsonins and liposomes: Their interaction and opsonophagocytosis. *Crit Rev Ther Drug Carrier Syst 9*, 39–90.

Patil, S., Sandberg, A., Heckert, E., Self, W., and Seal, S. 2007. Protein adsorption and cellular uptake of cerium oxide nanoparticles as a function of zeta potential. *Biomaterials 28*, 4600–4607.

Rodriguez, P.L., Harada, T., Christian, D.A., Pantano, D.A., Tsai, R.K., and Discher, D.E. 2013. Minimal "self" peptides that inhibit phagocytic clearance and enhance delivery of nanoparticles. *Science 339*, 971–975.

Rothen-Rutishauser, B.M., Schurch, S., Haenni, B., Kapp, N., and Gehr, P. 2006. Interaction of fine particles and nanoparticles with red blood cells visualized with advanced microscopic techniques. *Environ Sci Technol 40*, 4353–4359.

Sarlo, K., Blackburn, K.L., Clark, E.D., Grothaus, J., Chaney, J., Neu, S., Flood, J., Abbott, D., Bohne, C., Casey, K. et al. 2009. Tissue distribution of 20 nm, 100 nm and 1000 nm fluorescent polystyrene latex nanospheres following acute systemic or acute and repeat airway exposure in the rat. *Toxicology 263*, 117–126.

Schurch, S., Gehr, P., Im Hof, V., Geiser, M., and Green, F. 1990. Surfactant displaces particles toward the epithelium in airways and alveoli. *Respir Physiol 80*, 17–32.

Semmler, M., Seitz, J., Erbe, F., Mayer, P., Heyder, J., Oberdorster, G., and Kreyling, W.G. 2004. Long-term clearance kinetics of inhaled ultrafine insoluble iridium particles from the rat lung, including transient translocation into secondary organs. *Inhal Toxicol 16*, 453–459.

Semmler-Behnke, M., Takenaka, S., Fertsch, S., Wenk, A., Seitz, J., Mayer, P., Oberdorster, G., and Kreyling, W.G. 2007. Efficient elimination of inhaled nanoparticles from the alveolar region: Evidence for interstitial uptake and subsequent reentrainment onto airway epithelium. *Environ Health Perspect 115*, 728–733.

Simonson, A.B., and Schnitzer, J.E. 2007. Vascular proteomic mapping *in vivo*. *J Thromb Haemost 5 Suppl 1*, 183–187.

Thornton, E.E., Looney, M.R., Bose, O., Sen, D., Sheppard, D., Locksley, R., Huang, X.Z., and Krummel, M.F. 2012. Spatiotemporally separated antigen uptake by alveolar dendritic cells and airway presentation to T cells in the lung. *J Exp Med 209*, 1183–1199.

Tonnis, W.F., Kersten, G.F., Frijlink, H.W., Hinrichs, W.L., de Boer, A.H., and Amorij, J.P. 2012. Pulmonary vaccine delivery: A realistic approach? *J Aerosol Med Pulm Drug Deliv 25*, 249–260.

von Garnier, C., Wikstrom, M.E., Zosky, G., Turner, D.J., Sly, P.D., Smith, M., Thomas, J.A., Judd, S.R., Strickland, D.H., Holt, P.G. et al. 2007. Allergic airways disease develops after an increase in allergen capture and processing in the airway mucosa. *J Immunol 179*, 5748–5759.

Wichmann, H.E., Spix, C., Tuch, T., Wolke, G., Peters, A., Heinrich, J., Kreyling, W.G., and Heyder, J. 2000. Daily mortality and fine and ultrafine particles in Erfurt, Germany part I: Role of particle number and particle mass. *Res Rep* 5–86; discussion 87–94.

Wiebert, P., Sanchez-Crespo, A., Falk, R., Philipson, K., Lundin, A., Larsson, S., Moller, W., Kreyling, W.G., and Svartengren, M. 2006. No significant translocation of inhaled 35-nm carbon particles to the circulation in humans. *Inhal Toxicol 18*, 741–747.

Zhang, J.S., Liu, F., and Huang, L. 2005. Implications of pharmacokinetic behavior of lipoplex for its inflammatory toxicity. *Adv Drug Deliv Rev 57*, 689–698.

Zrazhevskiy, P., Sena, M., and Gao, X.H. 2010. Designing multifunctional quantum dots for bioimaging, detection, and drug delivery. *Chem Soc Rev 39*, 4326–4354.

11

The Pulmonary Lymphatic System

Akira Tsuda

CONTENTS

11.1 Introduction

The fate of nanoparticles (NPs) after they deposit on the respiratory surface depends on which part of the lungs they have landed on (Oberdorster et al. 2005). NPs, which deposit on the conducting airways, are quickly cleared by a mucociliary escalator mechanism delivering them to the gastrointestinal (GI) tract (Brain et al. 1977; Stuart 1984; International Commission on Radiological Protection [ICRP] 1994; Kreyling and Scheuch 2000; Kreyling et al. 2007). In contrast, NPs, which deposit deeper in the gas-exchange region of the lungs where a mucociliary clearance mechanism does not exist, are either engulfed by the alveolar macrophages (AMs) (Brain et al. 1977; Corry et al. 1984; Stuart 1984; Harmsen et al. 1985; ICRP 1994; Kreyling and Scheuch 2000; Chono et al. 2006; Kreyling et al. 2007; Geiser et al. 2008) or are taken up by alveolar epithelial pneumocytes, depending on the availability as well as chemotactic and phagocytic activity of AMs (e.g., Lauweryns and Baert 1977; Warheit et al. 1988; Geiser et al. 2008; Liu et al. 2010). NPs engulfed by AMs may remain inside the cell for several months (Brain et al. 1977) or may be carried to the draining lymph node by the AMs via the pulmonary lymphatic system. AM-mediated clearance is extensively discussed in Chapter 6. Note that recent studies show that another immune cell type, the antigen-presenting dendritic cell (DC), may be also involved in NP clearance from the alveolar surface in collaboration with AM. DCs usually stay in the interstitial space beneath the epithelium and may receive NPs from AMs via the tight junctions of the epithelial monolayer (Blank et al. 2007, 2008; Rothen-Rutishauser et al. 2007). This is discussed in detail in Chapter 9. The uptake of NPs by alveolar pneumocytes is a complex process, and is dependent on the physicochemical characteristics (e.g., size, shape, surface charge, and hydrophobicity) of the particles. This is studied extensively (Unfried et al. 2007; Mühlfeld et al. 2008) and is discussed in Chapter 9. Once the NPs exit from the alveolar pneumocytes (exocytosis) into the interstitial space, the NPs have two choices: (i) they move toward the draining lymph nodes via the lymph and subsequently to the circulatory system, or (ii) they move directly into the circulation. This chapter deals with the former route.

11.2 Lymphatic System

The lymphatic system works together with the blood vascular system. Nutrients, gases, hormones, globulins, etc., circulating in the bloodstream leave the circulation system through leakage between the tight junctions of the blood vessel endothelium or direct transendothelial transport, caveolar movement, etc., to feed the tissues. However, the leaked-out excess fluid together with plasma proteins (which could be as much as 3 L daily at the capillary beds; Marieb and Hoehn 2013) residing in the tissue spaces cannot be retrieved by the circulation system because the permeability back into the blood capillaries is not high enough. The lymphatic system, along with the kidneys, is another contributor of maintaining the blood volume.

The draining lymphatic system is made of the lymphatic vessels, which may be categorized into four groups: lymphatic capillaries, precollectors, lymph collectors, and lymphatic trunks. Lymphatic drainage begins with the lymphatic capillaries, which are found between the tissue cells and blood capillaries (Figure 11.1). Unlike blood capillaries, the lymphatic capillaries have remarkably high permeability owing to the unique arrangement of their endothelial cell layer (Calnan et al. 1967; Casley-Smith 1967); the endothelial cells of the walls of lymphatic capillaries are not tightly joined, forming flap-like minivalves (Figure 11.2). Since the loosely arranged endothelial cell layers work as one-way valves (Tranpnell 1963), high pressure in the interstitial space resulting from increased interstitial fluid volume allows the excess fluid into the lymphatic capillaries without any backflow. The lymph (the interstitial fluid is called lymph once it is in the lymphatic vessels) contains particles (proteins, cell debris, pathogens, etc.) and plasma fluid collected in the lymphatic capillaries. Lymph travels through a network of lymphatic precollecting vessels, collecting vessels, and lymphatic trunks. All of the lymphatic trunks merge into two collecting ducts, namely the thoracic duct and the right lymphatic duct, bringing lymph back into the blood circulation (Figure 11.3). The thoracic duct empties into the left

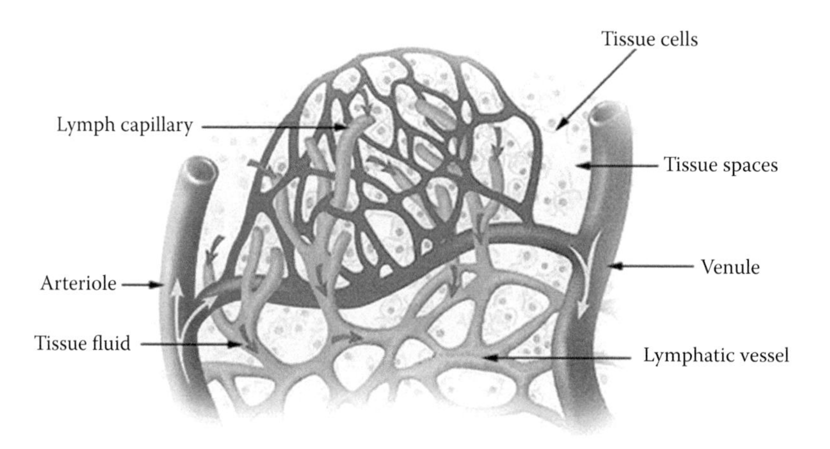

FIGURE 11.1
Lymphatic capillaries. Schematic shows intertwined lymph capillaries between tissue cells and blood capillaries, collecting the excess protein–containing interstitial fluid and waste in connective tissues. (Adapted from http://en.wikipedia.org/wiki/Lymph_capillary.)

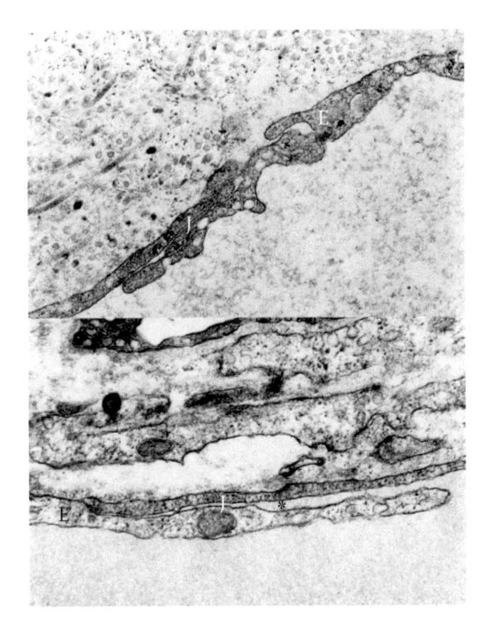

FIGURE 11.2
Electron micrographs showing extensive overlaps of adjacent endothelial cells (E) of rat lung lymphatic vessels at intercellular junctions (J). The width of the intercellular cleft (*) is variable. Both ×56,000. (Adapted from Figure 5 of Leak LV, *Environ. Health Perspect.*, 35, 55, 1980.)

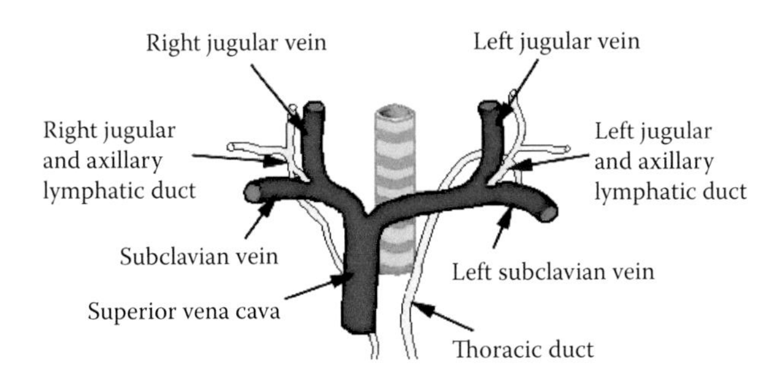

FIGURE 11.3
Lymphatic duct.

subclavian vein, and drains the lymph from the majority of the body (legs, abdomen, left side of head, left arm, and left thorax). The right lymphatic duct empties into the right subclavian vein and drains much less of the body's lymph (only from the right arm, right thorax, and right side of the head).

The lymphatic system has no central pump, similar to the heart in the circulatory system. The lymph moves in the lymphatic vessels by contraction of the surrounding skeletal muscles or by pressure changes during breathing; the larger lymphatic vessels (the lymphatic collector and trunks) produce their own propulsion with a network of smooth musculature located in the walls, but the smaller lymphatic vessels lack any muscle. In addition, large lymphatic vessels contain one-way valves, preventing any backflow of

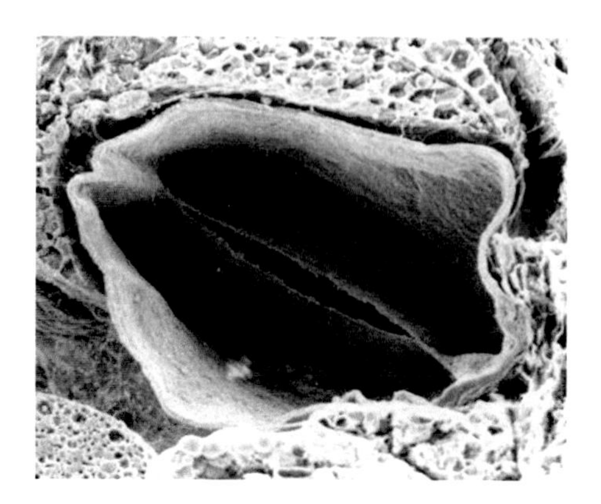

FIGURE 11.4
Scanning electron micrograph of a valve of a rat lung lymphatic duct, ×240. (Adapted from Figure 11 of Leak LV, *Environ. Health Perspect.*, 35, 55, 1980.)

the lymph (Figure 11.4). On its way to blood, lymphatic fluid travels through a successive number of lymph nodes, which filter out impurities from the lymph.

11.3 Anatomy of the Pulmonary Lymphatic System

The lymphatic systems in the lungs originate from two plexuses: a surface (pleural) plexus and a deep (peribronchial/perivascular) plexus (Figure 11.5; Tobin 1957; Leak 1980). The

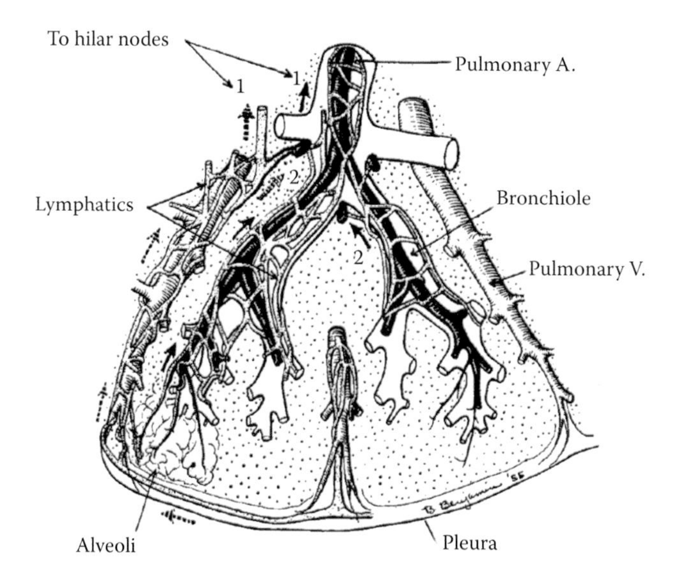

FIGURE 11.5
Diagram of the lymphatic drainage from two lobules in a segment of the lung. The course of flow in the lymphatics around the pulmonary artery and airways (indicated by solid arrows) is in a centripetal direction toward the hilar nodes (1) and bronchial nodes (2). The flow in the lymphatics around the pulmonary vein is from the pleura, peripheral part of the lobules, and the septa (cross-striped arrows). (Adapted from Figure 1 of Tobin CE, *Anat. Rec.*, 127, 611, 1957.)

surface lymphatic vessels are located beneath the pulmonary (visceral) pleura and drain lymph from the lung tissue and the visceral pleura. The surface lymphatic system drains into the bronchopulmonary lymph nodes (in the hilus of each lung). The deep plexus, which consists of the periarterial, perivenous, and peribronchiolar lymph vessels, are in the mucosal layers and connective tissue of the bronchi (note that while alveolar septa lacks lymphatics [Schraufnagel 2010], the interlobular septa are rich in lymphatics [Tobin 1957; Leak 1980]). These lymphatic vessels drain the roots of the lung and pass that lymph into the pulmonary lymph nodes.

There are several well-defined lymph nodes in the respiratory system (Figure 11.6). The bronchopulmonary or bronchial nodes are located near the first major division of the lobar bronchi. The lung hilar region is rich in (hilar) nodes; they are called the tracheobronchial lymph nodes. There is also a chain of lymph nodes on either side of the trachea, called the tracheal or paratracheal lymph nodes.

All lymph from the lungs drains into the tracheobronchial lymph nodes, which are therefore considered the most important for pulmonary drainage. From there, the lymph drains to the right and left bronchomediastinal lymph trunks, which end near the junction of the subclavian and internal jugular veins. The right bronchomediastinal lymph trunk merges with other trunks to form the right lymphatic duct, while the left bronchomediastinal lymph trunk ends at the thoracic duct (Figure 11.3). (Note that all lung lymph enters the bloodstream from the right lymph duct, with the exception of the upper lobule of the left lung, which drains into the thoracic duct; Morrow 1972.)

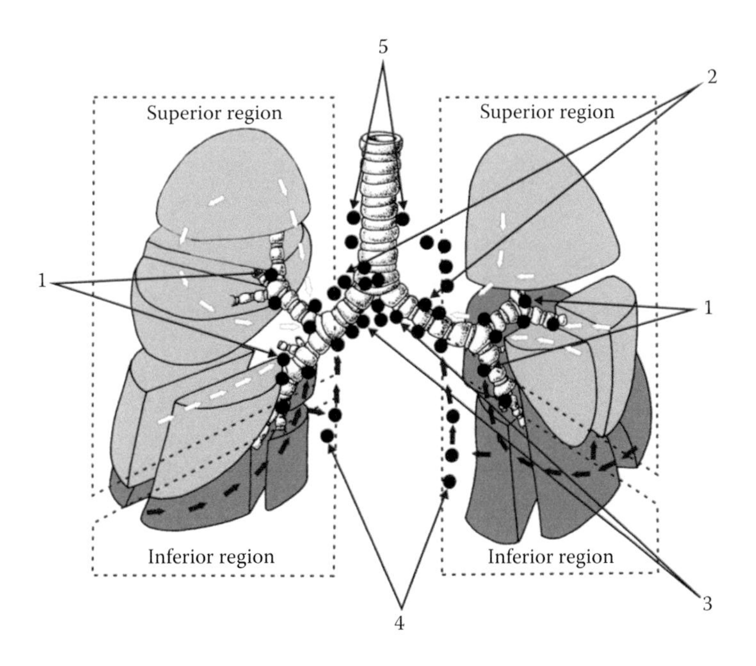

FIGURE 11.6
Diagram of the lymphatic drainage of the superior and inferior regions of the lung: 1—bronchopulmonary lymph nodes, 2—superior tracheobronchial lymph nodes, 3—inferior tracheobronchial lymph nodes, 4—pulmonary ligament lymph nodes, 5—paratracheal lymph nodes. (Adapted with permission from Figure 1 of Topol M, Maslon A, *Ann. Anat.*, 191, 568, 2009.)

11.4 Lymphatic Clearance of Particles

Inhaled NPs deposited on the lung surface first encounter a surfactant layer that forms the top layer of the epithelial lining fluid. The interaction of NPs with surfactant (Schürch et al. 1990; Geiser et al. 2003; Mijailovich et al. 2010) and the alveolar liquid layer (Nel et al. 2009) will determine their subsequent fate. These interactions are the subjects of other chapters (Chapters 4 and 9). This chapter deals with the lymphatic clearance of insoluble NPs, which traverse the type I pneumocyte layer either by the transcellular route (Gumbleton 2001; Spiekermann et al. 2002; Bitonti et al. 2004; Rejman et al. 2004) or the paracellular route (Patton 1996).

The function of the pulmonary lymphatic system is to keep the lungs in homeostatic fluid balance and to defend the body from airborne pathogens (particulates, microbes, bacteria, and viruses), which enter the pulmonary interstitial space. It may be noteworthy to mention that although there is no dispute in the importance of particle clearance via the pulmonary lymphatic drainage system (Green 1973; Corry et al. 1984; Leak and Ferrans Lee 1991; McIntire et al. 1998; Videira et al. 2002, 2006; Yang et al. 2008), the classical studies and recent studies differ in describing this process. The old studies describe this clearance as very slow. For instance, Ferin (1976) found very little accumulation of TiO_2 particles at 25 days after exposure in the rat pulmonary lymphatics; Sorokin and Brain (1975) reported gradual accumulation of Fe_2O_3 (submicron to micron size range) in regional lymphatics nearly a year after aerosol exposure of mice. These and other older studies are described in Morrow (1972), Lippmann et al. (1980), and Stuart (1984). In contrast, recent studies (Videira et al. 2002, 2006; Choi et al. 2010; Blank et al. 2013) indicated that particle transport in the pulmonary lymphatic drainage system occurs rather rapidly. Videira et al. (2002, 2006) reported that endotracheally administered lipid NPs were quickly (within a few minutes) disappearing from rat lungs and significantly accumulated in the regional lymph nodes. The particles that Videira et al. studied had a mean diameter of 200 nm measured by photon correlation spectroscopy; their surface charge was negative (ζ potential of –17.2 mV), and their shape was mostly spherical (confirmed uniformity of diameter by scanning electron microscopy). The reason for discrepancy between the classical and recent studies might be the different techniques or the type of particles used. Recent rapid advances in nanotechnology and biotechnology (e.g., new molecular imaging technology with new contrast agents utilizing NPs; Gibbs-Flournoy et al. 2011; Misri et al. 2012; Gong et al. 2013; Shibu et al. 2013; Yen et al. 2013a,b; Fang et al. 2014) enable imaging NP translocation to the regional lymph nodes more clearly than before.

The amount of NP translocation from the lung surface to secondary organs, potentially via pulmonary lymphatics and circulation, depends on the chemical composition of the particle. While rapid and substantial translocation of ^{13}C NP (26 nm) to the liver after inhalation exposure in a rat model (Oberdorster et al. 2002) has been reported, only <1% translocation of similar-size iridium NPs (15–20 nm in diameter) to the rat liver has been found (Kreyling et al. 2002).

11.5 NPs and Lung Lymphatics (Size and Surface Charge Dependence)

We have recently conducted a study focusing on NP translocation across the air–blood barrier to the mediastinal lymph nodes (Choi et al. 2010). The aim of this study was to investigate the effect of particle size and surface charge on NP translocation across the alveolar epithelium from the alveolar air into the bloodstream in a rat model. We have created a series of near-infrared (NIR) fluorescent NPs by systematically varying their size (5–320 nm) and surface charge (zwitterionic, anionic, cationic, and polar). Then, we instilled these NPs into the right lung using a specially designed catheter, and performed real-time imaging of their transport to the mediastinal lymph nodes (Figure 11.7a) using a fluorescence-assisted resection and exploration imaging system. This system is equipped with two independent channels of NIR fluorescence so that two NIR wavelengths (700 and 800 nm) can be acquired simultaneously.

The chest of the animal was opened, and both lungs were mechanically ventilated via the double-lumen catheter (Figure 11.7b). The tip of the catheter was placed in the right main stem bronchus, and a balloon at the tip of catheter was inflated to isolate the right lung from the rest. This balloon blockage had two functions: (i) it prevented mucociliary clearance of the administered NPs to the GI tract, so that any translocation found in this study originated from the right lung, and (ii) it prevented backflow of the administered NPs to the trachea to avoid high background signals interfering with clear imaging of the mediastinal lymph nodes.

The principal findings of this study are as follows. The most important parameter for NPs determining whether they can rapidly (within a few minutes) cross the air–blood barrier is particle size; particles larger than a hydrodynamic diameter (HD) of ~34 nm do not go across the alveolar wall to the bloodstream regardless of their surface charge (Figure 11.8a). On the other hand, when the HD of the NPs is smaller than ~34 nm, the surface charge becomes important. Anionic, polar, and zwitterionic NPs can go through the alveolar epithelium in this size range; however, cationic NPs cannot rapidly appear at the mediastinal

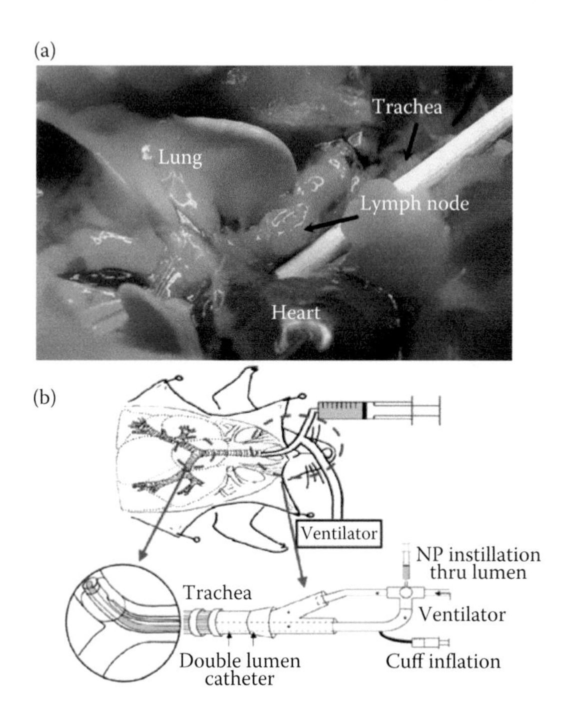

FIGURE 11.7
Image sites and experimental setup. (a) Before the administration of NPs, the chest of the animal was opened and target mediastinal lymph nodes were identified. (b) A specially designed double-lumen catheter with a balloon was inserted into the right main stem bronchus and used to instill NPs into the right lung, and the lungs were simultaneously ventilated. (From Choi HS et al., *Nat. Biotechnol.*, 28, 1300, 2010.)

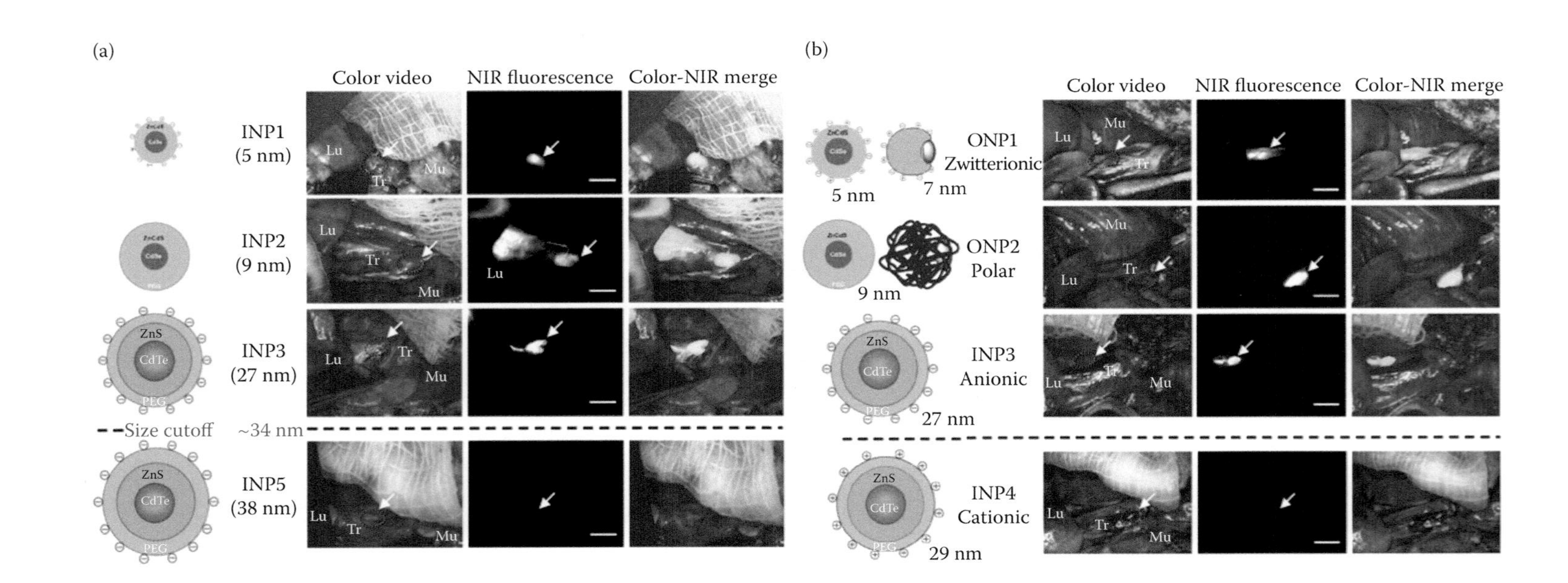

FIGURE 11.8
Size-dependent (a) and surface charge–dependent (b) translocation of NPs via the alveolar air–blood barrier. Color video image (left) and NIR fluorescence image (middle) are merged (right). Lu, lung; Mu, muscle; Tr, trachea. (Adapted from Figure 1 of Choi HS et al., *Nat. Biotechnol.*, 28, 1300, 2010.)

lymph nodes (Figure 11.8b). Figure 11.9 shows that small noncationic NPs (HD <34 nm) translocate rapidly from the lung to mediastinal lymph nodes.

Once the NPs enter the bloodstream, they can eventually go anywhere in the body. However, after entering the interstitial space in the lungs, which route NPs take into the bloodstream, whether NPs move directly into the pulmonary capillaries of the blood flow or they move toward the draining lymph nodes via lymphatic vessels and subsequently get into the circulation system, is not well understood, although it is potentially important in terms of lung injury and immune responses. We compared how long it takes for small-molecule sodium pertechnetate $\left(TcO_4^-\right)$ and our insoluble NPs to appear in the bloodstream after their administration into the lungs. TcO_4^- was used as a control because it is known to be rapidly taken up by blood. The results (Figure 11.10) show that while TcO_4^- appears in the bloodstream immediately after its administration, NPs exhibit some delay, suggesting that NPs are cleared via the pulmonary draining lymphatic pathway before entering the bloodstream.

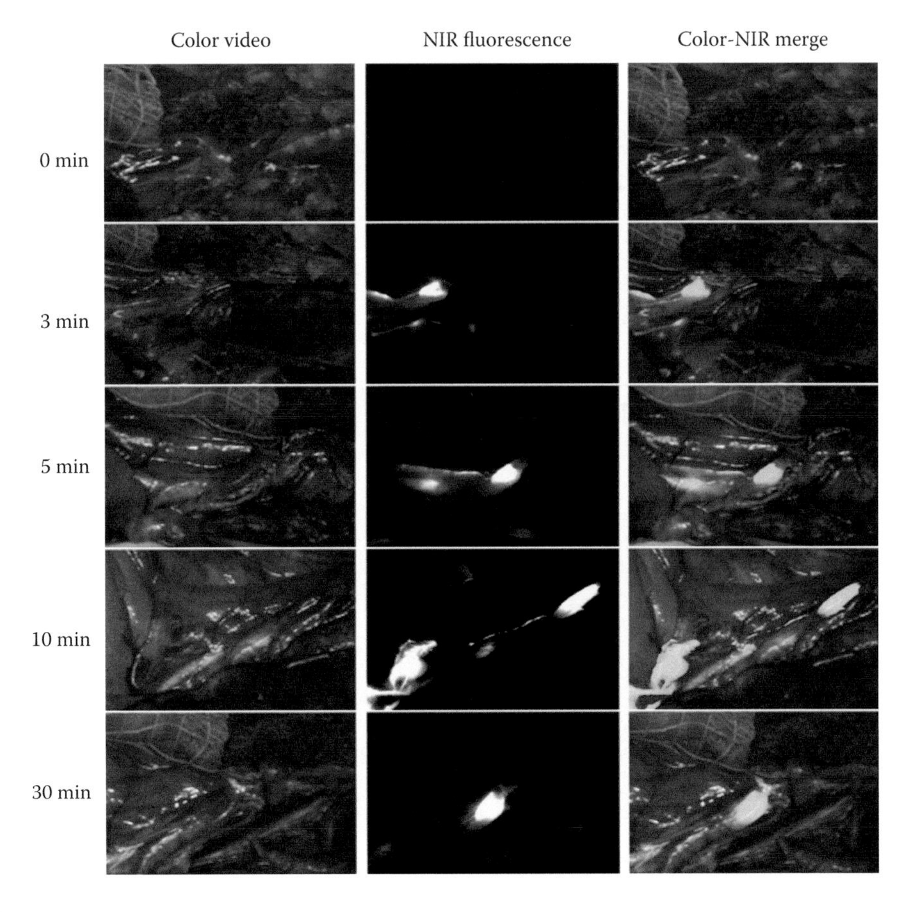

FIGURE 11.9
Anionic NPs (HD = 27 nm, CdTe(ZnS) inorganic core/PEG-COOH organic coating hybrid) instilled in the right lung appeared in the mediastinal lymph nodes within a few minutes. A bright lymphatic vessel leading to the lymph node is clearly seen. HD in PBS is 16 nm; HD in serum is 27 nm, determined by a combination of gel filtration chromatography and transmission electron microscope. Emission maximum: 800 nm. (See Choi HS et al., *Nat. Biotechnol.*, 28, 1300, 2010, for details.)

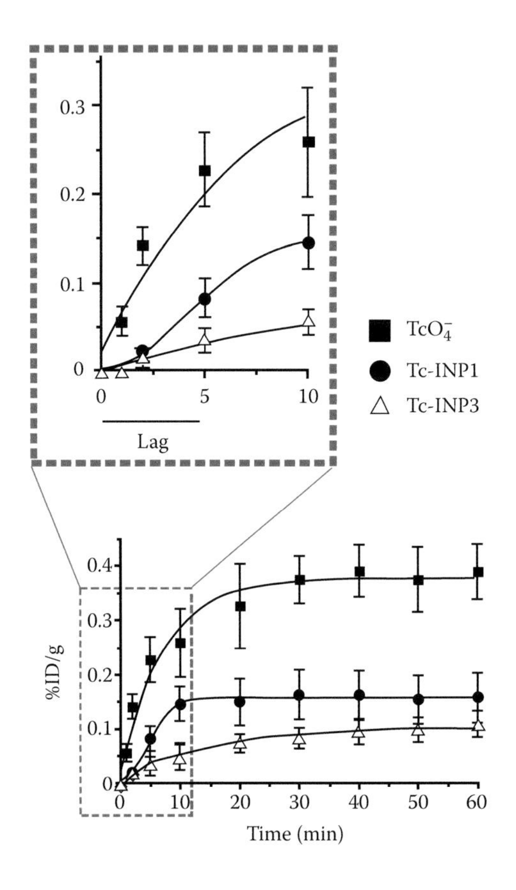

FIGURE 11.10
A few minutes after administration into the lung, both small-molecule TcO_4^- and insoluble NPs appeared in the bloodstream; however, NPs exhibited some delay, suggesting that NPs took the pulmonary draining lymphatic pathway before entering the bloodstream. (Adapted from Figure 2c of Choi HS et al., *Nat. Biotechnol.*, 28, 1300, 2010.)

Finally, the spread and accumulation of small NPs (HD <34 nm) in the body (beyond the pulmonary lymph nodes and the bloodstream) appear to be size and surface charge dependent. While both zwitterionic and anionic NPs <34 nm rapidly appeared in the bloodstream, zwitterionic NPs of an HD of 5 nm rapidly accumulated in the kidneys, and were subsequently cleared with urine at 30 min after administration (Figure 11.11a). On the other hand, anionic NPs of an HD of 27 nm slowly accumulated in the kidneys. However, they were not cleared by the kidneys (Figure 11.11b), and thus remained in the body.

Regarding NPs and pulmonary draining lymph nodes, we can draw several conclusions based on our recent results that are relevant to both NP-based drug delivery and air pollution. If the lung surface is targeted to treat lung disease, large (≥34 nm) cationic NPs would be suitable because they exhibit low systemic absorption. They would remain in the pulmonary system and would increase drug efficacy for lung disease. If the pulmonary lymph nodes are targeted to treat lung infections, disease, or tumor metastases, medium-size-range (5 nm ≤ HD ≤ 34 nm) noncationic NPs would be suitable because they could quickly reach the draining lymph nodes and deliver antibiotics, anti-inflammatory, or tumoricidal drugs to the lymph nodes within a minute. These NP-based drugs eventually enter the bloodstream and remain in the tissues of the body. Finally, if a quick and

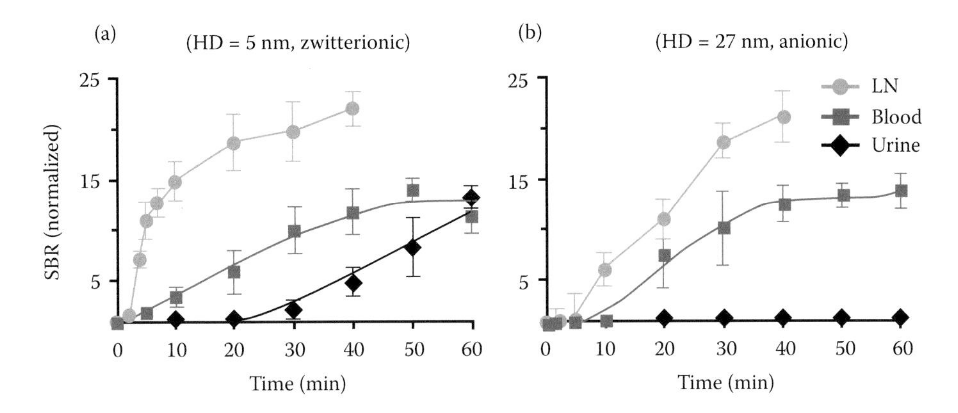

FIGURE 11.11
Biodistribution is size/surface charge dependent. Five-nanometer zwitterionic NPs were quickly cleared by the kidneys (a), while 27-nm anionic NPs remained in the tissues of the body (b). SBR, signal-to-background ratio. (Adapted from Figure 2a of Choi HS et al., *Nat. Biotechnol.*, 28, 1300, 2010.)

short systemic drug delivery is desired, very small size (≤5 nm) zwitterionic NPs would be suitable because they can enter the bloodstream rapidly with high efficacy but are quickly cleared by the kidneys (see Figure 11.11a).

The other side of the picture involves the adverse health effects of inhaled NPs, which are cleared by the pulmonary draining lymphatic pathway. If the size of inhaled pollutant NPs is relatively large (≥34 nm) and/or their surface is positively charged, they can be trapped in the pulmonary system, and the mucociliary escalator and/or macrophage-mediated clearance mechanisms may remove them from the body. On the other hand, if the inhaled pollutant NPs are noncationic and <34 nm, they may pose a danger to health because they can rapidly traverse the epithelial barrier. They may cause various adverse health effects by spreading throughout the body, including inflammation and cardiovascular disease; they may also contribute to carcinogenesis. Very small pollutant NPs (≤5 nm), unless they are cleared rapidly by the kidneys, are of special concern because they are potentially capable of reaching anywhere in the body, and get into cells once they reach the bloodstream.

Acknowledgments

This work was supported in part by research grants from the National Institutes of Health (HL054885, HL070542, HL074022, HL094567, and ES000002).

References

Bitonti AJ, Dumont JA, Low SC, Peters RT, Kropp KE, Palombella VJ, Stattel JM, Lu Y, Tan CA, Song JJ, Garcia AM, Simister NE, Spiekermann GM, Lencer WI, Blumberg RS. 2004. Pulmonary delivery of an erythropoietin Fc fusion protein in non-human primates through an immunoglobulin transport pathway. *Proc. Natl. Acad. Sci. U. S. A.* 101:9763–9768.

Blank F, Rothen-Rutishauser B, Gehr P. 2007. Dendritic cells and macrophages form a transepithelial network against foreign particulate antigens. *Am. J. Respir. Cell Mol. Biol.* 36:669–677.
Blank F, Von Garnier C, Rothen-Rutishauser B, Obregon C, Gehr P, Nicod L. 2008. The role of dendritic cells in the lung: What do we know from *in vitro* models, animal models and human studies? *Exp. Rev. Respir. Med.* 2:215–233.
Blank F, Stumbles PA, Seydoux E, Holt PG, Fink A, Rothen-Rutishauser B, Strickland DH, von Garnier C. 2013. Size-dependent uptake of particles by pulmonary APC populations and trafficking to regional lymph nodes. *Am. J. Respir. Cell Mol. Biol.* 49(1):67–77.
Brain JD, Proctor DF, Reid LM. 1977. *Respiratory Defense Mechanisms: Vols. 1 and 2 (Lung Biology in Health and Disease).* New York: Marcel Dekker.
Calnan JS, Rivero OR, Fillmore S, Mercurius-Taylor L. 1967. Permeability of normal lymphatics. *Br. J. Surg.* 54:278–285.
Casley-Smith R. 1967. The functioning of the lymphatic system under normal and pathological conditions: Its dependence on the time-structures and permeability of the vessels. In: Ruttiman A, ed. *Progress in Lymphology*. New York: Hafner, pp. 348–359.
Choi HS, Ashitate Y, Lee JH, Kim SH, Matsui A, Insin N, Bawendi MG, Semmler-Behnke M, Frangioni JV, Tsuda A. 2010. Rapid translocation of nanoparticles from the lung airspaces to the body. *Nat. Biotechnol.* 28(12):1300–1303.
Chono S, Tanino T, Seki T, Morimoto K. 2006. Influence of particle size on drug delivery to rat alveolar macrophages following pulmonary administration of ciprofloxacin incorporated into liposomes. *J. Drug Target.* 14:557–566.
Corry D, Kulkarni P, Lipscomb MF. 1984. The migration of bronchoalveolar macrophages into hilar lymph nodes. *Am. J. Pathol.* 115:321–328.
Fang J, Chandrasekharan P, Liu XL, Yang Y, Lv YB, Yang CT, Ding J. 2014. Manipulating the surface coating of ultra-small Gd_2O_3 nanoparticles for improved T1-weighted MR imaging. *Biomaterials* 35(5):1636–1642.
Ferin J. 1976. Lung clearance of particles. In: Aharonson EF, Ben-David A, Klingberg MA, eds. *Air Pollution and the Lung*. Jerusalem: Halsted Press-John Wiley, pp. 64–78.
Geiser M, Schurch S, Gehr P. 2003. Influence of surface chemistry and topography of particles on their immersion into the lung's surface-lining layer. *J. Appl. Physiol.* 94:1793–1801.
Geiser M, Casaulta M, Kupferschmid B, Schulz H, Semmler-Behnke M, Kreyling W. 2008. The role of macrophages in the clearance of inhaled ultrafine titanium dioxide particles. *Am. J. Respir. Cell Mol. Biol.* 38:371–376.
Gibbs-Flournoy EA, Bromberg PA, Hofer TP, Samet JM, Zucker RM. 2011. Darkfield–confocal microscopy detection of nanoscale particle internalization by human lung cells. *Part. Fibre Toxicol.* 8(1):2–13.
Gong T, Olivo M, Dinish US, Goh D, Kong KV, Yong KT. 2013. Engineering bioconjugated gold nanospheres and gold nanorods as label-free plasmon scattering probes for ultrasensitive multiplex dark-field imaging of cancer cells. *J. Biomed. Nanotechnol.* 9(6):985–991.
Green GM. 1973. Alveolobronchiolar transport mechanisms. *Arch. Intern. Med.* 131:109–114.
Gumbleton M. 2001. Caveolae as potential macromolecule trafficking compartments within alveolar epithelium. *Adv. Drug Deliv. Rev.* 49:281–300.
Harmsen AG, Muggenburg BA, Snipes MB, Bice DE. 1985. The role of macrophages in particle translocation from lungs to lymph nodes. *Science* 230:1277–1280.
International Commission on Radiological Protection. 1994. ICRP publication 66: Human respiratory tract model for radiological protection. A report of a task group of the International Commission on Radiological Protection. *Ann. ICRP* 24(1–3):1–482.
Kreyling W, Scheuch G. 2000. Clearance of particles deposited in the lungs. In: Heyder J, Gehr P, eds. *Particle Lung Interactions*. New York: Marcel Dekker, pp. 323–376.
Kreyling WG, Semmler M, Erbe F, Mayer P, Takenaka S, Schulz H, Oberdorster G, Ziesenis A. 2002. Translocation of ultrafine insoluble iridium particles from lung epithelium to extrapulmonary organs is size dependent but very low. *J. Toxicol. Environ. Health A* 65:1513–1530.
Kreyling WG, Möller W, Semmler-Behnke M, Oberdorster G. 2007. Particle dosimetry: Deposition and clearance from the respiratory tract and translocation towards extra-pulmonary sites. In:

Donaldson K, Borm P, eds. *Particle Toxicology*. Boca Raton, FL: CRC Press, Taylor & Francis Group, pp. 47–74.
Lauweryns JM, Baert JH. 1977. Alveolar clearance and the role of the pulmonary lymphatics. *Am. Rev. Respir. Dis.* 115:625–683.
Leak LV. 1980. Lymphatic removal of fluids and particles in the mammalian lung. *Environ. Health Perspect.* 35:55–76.
Leak LV, Ferrans Lee VJ. 1991. Lymphatics and lymphoid tissue. In: Crystal RG, West JB, Cherniack NS, Weibel ER, eds. *The Lung: Scientific Foundations*. New York: Raven Press, pp. 779–786.
Lippmann M, Yeates DB, Albert RE. 1980. Deposition, retention, and clearance of inhaled particles. *Br. J. Ind. Med.* 37:337–362.
Liu R, Zhang X, Pu Y, Yin L, Li Y, Zhang X, Liang G, Li X, Zhang J. 2010. Small-sized titanium dioxide nanoparticles mediate immune toxicity in rat pulmonary alveolar macrophages *in vivo*. *J. Nanosci. Nanotechnol.* 10(8):5161–5169.
Marieb EN, Hoehn K. 2013. *Human Anatomy & Physiology*, 9th ed. Upper Saddle River, NJ: Pearson Education Inc., publishing as Benjamin Cummings.
McIntire GL, Bacon ER, Toner JL, Cornacoff JB, Losco PE, Illig KJ, Nikula KJ, Muggenburg BA, Ketai L. 1998. Pulmonary delivery of nanoparticles of insoluble, iodinated CT X-ray contrast agents to lung draining lymph nodes in dogs. *J. Pharm. Sci.* 87:1466–1470.
Mijailovich SM, Kojic M, Tsuda A. 2010. Particle-induced indentation of the alveolar epithelium caused by surface tension forces. *J. Appl. Physiol.* 109:1179–1194.
Misri R, Saatchi K, Häfeli UO. 2012. Nanoprobes for hybrid SPECT/MR molecular imaging. *Nanomedicine* 7(5):719–733.
Morrow PE. 1972. Lymphatic drainage of the lung in dust clearance. *Ann. N. Y. Acad. Sci.* 200:46–65.
Mühlfeld C, Gehr P, Rothen-Rutishauser B. 2008. Translocation and cellular entering mechanisms of nanoparticles in the respiratory tract. *Swiss Med. Wkly.* 138(27–28):387–391.
Nel AE, Mädler L, Velegol D, Xia T, Hoek EM, Somasundaran P, Klaessig F, Castranova V, Thompson M. 2009. Understanding biophysicochemical interactions at the nano–bio interface. *Nat. Mater.* 8(7):543–557.
Oberdorster G, Sharp Z, Atudorei V, Elder A, Gelein R, Lunts A, Kreyling WG, Cox C. 2002. Extrapulmonary translocation of ultrafine carbon particles following whole-body inhalation exposure of rats. *J. Toxicol. Environ. Health A* 65:1531–1543.
Oberdorster G, Oberdorster E, Oberdorster J. 2005. Nanotoxicology: An emerging discipline evolving from studies of ultrafine particles. *Environ. Health Perspect.* 113:823–839.
Patton JS. 1996. Mechanisms of macromolecule absorption by the lungs. *Adv. Drug Deliv. Rev.* 19: 3–36.
Rejman J, Oberle V, Zuhorn IS, Hoekstra D. 2004. Size-dependent internalization of particles via the pathways of clathrin- and caveolae-mediated endocytosis. *Biochem. J.* 377:159–169.
Rothen-Rutishauser B, Muhlfeld C, Blank F, Musso C, Gehr P. 2007. Translocation of particles and inflammatory responses after exposure to fine particles and nanoparticles in an epithelial airway model. *Part. Fibre Toxicol.* 4:9.
Schraufnagel DE. 2010. Lung lymphatic anatomy and correlates. *Pathophysiology* 17:337–343.
Schürch S, Gehr P, Im Hof V, Geiser M, Green F. 1990. Surfactant displaces particles toward the epithelium in airways and alveoli. *Respir. Physiol.* 80:17–32.
Shibu ES, Sugino S, Ono K, Saito H, Nishioka A, Yamamura S, Sawada M, Nosaka Y, Biju V. 2013. Singlet-oxygen-sensitizing near-infrared-fluorescent multimodal nanoparticles. *Angew. Chem. Int. Ed. Engl.* 52(40):10559–10563.
Sorokin SP, Brain JD. 1975. Pathways of clearance in mouse lungs exposed to iron oxide aerosols. *Anat. Rec.* 181:581–625.
Spiekermann GM, Finn PW, Ward ES, Dumont J, Dickinson BL, Blumberg RS, Lencer WI. 2002. Receptor-mediated immunoglobulin G transport across mucosal barriers in adult life: Functional expression of FcRn in the mammalian lung. *J. Exp. Med.* 196:303–310.
Stuart BO. 1984. Deposition and clearance of inhaled particles. *Environ. Health Perspect.* 55:369–390.
Tobin CE. 1957. Human pulmonic lymphatics: An anatomic study. *Anat. Rec.* 127:611–633.

Topol M, Maslon A. 2009. Some variations in lymphatic drainage of selected bronchopulmonary segments in human lungs. *Ann. Anat.* 191:568–574.

Tranpnell DH. 1963. The peripheral lymphatics of the lungs. *Br. J. Radiol.* 36:660–672.

Unfried K, Albrecht C, Klotz LO, von Mikecz A, Grether-Beck S, Schins RPF. 2007. Cellular responses to nanoparticles: Target structures and mechanisms. *Nanotoxicology* 1:1–20.

Videira MA, Botelho MF, Santos AC, Gouveia LF, de Lima JJ, Almeida AJ. 2002. Lymphatic uptake of pulmonary delivered radiolabelled solid lipid nanoparticles. *J. Drug Target.* 10(8):607–613.

Videira MA, Gano L, Santos C, Neves M, Almeida AJ. 2006. Lymphatic uptake of lipid nanoparticles following endotracheal administration. *J. Microencapsul.* 23(8):855–862.

Warheit DB, Overby LH, George G, Brody AR. 1988. Pulmonary macrophages are attracted to inhaled particles through complement activation. *Exp. Lung Res.* 14(1):51–66.

Yang W, Peters JI, Williams RO III. 2008. Inhaled nanoparticles—A current review. *Int. J. Pharm.* 356:239–247.

Yen SK, Jańczewski D, Lakshmi JL, Dolmanan SB, Tripathy S, Ho VH, Vijayaragavan V, Hariharan A, Padmanabhan P, Bhakoo KK, Sudhaharan T, Ahmed S, Zhang Y, Tamil Selvan S. 2013a. Design and synthesis of polymer-functionalized NIR fluorescent dyes—Magnetic nanoparticles for bioimaging. *ACS Nano* 7(8):6796–6805.

Yen SK, Padmanabhan P, Selvan ST. 2013b. Multifunctional iron oxide nanoparticles for diagnostics, therapy and macromolecule delivery. *Theranostics* 3(12):986–1003.

12

Translocation and Accumulation in the Body

Wolfgang G. Kreyling

CONTENTS

12.1 Deposition of Inhaled NPs and Interactions with the Lung Epithelium

Deposition of inhaled nanoparticles (NPs) is governed by their diffusivity, leading to a rather homogeneous deposition density on the epithelium of the respiratory tract. This results in a major deposited fraction in the alveolar region of the lungs, with a maximum of 30%–40% of the inhaled aerosol with a size of 20 nm (International Commission on Radiological Protection 1994; MPPD software). Below that size, increasing fractions deposit in the airways of the head and thorax according to their increasing diffusivity with decreasing size, such that fewer NP reach the distal alveolar region. Once deposited in the peripheral lungs, NP are not only subject to phagocytosis by alveolar macrophages but also to endocytosis by epithelial cells. It appears plausible that macrophages take up all NP deposited in their immediate vicinity, while distant NP are not recognized (Kreyling et al. 2013). As a result, a major fraction (>80%) of the NP in the peripheral rodent lung enter the epithelium and penetrate into interstitial spaces (Semmler-Behnke et al. 2007a).

12.2 Clearance Pathways Out of the Lungs

Hence, NP behave differently compared with micron-sized particles, which are retained on the rodent epithelium, leaving the lungs by macrophage-mediated clearance at a rate of 2%–3% per day (Kreyling 1990; Kreyling and Scheuch 2000). Surprisingly, NP are cleared by the same clearance rate via this macrophage-mediated transport process, indicating relocation of the NP from the interstitial spaces and epithelium back on top of the epithelium (Semmler-Behnke et al. 2007a). Therefore, macrophage-mediated clearance is the most prominent long-term clearance mechanism in the peripheral lung of rodents. In fact, while the NP are retained in the interstitium close to lymphatic drainage and blood vessels, only rather small fractions are removed via these two pathways. Particle clearance from the peripheral lungs of humans and from those of dogs and monkeys differ from that in rodents: in the three large species, even micron-sized particles enter the epithelium and penetrate into interstitial spaces, and macrophage-mediated clearance occurs at a rate that is one order of magnitude lower than that of rodents (Nikula et al. 1997; Kreyling and Scheuch 2000; Kreyling and Geiser 2010; Kreyling et al. 2013). Yet, since micron-sized particles penetrate into interstitial spaces of the human, canine, and simian lungs, it appears plausible that NP also follow this pathway as we observed in rat and mouse lungs. Unfortunately, for humans, neither macrophage-mediated long-term clearance kinetics data nor translocation data of NP into the circulation are available. An article by Nemmar and colleagues (2002) has been widely cited as direct evidence for the translocation of inhaled NP from the human lung epithelium into the circulation. However, subsequent work has failed to confirm their findings. In their study, Nemmar and colleagues exposed young volunteers to an aerosol of primary carbon particles of about 10 nm, radiolabeled with ^{99m}technetium (^{99m}Tc). They detected the ^{99m}Tc tracer in blood within minutes after the exposure and in the liver and stomach within an hour. However, earlier, Brown et al. (2002) did not confirm these findings using similar carbonaceous NP, and more recently other investigators (Mills et al. 2006; Wiebert et al. 2006a,b) have repeated these studies with similarly sized carbon NP and were unable to find evidence for particle translocation into blood. Mills and colleagues (2006) found that radioactive technetium leached off 4–20 nm carbon NP when they deposited on the epithelium, and that the radioactive moiety, rather than the particle itself, was detected rapidly in circulating blood. In addition, Möller et al. (2006, 2008) clearly showed that the radiolabel is rapidly leached off, unless it has been stabilized by several necessary measures during the production of the carbon NP. Hence, existing data suggest that the translocation fractions of NP must be <1% of the deposited NP according to the lower limit of experimental detection, as no accumulation was observed in any secondary organ (Möller et al. 2008). An important underlying mechanism of the special behavior of NP may be selected binding of proteins and NP, affecting the biokinetics fate of the NP. In other words, coating of NP with selected proteins can influence their uptake and distribution and direct them to secondary organs and tissues (Lynch et al. 2007; Lundqvist et al. 2008).

12.3 Translocation into Circulation and Accumulation in Secondary Organs and Tissues

The cardiovascular effects observed in epidemiological studies triggered the discussion on enhanced translocation of ultrafine particles from the respiratory epithelium toward the circulation and subsequent organs, such as the heart, liver, spleen, and brain, eventually causing adverse effects on cardiac function and blood coagulation, as well as on functions of the central nervous system. There is clear evidence that NP can cross body membranes and reach the above-mentioned secondary organs and accumulate there.

To determine the accumulated fractions in such organs, the ultimate aim is to quantitatively balance the fractions of NP in all relevant organs and tissues of the body and include the remaining body and total excretion collected between application and autopsy. Otherwise, substantial uncertainty remains if only selected organs are analyzed (Figure 12.1). A concept of the assessment of quantitative particle biokinetics is shown in Figure 12.1.

On the basis of quantitative biokinetics analysis after NP application to the lungs of a rat model, small fractions of NP (iridium, carbon, gold, and preliminary results on titanium dioxide [TiO_2]) were found in all secondary organs studied, including the brain, heart, and even in the fetus (Kreyling et al. 2002, 2009; Semmler et al. 2004; Semmler-Behnke et al. 2007a,b, 2008). All NP were radiolabeled, with the label firmly fixed to the particle core usually using a radioisotope of the same chemical element as the NP core. In addition, auxiliary studies were performed to estimate the minute amount of NP dissolution and radiotracer leaching for subsequent data correction. Fractions in each of the secondary organs were usually well below 0.5% of the administered dose to the lungs but depended strongly on particle size in an inverse fashion. However, NP fractions in soft tissue and skeleton (without blood content) were usually larger than in individual organs and increased the totally translocated fraction to 5%–10% of the administered dose depending on the physicochemical NP properties. Particularly, the accumulation in the skeleton is remarkable and yet widely ignored in the open literature. Since the NP have entered the skeleton via blood circulation, they are likely being retained next to bone marrow cells with a potential of affecting those. Also, negatively ionic surface charged NP translocated more rapidly than positively charged NP of the same size. In addition, strong differences of the totally translocated fractions between chain-aggregated/agglomerated iridium, TiO_2, and carbon NP versus gold spheres of same size highlight the importance of NP material, morphological, and/or surface properties (Kreyling et al. 2013). Note that these freshly produced aerosols of 20-nm (median-sized) NP being inhaled within about 10 s after production exhibited the naked NP surface without any surface modification. Furthermore, NP accumulation in the rat brain results from both pathways: via the olfactory bulb versus circulation as was shown by differential inhalation studies exposing either the

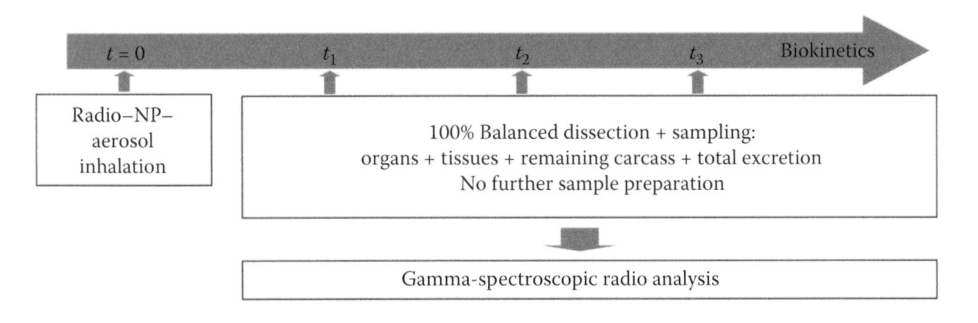

FIGURE 12.1
Concept of quantitative particle biokinetics assessment. Particles are administered at time $t = 0$. From this time point on, the entire urinary and fecal excretions are collected separately. Animals are euthanized at times t_1, t_2, t_3, etc., and organs and tissues of interest and the entire remaining carcass are sampled as well as the entire excretion (100% balanced sampling). When particles are radiolabeled, total organ and tissue samples will be analyzed directly using gamma spectrometry without any further chemical preparation.

entire respiratory tract of rats including nasal passages, or only the lungs of intratracheally intubated and ventilated rats using radiolabeled, 20-nm iridium NP aerosols.

12.4 Some Aspects of Other Studies of Biokinetics

There are numerous articles reporting biokinetics data after inhalation of various NP aerosols. There are even more reports providing biokinetics data after instillation or other artificial forms of application, which is not the subject of this chapter. Generally, most of these data lack the quantitative and balanced approach that leads to several uncertainties:

- Without excreta data over time, neither can the deposited fraction of NP in the conducting airways be estimated from the determination of fast mucociliary clearance nor can the long-term macrophage-mediated, cleared particle fraction be evaluated with sufficient accuracy as a function of time.
- Without determination of the retained NP fraction in soft tissue and the skeleton, two major sites of NP accumulation remain unknown. Although NP retention in soft tissues is most likely of no important toxicological relevance, it is the major sink of most biopersistent NP in the body, which is important for a proper quantitative balance. In contrast, NP accumulation in the skeleton may be lower than or in the same range as in secondary organs such as the liver, and this retention may well be of toxicological relevance. Note that the NP reach the skeletal site by the circulation, and it appears plausible that NP may interact with bone marrow cells.

Early pioneering studies in the nineties have been performed at the University of Rochester by Günter Oberdörster, Jurai Ferin, and their colleagues (Ferin et al. 1991; Oberdörster et al. 1994a,b). They were the first to demonstrate detectable translocation of nano-sized TiO_2 particles inhaled as aerosols of submicrometer-sized TiO_2 agglomerates/aggregates into the lung interstitium as well as into secondary organs. Meanwhile, a number of inhalation studies on rodents have been performed with endpoints of biokinetics and toxicological responses. Just very recently, a comprehensive list of these inhalation studies and instillation studies using various NP materials has been published by Balasubramanian and colleagues (2013, their Table 1). There is consensus that only small fractions of inhaled NP are translocated across the air–blood–barrier, leading to accumulation in secondary organs, including the liver and spleen as the most commonly reported organs. Notably, Balasubramanian and colleagues (2013) performed a rat inhalation study using aerosol consisting of agglomerated gold NP of a median size of 40–50 nm, which consisted of either 7- or 20-nm primary gold particles. The inhalation of the same-sized agglomerated gold NP led to the same lung deposition, and the deposited NP dose was also adjusted to be the same. Under these rather controlled conditions, the authors showed that not only on a number basis but also on an NP mass basis that significantly more of the 7-nm primary gold particles were translocated and accumulated in secondary organs than the 20-nm gold particles. Hence, they demonstrated that the deposited agglomerated gold NP disagglomerated into their primary particles to a substantial extent, giving rise to enhanced translocation of the smaller 7-nm gold NP when compared with the 20-nm gold NP. Interestingly, Ferin et al. (1991) had already postulated such a disagglomeration of their inhaled agglomerated TiO_2 particles. While these early findings led to much speculation without much experimental evidence, the new study confirms that disagglomeration not only occurred with TiO_2 agglomerates but also with gold agglomerates. Material dependence is likely to play a pivotal role as the weak forces between the agglomerates need to be overcome by the surface tension of the epithelial lining fluid.

The inhalation study using 20-nm iridium NP was extended to follow the fate of the NP during 6 months after a single 1-h inhalation, and yielded significant retention in secondary organs such as the liver, spleen, kidneys, heart, and brain (Semmler et al. 2004; Semmler-Behnke et al. 2007a). In this study, we found evidence for considerable particle relocation within the pulmonary tissue during the 6-month period. Combining exhaustive bronchoalveolar lavages (BAL) with our long-term biokinetics studies, we observed that 3 days after inhalation, only 10%–20% NP were accessible to BAL and >80% had already been relocated into epithelial and interstitial tissue. Even more surprising, these interstitially retained NP were predominantly cleared by macrophage mediation back to the luminal side of the epithelium and toward the mucociliary escalator of the bronchiole and bronchi to the larynx (from where they were swallowed and excreted) (Semmler-Behnke et al. 2007a). Pathways of relocation of inhaled NP in rodent lungs are schematically sketched in Figure 12.2.

Translocation to the lymphatic drainage and to blood circulation remained to be rather low, although the NP were retained rather closely to the lymphatic system and the blood vessels.

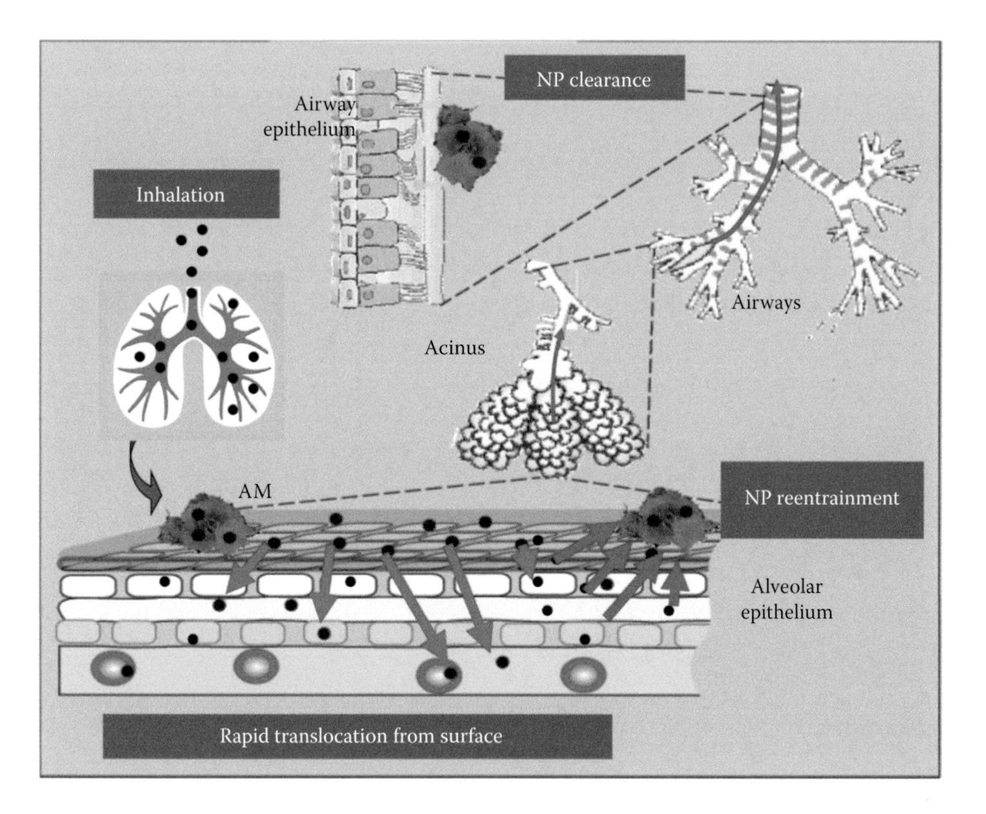

FIGURE 12.2
Pathways of relocation of inhaled NPs after deposition on the alveolar epithelium of the rodent lungs: rapid transport into the epithelium and interstitial spaces for long-term retention; upon clearance, NPs can reentrain back onto the epithelium for macrophage-mediated transport toward ciliated airways and the larynx. (From Semmler-Behnke et al. 2007a; Kreyling et al. 2013. With permission from Accounts of Chemical Research.)

12.5 Benefit of Complementing a Biokinetics Study with a Morphometric Study

While these macroscopic studies do not provide insight into which cells or cell compartments the NP are retained or relocated within an organ, complementary microscopic studies are required to fill this gap. Recently, we performed a microscopic study directly after and 24 h after inhalation of 20-nm chain-aggregated/agglomerated TiO_2 NP showing rapid penetration of NP into the epithelial cell layer, as well as translocation into interstitial spaces and the vascular endothelium (Kapp et al. 2004; Geiser et al. 2005, 2008; Mühlfeld et al. 2007). However, at that time, no macroscopic inhalation study was available using the same TiO_2 NP since the radiolabeling technology of the aerosolized TiO_2 NP was not yet developed. This has now been achieved, and the first results indicate that modest NP accumulation in all secondary organs occurs but that the total 24-h translocation of these chain-aggregated/agglomerated TiO_2 NP is significantly lower by a factor of 5 from that of similar 20-nm iridium NP. The aerosolized TiO_2 NP used were generated by spark ignition, very similar to the production of the TiO_2 NP for the morphometric study, and are currently carefully characterized showing 20-nm chain aggregates/agglomerates of polycrystalline primary anatase TiO_2 particles of 3–5 nm size (Kreyling et al. 2011). As a result, they are different from commercially available TiO_2 NP. However, they are designed to challenge the nanoscale size to the lower limits and to study the biokinetics of inhaled 20-nm NP. Commercially available TiO_2 NP powders cannot be redispersed to such small entities, yet during their initial production, for example, in a flame reactor, they used to be in the 10-nm range and below. Furthermore, disagglomeration of commercial submicrometer TiO_2 NP agglomerates has been shown (Ferin et al. 1991) and cannot be excluded neither in nanotechnological processes leading to human exposure nor after incorporation of those nanomaterials in the body.

The presented data suggest that NP parameters such as material, size, morphology, hydro-/lipophilicity, surface charge, surface ligands, and their possible exchange in various body fluids need to be considered. Unfortunately, quantitative biokinetics studies are not possible in human subjects. Existing data only confirm that translocated fractions to secondary organs do not exceed the fractions found in the rat model. However, reliable precise fractions in humans are still lacking.

Currently, no quantitative biokinetics data exist on carbon nanotubes or nanowires, while there are a few data on carbon black NP; yet, there is considerable data on the toxicological effects of carbon black NP *in vitro* and *in vivo*, as a result of the scientific interest on the effects of ambient ultrafine aerosol particles in air pollution (Donaldson et al. 2005, 2006; Poland et al. 2008). However, there is growing evidence that inhalation exposure to carbon nanotubes at elevated doses induce oxidative stress and pro-inflammatory reactions in mouse models (Shvedova et al. 2008a,b, 2009).

12.6 Predominant NP Lung Retention and Limited Translocation toward the Circulation

As discussed previously, insoluble NP in rodent lungs are predominantly retained long-term in interstitial spaces of the alveolar region. Yet, there is only limited translocation toward the circulation. This is indeed very surprising since we found significant fractions

TABLE 12.1

Translocated Percentages of Alveolar NP Deposits toward Blood and Subsequent Organs and Tissues after 24 h

Ir	TiO_2	EC	Au
7.96 ± 0.47	6.95 ± 0.14	2.18 ± 0.31	1.79 ± 0.39

Source: Iridium—Kreyling WG et al., *J. Toxicol. Environ. Health* 65, 2002, 1513; Semmler M et al., *Inhal. Toxicol.*, 16, 453, 2004. Elemental carbon labeled with ^{192}Ir: Kreyling et al. 2009; titanium dioxide + gold: Kreyling, personal communication.

Note: Four different materials (iridium, elemental carbon, TiO_2, and gold) were inhaled as freshly generated 20-nm NP aerosols for 1–2 h by healthy adult female Wistar–Kyoto rats. The iridium NP translocation was significantly higher than those of elemental carbon and TiO_2 NP.

of 20-nm TiO_2 NP in endothelial cells of rat lungs within 24 h after TiO_2 NP inhalation (Geiser et al. 2005). Unfortunately, there is no direct proof for this NP accumulation in endothelial cells of large animal species and in humans; however, the study of Nikula and coworkers (1997) provides indirect evidence as discussed previously.

Yet, the limited NP translocation in rat lungs strongly depends on the physicochemical NP properties. For instance, after a 1–2-h inhalation of 20-nm NP of different materials (iridium, elemental carbon, TiO_2, and gold) by healthy adult rats, the totally translocated NP fraction across the air–blood–barrier after 24 h varies significantly by about one order of magnitude, from almost 10% for iridium NP to about 7% for TiO_2 NP and to 2% for elemental carbon NP and gold NP (see Table 12.1).

12.7 Possible Mechanisms of Translocation across the Air–Blood–Barrier

In the following, the mechanisms of mediating NP transport across the air–blood barrier are addressed; yet, it needs to be emphasized that there are still major knowledge gaps:

- According to the limited phagocytosis of NP deposited rather homogeneously on the alveolar epithelium, it is unlikely that alveolar macrophages play a major role in short-term translocation (Geiser et al. 2005, 2008, 2013; Takenaka et al. 2012) toward the interstitium, while this transport cannot be excluded during long-term retention.
- Endocytosis, including macropinocytosis of epithelial cells, appears to be an important mechanism particularly immediately after NP deposition, depending on the protein coating within the epithelial lining fluid. Mostly, NP will be taken up in endosomal vesicles, which are processed from early to late endosomes and by lysosomal fusion. Since epithelial cells are oriented from their luminal to the interstitial side, exocytosis of late endosomes appears to be an important pathway into interstitial spaces (Conner and Schmid 2003). It seems likely that the initial protein coating may have been modulated during the cellular passage.
- There is little evidence on paracellular NP transport unless the NP are very small (<5 nm) and/or the NP may have been coated by proteins favoring paracellular transport.
- After entering the interstitium, NP are obviously taken up mostly endocytotically by several cell types, including interstitial macrophages and endothelial cells of the vasculature (Geiser et al. 2005, 2013).

- Lymphatic drainage of micron-sized particles is well known to be rather limited in rodents, in contrast to the human lungs, for example. The same negligible drainage also holds for NP in the rodent lungs (Geiser and Kreyling 2010).
- It is still a surprise that NP translocation from endothelial cells to blood circulation is limited. Since endothelial cells are also oriented from the vascular to the interstitial side, this may modulate exocytosis to the blood lumen and requires more research.
- The other surprise is the long-term clearance of NP from the air–blood–barrier back onto the epithelium for macrophage-mediated transport toward the larynx. Preliminary analyses of this transport of inhaled 20-nm TiO_2 NP and gold NP confirm those data obtained for inhaled iridium NP as mentioned previously. It remains unclear how the NP reappear back on the epithelium: either by macrophage transport from the interstitium to the alveolar epithelium and/or through broncho-associated lymphoid tissue. In any case, it remains remarkable that the long-term clearance kinetics of this macrophage-mediated transport toward the larynx occurs with the same rate as micron-sized particles, which were shown to reside on the epithelium taken up by macrophages until the latter find their way out toward the mucociliary escalator.

12.8 What Are the Consequences of NP Translocation—A Work-in-Progress Statement

Acute effects resulting directly from translocated NP in secondary organs are likely to be rather low because of the estimated low accumulation fractions of NP tested thus far. However, chronic exposure will lead to cumulative accumulation of insoluble NP in secondary organs, which may well mediate adverse health effects including pro-inflammatory responses, eventually progressing into inflammatory diseases in those secondary organs. In addition, beyond the direct effect of translocated NP, it appears worthwhile to investigate the effects caused by (i) mediators released from the lungs as the primary organ of intake to blood as a result of the interaction of freshly inhaled NP with lung tissues even after short-term exposures or by (ii) alterations of the autonomous nervous system as a result of the inhalation of engineered NP.

12.9 Gaps of Knowledge

Hence, there are gaps of knowledge that need to be addressed in due course for a comprehensive risk assessment. The following is a limited list of issues:

1. Mechanisms determining the transport of NP through cell membranes and biological membranes such as the air–blood barrier of the lungs, the epithelium of the intestine, the blood–brain–barrier, the placental barrier, etc., and the role of NP properties and appropriate metrics
2. Toxicological responses of cells and membranes to NP using well-defined and relevant NP doses

3. Repeated or chronic NP exposure studies with subsequent accumulation of NP in secondary organs for dose estimates to plan subsequent toxicological testing
4. Mediator release in circulation after NP exposure to the lungs and their role in triggering adverse cardiovascular effects
5. Dose–response studies in case of toxic outcomes in primary or secondary organs

Acknowledgments

This work was partially supported by European Union (EU) FP5 FIGD-CT-2000-00053, EU FP6 PARTICLE_RISK 012912 (NEST), EU FP7 NeuroNano (NMP4-SL-2008-214547) and ENPRA (NMP4-SL-2009- 228789), and US-NIH RO1HL070542.

Parts of this chapter have been published within the monograph "Nanomaterials Commission for the Investigation of Health Hazards of Chemical Compounds in the Work Area of the German Research Foundation, DFG." Wiley-VCH, Weinheim 2013.

References

Balasubramanian, S. K., Poh, K. W., Ong, C. N., Kreyling, W. G., Ong, W. Y. and Yu, L. E. 2013. The effect of primary particle size on biodistribution of inhaled gold nano-agglomerates. *Biomaterials* **34**(22), 5439–52, 10.1016/j.biomaterials.2013.03.080.

Brown, J. S., Zeman, K. L. and Bennett, W. D. 2002. Ultrafine particle deposition and clearance in the healthy and obstructed lung. *Am J Respir Crit Care Med* **166**(9), 1240–7.

Conner, S. D. and Schmid, S. L. 2003. Regulated portals of entry into the cell. *Nature* **422**(6927), 37–44, 10.1038/nature01451.

Donaldson, K., Aitken, R., Tran, L., Stone, V., Duffin, R., Forrest, G. and Alexander, A. 2006. Carbon nanotubes: A review of their properties in relation to pulmonary toxicology and workplace safety. *Toxicol Sci* **92**, 5–22.

Donaldson, K., Tran, L., Jimenez, L. A., Duffin, R., Newby, D. E., Mills, N., Macnee, W. and Stone, V. 2005. Combustion-derived nanoparticles: A review of their toxicology following inhalation exposure. *Part Fibre Toxicol* **2**, 10.

Ferin, J., Oberdörster, G., Soderholm, S. C. and Gelein, R. 1991. Pulmonary tissue access of ultrafine particles. *J Aerosol Med* **4**, 57–68.

Geiser, M., Casaulta, M., Kupferschmid, B., Schulz, H., Semmler-Behnke, M. and Kreyling, W. 2008. The role of macrophages in the clearance of inhaled ultrafine titanium dioxide particles. *Am J Respir Cell Mol* **38**, 371–6.

Geiser, M. and Kreyling, W. G. 2010. Deposition and biokinetics of inhaled nanoparticles. *Part Fibre Toxicol* **7**(1), 2, 10.1186/1743-8977-7-2.

Geiser, M., Quaile, O., Wenk, A., Wigge, C., Eigeldinger-Berthou, S., Hirn, S., Schaffler, M., Schleh, C., Moller, W., Mall, M. A. and Kreyling, W. G. 2013. Cellular uptake and localization of inhaled gold nanoparticles in lungs of mice with chronic obstructive pulmonary disease. *Part Fibre Toxicol* **10**(1), 19, 10.1186/1743-8977-10-19.

Geiser, M., Rothen-Rutishauser, B., Kapp, N., Schurch, S., Kreyling, W., Schulz, H., Semmler, M., Im Hof, V., Heyder, J. and Gehr, P. 2005. Ultrafine particles cross cellular membranes by nonphagocytic mechanisms in lungs and in cultured cells. *Environ Health Perspect* **113**, 1555–60.

International Commission on Radiological Protection 1994. Human respiratory tract model for radiological protection. A report of a Task Group of the International Commission on Radiological Protection. *Ann ICRP* **24**(1–3), 1–482.

Kapp, N., Kreyling, W., Schulz, H., Im Hof, V., Gehr, P., Semmler, M. and Geiser, M. 2004. Electron energy loss spectroscopy for analysis of inhaled ultrafine particles in rat lungs. *Microsc Res Tech* **63**, 298–305.

Kreyling, W. G. 1990. Interspecies comparison of lung clearance of "insoluble" particles. *J Aerosol Med* **3** Suppl. 1, S93–110.

Kreyling, W. G., Biswas, P., Messing, M. E., Gibson, N., Geiser, M., Wenk, A., Sahu, M., Deppert, K., Cydzik, I., Wigge, C., Schmid, O. and Semmler-Behnke, M. 2011. Generation and characterization of stable, highly concentrated titanium dioxide nanoparticle aerosols for rodent inhalation studies. *J Nanopart Res* **13**, 511–24, 10.1007/s11051-010-0081-5.

Kreyling, W. G. and Geiser, M. 2010. Dosimetry of inhaled nanoparticles. In: *Nanoparticles in Medicine and Environment, Inhalation and Health Effects*. (Eds. Marijnssen, J., Gradon, L.) Springer, Berlin, 145–72.

Kreyling, W. G. and Scheuch, G. 2000. Clearance of particles deposited in the lungs. In: *Particle–Lung Interactions*. (Eds. Heyder, J., Gehr, P.) Marcel Dekker, New York, 323–76.

Kreyling, W. G., Semmler, M., Erbe, F., Mayer, P., Takenaka, S., Schulz, H., Oberdörster, G. and Ziesenis, A. 2002. Translocation of ultrafine insoluble iridium particles from lung epithelium to extrapulmonary organs is size dependent but very low. *J Toxicol Environ Health* **65**, 1513–30.

Kreyling, W. G., Semmler-Behnke, M., Seitz, J., Scymczak, W., Wenk, A., Mayer, P., Takenaka, S., and Oberdorster, G. 2009. Size dependence of the translocation of inhaled iridium and carbon nanoparticle aggregates from the lung of rats to the blood and secondary target organs. *Inhalation Toxicology* **21**, 55–60.

Kreyling, W. G., Semmler-Behnke, M., Takenaka, S. and Moller, W. 2013. Differences in the biokinetics of inhaled nano- versus micrometer-sized particles. *Acc Chem Res* **46**, 714–22.

Lundqvist, M., Stigler, J., Elia, G., Lynch, I., Cedervall, T. and Dawson, K. A. 2008. Nanoparticle size and surface properties determine the protein corona with possible implications for biological impacts. *Proc Natl Acad Sci USA* **105**(38), 14265–70, 10.1073/pnas.0805135105.

Lynch, I., Cedervall, T., Lundqvist, M., Cabaleiro-Lago, C., Linse, S. and Dawson, K. A. 2007. The nanoparticle–protein complex as a biological entity; a complex fluids and surface science challenge for the 21st century. *Adv Colloid Interface Sci* **134–135**, 167–74, 10.1016/j.cis.2007.04.021.

Mills, N. L., Amin, N., Robinson, S. D., Anand, A., Davies, J., Patel, D., de la Fuente, J. M., Cassee, F. R., Boon, N. A., Macnee, W., Millar, A. M., Donaldson, K. and Newby, D. E. 2006. Do inhaled carbon nanoparticles translocate directly into the circulation in humans? *Am J Respir Crit Care Med* **173**(4), 426–31.

Möller, W., Felten, K., Seitz, J., Sommerer, K., Takenaka, S., Wiebert, P., Philipson, K., Svartengren, M. and Kreyling, W. G. 2006. A generator for the production of radiolabelled ultrafine carbonaceous particles for deposition and clearance studies in the respiratory tract. *J Aerosol Sci* **37**(5), 631–44.

Möller, W., Felten, K., Sommerer, K., Scheuch, G., Meyer, G., Meyer, P., Haussinger, K. and Kreyling, W. G. 2008. Deposition, retention, and translocation of ultrafine particles from the central airways and lung periphery. *Am J Respir Crit Care Med* **177**(4), 426–32, 10.1164/rccm.200602-301OC.

Mühlfeld, C., Geiser, M., Kapp, N., Gehr, P. and Rothen-Rutishauser, B. 2007. Re-evaluation of pulmonary titanium dioxide nanoparticle distribution using the "relative deposition index": Evidence for clearance through microvasculature. *Part Fibre Toxicol* **4**, 7, 10.1186/1743-8977-4-7.

Nemmar, A., Hoet, P. H., Vanquickenborne, B., Dinsdale, D., Thomeer, M., Hoylaerts, M. F., Vanbilloen, H., Mortelmans, L. and Nemery, B. 2002. Passage of inhaled particles into the blood circulation in humans. *Circulation* **105**(4), 411–4.

Nikula, K. J., Avila, K. J., Griffith, W. C. and Mauderly, J. L. 1997. Sites of particle retention and lung tissue responses to chronically inhaled diesel exhaust and coal dust in rats and cynomolgus monkeys. *Environ Health Perspect* **105** Suppl. 5, 1231–4.

Oberdörster, G., Ferin, J. and Lehnert, B. E. 1994b. Correlation between particle size, *in vivo* particle persistence, and lung injury. *Environ Health Perspect* **102** Suppl. 5, 173–9.

Oberdörster, G., Ferin, J., Soderholm, S. C., Gelein, R., Cox, C., Baggs, R. and Morrow, P. E. 1994a. Increased pulmonary toxicity of inhaled ultrafine particles: Due to lung overload alone? *Ann Occup Hyg* **38** Suppl. 1, 295–302.

Poland, C. A., Duffin, R., Kinloch, I., Maynard, A., Wallace, W. A. H., Seaton, A., Stone, V., Brown, S., MacNee, W. and Donaldson, K. 2008. Carbon nanotubes introduced into the abdominal cavity of mice show asbestos-like pathogenicity in a pilot study. *Nat Nanotechnol* 3(7): 423–8.

Semmler, M., Seitz, J., Erbe, F., Mayer, P., Heyder, J., Oberdorster, G. and Kreyling, W. G. 2004. Long-term clearance kinetics of inhaled ultrafine insoluble iridium particles from the rat lung, including transient translocation into secondary organs. *Inhal Toxicol* **16**, 453–9.

Semmler-Behnke, M., Fertsch, S., Schmid, O., Wenk, A. and Kreyling, W. G. 2007b. Uptake of 1.4 mm versus 18 mm gold particles by secondary target organs is size dependent in control and pregnant rats after intertracheal or intravenous application. In: *Euro Nanoforum 2004*. Proceedings of Euro Nanoforum - Nanotechnology in Industrial Applications pp. 102–4. http://www.euronanoforum2007.de/download/Proceedings%20ENF2007.pdf.

Semmler-Behnke, M., Kreyling, W., Lipka, J., Fertsch, S., Wenk, A., Takenaka, S., Schmid, G. and Brandau, W. 2008. Biodistribution of 1.4- and 18-nm gold particles in rats. *Small* **4**, 2108–11.

Semmler-Behnke, M., Takenaka, S., Fertsch, S., Wenk, A., Seitz, J., Mayer, P., Oberdorster, G. and Kreyling, W. G. 2007a. Efficient elimination of inhaled nanoparticles from the alveolar region: Evidence for interstitial uptake and subsequent reentrainment onto airways epithelium. *Environ Health Perspect* **115**, 728–33.

Shvedova, A. A., Fabisiak, J. P., Kisin, E. R., Murray, A. R., Roberts, J. R., Tyurina, Y. Y., Antonini, J. M., Feng, W. H., Kommineni, C., Reynolds, J., Barchowsky, A., Castranova, V. and Kagan, V. E. 2008b. Sequential exposure to carbon nanotubes and bacteria enhances pulmonary inflammation and infectivity. *Am J Respir Cell Mol Biol* **38**(5), 579–90, 10.1165/rcmb.2007-0255OC.

Shvedova, A. A., Kisin, E. R., Murray, A. R., Kommineni, C., Castranova, V., Fadeel, B., and Kagan, V. E. 2008a. Increased accumulation of neutrophils and decreased fibrosis in the lung of NADPH oxidase-deficient C57BL/6 mice exposed to carbon nanotubes. *Toxicol Appl Pharmacol* **231**, 235–40.

Shvedova, A. A., Kisin, E. R., Porter, D., Schulte, P., Kagan, V. E., Fadeel, B. and Castranova, V. 2009. Mechanisms of pulmonary toxicity and medical applications of carbon nanotubes: Two faces of Janus? *Pharmacol Ther* **121**(2), 192–204.

Takenaka, S., Moller, W., Semmler-Behnke, M., Karg, E., Wenk, A., Schmid, O., Stoeger, T., Jennen, L., Aichler, M., Walch, A., Pokhrel, S., Madler, L., Eickelberg, O. and Kreyling, W. G. 2012. Efficient internalization and intracellular translocation of inhaled gold nanoparticles in rat alveolar macrophages. *Nanomedicine (Lond)* 7(6), 855–65, 10.2217/nnm.11.152.

Wiebert, P., Sanchez-Crespo, A., Falk, R., Philipson, K., Lundin, A., Larsson, S., Moeller, W., Kreyling, W. and Svartengren, M. 2006a. No significant translocation of inhaled 35-nm carbon particles to the circulation in humans. *Inhal Toxicol* **18**(10), 741–7.

Wiebert, P., Sanchez-Crespo, A., Seitz, J., Falk, R., Philipson, K., Kreyling, W. G., Moller, W., Sommerer, K., Larsson, S. and Svartengren, M. 2006b. Negligible clearance of ultrafine particles retained in healthy and affected human lungs. *Eur Respir J* **28**(2), 286–90.

Section V

Drug Delivery to the Respiratory Tract

13

Practical Considerations for Drug Delivery to the Respiratory Tract

John S. Patton

CONTENTS

13.1 Initial Comments

There is tremendous excitement today about nanotechnology. The nanometer size range spans the diameters of individual soluble protein molecules (2–16 nm; http://www.ivdtechnology.com/article/measuring-hydrodynamic-radius-nanoparticle-formulations), to the thickness of the type 1 alveolar epithelial cells of the lungs (100 nm) (Patton 1996), to the diameters of viruses (30–300+ nm) and bacteria (250–3000 nm) (http://www.ionizers.org/Sizes-of-Bacteria.html). Human cells are amazing assemblages of exquisite nanostructures that make up the living system (bilayer membranes, ribosomes, endosomes, mitochondria, chloroplasts, Golgi apparatus, chromosomes, tight junctions, etc.). Biochemists, cell biologists, and drug developers have been examining, isolating, characterizing, and making some types of nano-sized structures for nearly a hundred years (subcellular organelles, micelles, crystals, colloids, liposomes, etc.). Why has this particular size range become so exciting lately? It is because we are learning how to precisely make complex nanostructures and how to manipulate their properties, and discovering their interesting and powerful applications in medicine and in all fields of science. It is the size range just above the size of a small molecule (the water molecule has a diameter of ~0.3 nm), partly visible by electron microscopy and at an order level in which an almost infinite number of complex, multifunctional physical chemical structures relevant to humans can be designed.

13.2 Optimum Aerosol Particle Size for Efficient Lung Deposition

There is no medical device thus far that can produce dense clouds of nanoparticles (NPs) that can conveniently deliver the required masses of typical inhaled medicines in a few breaths. If there were, it would not be very efficient because most of the mass of inhaled NPs would remain suspended in the inhaled air and then be exhaled (Al-Hallak et al. 2011; Carvalho et al. 2011; Londahl et al. 2013). Thus, NPs must be placed into larger aerosol particles (~1–5 micron diameters) that can be efficiently inhaled and deposited in the lungs so that an adequate drug dose can be conveniently delivered to patients within a reasonable time.

Inhaled medicines are typically delivered to the respiratory tract in clouds of polydisperse aerosol particles that average in size from about ~1 to 5 microns in aerodynamic diameter, and number in the hundreds of millions of particles per inhalation. Within these clouds of inhaled medicines, there are some smaller particles in the nanometer size range that make up a significant percentage of the mass and some larger particles and clumps >10 microns in diameter. Within this range, all particles, but especially the larger ones, have the possibility of impacting in the mouth and throat if they are inhaled too rapidly, and thus they may never reach the lungs. Also within this range, the smaller particles particularly (i.e., <0.5 microns [500 nm]) have a possibility of staying suspended in the inhaled air and being exhaled. Therefore, the large and small ends of the size distribution of medical aerosols are mostly wasted with most of the aerosol mass deposited in the lungs being within 1–5 microns. The average particle size of cigarette smoke is about 300 nm (Becquemin et al. 2007), which is too small for high-efficiency therapeutic drug deposition, although obviously efficient enough to deliver the potent nicotine molecule even with a high percentage of exhaled smoke.

There is a fascinating phenomenon with NPs below the size of ~200–300 nm—their lung deposition starts to increase to a certain point. This may contribute to the efficiency of nicotine absorption from combustion smoke. According to Londahl et al. (2013), the alveolar deposition fraction peaks at about 30% with NP diameters of ~30 nm and tracheobronchial deposition peaks at about 35% with NP diameters of about 6–7 nm. The implications for this phenomenon for therapeutic drug delivery are not clear. Could therapeutic proteins be molecularly dispersed in air and inhaled as single molecules or clumps of a few molecules?

Like exhaled smoke, exhaled medicines are not desired. Thus, unless one wants to deal with filters to capture exhaled aerosols, the most useful and efficient average aerodynamic size for an inhaled cloud of therapeutic drug particles is again around ~1–5 microns in diameter—a size range where little mass is exhaled (Clark 2012). Moreover, if slowly inhaled (the secret to efficient lung deposition) (Brand et al. 1999, 2003), a size range where a high percentage of the mass can be deposited in the lungs (>80% of the dose emitted from some inhalers) is needed. However, for most commercial medical inhalers, it is fortunate to get 20%–30% of the loaded dose emitted from the device into the lungs because of the high-velocity exit of the aerosols from the device (and the subsequent high deposition in the mouth and throat) caused by either the device ejecting the aerosol out too fast or the patient inhaling too fast (Borgstrom et al. 2006; Clark 2012). By using different breathing maneuvers, the lung deposition site can be partially targeted (Brand et al. 1999, 2003). When considering inhaled NP gene delivery vectors, one needs to consider that any inhaler delivery system will leave some material in the back of the throat (10%–90%; see Section 13.4); therefore, in addition to the lungs, gene expression is a possibility in the throat and esophagus and further down the gastrointestinal tract. Since gene delivery may afford a therapeutic effect for a long time, the need for short inhalation times may not be required and subjects may be happy to inhale for a period of many hours to receive more targeted pulmonary gene delivery.

However, if we put aside the deposition site, delivery time, and the overall efficiency of aerosol delivery for the moment, it is important to visualize how many NPs could fit into a typical aerosol drug delivery particle that is being inhaled with millions of other similar-sized aerosol particles. Many would say that a 2-micron particle is one of the best sizes for efficient aerosol drug delivery, assuming the particle does not agglomerate too much as it comes out of the inhaler and the aerosol is slowly inhaled. A 2-micron-diameter liquid aerosol particle is shown in Figure 13.1, next to some well-known NPs drawn to scale: a monoclonal antibody (hydrodynamic diameter, 8–16 nm); a unilamellar liposome with a diameter of 100 nm (also similar to the diameter of an influenza virus); and a very small bacteria, *Mycoplasma pneumoniae*, with a diameter of 250 nm.

It is clear that the best way to deliver a significant mass of NPs into the lungs in a few breaths is to place them into the ~1–5-micron particles. Note in Figure 13.1 how many NPs could be tightly packed into the volume of a 2-micron-diameter aqueous spherical aerosol

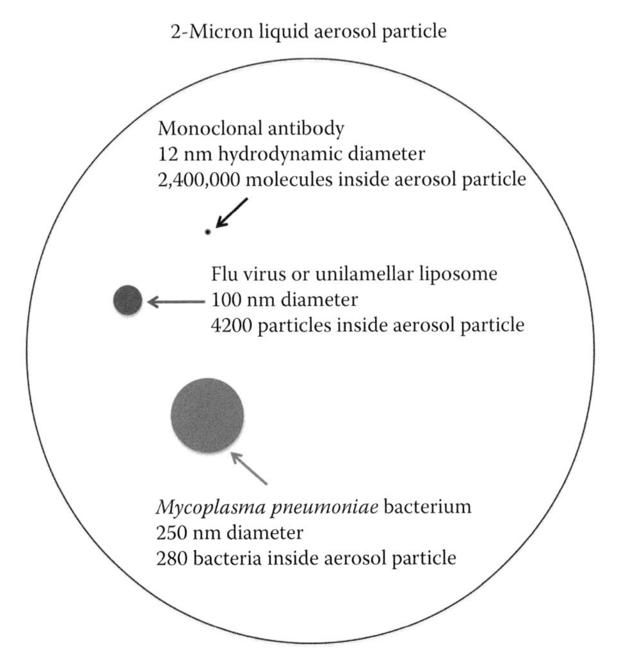

FIGURE 13.1
Scale drawing of a typical optimum-sized inhalable liquid aerosol particle with representative NPs shown to scale.

particle (assuming that each spherical NP has a packing volume of a cube, which leaves some room for additional excipient). Dry powder particles are usually not spherical; however, with new particle engineering techniques, they can be made to be roughly spherical under certain conditions and with certain excipients (Edwards and Dunbar 2002; Weers et al. 2007). This author is not aware of any particular advantage of the spherical shape for dispersion and lung deposition of aerosol particles. However, it is in the porosity of particles where some real advantages can be seen (Edwards and Dunbar 2002; Weers et al. 2007).

A typical breath of an inhaled medicine will contain hundreds of millions of aerosol particles. A single 2-micron-diameter dry aerosol particle can contain several hundreds of million insulin molecules (diameter 2 nm)—not shown in Figure 13.1. If one inhales dry powder aerosol insulin particles in the amorphous dry state or an aerosol of water particles with the insulin dissolved in them, the absorption of the inhaled insulin occurs at the same rate (Heinemann and Heise 2004). Amorphous insulin is highly water soluble. The individual molecules are quickly released from the dry 1–2 μm particle in the pulmonary fluids and behave like the molecules delivered in the aqueous aerosol. Larger macromolecules like the therapeutically important antibodies with hydrodynamic diameters of about 8–16 nm are NPs all by themselves (Figure 13.1). For details of the determinants of protein absorption, see Section 13.6 and Patton (1996).

13.3 Practical Issues of Inhaled Mass and Irritation

There is a limit to how much mass can be chronically inhaled by humans without causing coughing, irritation, bronchospasm, and inflammation (Brannan et al. 2005). Even the most benign aerosol will cause physical irritation and coughing if the mass is too high. Furthermore, some aerosols will be more irritating than others. I have saved very large amounts of development dollars by first inhaling a new experimental aerosol before embarking on very expensive toxicological studies. Everything can be perfect on the bench top—chemical and physical stability, particle size, dispersion, and delivered dose—but when you inhale it, it can make your lungs erupt with pain.

While one can chronically deliver grams of therapeutics and their excipients (additives) orally, the practical upper limit of daily pulmonary mass delivery (to subjects without asthma, where the safe daily mass permissible would be much lower) is probably about a total of 100–200 mg of aerosol dry mass (drugs plus excipients) per single dose (Konstan et al. 2010, 2011). One would want a lower amount than this, if possible, for the patient's comfort. If one extrapolates to animals, based purely on the dry mass of aerosol/weight of animal, this would be equivalent to about 0.3–0.6 mg maximum inhaled dry mass for a 250-g rat or about 25–50 μg inhaled dry mass per a 20-g mouse.

If an aerosol particle contains 10% drug and 90% excipient, then the maximum drug mass one should give to animals would be about 30–60 μg for a 250-g rat or 2.5–5 μg for a 20-g mouse. These simple pulmonary mass limitations need to be considered when designing animal studies—investigators need to keep these limits in mind, which are independent of the drug. Obviously, more potent drugs have a better chance of staying under the mass limits; however, adding a high percentage of excipient can destroy the potency advantage.

Extremely potent drugs have dominated inhaled drug delivery historically, and until recently, delivering tens or hundreds of milligrams was not possible unless one inhaled from a nebulizer solution for long periods. Today, the three most popular classes of inhaled drugs that are used singly or in combination to treat asthma and chronic obstructive pulmonary disease (COPD, primarily caused by smoking, called "smokers lungs") are the bronchodilators, bronchosteroids, and anticholinergics. These drugs are some of the most potent drugs in the physician's medicine chest and require doses delivered to the lungs of only 100–200 µg or less. Representatives from these classes were available as oral medicines back in the 1950s and have since all been converted to inhaled forms to provide fast action, lower the whole-body dose, and reduce systemic side effects.

13.4 Inhaler Devices

The inhalers that were invented during the last 60 years to deliver these potent drugs have not had to be very efficient since the drugs are relatively inexpensive, and as long as 10%–20% of the aerosol emitted from the device reach the lungs, people were generally satisfied. The 80%–90% of the drug that was "lost" was usually stuck in the mouth and the back of the throat and swallowed (Clark 2012). Thus, with 10% efficiency and a lung deposition requirement of 100 µg, one would need to have an inhalation device that emits about 1 mg of drug aerosol. Because of the great potency of the most important inhaled drugs (and despite the device's poor delivery efficiencies), the payload capabilities (mass of drug delivered to the lungs per inhalation) of the inhalers developed in the last 60 years have been limited. Unfortunately, most of these inhalers developed have also suffered from reliability of dosing issues; however, today's inhalers have become much more reliable (Borgstrom et al. 2006). The following are descriptions of the three classes of inhaled drug delivery devices and their payload capabilities, and some features with regard to NPs.

13.4.1 Nebulizers

These electronic devices that aerosolize liquid formulations (usually aqueous) have been around for a long time. Their demise was often predicted because they were large and cumbersome, and it could take 20–30 min to dispense a few milliliters of fluid. However, their gentle soft mist and ease and reliability of dosing have kept them as popular as ever among a subset of asthmatics and other users, including children, who could wear a mask to enable both mouth and nose breathing. If a drug has good water solubility, such as tobramycin, then high payloads (100 mg) dissolved in 2 to 4 ml of fluid can be given during 20–30 min. For the researcher, they represent the least expensive and easiest devices for exposing animals and obtaining human data. Now some companies, to make them discreet and even more acceptable, are miniaturizing those devices (e.g., www.Dancebiopharm.com). They come in two primary forms: the jet nebulizer and the vibrating mesh. These operate by two very different mechanisms.

- *Jet nebulizers*: These operate by directing a high-velocity air–liquid jet of solution through a venturi nozzle against a plastic ball in a semicontained space, where only the smallest particles generated are emitted into the breathing air stream, while the majority of fluid falls back and is redirected against the ball again and again through the venturi nozzle until eventually the fluid volume runs low and the device starts to sputter. Soluble small molecules are usually fine with this process, unless they foam, and some proteins also do well (i.e., Pulmozyme); however, the process can denature some proteins, causing them to aggregate and precipitate (i.e., growth hormone and interferons) and can cause the water-soluble contents of liposomes to be released (Niven and Schreier 1990; Niven et al. 1991, 1992). However, it is difficult to predict what preparations will be adversely affected by jet nebulization. Fortunately, it is easy to test on the laboratory bench. Although not reduced to practice commercially, nebulization with subsequent particle drying could be used to produce very fine particles, potentially in the nanometer range. Depending on the chemical composition, a dry aerosol may take up moisture in the airways and grow in size.

- *Vibrating mesh nebulizers*: These newer devices utilize a piezo-actuator to vibrate metal or plastic meshes with small holes in them (3–5 μm diameter) up to 120,000 cycles/s to produce droplets in the 3–5 μm size range. The solution is not subjected to constant recycling and is sucked through the mesh in a single pass. Pulmicourt suspension is an example of an aqueous suspension of drug formulation that works well with the vibrating mesh. Although clogging may occur if one tries to use large suspended particles, the vibrating mesh could be ideal for delivering suspensions of insoluble NPs, especially the highly engineered virus-like particles that have a stable structure and small size (<500 nm).

13.4.2 Metered Dose Inhalers (MDIs)

These famous "asthma puffers" were invented by Charlie Thiel at 3M in the 1950s and became the most popular inhaler in the world (Rau 2005). Drug particles are usually suspended in inert propellants, namely chlorofluorocarbons (CFCs—now banned in many countries for destroying atmospheric ozone) and hydrofluoroalkanes (HFAs—a potent greenhouse gas whose use may eventually also be banned), in tightly sealed canisters that have a valve that releases specific volumes of fluid suspension (28–100 μl) when one presses the canister down on the actuator. The very low boiling propellants rapidly evaporate once out of the device, leaving the particles as an aerosol, albeit a high-velocity one. Few molecules are soluble in HFAs; thus, one must make suspensions of solid drug particles that do not aggregate. These devices are notorious for being difficult for the patient to coordinate inhalation of the aerosol with pressing the canister. Add-on spacer devices have been developed to help minimize this problem as well as a breath-actuated MDI, the Autohaler. Because of the very high velocity of the aerosol, without a spacer, most of it (up to 90%) becomes impacted in the back of the throat and is swallowed (Rau 2005). In addition to these reliability issues, the payload from MDIs is very small, a maximum of about 1–2 mg per actuation, of which only a small fraction (~5–10%) reaches the lungs. Formulating in MDIs is notoriously finicky and difficult. Making particles that do not aggregate and change character over time (Oswald ripening) is difficult, and the number of excipients that are compatible with HFAs is extremely limited. Expensive specialized equipment is needed to fill the canisters.

There is one highly unique formulation with MDIs that created very fine particles (QVar) with high lung deposition efficiency (~65%). Ethanol is one bio-acceptable molecule (at least in tiny doses) that can be dissolved in HFAs up to about 15%. 3M dispersed beclomethasone, a lipophilic broncho-steroid, in ethanol to make the first MDI formulation that created very fine particles when the propellant evaporates (Leach 2002). Chiesi also has what they call extrafine particles (1.4–1.5-micron aerodynamic diameters), and their technique is called Modulite (http://www.future-science-group.com/_img/pics/Beclomethasone_diprorionate_formoterol_4.pdf).

In general, we do not recommend that researchers start with MDIs if they are considering NP delivery to the lungs because the equipment, materials, and testing are cumbersome and expensive. However, for very potent and stable preparations, they may be suitable.

13.4.3 Dry Powder Inhalers

Because of the phase out of the CFCs and uncertainty about HFAs, there has been an explosion in the development of dry powder inhalers during the last 25 years (Edwards and Dunbar 2002; White et al. 2005; Harper et al. 2007; Weers et al. 2007; Sadrzadeh et al. 2010). These have their own set of advantages and problems, and historically have been tricky and expensive to develop; however, many advances have been made, and making a dispersible dry powder system is becoming much easier (Weers et al. 2007). Powders can be quite stable, and new processing methods (i.e., spray drying) are a dream for the formulator of NPs because so many process variables can be manipulated in designing particles, such as soluble drug and excipient concentrations, spray droplet particle size, suspension mixing with soluble excipients, drying temperature, spray flow rate, and solvent ratios (water-to-ethanol).

Spray drying was first introduced into the aerosol formulation design by Robert Platz in the late 1980s at SRI and later commercialized in inhaled insulin (Exubera) at Inhale/Nektar where Mr. Platz was a co-founder (there are 39 spray drying patents with his name). Spray drying has not yet been widely adopted by pharmaceutical companies but it will be. There are small bench-top spray dryers (Buchi) for development work, and innovative small companies are adopting spray drying for formulating medical aerosols (i.e., http://flurrypowders.com). Although spray drying has thus far been used to make particles with an average diameter of 1–3 μm, it should be possible to go into the nanometer range. More efficient collecting methods will have to be developed to capture spray-dried NPs as they tend to stay suspended in the exhaust air in the typical cyclone precipitators that are used to collect particles in the 1–5 micron range. However, this is not necessarily required because suspensions of NPs can be co-spray-dried with water-soluble excipients to make optimal inhalable particles (1–3 μm) that once impacted in the lungs can "dissolve" to release the free NPs into the lung lining fluids (Li and Zhang 2012).

13.4.3.1 Two Major Challenges with Dry Powders: Agglomeration and Moisture Protection

One common and sometimes daunting problem with dry powder aerosols is agglomeration (Wong et al. 2010), and this can be an issue with NPs, which also makes them stick to packaging and increases their aerodynamic size and tendency to be impacted in the mouth, throat, and upper airways. The tendency of particles to stick to each other can be reduced by altering the surface chemistry and structure of the particles (Weers et al. 2007). The "dispersibility" of dry powders has thus become their key feature for inhalation purposes. It is not good enough to make the right-sized particles; they also need to be dispersible. In general, uncharged particle surfaces with slightly hydrophobic character are the most dispersible.

Agglomeration's evil brother is moisture sensitivity, which can be death to a good dispersible powder. Some spray-dried powders are so hygroscopic that it is impossible to weigh them in an open laboratory as they suck moisture from the room and turn to goo. Thus, one of the most challenging aspects of developing a dry powder aerosol is the need for moisture control and protection during processing, storage, and packaging. Dry powders for inhalation have to be handled in humidity-controlled glove boxes and double foil packaged to prevent moisture uptake as products. Fortunately, technologies have been developed and are available to enable companies to develop and package dry powder aerosol products (White et al. 2005; Harper et al. 2007; Sadrzadeh et al. 2010; http://flurry powders.com).

13.4.4 SUPRAER Technology

Donovan Yeates, Tony Hickey, and their collaborators have developed a novel inhalation device that they claim enables the gentle delivery of labile biologics, microorganisms, and cells. In addition, the device heats liquid aerosol particles to turn them into dry powders, thus reducing their size (http://www.kaerbio.com/supreaertechnology .htm).

13.5 Events at the Air–Blood Tissue Barrier: Regional Deposition and Clearance of Nanoparticles

Inhaled NPs do not readily cross from the lungs into the systemic circulation. Thus, they can potentially provide controlled release of therapeutic entities in the lungs. If deposited, they stick in the lungs and if not metabolized, may gradually over hours, days, weeks, months, or years migrate into other parts of the body. Multiple mechanisms are in place to hold, dissolve, destroy, and move them out of the lungs into the gastrointestinal tract or into the foreign particle–processing mechanisms of the immune system. A recent study with biopersistent 20-nm radiolabeled gold particles in mice found that about 1% of the particles reached the circulation after 2 h, with >70% associated with lung tissue and lavage cells (Schleh et al. 2013).

The two major regions of the lungs, conducting airways (trachea, bronchi, and bronchioles) and alveoli, are composed of different cells and possess very different barrier properties and clearance mechanisms to keep the lungs clean (Patton 1996; Ganesan et al. 2013). The critical mucociliary escalator operates in the airways to sweep deposited inhaled particles up and out of the lungs during a period of many hours such that the material is eventually swallowed. This cilia-driven clearance mechanism is absent from the alveolar region, where surfactant envelopment and opsonization of particles by pulmonary "collectins" (Atochina-Vasserman et al. 2010) (the corona effect), and subsequent macrophage/dendritic cell particle engulfment and digestion (Ruge et al. 2012) along with dispersion, metabolism, and absorption of soluble molecules, are the major clearance mechanisms for molecules and particles. Lung surfactant molecules spread over all of the air/lung interphase and will tend to envelope all inhaled foreign particles, and targeting aerosols to general regions of the lungs (i.e., peripheral [alveoli] vs. central [airways]) is partially possible by manipulating the breathing maneuver and particle size (Brand et al. 1999, 2003; Borgstrom et al. 2006; Clark 2012); however, very specific regional targeting is virtually impossible with conventional aerosols.

Thus, many partially soluble or insoluble nano- to micron-sized particles tend to eventually be scavenged by airway and alveolar dendritic cells and macrophages that have evolved to sense, engulf, and destroy viral, fungal, and bacterial cells and structures that mimic or suggest such entities (Ruge et al. 2012). Thus, there can be a relatively high efficiency uptake of nano- and micron-sized particles by these types of immune-processing cells (Lombry et al. 2004).

In the airways, just the penetration of the mucus layer can be challenging for NPs. Justin Hanes and colleagues have studied NP penetration of airway mucus extensively (Schuster et al. 2013) and have developed highly compacted DNA NPs that overcome the mucus barrier (da Silva et al. 2014; Suk et al. 2014).

There are many deadly and often incurable lung diseases with unique pathologies (i.e., idiopathic pulmonary fibrosis, primary pulmonary hypertension, lung cancer, pneumonia, adult respiratory distress syndrome, etc.). The specific diseased tissue is often poorly vascularized and difficult to reach by aerosol or from the blood side. The science of targeting these complex diseased regions in the lung is embryonic and an extremely fertile region for new nanomedical research (Patton et al. 2010).

13.6 Events at the Air–Blood Tissue Barrier: Systemic Absorption and Metabolism of Nanomolecules (i.e., Proteins)

Although there are many DNA, RNA, and other nano-sized biological molecules in development, the class with the most commercial impact and attention today are proteins. This exciting class of innovative therapeutics produced by recombinant DNA technology, many with hydrodynamic diameters of 8–16 nm (the antibodies and conjugates), includes many of the top 25 best-selling drugs of 2013. Eleven of the top sellers are proteins with collective revenues of $73.4B, including the top 3 (Humira, Remicade, and Rituxin) (*Genetic Engineering News*, March 3, 2014). The top two antibodies are anti-inflammatory proteins and could potentially be inhaled to control asthma without the systemic side effects of these proteins, which can be significant. We have spent almost 30 years studying the absorption of insulin and other macromolecules by the lungs (Wolff 1998; Patton et al. 2004b; Fineberg et al. 2005; Ceglia et al. 2006; Patton and Byron 2007; McElroy et al. 2013). The development of inhaled insulin has attracted many billions of investor dollars, and a new inhaled insulin product, Afrezza, was approved by the FDA on June 27 of this year (Neumiller et al. 2010). Protein molecules are generally freely soluble in water and physiological fluids. The lungs are the only natural noninvasive port of entry into the body for pharmaceutical quantities of therapeutic macromolecules. The nasal, dermal, oral, and other routes of noninjection delivery present formidable challenges for proteins. Very poorly permeable epithelium (all routes but pulmonary), very strong digestive enzyme barriers (the gastrointestinal tract but also other routes to a lesser degree), and limited residence time (oral and nasal) make these routes of systemic delivery extremely difficult.

13.6.1 Events at the Air–Blood Tissue Barrier: Features of Systemic Pulmonary Absorption and Processing of Proteins

Just as with absorption from the subcutaneous injection site, absorption from the lungs is molecular weight dependent with small peptides being absorbed within minutes to tens of minutes and larger molecules within hours or days (Patton et al. 1994, 1999). Despite the many years of studying the absorption of insulin and other protein molecules, the details of how they are absorbed are still poorly understood. The most thoughtful recent discussion of the science on the pulmonary absorption of insulin and other biotherapeutics is that by Sakagami (2013). In general, we have learned some rough unifying principles about the systemic absorption of peptides and proteins once they have been deposited into the lungs (Patton 1996; Patton et al. 1998, 2004a,b; Hastings et al. 2004).

- Small natural sequence peptides up to about 30 amino acids are almost completely degraded by lung enzymes before they can reach the systemic circulation and have systemic bioavailabilities <1%.
- Small chemically blocked (to inhibit peptidase degradation) versions of the natural small peptides are rapidly absorbed within 10–20 min by the lungs, and with high bioavailabilities (50%–95%) and with peak levels in the blood within 30 min.
- Intermediate-sized peptides and proteins of 50–200 amino acids, typically the famous hormones and cytokines such as insulin, growth hormone, interferons, interleukins, colony-stimulating factors, and erythropoietins with molecular weights in the 5–35 kDa range, are absorbed with clinically acceptable but small to modest bioavailabilities of 5%–20% and during a period of several hours with peak levels in the first hour. They are cleared from the systemic circulation even faster.
- Larger proteins, typical of the long-lasting plasma proteins that circulate for days in the systemic circulation and avoid kidney clearance by virtue of their size, such as albumin (57 kDa) and immunoglobulin (150 kDa), are slowly absorbed from the lower airways with low systemic bioavailabilities of a few percent, if any. But see below on this page for antibody absorption from the upper airways.
- Of the many potential ways to retain and/or protect proteins in the lungs after inhalation, such as microencapsulation in liposomes, incorporation into polymeric matrices (Sakagami and Byron 2005), precipitations into insoluble or slowly soluble particles, or chemical modification (i.e., PEGylation) to increase molecular weight, PEGylation seems to be the most promising (Niven et al. 1995; Leach et al. 2004). Particle engineering (as opposed to the chemical engineering such as PEGylation) led to low encapsulation efficiencies, difficult manufacturing processes, and large uncontrolled release of the drug upon the particles first being deposited in the lungs—the difficult burst effect. PEGylation avoided or reduced all of these issues and probably also led to greater chemical stability of the drug as a polyethylene glycol conjugate.

The use of specific cell membrane ligands (receptors, transporters, etc.) as molecular and NP binding and targeting sites and transporters is in its infancy but is sure to become an important field. For example, by using the Fc transporter system that is located primarily in the airways, Bitonti et al. (2004), Dumont et al. (2005, 2006), and Bitonti and Dumont (2006) and were able to deliver Fc–cytokine conjugates selectively to the systemic circulation with good bioavailabilities. They found that with shallow tidal breathing, in which their aerosol particles deposited primarily in the upper airways, they obtained higher bioavailabilities than with deep inhalation that led to greater alveolar deposition where the concentration of immunoglobulin transports is less.

13.7 Summary

There are many exciting possibilities for the formulation and delivery of therapeutic NPs to the lungs. Improvements in devices, packaging, and, particularly, aerosol and NP formulation methods should help make the development of new inhaled NP therapies in the future a reality.

References

Al-Hallak M.K.D.K., M.K. Sarfraz, S. Azarmi, W.H. Roa, W.H. Finlay and R. Löbenberg. 2011. Pulmonary delivery of inhalable nanoparticles: Dry powder inhalers. *Ther Deliv.* 2: 1–12.

Atochina-Vasserman E.N., M.F. Beers and A.J. Gow. 2010. Review: Chemical and structural modifications of pulmonary collectins and their functional consequences. *Innate Immun.* 16: 175–182.

Becquemin M.H., J.F. Bertholon, M. Attoui, F. Roy, M. Roy and B. Dautzenberg. 2007. Particle size in the smoke produced by six different types of cigarette. *Rev Mal Respir.* 24: 845–852.

Bitonti A.J. and J.A. Dumont. 2006. Pulmonary administration of therapeutic proteins using an immunoglobulin transport pathway. *Adv Drug Deliv Rev.* 58: 1106–1118.

Bitonti A.J., J.A. Dumont, S.C. Low, R.T. Peters, K.E. Kropp, V.J. Palombell, J.M. Stattel, Y. Lu, C.A. Tan, J.J. Song, A.M. Garcia, N.E. Simister, G.M. Spiekermann, W.I. Lencer and R.S. Blumberg. 2004. Pulmonary delivery of an erythropoietin Fc fusion protein in non-human primates through an immunoglobulin transport pathway. *Proc Natl Acad Sci U S A.* 101: 9763–9768.

Borgstrom L., B. Olsson and L. Thorsson. 2006. Degree of throat deposition can explain the variability in lung deposition of inhaled drugs. *J Aerosol Med.* 19: 473–483.

Brand P., H. Beckmann, M. Mass Enriquez, T. Meyer, B. Mullinger, K. Sommerer, N. Weber, T. Weuthen and G. Schuech. 2003. Peripheral deposition of α1-protease inhibitor using commercial inhalation devices. *Eur Respir J.* 22: 263–267.

Brand P., K. Häussinger, T. Meyer, G. Scheuch, H. Schulz, T. Selzer and J. Heyder. 1999. Intrapulmonary distribution of deposited particles. *J Aerosol Med.* 12: 275–284.

Brannan J.D., S.D. Anderso, C.P. Perry, R. Freed-Martens, A.R. Lassig and B. Charlton; Aridol Study Group. 2005. The safety and efficacy of inhaled dry powder mannitol as a bronchial provocation test for airway hyper-responsiveness: A phase 3 comparison study with hypertonic (4.5%) saline. *Respir Res.* 6: 144.

Carvalho T.C., J.I. Petters and R.O. Williams 3rd. 2011. Influence of particle size on regional lung deposition—What evidence is there? *Int J Pharm.* 406: 1–10.

Ceglia L., J. Lau and A.G. Pittas. 2006. Meta-analysis: Efficacy and safety of inhaled insulin therapy in adults with diabetes mellitus. *Ann Intern Med.* 145: 665–675.

Clark A.R. 2012. Understanding penetration index measurements and regional lung targeting. *J Aerosol Med Pulm Drug Deliv.* 25: 179–187.

da Silva A.L., S.V. Martini, S.C. Abreu, S. Cdos Samary, B.L. Diaz, S. Fernezlian, V.K. de Sá, V.L. Capelozzi, N.J. Boylan, R.G. Goya, J.S. Suk, P.R. Rocco, J. Hanes and M.M. Morales. 2014. DNA nanoparticle-mediated thymulin gene therapy prevents airway remodeling in experimental allergic asthma. *J Control Release.* 180: 125–133.

Dumont J.A., A.J. Bitonti, D. Clark, S. Evans, M. Pickford and S.P. Newman. 2005. Delivery of an erythropoietin–Fc fusion protein by inhalation in humans through an immunoglobulin transport pathway. *J Aerosol Med.* 18(3): 294–303.

Dumont J.A., S.C. Low, R.T. Peters and A.J. Bitonti. 2006. Monomeric Fc fusions: Impact on pharmacokinetic and biological activity of protein therapeutics. *BioDrug.* 20: 151–160.

Edwards D.A. and C. Dunbar. 2002. Bioengineering of therapeutic aerosols. *Annu Rev Biomed Eng.* 4: 93–107.

Fineberg S.E., T. Kawabata, D. Finco-Kent, C. Liu and A. Kranser. 2005. Antibody response to inhaled insulin in patients with type 1 or type 2 diabetes. An analysis of initial phase II and III inhaled insulin (Exubera) trials and a two-year extension trial. *J Clin Endocrinol Metab.* 90: 3287–3294.

Ganesan S., A.T. Comstock and U.S. Sajjan. 2013. Barrier function of airway tract epithelium. *Tissue Barriers.* 1: e24997.

Genetic Engineering News. March 3, 2014. "The Lists": The Top 25 Best Selling Drugs of 2013.

Harper N.J., S. Gray, J. De Groot, J.M. Parker, N. Sadrzadeh, C. Schuler, J.D. Schumacher, S. Seshadri, A.E. Smith, G.S. Steeno, C.L. Stevenson, R. Taniere, M. Wang and D.B. Bennett. 2007. The design and performance of the Exubera® pulmonary insulin delivery system. *Diabetes Technol Ther.* 9: S-16–S-27.

Hastings R.H., H.G. Folkesson and M.A. Matthay. 2004. Mechanisms of alveolar protein clearance in the intact lung. *Am J Physiol Lung Cell Mol Physiol.* 286: L679–L689.

Heinemann L. and T. Heise. 2004. Current status of the development of inhaled insulin. *Br J Diabetes Vasc Dis.* 4: 295–301.

Konstan M.W., P.A. Flume, M. Kappler, R. Chiron, M. Higgins, F. Brockhaus, J. Zhang, G. Angyalosi, E. He and D.E. Geller. 2010. Safety, efficacy and convenience of tobramycin inhalation powder in cystic fibrosis patients: The EAGER trial. *J Cyst Fibros.* 10: 54–61.

Konstan M.W., D.E. Geller, P. Minic, F. Brockhaus, J. Zhang and G. Angyalosi. 2011. Tobramycin inhalation powder for *P. aeruginosa* infection in cystic fibrosis: The EVOLVE trial. *Pediatr Pulmonol.* 46: 230–238.

Leach C.L. 2002. Lung deposition of hydrofluoroalkane-134*a* beclomethasone is greater than that of chlorofluorocarbon fluticasone and chlorofluorocarbon beclomethasone: A cross-over study in healthy volunteers. *Chest.* 122: 510–516.

Leach C., M.-C. Kuo, B. Beuche, S. Fishburn, T. Vegas, M. Bossard, L. Guo, M.D. Bentley, C.H. Hobbs. A.D. Cherrington and J.S. Patton. 2004. Modifying the pulmonary absorption and retention of proteins through PEGylation. In: *Respiratory Drug Delivery*, IX Proceedings (R.N. Dalby, P.R. Byron, J. Peart, J.D. Suman and S.J. Farr, eds.) pp. 69–77. Davis Healthcare Publishing, River Grove, IL.

Li H.Y. and F. Zhang. 2012. Preparation of nanoparticles by spray drying and their use for efficient pulmonary drug delivery. *Methods Mol Biol.* 906: 295–301.

Lombry C., D.A. Edwards, V. Preat and R. Vanbever. 2004. Alveolar macrophages are a primary barrier to pulmonary absorption of macromolecules. *Am J Physiol Lung Cell Mol Physiol.* 286: L1002–L1008.

Londahl J., W. Moller, J.H. Pagels, W.G. Kreyling, E. Swietlicki and O. Schmid. 2013. Measurement techniques for respiratory tract deposition of airborne nanoparticles: A critical review. *J Aerosol Med Pulm Drug Deliv.* 26: 1–26.

McElroy E.C., C. Kirton, D. Gliddon and R.K. Wolff. 2013. Inhaled biopharmaceutical drug development: Nonclinical consideration and case studies. *Inhal Toxicol.* 25: 219–232.

Neumiller J.J., R.K. Campbell and L.D. Wood. 2010. A review of inhaled technosphere insulin. *Ann Pharmacother.* 44: 1231–1239.

Niven R.W., T. Carvajal and H. Schreier. 1992. Nebulization of liposomes. III. The effects of operating conditions and local environment. *Pharm Res.* 9: 515–520.

Niven R.W. and H. Schreier. 1990. Nebulization of liposomes. I. Effects of lipid composition. *Pharm Res.* 7: 1127–1133.

Niven R.W., M. Speer and H. Schreier. 1991. Nebulization of liposomes. II. The effects of size and modeling of solute release profiles. *Pharm Res.* 8: 217–221.

Niven R.W., K.L. Whitcomb, L. Shaner, A.Y. Ip and O.B. Kinstler. 1995. The pulmonary absorption of aerosolized and intra-tracheally instilled rhG-CSF and mono-PEGylated rhG-CSF. *Pharm Res.* 12: 1343–1349.

Patton J. 1996. Mechanisms of macromolecule absorption by the lungs. *Adv Drug Deliv Rev.* 19: 3–36.

Patton J.S., J.D. Brain, L.A. Davies, J. Fiegel, M. Gumbleton, K. Kim, M. Sakagami, R. Vanbever and C. Ehrhardt. 2010. The particle has landed—Characterizing the fate of inhaled pharmaceuticals. *J Aerosol Med Pulm Drug Deliv*. 23 Suppl 2: S71–S87.

Patton J.S., J.G. Bukar and M.A. Eldon. 2004a. Clinical pharmacokinetics and pharmacodynamics of inhaled insulin. *Clin Pharmacokinet*. 43: 781–801.

Patton J., J. Bukar and S. Nagarajan. 1999. Inhaled insulin. *Adv Drug Deliv Rev*. 35: 235–247.

Patton J.S. and P.R. Byron. 2007. Inhaling medicines: Delivering drugs to the body through the lungs. *Nat Rev Drug Discov*. 6: 67–74.

Patton J.S., C.S. Fishburn and J.G. Weers. 2004b. The lungs as a portal of entry for systemic drug delivery. *Proc Am Thorac Soc*. 1: 338–344.

Patton J., S. Nagarajan and A. Clark. 1998. Pulmonary absorption and metabolism of peptides and proteins. *Respir Drug Deliv VI*. 1: 17–24.

Patton J., P. Trinchero and R. Platz. 1994. Bioavailability of pulmonary delivered peptide and protein: Interferon alpha, calcitonin and parathyroid hormones. *J Control Release*. 28: 79–85.

Rau, J.I. 2005. The inhalation of drugs: Advantages and problems. *Respir Care*. 50: 367–382.

Ruge C.A., U.F. Schaefer, J. Herrmann, J. Kirch, O. Canadas, M. Echaide, J. Perez-Gil, C. Casals, R. Muller and C. Lehr. 2012. The interplay of lung surfactant proteins and lipids assimilates the macrophage clearance of nanoparticles. *PLoS One*. 7(7): e40775.

Sadrzadeh N., D.P. Miller, D. Lechuga-Ballesteros, N.J. Harper, C.L. Stevenson and D.B. Bennett. 2010. Solid-state stability of spray-dried insulin powder for inhalation: Chemical kinetics and structural relaxation modeling of Exubera above and below the glass transition temperature. *J Pharm Sci*. 99: 3698–3710.

Sakagami M. 2013. Systemic delivery of biotherapeutics through the lung: Opportunities and challenges for improved lung absorption. *Ther Deliv*. 4: 1511–1525.

Sakagami M. and P.R. Byron. 2005. Respirable microspheres for inhalation: The potential of manipulating pulmonary disposition for improved therapeutic efficacy. *Clin Pharmacokinet*. 44: 263–277.

Schleh C., U. Holzwarth, S. Hirn, A. Wenk, F. Simonelli, M. Schaffler, W. Moller, N. Gibson and W.G. Kreyling. 2013. Biodistribution of inhaled gold nanoparticles in mice and the influence of surfactant protein D. *J Aerosol Med Pulm Drug Deliv*. 26: 24–30.

Schuster B.S., J.S. Suk, G.F. Woodworth and J. Hanes. 2013. Nanoparticle diffusion in respiratory mucus from humans without lung disease. *Biomaterials*. 34: 3439–3446.

Suk J.S., A.J. Kim, K. Trehan, C.S. Schneider, L. Cebotaru, O.M. Woodward, N.J. Boylan, M.P. Boyle, S.K. Lai, W.B. Guggino and J. Hanes. 2014. Lung gene therapy with highly compacted DNA nanoparticles that overcome the mucus barrier. *J Control Release*. 178: 8–17.

Weers J.G., T.E. Tarara and A.R. Clark. 2007. Design of fine particles for pulmonary drug delivery. *Expert Opin Drug Deliv*. 4: 1–17.

White S., D.B. Bennett, S. Cheu, P.W. Conley, D.B. Guzek, S. Gray, S. Howard, R. Malcolmson, J.N. Parker, P. Roberts, N. Sadrzadeh, J.D. Schumacher, S. Seshadri, G.W. Sluggett, C.L. Stevenson and N.J. Harper. 2005. Exubera: Pharmaceutical development of a novel product for pulmonary delivery of insulin. *Diabetes Technol Ther*. 7: 896–906.

Wolff R.K. 1998. Safety of inhaled proteins for therapeutic use. *J Aerosol Med Pulm Drug Deliv*. 11: 197–219.

Wong W., D.F. Fletcher , D. Traini, H.K. Chan, J. Crapper and P.M. Young. 2010. Particle aerosolisation and break-up in dry powder inhalers 1: Evaluation and modeling of venturi effects for agglomerated systems. *Pharm Res*. 27: 1367–1376.

14

Drug and Gene Delivery in the Lungs

Satoshi Uchida, Keiji Itaka, and Kazunori Kataoka

CONTENTS

14.1 Basic Design of a Polyplex Nanomicelle for Inhalation Delivery

14.1.1 NP-Based Drug Delivery to the Lungs and Polyplex Nanomicelle Application

Inhalation drug delivery is an attractive approach because it is less invasive than other treatment methods. For treating respiratory diseases, inhalation allows direct accessibility to target cells, e.g., β-agonist inhalation for the treatment of asthma. This approach is also appealing for systemic drug delivery because the lungs have a large epithelial surface area, which permits high drug permeability, and low drug-metabolizing activity than other administration routes such as the gastrointestinal tracts (Patton and Byron 2007). The inhaled insulin product Exubera is one of the most successful examples of systemic drug delivery. However, the clinical application of inhalation drug delivery remains limited. A major problem associated with conventional inhalation therapy is the rapid clearance of drugs from lung tissues, necessitating repetitive administration to obtain a sustained therapeutic effect. One promising solution to inhibit rapid clearance is increasing the molecular size of the drug by encapsulating it into functional particles. Nanoparticles (NPs) with the size of several tens of nanometers are effective for this purpose because they are advantageous for efficient cellular uptake by endocytosis, wide distribution, and improved tissue retention (Azarmi et al. 2008; Rytting et al. 2008; Bailey and Berkland 2009). Indeed, adenoviruses with smart mechanisms to internalize into target cells possess the scale of 70–100 nm (Kennedy and Parks 2009).

Despite their advantages of avoiding rapid clearance, a significant problem remains to be solved before inhalation delivery of NPs into the lungs can be realized. Because respiratory organs are always exposed to foreign materials, they possess sophisticated systems to clear these materials, for example, mucosal excretion and phagocytosis by alveolar macrophages, which may impair the efficient delivery of NPs to target cells. In addition, the exposure to foreign materials can evoke significant inflammatory responses, leading to tissue damage, although these responses are dependent on the physicochemical property of the materials. Interactions with inhaled NPs have been vigorously investigated from the standpoint of toxicity. Studies of carbon nanotubes are well known in the field of nanotoxicology. Inhalation or intraperitoneal (i.p.) administration of carbon nanotubes (4 mg/kg for inhalation and 2 mg/kg for i.p. administration) caused asbestos-like pathogenesis in mice, including inflammation and the formation of granulomas, presumably because of their needle-shaped structure and nondegradable nature (Poland et al. 2008; Ryman-Rasmussen et al. 2009). The distinguishing characteristic of NPs is a relatively large surface area per unit volume, exhibiting increased interaction with surrounding biomolecules in the body and subsequent unfavorable responses (Nel et al. 2006; Li et al. 2010). Among these responses, oxidative stress was most frequently reported, which causes activation of proinflammatory signaling cascades, and cell death. To avoid these surface-mediated toxicities, optimization of surface property is particularly important.

Modification of the surface with biocompatible polymers such as poly(ethylene glycol) (PEG) is a good approach to reduce the toxicity of NPs. The toxicity of the cationic polyplex from polyethyleneimine (PEI) was significantly reduced by surface modification with high-density PEG (Beyerle et al. 2010a,b). The PEG layer on the surface of NPs is thought to inhibit the interaction of surrounding biomolecules such as serum proteins with the NPs. Thus, the PEG surface provides a "stealth" property to NPs by avoiding recognition by the host immune system.

In contrast, the surface charge has a large impact on inflammatory responses when PEGylation is not enough to mask the surface charge. Intratracheal administration of cationic stearylamine-based PEG–polylactic acid (PLA) NPs with a ζ-potential of +32 mV and a diameter of 150 nm (7.5 mg/kg/day for 5 consecutive days) induced a higher level of inflammatory responses in mouse lung than similarly sized anionic PEG–PLA NPs without stearylamine with a ζ-potential of −31 mV (Harush-Frenkel et al. 2010).

Particle size is another important factor that can be used to regulate the toxicity of NPs. The influence of particle size on the uptake of the particle by alveolar macrophage was investigated after pulmonary administration of ciprofloxacin (CPFX)-loaded liposome with various sizes, composed of hydrogenated soybean phosphatidylcholine, cholesterol, and dicetylphosphate (193 μg CPFX/kg) (Chono et al. 2006). The amount of liposome uptake by alveolar macrophages showed a three to four times increase with an increase in the size in the range of 100–1000 nm.

Biodegradability is also a critical factor influencing the toxicity of NPs. In intratracheal administration to mice, a significant decrease in the inflammatory response in lung tissues was observed using 80-nm-sized polymeric NPs composed of biodegradable poly(lactic-co-glycolic acid) (10 mg/kg) compared with that using a similarly sized control NP composed of nonbiodegradable polystyrene (10 mg/kg) (Dailey et al. 2006).

On the basis of these toxicity issues, we propose a polyplex nanomicelle composed of a PEG shell and a drug-loading core as a good candidate for inhalation drug delivery

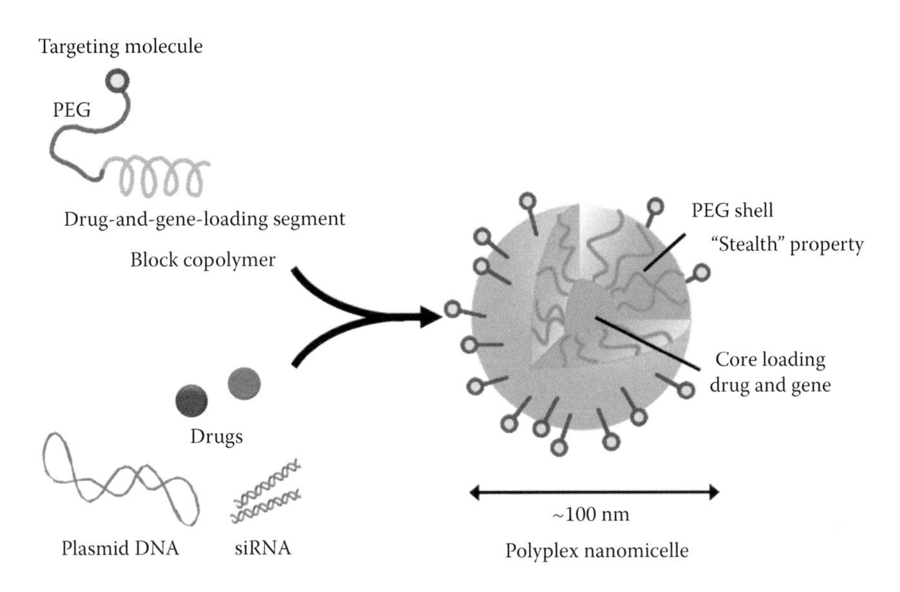

FIGURE 14.1
Basic design of polyplex nanomicelles.

(Figure 14.1) (Kataoka et al. 2001). The nanomicelle is prepared by self-assembly of diblock copolymers containing PEG and a drug-and-gene-loading segment. Various drugs and nucleic acids can be incorporated into the nanomicelle through various driving forces of micelle formation, such as hydrophobic interaction (for hydrophobic anticancer drugs), ligand-exchange reaction (for metal complex drugs, e.g., cisplatin), and electrostatic interaction (for nucleic acids) (Figure 14.2). In general, the nanomicelle has a relatively small size (<100 nm) with a PEG surface, showing excellent stability under physiological conditions. Because it can avoid recognition by the reticuloendothelial system in the body, the nanomicelle architecture is advantageous for reducing acute immune responses. In addition, the use of biodegradable polymers significantly decreases the risk of cumulative toxicity that may affect cell homeostasis in a time-dependent manner.

14.1.2 Gene Delivery into the Lungs and Application of Polyplex Nanomicelles

Among various pulmonary diseases, cystic fibrosis (CF) is one of the most significant targets of inhalation therapy. CF is a single-gene disorder of the CF transmembrane conductance regulator (CFTR), with high mortality at young ages. To treat this disease, gene introduction for the replacement of CFTR is a promising strategy, and various methods to introduce genes into the lungs have been vigorously examined. Among these methods, viral gene vectors such as adenovirus and adeno-associated virus have been most frequently used to obtain high transfection efficiency, and several clinical trials have already been performed (Griesenbach and Alton 2009; Davies and Alton 2010). However, although administration of these viral vectors did not cause significant problems in phase I studies, outcomes of larger-scale studies were unsatisfactory. The risk of viral vectors to cause severe immunogenicity hampered the dose and frequency of virus administration necessary to achieve a sufficient therapeutic effect. Indeed, in the clinical trial of ornithine transcarbamylase deficiency, direct injection of an adenovirus vector into the

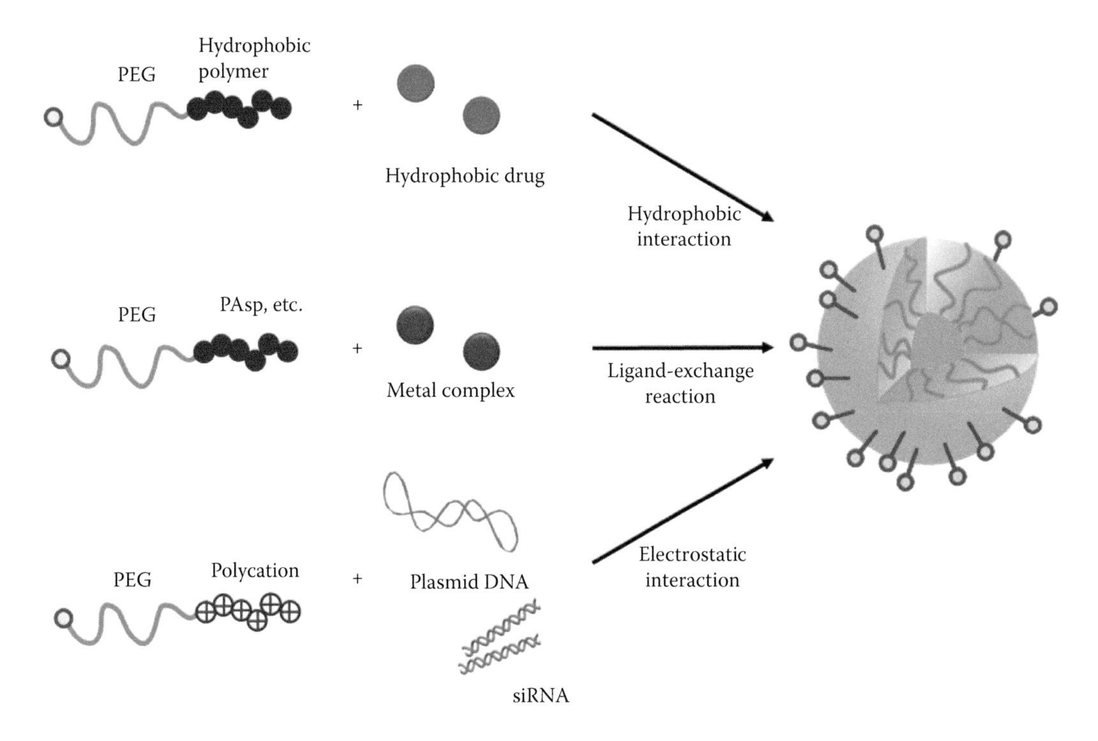

FIGURE 14.2
Incorporation of various drugs and nucleic acids into nanomicelles.

hepatic artery caused the death of a patient by systemic inflammatory response syndrome (Marshall 1999).

As established in the CF gene therapy trials, safety issues should be highlighted during the development of NPs for inhalation delivery into the lungs. The polyplex nanomicelle has a high potential for inhalation gene delivery because of its "stealth" property, which allows it to avoid tissue damage and immune responses. In Sections 14.2, 14.3.1, and 14.3.2, the development of polyplex nanomicelles for gene delivery and their applications into the lungs will be briefly summarized.

14.2 Development of Polyplex Nanomicelles for Gene Delivery

Polyplex nanomicelles are organized by self-assembly of amphiphilic block copolymers composed of PEG and another segment (Kataoka et al. 2001). For plasmid DNA (pDNA), PEG–polycation block copolymers are used to form a core–shell micellar structure (Figure 14.2). Using PEG with a molecular weight (Mw) of 12,000, the nanomicelles possess dense PEG palisades on their surfaces to effectively compartmentalize condensed pDNA in the core. Detailed analyses of the physicochemical properties and *in vitro* transfection efficiencies have been performed for PEG–poly(L-lysine) (PEG–PLys) block copolymers. Under optimal conditions, mixing PEG–PLys and pDNA induced the homogeneous formation of polyplex nanomicelles with a diameter of 80–100 nm, possessing a single molecule of pDNA in the core (Itaka et al. 2003). This distinct core–shell structure provided high solubility and improved the nuclease resistance of incorporated pDNA (Katayose and Kataoka 1997, 1998). The practical advantage of this composition is the excellent stability obtained under physiological conditions, such as those encountered in serum-containing medium, leading to high transfection efficiency in the presence of serum (Itaka et al. 2002, 2003).

However, although PLys has a biocompatible nature as a ubiquitous amino acid in the body, in particular, within nuclei as the chief component of histone proteins, it generally exhibits low transfection efficiency. The restricted release of pDNA from polyplexes within cells is responsible for its low gene introduction capacity (Goncalves et al. 2002; Itaka et al. 2004). In addition, PLys is not capable of facile endosome escape, unlike PEI, which has a characteristic so-called proton sponge effect (Boussif et al. 1995). Because of the presence of unprotonated amines at neutral pH, PEI is considered to induce H^+ influx into the endosome, thereby reducing pH and raising the internal osmotic pressure, resulting in osmotic rupture of the endosome. However, PEI induces the damage of cell membrane within an hour after transfection ("acute" toxicity), and also perturbs the cell homeostasis in a time-dependent manner ("cumulative" toxicity), leading to cell death (Godbey et al. 2001; Moghimi et al. 2005). These toxicological problems of PEI motivated us to develop the next generation of polyplex nanomicelles possessing a high capacity of endosomal escape. The synthesis of cationic polymers was based on the finding that the flanking benzyl ester groups of poly(β-benzyl L-aspartate) (PBLA) could undergo a quantitative aminolysis reaction with various polyamine compounds under mild anhydrous conditions below room temperature (Kanayama et al. 2006; Pittella et al. 2011). By this strategy, an *N*-substituted poly(aspartamide) (PAsp) derivative library possessing various cationic side chains from a single platform of PBLA was prepared. A series of transfection and cytotoxicity assays revealed that a diamine structure with two distinct pKa values, primary and secondary amino groups in the side chain, was found to be effective for internalizing the polyplexes into cells.

Among the library, a 1,2-diaminoethane structure at the side chain of the cationic segment, PEG–*b*-P[Asp(DET)] (Figure 14.3), showed remarkably high transfection efficiency. An important point is that the cationic polymer P[Asp(DET)] exhibited remarkably lower toxicity. It is assumed that there are two major reasons for the low toxicity of P[Asp(DET)]. One reason is the pH-selective membrane destabilization of P[Asp(DET)] (Miyata et al. 2008), which contributes to reduced "acute" toxicity. A leakage assay of the cytoplasmic enzyme lactate dehydrogenase (LDH) and confocal laser scanning microscopy revealed that P[Asp(DET)] induced minimal membrane destabilization at physiological pH; yet,

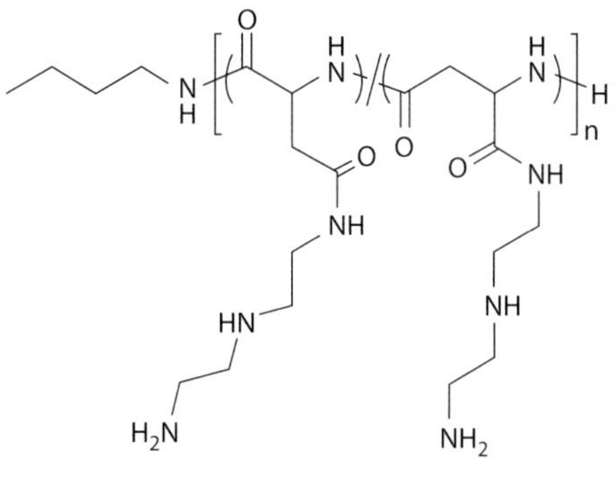

FIGURE 14.3
Molecular structure of P[Asp(DET)].

there was a significant increase in destabilization under acidic conditions (pH ~5). This pH-selective membrane destabilization profile corresponded to a protonation change in the flanking diamine unit, that is, the monoprotonated gauche form at physiological pH and the deprotonated anti form at acidic pH. Eventually, P[Asp(DET)] effectively reduced unfavorable interactions with plasma membranes and/or cytoplasmic organelles at physiological pH, and when residing in endosomes or lysosomes the membrane destabilization induced at acidic pH enabled the smooth translocation of polyplexes into the cytoplasm.

Another reason for the low toxicity of P[Asp(DET)] is biodegradability. This is strongly related to "cumulative" cytotoxicity, which may perturb cellular homeostasis in a time-dependent manner. Indeed, in the case of PEI, even after obtaining good transgene expression, cell viability gradually decreased with fluctuations in the expression of endogenous genes (Masago et al. 2007). In contrast, P[Asp(DET)] showed high efficacy for inducing cell differentiation after transfection of transcription factor–expressing pDNA. Pharmacogenomic analysis suggested that P[Asp(DET)] maintained cellular homeostasis after transfection (Itaka et al. 2007; Masago et al. 2007). A key feature is that the degradation products of the Asp(DET) monomer showed almost no cytotoxicity even with an extremely high amine concentration (1 mM of amine in the culture medium). Thus, it is likely that once internalized into cells, P[Asp(DET)] will be safely degraded to a nontoxic form, thereby minimizing time-dependent effects on cellular homeostasis. These aspects of excellent biocompatibility are crucial for practical applications, in particular, for inhalation gene delivery into the lungs (Itaka et al. 2010).

14.3 Inhalation Gene Delivery into the Lungs Using Polyplex Nanomicelles

14.3.1 Application of Nanomicelles to an Animal Model of Disease

The feasibility of using nanomicelles for clinical applications was assessed using an animal model of idiopathic pulmonary arterial hypertension (PAH) (Harada-Shiba et al. 2009). PAH is a life-threatening disease characterized by a progressive increase in pulmonary vascular resistance, leading to right heart failure. Adrenomedulline (AM), a peptide isolated from human pheochromocytoma, shows substantial therapeutic effects because of its strong vasodilatory effect, and the abundance of AM receptors in the lung (Nagaya et al. 2005). Indeed, the inhalation of AM peptide was reported to ameliorate PAH in an animal study as well as in a clinical trial (Nagaya et al. 2003, 2004). However, in the clinical trial, the effect of reducing vascular resistance was sustained for only an hour, resulting in limited outcomes. Thus, gene introduction of AM should provide a promising approach to obtain satisfactory results with prolonged reduction of vascular resistance.

PAH was induced in rats by subcutaneous injection of monocrotaline, and AM-expressing pDNA was intratracheally administered using a microspray. Therapeutic effects were assessed by evaluating the reduction of right ventricle pressure after treatment. AM gene introduction using PEG–P[Asp(DET)]/pDNA nanomicelles yielded a significant therapeutic effect, whereas the introduction using the LPEI/pDNA polyplex or naked pDNA showed no effect. Transgene expression was detectable until day 14, with minimal neutrophilic infiltration in lung tissues and minimal induction of pro-inflammatory cytokines on day 7. These observations demonstrated the high potential of nanomicelles for use in pulmonary gene therapy.

14.3.2 Toxicologic Studies of Nanomicelles in the Lungs

As mentioned previously, the safety of inhaled material is a critical issue because the respiratory system is highly sensitive to foreign materials. To evaluate the safety of polyplex nanomicelles, immunological studies were performed on day 7 after intratracheal administration of the nanomicelle and LPEI/pDNA polyplex containing 10 μg pDNA/kg to mouse lung, by quantitative real-time polymerase chain reaction measurement of inflammatory molecules (interleukin [IL]-6, tumor necrosis factor [TNF]-α, IL-10, and cyclooxygenase [Cox]-2) and review of histopathologic sections (Harada-Shiba et al. 2009). Even a single administration of LPEI/pDNA induced a high expression of the inflammatory molecules as well as significant infiltration of neutrophils to terminal bronchioles and alveoli. In contrast, a single administration of the nanomicelles with the same pDNA dose as LPEI/pDNA did not increase the expression of inflammatory molecules compared with untreated controls, and resulted in nearly intact histological findings. The high "cumulative" toxicity of LPEI derived from its nonbiodegradable nature was avoided using biodegradable polycation P[Asp(DET)].

"Acute" toxicity of nanomicelles following intratracheal gene administration was measured by LDH leakage into bronchoalveolar lavage fluid (BALF) in mice (Uchida et al. 2011). Nanomicelles containing 0.4 mg pDNA/kg induced some leakage of LDH at 30 min after administration. *In vitro* analyses of cell membrane damage showed that the "acute" toxicity of nanomicelle was induced by free polycations, which were not attached to nanomicelles, although the toxicity of free PEG–P[Asp(DET)] was much lower than that of LPEI. In most of the polycation-based gene delivery systems, an excess ratio of cationic polymers to pDNA is required to obtain efficient transfection. Thus, a substantial amount of polycations exist in the free state. To reduce the toxicity of free PEG–P[Asp(DET)], an anionic polycarbohydrate chondroitin sulfate (CS) was added to the nanomicelle solution to inhibit binding of the free polymers to cell membranes or other anionic biomolecules. After adding CS at a CS/PEG–PAsp(DET) charge ratio of 10:1, leakage of LDH into BALF was suppressed to the level of that in untreated controls, whereas the transfection efficiency of the nanomicelles remained high.

Another method to alleviate the toxicity of free polycations is to reduce the charge ratio of polycations to pDNA, although the transfection efficiency is also reduced. To obtain sufficient expression at a low charge ratio, the surface PEG density of nanomicelles should be optimized because high PEG density prevents cellular uptake and hampers intracellular processing of nanomicelles (Takae et al. 2008). Nanomicelles with various PEG densities were prepared by the combined use of the PEG–P[Asp(DET)] block copolymer (B) and non-PEGylated P[Asp(DET)] homopolymer (H) at various mixture ratios (B–H formulation) (Uchida et al. 2012). At the optimal B–H ratio, the nanomicelle yielded very high transgene expression with no detectable toxicity, as evaluated by the expression of inflammatory molecules (IL-6, TNF-α, IL-10, and Cox-2) and histopathological sections 4 h after the transfection.

Interestingly, the P[Asp(DET)]/pDNA polyplex without PEGylation induced high inflammatory responses even at a low polycation-to-pDNA charge ratio, leading to inefficient transgene expression (Uchida et al. 2012). Because the P[Asp(DET)] free polymer did not induce significant toxicity at this charge ratio, some properties of the unPEGylated polyplex triggered the unfavorable response. The unPEGylated polyplex showed rapid aggregation after incubation in BALF, whereas the PEGylated nanomicelle did not aggregate even after 90 min of incubation (Figure 14.4). These results are consistent with the immunohistochemical observation that the unPEGylated polyplex was taken up by macrophages to a significantly greater extent than the PEGylated nanomicelle. Thus, these findings strongly suggest that rapid aggregation of the unPEGylated polyplex in lung tissues caused the activation of macrophages and subsequent strong inflammatory responses.

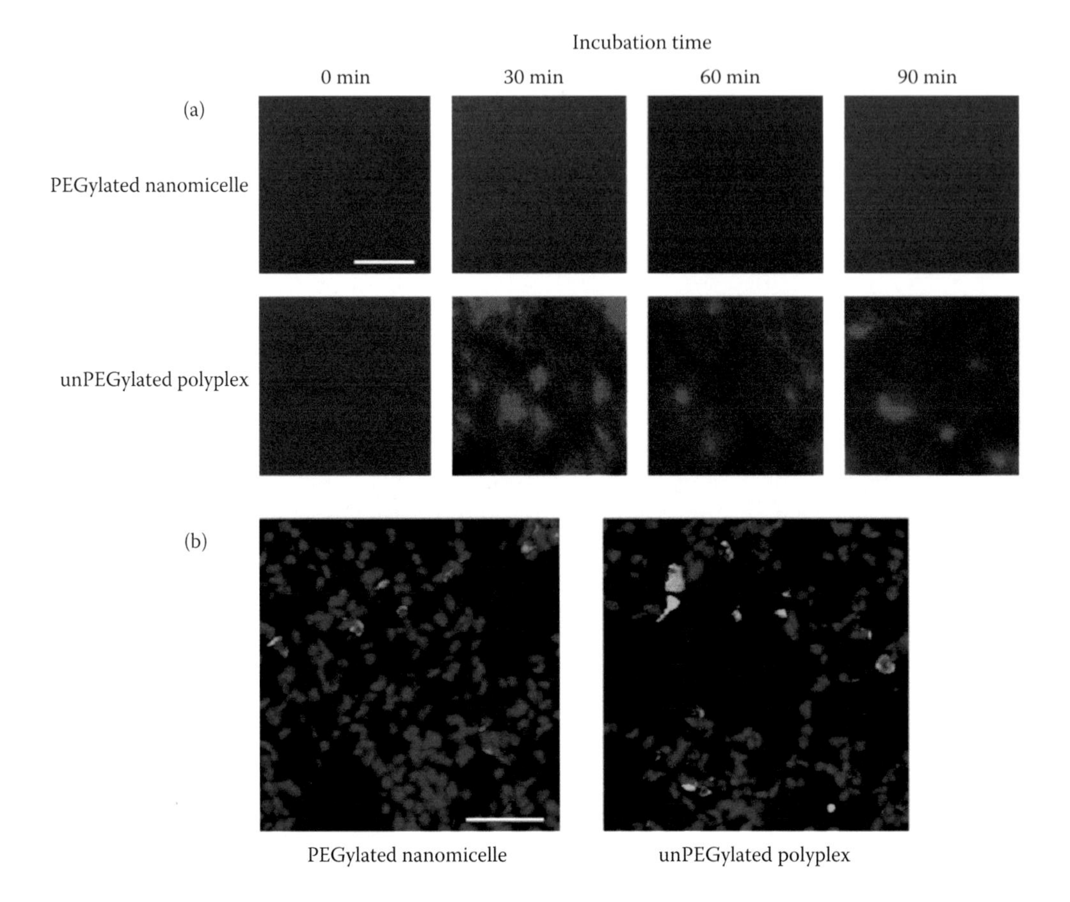

FIGURE 14.4
Macrophage activation in lung tissues induced by aggregation of polyplexes. (a) Aggregation of polyplex during incubation in BALF. Nanomicelle with B–H formulation (PEGylated nanomicelle) and the P[Asp(DET)]/pDNA polyplex (unPEGylated polyplex) were prepared from Cy5-labeled pDNA and observed by fluorescence microscopy after adding BALF. Bar, 50 μm. (b) Uptake of polyplexes by macrophages. Fluorescence microscopic images of the lung were taken at 4 h after administration of polyplexes. Polyplexes were prepared from Cy5-labeled pDNA (red). Macrophages were immunostained using anti-F4/80 antibodies (green). Cell nuclei were stained with Hoechst 33342 (blue). Bar, 50 μm. (Reprinted by permission from Macmillan Publishers Ltd. *Mol Ther*, Uchida et al. 2012, copyright 2012.)

Collectively, these observations displayed three determinant factors of toxicity caused by intratracheally administered NPs: (i) the biodegradability of polycations and/or cationic polyplexes, (ii) amount of free polycations, and (iii) intravital aggregation behavior of polyplexes. Considering these factors, nanomicelles with optimized PEG shielding and CS-treated nanomicelles are good candidates for practical pulmonary gene therapy in future clinical settings.

14.3.3 Other Nonviral Carriers for Gene Transfection to the Lung

Aside from PEG–PAsp(DET)-based polyplex nanomicelles, various formulations of nonviral carriers were studied for gene transfection to the lung, including systems composed of cationic polymers and cationic lipids (Pringle et al. 2009). Among them, one of the most intensively studied systems is the GL67A lipoplex, which is composed of GL67 cationic lipid, dioleoyl phosphatidylethanolamine, and dimyristoyl phosphatidylethanolamine–PEG at a mixing ratio of 1:2:0.05. This system has already been used for large animals, where the GL67A lipoplex (52.8 mg pDNA/sheep) showed remarkably high efficiency of CFTR transfection by lung delivery (McLachlan et al. 2011). In addition, a clinical trial for human CF patients has already been performed, where the GL67A loaded with CFTR-expressing pDNA (42.2 mg pDNA/patient) was administered by nebulization and a significant functional recovery of chloride channel in the airway was obtained compared with a placebo group (Alton et al. 1999). However, several adverse effects accompanied the treatment with this system. The patients showed several clinical signs such as reduction of respiratory functions, inflammatory responses including influenza-like symptoms, and a significant increase in serum inflammation markers (C-reactive proteins and IL-6) and white blood cell counts. Although a part of these reactions can be attributed to the immunogenicity of pDNA molecules as mentioned in Section 14.3.4, a placebo group treated with GL67A without containing pDNA also showed such adverse effects, suggesting the toxicity of the lipid particles itself. Because lung gene therapy is desired for patients with impaired pulmonary function, the system still needs to overcome the toxicity issues for further applications.

In the studies, polyplex-based systems such as branched PEI (BPEI; 8 mg pDNA/sheep) and nanomicelles prepared from pDNA and PEG(10k)-PLys(30-mer) block copolymer (40 mg pDNA/sheep) were also evaluated in the lung of mice and sheep (Ziady et al. 2003; McLachlan et al. 2011). Although the transfection efficiency of these polyplex-based systems was lower than that of GL67A, the immunogenicity of PEG–PLys-based pDNA nanomicelles was also significantly lower than the lipoplex-based systems such as GL67A and lipofectin. Likewise, in a clinical trial for CF patients, intranasal delivery of the nanomicelles loaded with CFTR-expressing gene (0.8, 2.67, and 8.0 mg pDNA/patient) did not induce significant increase in inflammatory markers in serum and nasal washing, with a partial functional recovery of chloride channel (Konstan et al. 2004). Collectively, these studies demonstrated that the use of nanomicelle formulation with PEG shielding is an effective strategy for reducing immunogenicity of nonviral carriers in pulmonary delivery.

14.3.4 Immunogenicity of pDNA

CpG dinucleotides in pDNA are immunogenic to mammalian cells. Mammalian genomic DNA has much a lower frequency of CpG dinucleotides (1 of 50–100) than the expected value (1 of 16), and its cytosine is highly methylated. Thus, unmethylated CpG dinucleotides, which are uncommon in mammals, are recognized by Toll-like receptor 9 to evoke innate immune responses (Kawai and Akira 2010). Immunogenicity of CpG dinucleotides was evaluated in mouse lung after a single intranasal insufflation of the GL67A lipoplex (3.2 mg/kg pDNA) containing a series of pDNAs with various numbers of CpG dinucleotides (Hyde et al. 2008). This study showed that even a single CpG dinucleotide in pDNA was sufficient to induce significantly higher immune responses than completely CpG-free pDNA. The immunogenicity of CpG dinucleotides was confirmed using another nonviral transfection system, in which a single nebulization of BPEI/pDNA polyplex (2 mg pDNA/mouse) induced CpG-dependent activation of lung residual macrophages (Lesina et al. 2010). In addition, CpG dinucleotides, in particular those in promoter and enhancer regions, are responsible for transcriptional silencing after methylation in mammalian cells (Mitsui et al. 2009). Thus, the use of pDNA completely free of CpG dinucleotides is preferable to obtain prolonged transgene expression without evoking immune responses.

14.3.5 Aerosol Delivery

In previous studies of PEG–PAsp(DET)-based nanomicelles, pDNA solution was administered using a microspray (Microsprayer; Penn Century Inc., Philadelphia, PA) in the trachea after tracheotomy. This procedure is too invasive for clinical use because repeated administration is required for the treatment of chronic diseases. Thus, aerosolization of the nanomicelle solution for oral inhalation was attempted. The size of aerosol should be 1–5 μm for effective delivery to the lung tissue (Patton and Byron 2007). To generate such a fine aerosol, there are two types of procedures: liquid nebulization and dry powder formation.

For liquid nebulizers, aerosols are prepared using various forces such as compressed gas for jet nebulizers and high-frequency sound waves for ultrasonic nebulizers (Dolovich and Dhand 2011). During the course of nebulization, fragmentation of pDNA can be induced by shear stress (Lentz et al. 2006). This fragmentation is protected when the pDNA is in a highly condensed state, although condensation of pDNA shows an inhibitory effect on transgene expression under other transfection conditions. In nebulization (1 mg pDNA per mouse), BPEI, which induces a high degree of pDNA condensation, yielded high transgene expression in the lungs of mice (Rudolph et al. 2005), whereas BPEI showed relatively low efficiency for *in vitro* transfection (Itaka et al. 2004). Thus, to achieve efficient transfection using nanomicelles after nebulization, the cationic segment of the block copolymer should be further optimized to increase the condensation degree of pDNA.

Alternatively, aerosol PEG–P[Asp(DET)] nanomicelles can be prepared in a dry powder form using the spray freeze drying (SFD) technique. In SFD, a nanomicelle solution is sprayed into a cryogen such as liquid nitrogen, followed by sublimation of water moiety in a freeze dryer (Mohri et al. 2010). The resulting dry powder has a highly porous structure, which contributes to high dispersibility for inhalation. In our preliminary study, PEG–P[Asp(DET)] nanomicelles maintained their high transfection efficiency in the lungs of mice after dry powder formulation, suggesting its feasibility for inhalation treatment.

In the administration of nonviral and viral gene vectors, dry powder formulation has several advantages over liquid nebulization, such as relatively low drug loss during the administration, and improved long-term storage stability (Dolovich and Dhand 2011). Most nonviral vectors such as lypoplexes and polyplexes gradually lose their transfection capability during storage in an aqueous solution by mechanisms such as aggregation of these complexes and degradation of pDNA, whereas the dry powder formulation prevents these unfavorable events. Collectively, the nanomicelle system becomes more practical for clinical application after dry powder formulation in terms of the simplicity of administration and storage stability.

14.4 Summary

Inhalation drug delivery is attractive because of direct accessibility of target cells and noninvasive administration methods. However, the high immunogenicity of the respiratory system hampers effective drug performance after inhalation. The polyplex nanomicelle with a PEG shell and drug-loading core provides an excellent system for inhalation delivery because of its "stealth" property. The nanomicelle system can be applied to pDNA delivery using the PEG–polycation block copolymer. During optimization of the cationic block, P[Asp(DET)] was found to have both high transfection competency and high biocompatibility, which were directly attributed to two distinct properties of the polymer: biodegradability and pH-sensitive membrane destabilization. PEG–P[Asp(DET)]/pDNA nanomicelles showed safe and efficient gene introduction after intratracheal administration to lung tissues, leading to therapeutic effects in an animal model of idiopathic PAH. In contrast, the unPEGylated polyplex from polycation and pDNA exhibited rapid aggregation and subsequent uptake by macrophages, resulting in high inflammatory responses and impaired transgene expression. Nanomicelles can be prepared in the form of a dry powder, which permits oral inhalation. Thus, nanomicelles are a promising system for future clinical gene delivery to pulmonary tissues.

References

Alton, E. W., M. Stern, R. Farley, A. Jaffe, S. L. Chadwick, J. Phillips, J. Davies, S. N. Smith, J. Browning, M. G. Davies, M. E. Hodson, S. R. Durham, D. Li, P. K. Jeffery, M. Scallan, R. Balfour, S. J. Eastman, S. H. Cheng, A. E. Smith, D. Meeker, and D. M. Geddes. 1999. Cationic lipid-mediated CFTR gene transfer to the lungs and nose of patients with cystic fibrosis: A double-blind placebo-controlled trial. *Lancet* 353:947–54.

Azarmi, S., W. H. Roa, and R. Lobenberg. 2008. Targeted delivery of nanoparticles for the treatment of lung diseases. *Adv Drug Deliv Rev* 60 (8):863–75.

Bailey, M. M., and C. J. Berkland. 2009. Nanoparticle formulations in pulmonary drug delivery. *Med Res Rev* 29 (1):196–212.

Beyerle, A., M. Irmler, J. Beckers, T. Kissel, and T. Stoeger. 2010a. Toxicity pathway focused gene expression profiling of PEI-based polymers for pulmonary applications. *Mol Pharm* 7 (3):727–37.

Beyerle, A., O. Merkel, T. Stoeger, and T. Kissel. 2010b. PEGylation affects cytotoxicity and cell-compatibility of poly(ethylene imine) for lung application: Structure–function relationships. *Toxicol Appl Pharmacol* 242 (2):146–54.

Boussif, O., F. Lezoualc'h, M. A. Zanta, M. D. Mergny, D. Scherman, B. Demeneix, and J. P. Behr. 1995. A versatile vector for gene and oligonucleotide transfer into cells in culture and *in vivo*: Polyethylenimine. *Proc Natl Acad Sci U S A* 92 (16):7297–301.

Chono, S., T. Tanino, T. Seki, and K. Morimoto. 2006. Influence of particle size on drug delivery to rat alveolar macrophages following pulmonary administration of ciprofloxacin incorporated into liposomes. *J Drug Target* 14 (8):557–66.

Dailey, L. A., N. Jekel, L. Fink, T. Gessler, T. Schmehl, M. Wittmar, T. Kissel, and W. Seeger. 2006. Investigation of the proinflammatory potential of biodegradable nanoparticle drug delivery systems in the lung. *Toxicol Appl Pharmacol* 215 (1):100–8.

Davies, J. C., and E. W. Alton. 2010. Gene therapy for cystic fibrosis. *Proc Am Thorac Soc* 7 (6):408–14.

Dolovich, M. B., and R. Dhand. 2011. Aerosol drug delivery: Developments in device design and clinical use. *Lancet* 377 (9770):1032–45.

Godbey, W. T., K. K. Wu, and A. G. Mikos. 2001. Poly(ethylenimine)-mediated gene delivery affects endothelial cell function and viability. *Biomaterials* 22 (5):471–80.

Goncalves, C., C. Pichon, B. Guerin, and P. Midoux. 2002. Intracellular processing and stability of DNA complexed with histidylated polylysine conjugates. *J Gene Med* 4 (3):271–81.

Griesenbach, U., and E. W. Alton. 2009. Gene transfer to the lung: Lessons learned from more than 2 decades of CF gene therapy. *Adv Drug Deliv Rev* 61 (2):128–39.

Harada-Shiba, M., I. Takamisawa, K. Miyata, T. Ishii, N. Nishiyama, K. Itaka, K. Kangawa, F. Yoshihara, Y. Asada, K. Hatakeyama, N. Nagaya, and K. Kataoka. 2009. Intratracheal gene transfer of adrenomedullin using polyplex nanomicelles attenuates monocrotaline-induced pulmonary hypertension in rats. *Mol Ther* 17 (7):1180–6.

Harush-Frenkel, O., M. Bivas-Benita, T. Nassar, C. Springer, Y. Sherman, A. Avital, Y. Altschuler, J. Borlak, and S. Benita. 2010. A safety and tolerability study of differently-charged nanoparticles for local pulmonary drug delivery. *Toxicol Appl Pharmacol* 246 (1–2):83–90.

Hyde, S. C., I. A. Pringle, S. Abdullah, A. E. Lawton, L. A. Davies, A. Varathalingam, G. Nunez-Alonso, A. M. Green, R. P. Bazzani, S. G. Sumner-Jones, M. Chan, H. Li, N. S. Yew, S. H. Cheng, A. C. Boyd, J. C. Davies, U. Griesenbach, D. J. Porteous, D. N. Sheppard, F. M. Munkonge, E. W. Alton, and D. R. Gill. 2008. CpG-free plasmids confer reduced inflammation and sustained pulmonary gene expression. *Nat Biotechnol* 26 (5):549–51.

Itaka, K., A. Harada, K. Nakamura, H. Kawaguchi, and K. Kataoka. 2002. Evaluation by fluorescence resonance energy transfer of the stability of nonviral gene delivery vectors under physiological conditions. *Biomacromolecules* 3 (4):841–5.

Itaka, K., A. Harada, Y. Yamasaki, K. Nakamura, H. Kawaguchi, and K. Kataoka. 2004. In *situ* single cell observation by fluorescence resonance energy transfer reveals fast intra-cytoplasmic delivery and easy release of plasmid DNA complexed with linear polyethylenimine. *J Gene Med* 6 (1):76–84.

Itaka, K., T. Ishii, Y. Hasegawa, and K. Kataoka. 2010. Biodegradable polyamino acid-based polycations as safe and effective gene carrier minimizing cumulative toxicity. *Biomaterials* 31 (13):3707–14.

Itaka, K., S. Ohba, K. Miyata, H. Kawaguchi, K. Nakamura, T. Takato, U. I. Chung, and K. Kataoka. 2007. Bone regeneration by regulated *in vivo* gene transfer using biocompatible polyplex nanomicelles. *Mol Ther* 15 (9):1655–62.

Itaka, K., K. Yamauchi, A. Harada, K. Nakamura, H. Kawaguchi, and K. Kataoka. 2003. Polyion complex micelles from plasmid DNA and poly(ethylene glycol)-poly(L-lysine) block copolymer as serum-tolerable polyplex system: Physicochemical properties of micelles relevant to gene transfection efficiency. *Biomaterials* 24 (24):4495–506.

Kanayama, N., S. Fukushima, N. Nishiyama, K. Itaka, W. D. Jang, K. Miyata, Y. Yamasaki, U. I. Chung, and K. Kataoka. 2006. A PEG-based biocompatible block catiomer with high buffering capacity for the construction of polyplex micelles showing efficient gene transfer toward primary cells. *ChemMedChem* 1 (4):439–44.

Kataoka, K., A. Harada, and Y. Nagasaki. 2001. Block copolymer micelles for drug delivery: Design, characterization and biological significance. *Adv Drug Deliv Rev* 47 (1):113–31.

Katayose, S., and K. Kataoka. 1997. Water-soluble polyion complex associates of DNA and poly(ethylene glycol)-poly(L-lysine) block copolymer. *Bioconjug Chem* 8 (5):702–7.

Katayose, S., and K. Kataoka. 1998. Remarkable increase in nuclease resistance of plasmid DNA through supramolecular assembly with poly(ethylene glycol)-poly(L-lysine) block copolymer. *J Pharm Sci* 87 (2):160–3.

Kawai, T., and S. Akira. 2010. The role of pattern-recognition receptors in innate immunity: Update on toll-like receptors. *Nat Immunol* 11 (5):373–84.

Kennedy, M. A., and R. J. Parks. 2009. Adenovirus virion stability and the viral genome: Size matters. *Mol Ther* 17 (10):1664–6.

Konstan, M. W., P. B. Davis, J. S. Wagener, K. A. Hilliard, R. C. Stern, L. J. Milgram, T. H. Kowalczyk, S. L. Hyatt, T. L. Fink, C. R. Gedeon, S. M. Oette, J. M. Payne, O. Muhammad, A. G. Ziady, R. C. Moen, and M. J. Cooper. 2004. Compacted DNA nanoparticles administered to the nasal mucosa of cystic fibrosis subjects are safe and demonstrate partial to complete cystic fibrosis transmembrane regulator reconstitution. *Hum Gene Ther* 15 (12):1255–69.

Lentz, Y. K., T. J. Anchordoquy, and C. S. Lengsfeld. 2006. Rationale for the selection of an aerosol delivery system for gene delivery. *J Aerosol Med* 19 (3):372–84.

Lesina, E., P. Dames, and C. Rudolph. 2010. The effect of CpG motifs on gene expression and clearance kinetics of aerosol administered polyethylenimine (PEI)-plasmid DNA complexes in the lung. *J Control Release* 143 (2):243–50.

Li, J. J., S. Muralikrishnan, C. T. Ng, L. Y. Yung, and B. H. Bay. 2010. Nanoparticle-induced pulmonary toxicity. *Exp Biol Med (Maywood)* 235 (9):1025–33.

Marshall, E. 1999. Gene therapy death prompts review of adenovirus vector. *Science* 286 (5448):2244–5.

Masago, K., K. Itaka, N. Nishiyama, U. I. Chung, and K. Kataoka. 2007. Gene delivery with biocompatible cationic polymer: Pharmacogenomic analysis on cell bioactivity. *Biomaterials* 28 (34):5169–75.

McLachlan, G., H. Davidson, E. Holder, L. A. Davies, I. A. Pringle, S. G. Sumner-Jones, A. Baker, P. Tennant, C. Gordon, C. Vrettou, R. Blundell, L. Hyndman, B. Stevenson, A. Wilson, A. Doherty, D. J. Shaw, R. L. Coles, H. Painter, S. H. Cheng, R. K. Scheule, J. C. Davies, J. A. Innes, S. C. Hyde, U. Griesenbach, E. W. Alton, A. C. Boyd, D. J. Porteous, D. R. Gill, and D. D. Collie. 2011. Pre-clinical evaluation of three non-viral gene transfer agents for cystic fibrosis after aerosol delivery to the ovine lung. *Gene Ther* 18 (10):996–1005.

Mitsui, M., M. Nishikawa, L. Zang, M. Ando, K. Hattori, Y. Takahashi, Y. Watanabe, and Y. Takakura. 2009. Effect of the content of unmethylated CpG dinucleotides in plasmid DNA on the sustainability of transgene expression. *J Gene Med* 11 (5):435–43.

Miyata, K., M. Oba, M. Nakanishi, S. Fukushima, Y. Yamasaki, H. Koyama, N. Nishiyama, and K. Kataoka. 2008. Polyplexes from poly(aspartamide) bearing 1,2-diaminoethane side chains induce pH-selective, endosomal membrane destabilization with amplified transfection and negligible cytotoxicity. *J Am Chem Soc* 130 (48):16287–94.

Moghimi, S. M., P. Symonds, J. C. Murray, A. C. Hunter, G. Debska, and A. Szewczyk. 2005. A two-stage poly(ethylenimine)-mediated cytotoxicity: Implications for gene transfer/therapy. *Mol Ther* 11 (6):990–5.

Mohri, K., T. Okuda, A. Mori, K. Danjo, and H. Okamoto. 2010. Optimized pulmonary gene transfection in mice by spray-freeze dried powder inhalation. *J Control Release* 144 (2):221–6.

Nagaya, N., S. Kyotani, M. Uematsu, K. Ueno, H. Oya, N. Nakanishi, M. Shirai, H. Mori, K. Miyatake, and K. Kangawa. 2004. Effects of adrenomedullin inhalation on hemodynamics and exercise capacity in patients with idiopathic pulmonary arterial hypertension. *Circulation* 109 (3):351–6.

Nagaya, N., H. Mori, S. Murakami, K. Kangawa, and S. Kitamura. 2005. Adrenomedullin: Angiogenesis and gene therapy. *Am J Physiol Regul Integr Comp Physiol* 288 (6):R1432–7.

Nagaya, N., H. Okumura, M. Uematsu, W. Shimizu, F. Ono, M. Shirai, H. Mori, K. Miyatake, and K. Kangawa. 2003. Repeated inhalation of adrenomedullin ameliorates pulmonary hypertension and survival in monocrotaline rats. *Am J Physiol Heart Circ Physiol* 285 (5):H2125–31.

Nel, A., T. Xia, L. Madler, and N. Li. 2006. Toxic potential of materials at the nanolevel. *Science* 311 (5761):622–7.

Patton, J. S., and P. R. Byron. 2007. Inhaling medicines: Delivering drugs to the body through the lungs. *Nat Rev Drug Discov* 6 (1):67–74.

Pittella, F., M. Zhang, Y. Lee, H. J. Kim, T. Tockary, K. Osada, T. Ishii, K. Miyata, N. Nishiyama, and K. Kataoka. 2011. Enhanced endosomal escape of siRNA-incorporating hybrid nanoparticles from calcium phosphate and PEG-block charge-conversional polymer for efficient gene knockdown with negligible cytotoxicity. *Biomaterials* 32 (11):3106–14.

Poland, C. A., R. Duffin, I. Kinloch, A. Maynard, W. A. Wallace, A. Seaton, V. Stone, S. Brown, W. Macnee, and K. Donaldson. 2008. Carbon nanotubes introduced into the abdominal cavity of mice show asbestos-like pathogenicity in a pilot study. *Nat Nanotechnol* 3 (7):423–8.

Pringle, I. A., S. C. Hyde, and D. R. Gill. 2009. Non-viral vectors in cystic fibrosis gene therapy: Recent developments and future prospects. *Expert Opin Biol Ther* 9 (8):991–1003.

Rudolph, C., A. Ortiz, U. Schillinger, J. Jauernig, C. Plank, and J. Rosenecker. 2005. Methodological optimization of polyethylenimine (PEI)-based gene delivery to the lungs of mice via aerosol application. *J Gene Med* 7 (1):59–66.

Ryman-Rasmussen, J. P., M. F. Cesta, A. R. Brody, J. K. Shipley-Phillips, J. I. Everitt, E. W. Tewksbury, O. R. Moss, B. A. Wong, D. E. Dodd, M. E. Andersen, and J. C. Bonner. 2009. Inhaled carbon nanotubes reach the subpleural tissue in mice. *Nat Nanotechnol* 4 (11):747–51.

Rytting, E., J. Nguyen, X. Wang, and T. Kissel. 2008. Biodegradable polymeric nanocarriers for pulmonary drug delivery. *Expert Opin Drug Deliv* 5 (6):629–39.

Takae, S., K. Miyata, M. Oba, T. Ishii, N. Nishiyama, K. Itaka, Y. Yamasaki, H. Koyama, and K. Kataoka. 2008. PEG-detachable polyplex micelles based on disulfide-linked block catiomers as bioresponsive nonviral gene vectors. *J Am Chem Soc* 130 (18):6001–9.

Uchida, S., K. Itaka, Q. Chen, K. Osada, T. Ishii, M. A. Shibata, M. Harada-Shiba, and K. Kataoka. 2012. PEGylated polyplex with optimized PEG shielding enhances gene introduction in lungs by minimizing inflammatory responses. *Mol Ther* 20 (6):1196–203.

Uchida, S., K. Itaka, Q. Chen, K. Osada, K. Miyata, T. Ishii, M. Harada-Shiba, and K. Kataoka. 2011. Combination of chondroitin sulfate and polyplex micelles from poly(ethylene glycol)-poly{*N'*-[*N*-(2-aminoethyl)-2-aminoethyl]aspartamide} block copolymer for prolonged *in vivo* gene transfection with reduced toxicity. *J Control Release* 155 (2):296–302.

Ziady, A. G., C. R. Gedeon, O. Muhammad, V. Stillwell, S. M. Oette, T. L. Fink, W. Quan, T. H. Kowalczyk, S. L. Hyatt, J. Payne, A. Peischl, J. E. Seng, R. C. Moen, M. J. Cooper, and P. B. Davis. 2003. Minimal toxicity of stabilized compacted DNA nanoparticles in the murine lung. *Mol Ther* 8 (6):948–56.

15

Bio-Nanocapsules: Novel Drug Delivery

Shun'ichi Kuroda

CONTENTS

15.1 Introduction

Drug delivery systems (DDSs) represent an emerging technology for targeting necessary and sufficient amounts of therapeutic materials (small molecules or genes) to specific sites in a time-controlled manner. Recent developments in nanotechnology have led to the generation of new nanoparticles (NPs) that can function as nanocarriers, ranging in size from 50 to 150 nm in diameter. Composed of biocompatible materials, these NPs have been formulated into nanomedicines by incorporating therapeutic materials. Conventional nanocarriers have been classified into four groups (Kasuya and Kuroda 2009): (i) liposomes, nanocapsules consisting of a lipid bilayer; (ii) nanomicelles, NPs consisting of an amphiphilic diblock copolymer; (iii) polymers; and (iv) viruses.

Liposomes and nanomicelles are approximately 100-nm hollow nanocapsules that can incorporate and carry drugs and genes. As DDS nanocarriers, they can be further categorized as "passive" or "active" targeting particles. "Passive-targeting DDS nanocarriers" accumulate spontaneously at the sites of inflammation or in angiogenic tumors where the endothelium is hyperpermeable to >100-nm NPs; this has been described as the enhanced permeability and retention effect (Maeda et al. 2000). On the other hand, "active-targeting DDS nanocarriers" display tissue-specific molecules (e.g., transferrin and folic acid as tumor-specific molecules; Kolhatkar et al. 2011) and accumulate in the tissues of interest. Both passive- and active-targeting DDS nanocarriers show comparable rates of accumulation at sites of inflammation and angiogenic tumors (Torchilin 2010). Presently, tumors remain the primary target of passive-targeting DDS nanocarriers, whereas nontumor tissues are increasingly becoming the targets of active-targeting DDS nanocarriers.

Typical polymer-based DDS nanocarriers are composed of a cationic polymer that spontaneously forms a stable complex with anionic nucleic acids by ionic interactions. These complexes, called polyplexes, are net positively charged, potentially cytotoxic, and relatively large (>100 nm in diameter) (Neu et al. 2005); thus, polyplexes may not be suitable for *in vivo* delivery. Viruses can efficiently infect a wide variety of cells *in vitro* and *in vivo*; as a DDS nanocarrier, they can only deliver genes that have been incorporated along with their own genomes. The specificity of a virus for a particular target cell varies; some are highly specific, while others are less selective. However, the application of virus-mediated gene therapy in the clinic has been limited after a highly publicized case of a death associated with gene therapy (Savulescu 2001). Thus, viruses have been used as nanocarriers in a carefully controlled way (i.e., replication-deficient viruses).

15.2 Attributes Required for Future Nanocarriers

Figure 15.1 lists the essential characteristics that nanocarriers are required to have to maximize their therapeutic efficacy in treating a wide variety of diseases; these are in addition

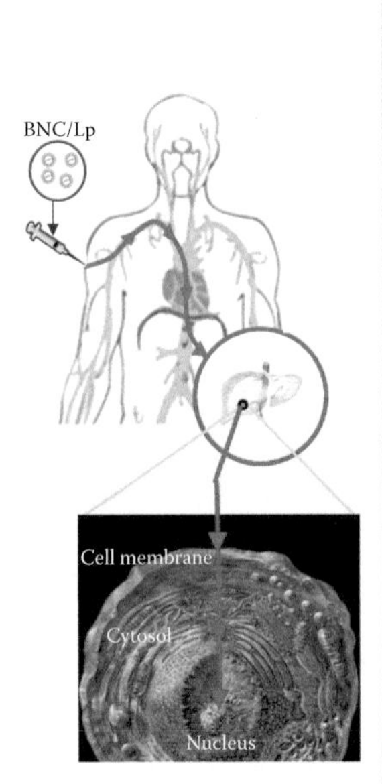

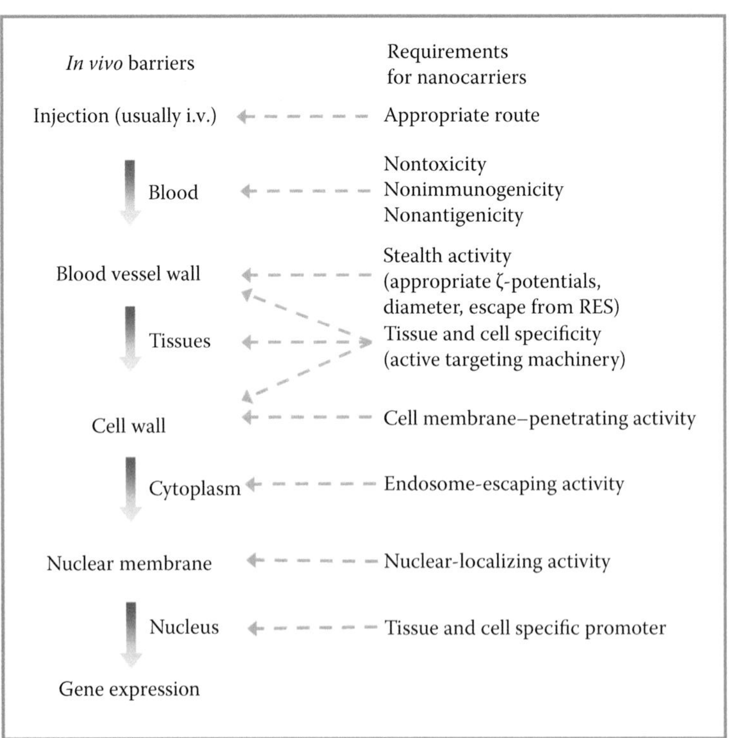

FIGURE 15.1
Essential characteristics required for nanocarriers to maximize their therapeutic efficacy in treating a wide variety of diseases.

to their "targeting activity." To reduce trapping by organs with a highly developed reticuloendothelial system (RES), such as the liver, lung, and spleen (phagocyte-rich organs), the surface of nanocarriers have been modified with polyethylene glycol (PEG). The formation of a hydrated phase minimizes recognition by the RES and maximizes the circulating levels, thus increasing the so-called stealth activity of nanocarriers (Matthews and McCoy 2004). However, the repetitive administration of PEGylated nanomedicines can elicit the production of anti-PEG IgM antibodies that can accelerate blood clearance (Ishida et al. 2006). Thus, new strategies need to be developed to incorporate stealth activity into nanocarriers while minimizing immunogenicity. Furthermore, since the subcellular targets of nanocarriers (e.g., interstitium, cell membrane, endosomes, cytoplasm, and nucleus) should be optimized for incorporated therapeutic materials to maximize their efficacy, the nanocarriers should possess "cell membrane–penetrating activity," "endosome-escaping activity," and "nuclear-localizing activity," concurrently within extremely small space without any interference.

15.3 Lessons from Viruses as Natural Nanocarriers

Viruses that infect humans harbor a nanostructure, circulate in the body, and adsorb onto specific target cells and tissues, which they then invade to establish infection. Through the mutation and adaptation that occurred during the evolution of viruses in the human body, viruses have successfully acquired the ability to target specific cells and tissues. Examples of viruses and their target cells include Japanese encephalitis virus and poliovirus targeting the brain, hepatitis B and C viruses targeting the liver, rotavirus and noroviruses targeting the intestine, human papilloma virus targeting the endocervix, and human immunodeficiency virus and Epstein–Barr virus targeting lymphocytes. Unlike nonviral nanocarriers, viruses are able to establish infection even at low titers, suggesting that they are innate and efficient nanocarriers. By analyzing the surface structures of viruses, we have been able to elucidate the molecular mechanisms underlying the following functions (see Figure 15.1) that are invaluable for their development as nanocarriers:

1. Optimal entry (administration) route
2. Stealth activity
3. Nonimmunogenicity, nonantigenicity, and nontoxicity
4. Active targeting to specific cells and tissues
5. Cell membrane–penetrating activity and endosome-escaping activity
6. Nuclear-localizing activity

In principle, nanocarriers displaying surface proteins of a particular virus could simulate the above viral activities *in vivo*. Liposomes displaying viral surface proteins—termed virosomes—were first used as immunogens for vaccines (Almeida et al. 1975). Since then, Kaneda (2000) demonstrated that virosomes composed of liposomes containing Sendai virus envelope proteins could deliver drugs and genes into cells and tissues. However, Sendai virus–derived virosomes have shown high immunogenicity and less targeting activity, thus hampering the development of these virosomes as specific DDS nanocarriers.

15.4 Bio-Nanocapsules: Hybrids of Human Liver–Specific Hepatitis B Virus and Liposomes

Hepatitis B virus (HBV) is a DNA virus of approximately 42 nm that specifically infects the liver of humans and chimpanzees. The highly infectious virions can be transmitted by body fluids, and even at low titers they can evade the RES to establish infection in hepatocytes. The HBV surface antigen (HBsAg) L protein, comprising pre-S1 (108 amino acids in HBV serotype y), pre-S2 (55 amino acids), and S (226 amino acid) regions, plays a central role in the infection machinery of HBV. When expressed in eukaryotic cells, L proteins self-aggregate to form hollow NPs of 30–40 nm in diameter stabilized by the S region, which contains three transmembrane segments (Figure 15.2) (Kuroda et al. 1992; Yamada et al. 2001; Jung et al. 2011). The N-terminal half of the pre-S1 region (amino acid residues 10–36) contains a human liver–specific binding site (Neurath et al. 1986), whereas the central domain of the pre-S2 region contains a receptor for polymerized human albumin (pHSA) (Itoh et al. 1992); both of these binding sites are exhibited on the surface of HBV and L protein particles. In 2003, we successfully produced large amounts of L protein particles in *Saccharomyces cerevisiae*, and we incorporated drugs and genes by electroporation into these particles. When administered systemically into a mouse xenograft model, the L protein NPs delivered their payloads specifically and efficiently to human hepatic tumors (Yamada et al. 2003), demonstrating liver-specific targeting. We have since showed that the uptake of L protein particles by human hepatic cells is comparable to that of HBV in magnitude, and it is pre-S1 dependent (Yamada et al. 2012). We have designated these nonviral human liver–specific NPs as bio-nanocapsules (BNCs).

The structure of the BNC is very similar to that of the immunogen for conventional hepatitis B vaccines, which have also been produced in *S. cerevisiae* during the last three

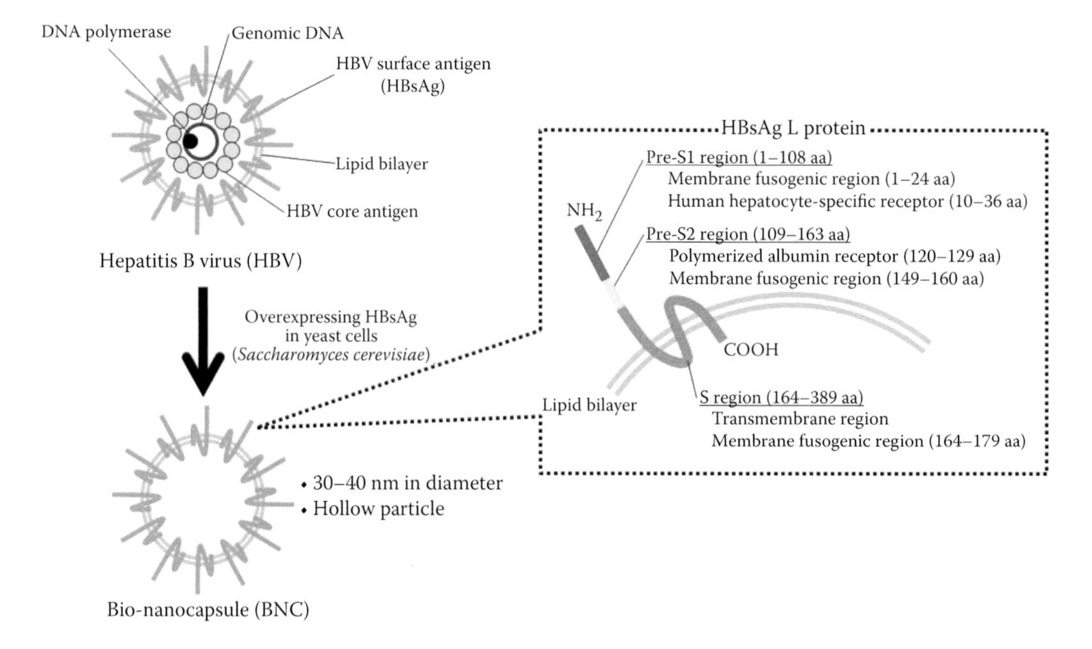

FIGURE 15.2
Structures of hepatitis B virus and bio-nanocapsule.

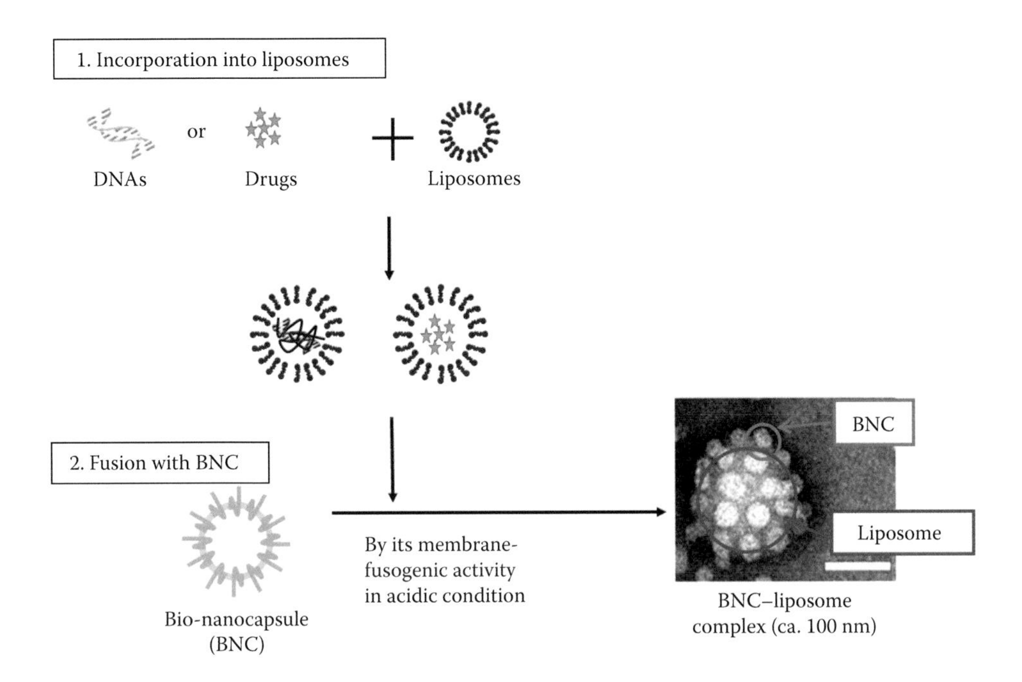

FIGURE 15.3
Process flowchart of the preparation of BNC–liposome complexes.

decades. The immunogenicity of BNC without adjuvant is not as high as that of the hepatitis B vaccine; nevertheless, studies are under way to reduce it further by engineering L protein mutations based on those identified in HBV escape mutants (see Section 15.5). Using good manufacturing practice (GMP) similar to that for recombinant hepatitis B vaccines, large amounts of BNC can be produced for clinical use.

Difficulties of using electroporation to produce BNC-based nanomedicines on a large scale led us to develop an alternative method for incorporating therapeutic agents into BNCs. In 2008, we discovered that the membrane fusogenic activity of BNCs facilitated the spontaneous formation of stable complexes of BNCs with liposomes (Jung et al. 2008). Transmission electron microscopy revealed that the liposomes are surrounded by 10–20 BNCs, suggesting that the complex has a similar structure to virosomes. We have shown that BNC–liposome complexes containing small molecules or genes, approximately 100 nm in diameter, are able to deliver their payloads to human hepatic cells and tumors *in vitro* and *in vivo* (Figure 15.3). The specificity and incorporation rate of BNC–liposome complexes into human hepatic cells were comparable to those of BNC alone and HBV.

15.5 Bio-Nanocapsules as Candidate Nanocarriers

Minimizing or eliminating the immunogenicity and antigenicity of nanocarriers is necessary for their long-term administration. This is particularly important in humans since viral proteins are highly immunogenic and antigenic. While liposomes can attenuate the immunogenicity and antigenicity of some conjugated proteins, this is still a major drawback of virosomes, including BNC–liposome complexes. Recently, we compared the amino acid sequences of various HBV escape mutants, which could propagate in recipients of the hepatitis B vaccine without being cleared by immune system in humans, and found two common mutations in the L protein. After repetitive injections of BNCs harboring the two mutations into mice (at 4-week intervals during 28 weeks), more than half of the mice did not develop anti-BNC antibodies (Jung et al. submitted). Thus, further study of HBV escape mutants as potential nanocarriers, as well as other mutants of viruses, is warranted.

Studies have shown that BNCs can achieve high systemic delivery. When BNC–liposome complexes containing the anticancer drug doxorubicin were injected into a mouse xenograft model, the circulating levels were comparable to those of PEGylated liposomes containing the same drug (Kasuya et al. 2009). These data suggest that coating liposomes with BNCs (i) increases the stability of liposomes in blood to levels comparable to the ones seen with PEGylated liposomes, and (ii) facilitates evasion from the RES. Recent evidence points to a role for the pHSA receptor in the pre-S2 region in conferring this "stealth" activity on BNC–liposome complexes, BNCs, and HBV. Specifically, HSA-coated polystyrene nanospheres injected intravenously into rats remained in the circulation longer than non-coated nanospheres and exhibited significantly reduced hepatic clearance (Ogawara et al. 2004). Ogawara et al. postulated that HSA prevented the binding of other serum proteins that would render the nanospheres more susceptible to opsonization. Moreover, in severe combined immunodeficiency mice harboring normal human liver tissues under the kidney capsule, BNC was demonstrated to accumulate only in the human liver tissue without being trapped by other tissues (Matsuura et al. 2011). Recently, we have shown that 100-nm polystyrene beads displaying a synthetic peptide functioning as a receptor for pHSA efficiently escape from the RES in a mouse xenograft model (Takagi et al. submitted).

Once endocytosed by cells, HBV requires a low pH within the endosomes to fuse with the endosome membrane and to release its natural payload (i.e., core proteins, genomic DNA, or polymerase) into the cytoplasm. Others have identified membrane fusogenic domains at the N-terminal half of the S region and the C-terminal half of the pre-S2 region (Rodríguez-Crespo et al. 1995; Oess and Hildt 2000); we have recently found a new membrane fusogenic domain at the N-terminal 20 amino acids of the pre-S1 region (Oeda et al. submitted). Deleting the pre-S1 region or blocking this region with anti-pre-S1 antibodies completely inhibited the formation of the BNC–liposome complex, suggesting that the membrane fusogenic domain in the pre-S1 region is indispensable for the movement of payloads of the BNC–liposome complex, BNC, and HBV from endosomes to the cytoplasm (endosomal escape).

We have shown in several studies using a mouse xenograft model that BNCs or BNC–liposome complexes are capable of delivering nucleic acids encoding and expressing a variety of genes. Specifically, intravenous injection of a BNC containing an expression vector for blood coagulation factor IX elicited the production of factor IX for at least 1 month at a level sufficient to treat moderate levels of hemophilia B (Yamada et al. 2003). In a separate study, the size of human hepatocyte–derived tumors, but not of human colon cancer–derived tumors, was significantly reduced by the intravenous injection of a BNC containing an expression vector for the herpes simplex virus–derived thymidine kinase (HSV-tk), along with the subcutaneous injection of ganciclovir, an inhibitor of HSV-tk (Iwasaki et al. 2007). Moreover, intravenously injected BNC–liposome complex containing doxorubicin (4 mg/kg) caused a significant reduction in the size of human hepatocyte-derived tumors; doxorubicin alone and PEGylated liposomes containing doxorubicin produced a lower effect at the same dose (Kasuya et al. 2009). These findings show that BNCs and BNC–liposomes possess increased functionality over other nanocarriers by their ability to introduce drugs and genes to be expressed.

15.6 Retargeting of Bio-Nanocapsules

Because the pre-S1 region represents a specific ligand for a receptor on hepatocytes, BNCs were originally developed to target liver cells. The liver is often a target of viral infections and subject to various metabolic disorders; therefore, the human liver-specific targeting ability of BNCs may have therapeutic value. Nevertheless, it is necessary to target BNCs (and BNC–liposome complexes) to other tissues and organs as well to expand their clinical utility. To this end, we and others have carried out studies to change the target tissue specificity of the BNC by replacing the pre-S1 region with other molecules through genetic recombination. Specifically, we modified the BNC to display a tandem form of the Z domain (IgG Fc-binding motif) from *Staphylococcus aureus* protein A, to which we could tether IgGs of interest; these modified BNCs were designated as ZZ-BNC (Iijima et al. 2011). Tsutsui et al. (2007) conjugated an anti-epidermal growth factor receptor (EGFR) IgG to the ZZ-BNC; this complex was injected intracerebroventricularly into mice harboring an EGFR-expressing glioma, where it was found to accumulate specifically in the tumor (Tsutsui et al. 2007). When ZZ-BNCs conjugated with anti-selectin IgG were mixed with liposomes containing either a green fluorescent protein expression vector or 100-nm fluorescent polystyrene beads and were applied intravenously in mouse models of either uveitis or arthritis, the BNCs were found localized to the retina or knee joint, respectively, suggesting successful targeting into the inflamed tissues. These complexes also delivered fluorescent dye to infarcted myocardium in a rat model of myocardial ischemia/reperfusion injury (Jung et al. submitted). Recently, we made ZZ-BNCs conjugated with anti-CD11c IgG and mixed them with liposomes containing a Japanese encephalitis virus (JEV)-derived envelope protein. Mice receiving two intravenous injections 4 weeks apart exhibited a higher titer of anti-JEV antibodies 2 weeks after the second injection compared with mice receiving control vaccines. This suggests that the anti-CD11c antibody can retarget the ZZ-BNC–liposome complex to CD11c$^+$ mouse splenic dendritic cells to elicit the efficient production of anti-JEV antibodies (Matsuo et al. 2012).

In addition to antibodies, other conjugates have also been tested for their ability to target BNCs to specific tissues. *Phaseolus vulgaris* agglutinin-L4 (PHA-L4) isolectin recognizes β1–6 branching *N*-acetylglucosamine (β1–6GlcNAc), which is abundantly expressed as a part of high-mannose glycans in various highly metastatic cancers. ZZ-BNCs were conjugated with PHA-L4 isolectins and mixed with a liposome complex containing fluorescent dyes or luciferase expression vector. Following injection of the isolectin–ZZ-BNC–liposome complex into mice harboring tumors with or without β1–6GlcNAc (i.e., malignant or benign tumors respectively), payloads were found only in malignant, but not benign, tumors (Kasuya et al. 2008). These data demonstrate that antibodies and lectins are useful for the *in vivo* pinpoint targeting of nanocarriers. Finally, Laakkonen et al. (2008) have reported on the use of homing peptides to specifically recognize various stages of lymphatic tumors. However, our laboratory has not been successful in using homing peptide-displaying BNCs for the *in vivo* targeted delivery of payloads. For accomplishing *in vivo* pinpoint delivery of BNC-based nanomedicines, it would be important to display targeting molecules that can specifically recognize the complicated structures on the surface of the targeted sites.

15.7 Future of Bio-Nanocapsules

To date, >30 patent applications have been filed in Japan related to BNCs; >20 have been filed internationally. Basic patents for BNCs have been registered in Japan, Korea, United States, and Europe. BNCs fulfill all of the requirements necessary to effectively deliver therapeutic materials, both small molecules and genes, to target cells, and thereby hold much promise as nanocarriers. However, since BNCs can mediate the movement of payloads only from extracellular space via endosomes to the cytoplasm, the machinery for the transport of genes from the cytoplasm to the nucleus must be installed for maximizing the transgene expression, which is a common problem for every nonviral nanocarrier (Hama et al. 2006). Furthermore, the BNC–liposome complex is composed of biologics and chemical products that complicate CMC (chemistry, manufacturing, and controls), retard the production of BNC-based nanomedicines under current GMP conditions, and increase the cost of the drug. Finally, the idea of BNCs and the BNC–liposome complex should not be limited to self-aggregating proteins from HBV but should be expanded to proteins from other viruses to yield a wider array of BNCs selective for a variety of tissues. Moreover, the use of nanocarriers other than liposomes needs to be explored. As demonstrated by the targeting of polyplex, a complex of polyethyleneimine and DNA (Somiya et al. 2012), BNCs could be used in combination with various nanocarriers.

Acknowledgments

This work was supported in part by KAKENHI (Grant-in-Aid for Scientific Research [A] [25242043]); the Science and Technology Research Promotion Program for Agriculture, Forestry, Fisheries and Food Industry; and the Health Labor Sciences Research Grant from the Ministry of Health Labor and Welfare.

References

Almeida, J. D., D. C. Edwards, C. M. Brand, and T. D. Heath. 1975. Formation of virosomes from influenza subunits and liposomes. *Lancet* 2: 899–901.

Hama, S., H. Akita, R. Ito, H. Mizuguchi, and H. Harashima. 2006. Quantitative comparison of intracellular trafficking and nuclear transcription between adenoviral and lipoplex systems. *Mol. Ther.* 13: 786–94.

Iijima, M., H. Kadoya, S. Hatahira et al. 2011. Nanocapsules incorporating IgG Fc-binding domain derived from *Staphylococcus aureus* protein A for displaying IgGs on immunosensor chips. *Biomaterials* 32: 1455–64.

Ishida, T., K. Atobe, X. Wang, and H. Kiwada. 2006. Accelerated blood clearance of PEGylated liposomes upon repeated injections: Effect of doxorubicin-encapsulation and high-dose first injection. *J. Control Release* 115: 251–8.

Itoh, Y., S. Kuroda, T. Miyazaki, S. Otaka, and Y. Fujisawa. 1992. Identification of polymerized-albumin receptor domain in the pre-S2 region of hepatitis B virus surface antigen M protein. *J. Biotechnol.* 23: 71–82.

Iwasaki, Y., M. Ueda, T. Yamada et al. 2007. Gene therapy of liver tumors with human liver-specific nanoparticles. *Cancer Gene Ther.* 14: 74–81.

Jung, J., M. Iijima, N. Yoshimoto et al. 2011. Efficient and rapid purification of drug- and gene-carrying bio-nanocapsules, hepatitis B virus surface antigen L particles, from *Saccharomyces cerevisiae*. *Protein Expr. Purif.* 78: 149–55.

Jung, J., T. Matsuzaki, K. Tatematsu et al. 2008. Bio-nanocapsule conjugated with liposomes for *in vivo* pinpoint delivery of various materials. *J. Control Release* 126: 255–64.

Kaneda, Y. 2000. Virosomes: Evolution of the liposome as a targeted drug delivery system. *Adv. Drug Deliv. Rev.* 43: 197–205.

Kasuya, T., J. Jung, H. Kadoya et al. 2008. *In vivo* delivery of bionanocapsules displaying *Phaseolus vulgaris* agglutinin-L4 isolectin to malignant tumors overexpressing *N*-acetylglucosaminyltransferase V. *Hum. Gene Ther.* 19: 887–95.

Kasuya, T., J. Jung, R. Kinoshita et al. 2009. Bio-nanocapsule–liposome conjugates for *in vivo* pinpoint drug and gene delivery. *Methods Enzymol.* 464: 147–66.

Kasuya, T., and S. Kuroda. 2009. Nanoparticles for human liver–specific drug and gene delivery systems: *In vitro* and *in vivo* advances. *Expert Opin. Drug Deliv.* 6: 39–52.

Kolhatkar, R., A. Lote, and H. Khambati. 2011. Active tumor targeting of nanomaterials using folic acid, transferrin and integrin receptors. *Curr. Drug Discov. Technol.* 8: 197–206.

Kuroda, S., S. Otaka, T. Miyazaki, M. Nakao, and Y. Fujisawa. 1992. Hepatitis B virus envelope L protein particles. Synthesis and assembly in *Saccharomyces cerevisiae*, purification and characterization. *J. Biol. Chem.* 267: 1953–61.

Laakkonen, P., L. Zhang, and E. Ruoslahti. 2008. Peptide targeting of tumor lymph vessels. *Ann. N. Y. Acad. Sci.* 1131: 37–43.

Maeda, H., J. Wu, T. Sawa, Y. Matsumura, and K. Hori. 2000. Tumor vascular permeability and the EPR effect in macromolecular therapeutics: A review. *J. Control Release* 65: 271–84.

Matsuo, H., N. Yoshimoto, N. Iijima et al. 2012. Engineered hepatitis B virus surface antigen L protein particles for *in vivo* active targeting of splenic dendritic cells. *Int. J. Nanomedicine* 7: 3341–50.

Matsuura, Y., H. Yagi, S. Matsuda et al. 2011. Human liver-specific nanocarrier in a novel mouse xenograft model bearing noncancerous human liver tissue. *Eur. Surg. Res.* 46: 65–72.

Matthews, S. J., and C. McCoy. 2004. Peginterferon alfa-2a: A review of approved and investigational uses. *Clin. Ther.* 26: 991–1025.

Neu, M., T. Fischer, and D. Kissel. 2005. Recent advances in rational gene transfer vector design based on poly(ethylene imine) and its derivatives. *J. Gene Med.* 7: 992–1009.

Neurath, A. R., S. B. Kent, N. Strick, and K. Parker. 1986. Identification and chemical synthesis of a host cell receptor binding site on hepatitis B virus. *Cell* 46: 429–36.

Oess, S., and E. Hildt. 2000. Novel cell permeable motif derived from the PreS2-domain of hepatitis-B virus surface antigens. *Gene Ther.* 7: 750–8.

Ogawara, K., K. Furumoto, S. Nagayama et al. 2004. Pre-coating with serum albumin reduces receptor-mediated hepatic disposition of polystyrene nanosphere: Implications for rational design of nanoparticles. *J. Control Release* 100: 451–5.

Rodríguez-Crespo, I., E. Núñez, J. Gómez-Gutiérrez et al. 1995. Phospholipid interactions of the putative fusion peptide of hepatitis B virus surface antigen S protein. *J. Gen. Virol.* 76: 301–8.

Savulescu, J. 2001. Harm, ethics committees and the gene therapy death. *J. Med. Ethics* 27: 148–50.

Somiya, M., N. Yoshimoto, M. Iijima et al. 2012. Targeting of polyplex to human hepatic cells by bio-nanocapsules, hepatitis B virus surface antigen L protein particles. *Bioorg. Med. Chem.* 20: 3873–9.

Torchilin, V. P. 2010. Passive and active drug targeting: Drug delivery to tumors as an example. *Handb. Exp. Pharmacol.* 197: 3–53.

Tsutsui, Y., K. Tomizawa, M. Nagita et al. 2007. Development of bionanocapsules targeting brain tumors. *J. Control Release* 122: 159–64.

Yamada, M., A. Oeda, J. Jung et al. 2012. Hepatitis B virus envelope L protein-derived bio-nanocapsules: Mechanisms of cellular attachment and entry into human hepatic cells. *J. Control Release* 160: 322–9.

Yamada, T., H. Iwabuki, T. Kanno et al. 2001. Physicochemical and immunological characterization of hepatitis B virus envelope particles exclusively consisting of the entire L (pre-S1 + pre-S2 + S) protein. *Vaccine* 19: 3154–63.

Yamada, T., Y. Iwasaki, H. Tada et al. 2003. Nanoparticles for the delivery of genes and drugs to human hepatocytes. *Nat. Biotechnol.* 21: 885–90.

Section VI

Special Issues

16

Physicochemical, Colloidal, and Transport Properties

Heinrich Hofmann, Lionel Maurizi, Marie-Gabrielle Beuzelin, Usawadee Sakulkhu, and Vianney Bernau

CONTENTS

16.1 Introduction

Following the definition from the European Commission, all engineered particles <100 nm in diameter are nanoparticles (NPs) (http://europa.eu, 2011). By this definition, many "nanoproducts" exist on the market, in which nanoscaled particles are the main building blocks. From a more scientific viewpoint, nanomaterials are defined by the change of important physicochemical properties with decreasing crystal or particle size (Brune et al. 2006). Such significant changes do not normally occur in particles >30 nm but mostly in particles <10 nm. The reasons for the changes in the material properties are the high surface/volume ratio and the changes in the electronic band structure (quantum confinement). The surface/volume ratio is not only a geometrical parameter that increases the reactivity per mass of a material but it also affects thermodynamic properties such as phase stability and equilibrium, including the melting temperature, homogeneity of the particles, and shape. Simultaneously, quantum confinement occurs, and the electrical and optical properties including the chemical reactivity are changed. A third origin of changes at the nanosize are properties such as ferromagnetism or ferroelectricity; they both depend on the interaction potential between the atoms (exchange energy), which decreases with a decreasing number of atoms per particle. If the thermal energy is larger than the interaction energy, the material loses its ordered arrangement of magnetic moments and becomes paramagnetic. Such properties depend not only on the temperature but also on the observation time. This short list demonstrates that NPs are indeed very complex materials and that a large number of properties should differ from the bulk behavior if the size of the particles decreases. It is important to be able to predict the behavior of NPs in a biological environment, such as cell cultures or *in vivo*. In this chapter, only aspects that are directly relevant for their behavior in a biological environment are discussed.

16.2 Physicochemical Properties of NPs

The size, morphology, crystalline phase, and crystal orientation of the crystal plane expressed at the surface determine the chemical and physical properties of NPs. Properties such as melting point, surface tension, crystalline phase, and catalytic activity are important characteristics for the application of the particles as well as for all aspects regarding interaction with living systems. In addition to the aforementioned physicochemical properties of nanoparticulate materials, Rivera-Gil et al. (2013) listed colloidal stability, purity, inertness, size, shape, charge, and their ability to adsorb environmental compounds such as proteins. These properties are again related to other basic physicochemical properties mostly in a complex manner, and the correlation of these properties with toxicity is not straightforward. The critical size, at which a significant deviation from the bulk properties occurs, depends strongly on the material but also on the property of interest (Barnard and Xu 2008). For example, a significant change (increase) in the band-gap energy, an important parameter for optical and electrical properties, and thus also for chemical properties, can be observed mostly for particles <10 nm. In contrast, melting temperatures decrease if the size of the particle is <5 nm. Another interesting example is the behavior of ferromagnetic materials. The change from permanent magnetism (e.g., ferromagnetism) to paramagnetic behavior occurs for oxides in the range of 20–30 nm and for metals <5 nm. These few examples clearly illustrate that the properties of NPs are not a constant value as for bulk materials; for each size, shape, and environment, we observe different properties. Titanium dioxide, a well-investigated nanomaterial that is also a bulk product used in various industrial applications, is a highly complex nanomaterial. Different crystalline phases are known (rutile, anatase, and brookite); however, the thermodynamic stability of the phases is strongly dependent on the size of the crystals. Below 5 nm, anatase is the thermodynamically stable phase, whereas for particles >33 nm, rutile is the stable phase. Regarding reactivity, not only are small titanium dioxide particles more photocatalytic because the surface/volume ratio is larger; in addition, a strong increase occurs because the amount of the more reactive anatase is also increasing with decreasing size, and the band-gap energy is increasing, which could have a significant effect on specific reactions. However, in reality, titanium dioxide NPs exhibit a size distribution, and therefore, the crystalline phase cannot be simply predicted from the mean sizes of the powder. Additionally, the thermodynamically preferred crystal phases also depend on the shape and the thermal and chemical environment, especially from H_2O adsorbed at anatase or OH^- adsorbed at rutile, as both stabilize the corresponding crystal structure. The effect of surface chemistry on phase stability leads to the coexistence of both phases over a large size range from 5 to 33 nm, and an abrupt phase change such as Gibb's phase rule is demanding is not observed (Barnard 2011). It is evident that the prediction of the properties of NPs, for example, their capacity for generation of reactive oxygen species, is very difficult and only possible with detailed modeling of the crystal and surface structure, the arrangement of atoms at the surface, and their interaction with the surrounding environment.

16.3 Particle–Particle Interactions

The interaction between two particles is generally determined by three main factors: attractive van der Waals forces, repulsive electrostatic forces, and steric hindrances. In addition to these forces, hydrophobic/hydrophilic interactions, depletion forces, and/or osmotic pressure induced by dissolved macromolecules may be observed. In a biologically relevant environment, we must address all these forces simultaneously. In water, the standard theory for interactions between small particles in colloids is the Derjaguin, Landau, Vervey, and Overbeek (DLVO) theory (Vervey and Overbeek 1948). The DLVO interaction potential is the sum of an effective repulsive electrostatic term and a van der Waals term, which is mostly attractive in nature. The van der Waals term is calculated as the integral of the interatomic dispersion interactions over the volume of both particles, while considering the dielectric properties of the medium between the particles (French 2000). Dispersion forces are the dipole interactions between molecules caused by the rapid change of the electron–nucleus arrangement, leading to changing dipoles that at any instant induce electrical interactions with another molecule/particle, resulting in interactive forces. van der Waals potentials are calculated using a constant (e.g., Hamaker constant) times a geometrical term, including the distance, shape, and size of particles. Later works (Lifshitz 1956; Parsegian and Ninham 1969; Hough and White 1980; Faure et al. 2011) have demonstrated that the Hamaker constant can be correctly calculated and predicted on the basis of spectroscopic measurements of the dielectric response of the material as a function of distance between particles and the medium (water, cell medium, air). Hamaker's van der Waals force description is widely accepted and valid for short-range interactions. For long-range interactions, other models exist (Vincent 1973), which also account for retardation effects because electromagnetic waves only have a limited velocity in the suspending medium. Only limited data regarding the dielectric function of nanosized materials exist, and therefore, bulk values must be used for the calculations. Electrostatic interactions result mainly from the overlapping of the ionic double layer of two neighboring particles. An exact analytical solution for the interaction energy of double layers can only be observed for noncorrelated point charges at charged infinite plates. The Debye–Hückel theory uses the linearized Poisson–Boltzmann (PB) distribution to describe the concentration of charges (cations and anions) as a function of the distance from the plate surfaces. Hogg et al. (1966) and Bell et al. (1970) modified the result for charged infinite plates for the application to curved surfaces. Figure 16.1a shows typical values for the interaction potential of van der Waals and electrostatic forces, as well the sum of these two forces according to the DLVO theory for 100-nm-diameter TiO_2 in water with a ζ-potential of 30 mV. A barrier of 12 k_BT (k_B is the Boltzmann constant) could clearly be observed at a distance of 4 nm from the particle surface. This value is just enough for the stabilization of a dilute suspension. When adding a polymer on the particle surfaces (also called a surfactant), the steric hindrance prevents attraction by van der Waals forces, and the suspension is colloidally stable.

Although the DLVO theory is well accepted and generally yields good predictions of the colloidal stability of suspensions of particles with diameters in the micrometer range, when the calculation is applied to nanosized objects (diameter <100 nm), the results do not always correlate well with experimental results. If we consider a suspension of 10-nm-sized maghemite NPs (γFe_2O_3) in an aqueous solution of nitric acid at 0.01 M and a ζ-potential of 20 eV, using the DLVO theory, we can calculate an energy barrier against a coagulation of 1–3 k_BT. This value would correspond to an aggregation time faster than 1 s. However, this

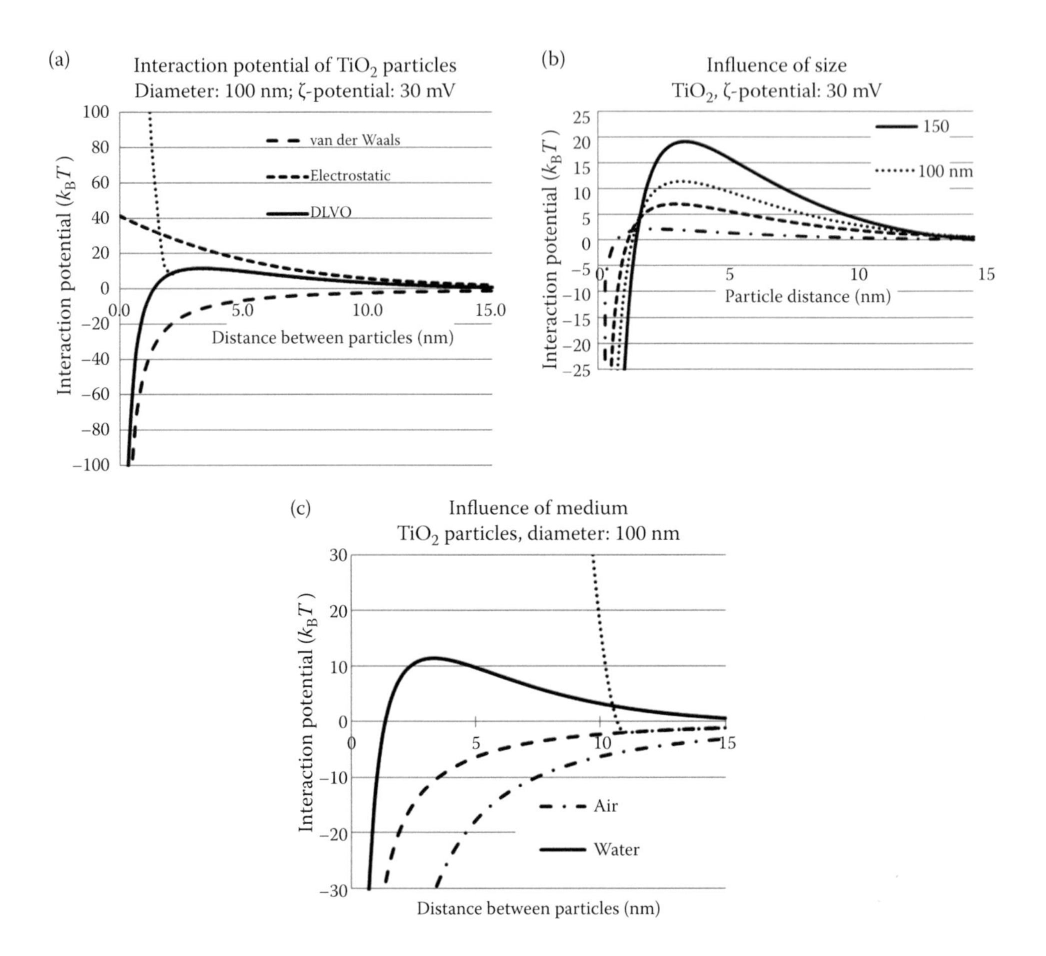

FIGURE 16.1
Interaction potential (DLVO) for titania particles of different sizes and in different media. (a) Attractive van der Waals, repulsive electrostatic potential, and total potential for 100-nm particles. The dotted line corresponds to particles with a strong bonded polymer or protein layer of 3-nm thickness. (b) DLVO potential of titania particles of different sizes. (c) DLVO potential of titania particles of 100 nm in different media. All values were calculated using Hamaker 2. (Free access from http://ltp.epfl.ch/page-35617-en.html.)

type of suspension can be maintained for >2 years at room temperature with no measurable aggregation. Figure 16.1b shows the decreasing DLVO potential for a TiO_2 in aqueous suspension (ζ-potential of 30 mV) with decreasing size. Only particles with a diameter >50 nm are colloidally stable in pure water. To compute the effective electrostatic component, the ions are described by point charges, and two approximations are made: the PB approximation (i.e., a mean-field treatment of ions) and an expansion of the charge density to linear order in the electrostatic potential. The interaction between highly charged particles in water is determined by the screening of the counterions and additional salt ions; the latter are highly present in biofluids. For strongly interacting concentrated particle suspensions, the effective interparticle forces exhibit a many-body character, which is induced by the nonlinear particle configuration dependent on counterion screening. The DLVO potential and its generic Yukawa expression for the potential of the mean force between particles, developed to describe interactions in highly diluted and weakly interacting colloidal suspensions (Dahirel and Jardat 2010), have been extensively used to describe concentrated suspensions with nanosized particles and high salt concentrations up to 0.15 M. Even recently, many experimental works on the suspensions of NPs such as proteins (Zhang et al. 2007) or iron oxides (Mériguet et al. 2004) could be interpreted assuming a Yukawa potential. These results suggest that these solutions of NPs maintain the dominant features of colloidal suspensions, even if the size of the solute particles corresponds to the lowest limit of the colloidal domain. It is important to note that nanosized particles can exhibit quantum confinement and, therefore, different optical properties such as the dielectric function. Faure et al. (2011) recalculated the Hamaker constant of superparamagnetic iron oxide in various media, taking advantage of the recent publication of the optical properties of iron oxides by Tepper et al. (2004). It was demonstrated that the Hamaker constant for 100-nm particles remains very similar to the bulk value; however, a more precise value is given. Knowing that iron oxide NPs for medical applications are in the size range of 7 nm, the nanosize will exhibit a nonnegligible effect on attractive forces. The Velegol group observed that van der Waals forces calculated using the classical approach are inaccurate for nearly all situations with nanocolloids (Kim et al. 2007). The assumptions of a uniform dielectric function (even up to the surface), the absence of discrete atom effects, or the assumption of interacting semi-infinite bodies all fail for the nanocolloid case. Another possible explanation for the discrepancy between the experimental and theoretical results is the breakdown of the assumption of a PB-like ionic double layer. Two distinct nanosize effects can be identified that might break this assumption for sufficiently small particle diameters:

- In the DLVO theory, the ionic distribution around a particle is considered to be similar to the ionic distribution close to an infinite plate with equal surface charge. However, when the size ratio between the double-layer thickness ($1/\kappa$) and the particle diameter is no longer negligible ($\kappa a < 10$), this assumption is no longer valid.
- In the DLVO theory, ions are approximated by point charges. However, when the size ratio between the solvated ionic species and the particle diameter becomes nonnegligible, ions can no longer be regarded as point charges but as finite size particles that cannot overlap. In other words, the maximal volume or surface charge density in the medium is limited by the ionic volume.

For particles suspended in blood, lung surfactant, or cell medium with serum for *in vitro* tests, the particles are not dispersed in pure water with a low salt content and a constant pH value. Biologically relevant fluids have high ionic strengths (0.15 M), have a pH of ≤7.4 (e.g., 4.5 in endosomes), and contain macromolecules such as proteins with similar sizes as the particles, for example, albumin with a diameter of 5.5 nm. Figure 16.1c shows the effect of the medium on the interaction potential of TiO_2 with a 100-nm diameter. It is clearly observed that the highest stability exists in pure water at low pH (high ζ-potential), whereas in cell medium at pH 7.4 and high ionic strengths, the particles agglomerate. After adding serum proteins that adsorb at the particle surface (Lundqvist et al. 2008), theoretically, steric stabilization should occur. An experimental investigation indicated that such conclusions are too simplified; the same particle can exhibit very fast agglomeration or very high stability depending on the type of coating, charge, and the presence of serum (see Figure 16.2) (Petri-Fink and Hofmann 2007). In contrast, polymer-coated superparamagnetic iron oxide NPs (SPIONs) were tested for colloidal stability and cell uptake. Although all the examined particles were colloidally stable in water and PBS buffer, these colloidal particles exhibited different stabilities in Dulbecco's modified Eagle's medium (DMEM) or Roswell Park Memorial Institute (RPMI) medium with and without

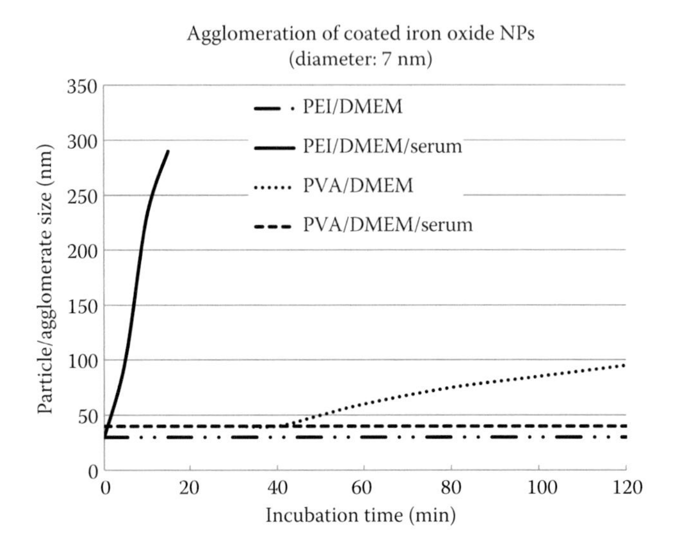

FIGURE 16.2
Agglomeration behavior of iron oxide NPs of 7 nm coated with PVA in different cell media with and without serum (10%).

fetal calf serum (FCS). Furthermore, the cell uptake (over various time intervals) depended on the cell medium and the presence/absence of serum. Systematic variations of any of the parameters, including the change of polymer (poly(vinyl-alcohol) [PVA], amino-PVA) or cell medium (DMEM or RPMI), and the presence/absence of serum, had an observable effect on agglomeration and uptake (Figure 16.3). It is clearly observed that particles in cell medium with 10% serum agglomerate less; however, the uptake is also lower than that of particles in cell medium without serum. The correlation between the agglomeration rate and uptake was 0.73 ($p = .02$). Replacement of PVA by amino-PVA, which means

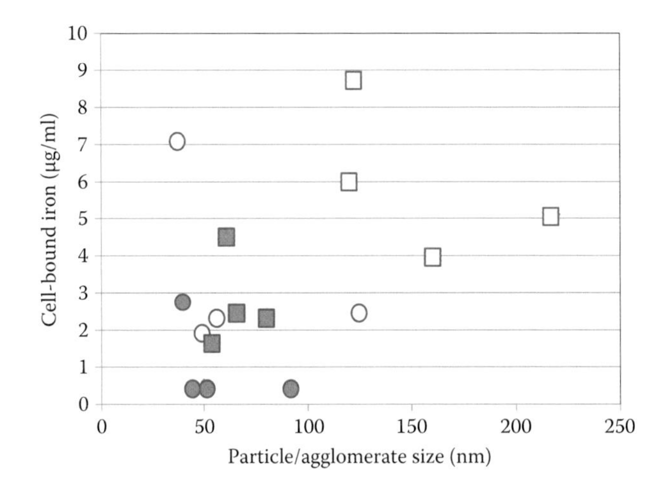

FIGURE 16.3
Cell uptake (HeLa cells) and agglomeration behavior of iron oxide NPs of 7 nm coated with PVA in different cell media with and without serum (10%). Full symbol, 1-h incubation time; empty symbol, 3-h incubation time; circles, with serum; squares, without serum.

increasing the positive charge from 5 to 25 mV, led to significantly faster agglomeration and slightly higher uptake. The reason for this observation must be the interaction of the particles with proteins in the serum. Additionally, the high stability of PVA–SPIONs in FCS-supplemented medium could be explained by depletion stabilization caused by the serum proteins that have a similar size as the NPs. Additionally, for particles >100 nm, sedimentation would be the dominant transport process, and therefore, the amount of particles in contact with the cells could be larger (details are given in Section 16.6).

16.4 Protein Adsorption

Pioneering work regarding protein adsorption on oxide NPs was performed by the Dawson group (Lundqvist et al. 2008). These researchers demonstrated that although many of the major highly abundant proteins could be detected in the corona independently of their size and surface charge, an entire range of other proteins form an important part of the corona. For a fixed material type, these researchers observed that the size of the particle and its surface modification are able to entirely change the nature of the biologically active proteins in the corona and thereby possibly also affect their biological function. Furthermore, Dawson and colleagues stated that the fundamental question is the format of presentation of these proteins and to what degree they are still able to present native epitopes when embedded in the corona that are still open. Additionally, he claimed that particles studied *in vitro* (at low serum dilutions) may bear little relation to those that exist *in vivo* (high protein concentration as, e.g., in blood), suggesting the need for a reevaluation of how such studies should be planned in the future and how *in vitro–in vivo* extrapolations could be performed. The nanoscale surface curvature strongly affects protein adsorption, such that the protein-binding affinities for NP surfaces are different from their analogous bulk material (Xin-Rui et al. 2010), and the coronas associated with NPs of the same material but of different size can vary in composition (Shang et al. 2009). Recent investigations suggest that the interaction between NPs and plasma proteins and other blood components is a determining factor for the cellular uptake of the particles. Certainly, *in vivo*, the adsorbed protein layer may affect trafficking (Oberdörster 2010) and particle biodistribution (Dobrovolskaia et al. 2008; Ehrenberg et al. 2009).

Dawson and collaborators have shown in a large number of experimental investigations that the surfaces of NPs in contact with a biological fluid are modified by the adsorption of biomolecules such as proteins or lipids, leading to an interface layer that will finally determine the interaction of the "core/shell" system with cells and organs (Monopoli et al. 2011). The so-called protein corona is not stable in composition during the first hours; it appears that, first, the most abundant proteins adsorb fast; however, in the long term, proteins at low concentration but high adsorption energy are observed preferentially at the surface (Vroman effect). Dawson's group called the first corona the soft corona and the latter the hard corona. Recently, it was demonstrated that for SPIONs, this concept failed, and no change of the protein corona over time was observed (Jansch et al. 2012). However, in all cases, it is increasingly accepted that the colloidal stability and overall scale of even nonspecific cell–particle interactions is determined by the degree of "screening" of the NP surface by the protein corona. In NPs, for example, the superparamagnetic iron oxide of 7 nm (SPION) coated with PVA (neutral), or modified PVA with amino groups (positive charge) or carboxylic groups (negative charge) in contact with serum, the net charge of each particle has a negative value (–5 to –15 mV). A comparison of ζ-potential values between SPIONs in PBS without and with serum reveals that the protein adsorptions to the surface of NPs could cause a significant change to the negative charges. By increasing the protein content, the ζ values would be further increased in the negatively charged direction (Wongsagonsup et al. 2005). After washing with PBS (30 times), the obtained charge differed strongly due to the protein movement from the surface of the SPIONs. The protein–NP interaction caused a significant increase in the hydrodynamic size of the NPs, which could be interpreted by the protein adsorption itself but additionally by the agglomeration. Severe washing of fixed NPs reduced these hydrodynamic sizes; however, tightly bound proteins (such as attached proteins in a stern layer) did not allow the particles to attain their primary size. Interestingly, at least 35 different proteins were detected in a corona of PVA-coated NPs. The composition of the corona must therefore be very different from particle to particle because for steric reasons, only 5–10 proteins could be adsorbed per particle. Additionally, it is obvious from the detection of proteins at the surface of differently charged PVA–SPION that the surface coating of the particles determines the type of proteins adsorbed (Figure 16.4). PVA-coated particles partially exhibit a very similar adsorption; however, some proteins could be observed only on one type of particle or on two. Silica-coated particles (values acquired from Jansch et al. 2012) exhibit a very different

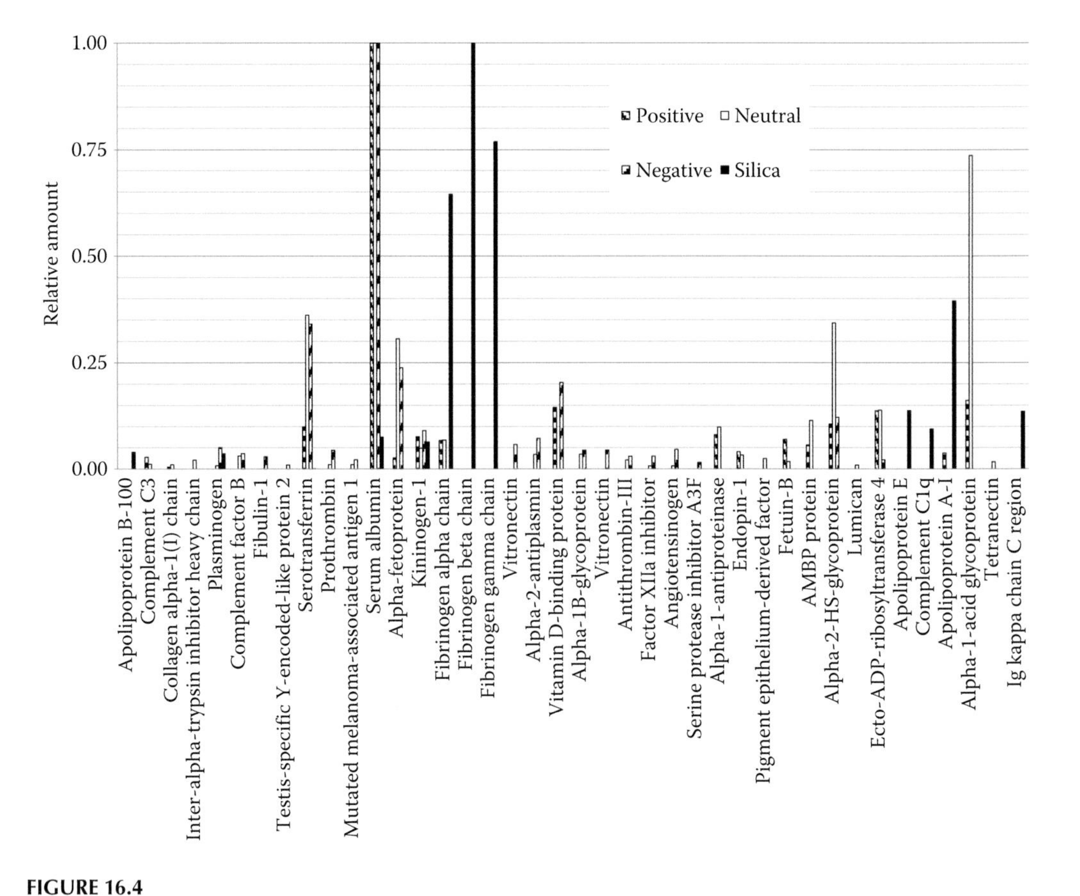

FIGURE 16.4
Relative amount of adsorbed proteins on NPs with different charges (positive, neutral, and negative) and coating materials (PVA and silica). (Values for silica obtained from Jansch M. et al., *Int. J. Pharm.*, 428, 125, 2012; values for iron oxide NPs obtained from Petri-Fink A., Hofmann H., *IEEE Trans. Nanobiosci.*, 6, 289, 2007.)

adsorption profile. This different protein adsorption led to different agglomeration behavior, as illustrated in Figure 16.3, which because of the unpredictable complex adsorption behavior is not yet predictable. It is evident that such a different protein coating would also have a large effect on the interaction with cells, such as particle adhesion at the cell membrane, uptake mechanism and biodistribution, and toxicity. For example, it has been demonstrated that positively charged PVA–SPIONs interact more strongly with HeLa cells than neutral or negatively charged PVA–SPIONs, which is consistent with the preferential uptake of cationic agents that has been widely described in the literature (Neuberger et al. 2005). However, as discussed previously, all the particles exhibited a negatively charged surface produced normally from the overall negative charge of albumin, such that direct electrostatic interaction of the original positively charged particle and the negatively charged membrane could be excluded as the mechanism for the attraction of particles by cells. In addition, the isoelectric point (IEP) of the proteins, or in other words, the overall charge of the protein, has no or only a nonsignificant effect on the adsorption. Most of the adsorbed proteins have an IEP < pH 7, which means that the proteins have a negative charge; however, we have also detected most of these proteins on negatively charged particles. It appears that local charges at the protein surface (Sugio et al. 1999) are more important for the protein–particle interaction than the overall charge.

16.5 Particle Agglomeration

Nel et al. (2009) provided a complete overview regarding the possible interaction of NPs with the biological environment, especially regarding the reactions and interactions at the particle surface. The authors clearly demonstrated that a large number of reactions could take place but that only a few are known in detail. These authors also claimed that, "It is well known that NPs agglomerate immediately upon addition to cell culture media." In their article, they mentioned that in addition to DLVO interactions, hydrophilic and hydrophobic interactions are responsible for this agglomeration; however, a quantitative prediction of the colloidal behavior is very difficult and not possible even today. Nel and coworkers developed a system for characterizing the agglomeration of NPs in cell medium and their effect on cell toxicity. The approach is based on fast-throughput size measurement using the dynamic light scattering of samples in well plates (HT-DLS; Dynapro Plate Reader, Wyatt Technology). These researchers measured mostly the fast agglomeration of NPs with agglomerate sizes up to several microns. Because sedimentation of agglomerated particles leads to incorrect measurements, the results produced by this method must be interpreted with care. Another method to detect agglomerates is the measurement of turbidity or light adsorbents in the UV/Vis region (Soos et al. 2009). However, using this method, the effects from sedimentation, agglomeration of proteins, adsorption of proteins, etc., also affect the estimated particle/agglomerate size. An interesting study regarding the effect of cell medium composition on the agglomeration behavior of titanium dioxide (P25, Degussa) was published by Ji et al. (2010). Six different cell culture media, including bronchial epithelial growth medium, DMEM, Luria–Bertani broth, tryptic soy broth, synthetic defined medium, and yeast extract peptone dextrose medium, were selected. From the published results, it is evident that the cell medium alone leads to a strong agglomeration (high ionic strengths), whereas the addition of bovine serum albumin leads to some stable suspension and fetal bovine serum to stable suspension in all the investigated combinations, which is explained by protein adsorption (steric stabilization). These results partially confirm the results presented in Figure 16.2 but are contradictory to the statement of Nel et al. cited above. It appears that conclusions that are too general, such as that all NPs agglomerate in suspensions, insufficiently describe the reality. In addition, the results presented by Ji et al. are not applicable to all inorganic particles. The results might be applicable for uncoated NPs; however, NPs with organic molecules as a shell would exhibit a different behavior.

A well-established and accepted method and theory exists for the characterization of the agglomeration process of NPs in well-defined solvent such as pure water with low ionic strength. Observations qualitatively agree with the trend from the DLVO predictions, although the predictions indicated that 32- and 12-nm particles were unstable in 10 mM NaCl. Experimental observations indicated that these particles were stable for extensive periods (He et al. 2008).

On the theoretical side, the agglomeration of particles in aqueous medium is well investigated, and models for the simulation and interpretation of the experimental results are available in the literature. Considering the significant developments that have occurred since the publication of the original article of Smoluchowski (Smoluchowski 1917), agglomeration and flocculation modeling could be described as an established field of research (Thomas et al. 1999). However, the successful application of these models is largely limited to idealized, artificial systems. In biological systems, which are physically and chemically heterogeneous, the correlation between simulated and experimental data is still poor.

16.6 Particle Transport

Particle transport processes to cells in cell culture during *in vitro* assays have an important effect on the cell doses. As expressed by Wittmaack (2011), most *in vitro* studies with nanomaterials would benefit from direct measures of cellular dose. However, the experimental measurement of cellular dose is often difficult or time consuming and, thus, is a considerable limitation of *in vitro* studies. Therefore, measurements of the target cell dose will often not be available in published articles, which makes the comparison of different investigations difficult, and only limited results could be gained from meta-studies. The dynamics of particles in liquids are well studied, and mathematical approaches for describing both transports by diffusion and gravitational settling have been developed by Mason and Weave (1924). On the basis of these approaches, Hinderliter et al. (2010) have developed a computational model of NP transport, that is, particokinetics by sedimentation and diffusion. This model can lead to a quantitative dosimetry for noninteracting (meaning no change in the size distribution during the *in vivo* experiment) spherical particles and their agglomerates in a common cell culture system.

The model clearly shows that for small particles (<50 nm), diffusion is the dominant transport mechanism, whereas for a particle diameter >150 nm, sedimentation is important. In between, a minimum for particle deposition on cell surfaces is observed because transport by diffusion decreases with increasing particle size, whereas particle deposition increases with increasing particle size due to sedimentation. Therefore, it is incorrect to develop conclusions from experimental observations regarding the effect of size, including this critical range on toxicity, without considering the very different amounts of particles in contact with cells. Additionally, the weaknesses of the model are that only noninteracting particles are simulated and that the particles size is constant. As discussed in the paragraphs above, interaction occurs between the particles, the size could change (increase), and for most of the particles the size range will move from the diffusion-controlled size area to the sedimentation-controlled regime. An additional difficulty is the density of the particles, which could decrease because the particles are not further primary particles but agglomerates with a maximum of random loose-packed configuration (54 vol% solid). The change in density will have an effect on the sedimentation velocity but not on the amount of particles transported by diffusion. Additionally, the boundary conditions of Mason's solution were selected such that all particles touching the cell membrane are taken up directly, meaning that the cell acts as an infinite sink. It is well known that this phenomenon does not occur in cell culture systems. After touching the cell membranes, the particles must activate an uptake mechanism; often, a complex signaling process is therefore necessary. If the uptake rate is lower than the transportation rate, particle layers are formed at the surface of the cell. Because the particles exhibit fast protein adsorption, and therefore a charge of −10 to −15 mV in cell culture, repulsion between the particles could theoretically occur, and only a monolayer will be formed. In cell culture, high ionic strengths of 0.15 mol will reduce this electrostatic repulsion and could permit the formation of thick particle layers, which can lead to a strongly reduced cell viability. To observe cytotoxicity depends, therefore, not only on the particle–cell interaction; it could also be caused by the type of particle–particle interaction and transport to the uptake ratio.

Various experimental approaches are available to monitor the transport rates of NPs to surfaces (adsorption). Optical waveguide lightmode spectroscopy and a quartz crystal microbalance (QCM) can be effective tools for *in situ* monitoring of the kinetics of the deposition of NPs onto planar surfaces. The QCM-D method has been used to determine the mass of NPs deposited on surfaces using the measurement of the energy dissipation factor or the decrease in the resonance frequency of a QCM-D crystal. For example, Chen and Elimelech (2008) studied the aggregation and deposition kinetics of fullerene NPs to silica surfaces. Their data revealed an increased rate of deposition with increasing salt concentration. Fatisson et al. (2009) studied the deposition of TiO_2 NPs onto silica surfaces over a broad range of solution conditions, for example, pH and ionic strength, illustrating the effect of electrostatic forces, interaction, or repulsion as a function of pH. Hinderliter et al. (2010) measured the transport rates of fluorescent polystyrene NPs by counting the number of particles reaching the bottom of cell culture dishes using total-internal-reflection fluorescence microscopy, where individual fluorescent NPs near the surface are selectively illuminated while the background fluorescence from the particle in suspension is efficiently suppressed. This method allows the amount of newly arrived particles to be determined as a function of time (manual counting). The rate of transport of NPs can also be indirectly determined from the cellular uptake rates. For example, Hinderliter et al. (2010) measured the rate of transport of iron oxide NPs of SPION taken up by RAW 264.7 macrophages as a function of time, using a magnetic particle detector, and quantified the results against standard curves using serial dilutions of SPION suspensions. Cho et al. (2011) calculated the number of gold particles taken up by cells using a UV/Vis spectroscopic method based on extinction spectra versus calibration curves and compared the results with those of other conventional quantification methods such as inductive coupled plasma mass spectrometry. Cho and coworkers have also developed experiments to assess the effects of sedimentation on cellular uptake during *in vitro* tests by comparing inverted cell culture configurations to usual upright configurations of cells cultured at the bottom of wells. Table 16.1 provides a brief overview of the existing methods for the quantitative estimation of the transport of NPs to cell membranes in cell cultures.

TABLE 16.1

Techniques Available for *In Situ* NP Adsorption Kinetic Studies on Planar Surfaces

Technique	Description	Model Particles	Reference
Optical waveguide lightmode spectroscopy (OWLS)	Total internally reflected light within a high refractive index transparent layer.	SiO_2	Huwiler et al. 2007
Quartz crystal microbalance (QCM)	An alternating field is applied between two metal electrodes attached to the surface of a piezoelectric crystal of a certain frequency. Adsorption of particle shift frequency.	SiO_2, TiO_2, fullerene	Huwiler et al. 2007; Chen and Elimelech 2008; Fatisson et al. 2009
Total internal reflection fluorescence microscopy (TIRFM)	Requires fluorescently labeled NPs. Fluorescent molecules on adsorbed NPs can be excited and monitored by using a microscope focused on the surface.	Polystyrene	Hinderliter et al. 2010

The amount of particles transported to cells in cell cultures with fixed cells can also be detected by chemical analysis. Figure 16.5 shows the amount of PVA-coated iron oxide NPs (SPIONs) observed after 1 and 3 h in close contact or inside the cells. Different cell culture media were used: DMEM and RPMI with or without 10% fetal bovine serum. Details about the experiment and the results are given in Petri-Fink and Hofmann (2007). Because the size of the particles was also measured during the same period under similar conditions (lower concentration), it was possible to calculate the transport of NPs by diffusion and sedimentation using a simplified method that calculated the diffusion and

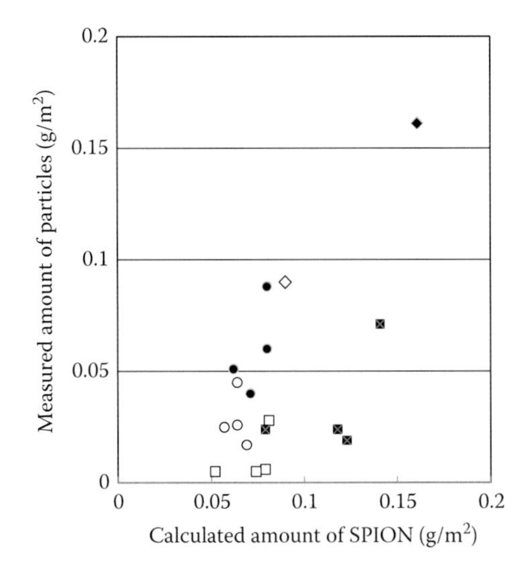

FIGURE 16.5
Comparison of measured and calculated (diffusion as main transport mechanism) amount of iron oxide NPs. Circles, cell medium without serum; squares, cell medium with serum; filled symbol, 1-h incubation time; empty symbol, 3-h incubation time. (Values obtained from Petri-Fink A., Hofmann H., *IEEE Trans. Nanobiosci.*, 6, 289, 2007.)

sedimentation separately. This calculation was possible because under the given conditions (the density of the loosely packed PVA–SPION agglomerates was 1.05 g/cm^3), the transport by sedimentation was only 2–5% of the overall transport. Figure 16.5 demonstrates that a relatively good correlation exists between the measured and calculated amount of particles in close contact with cells, which means that the transport of particles under the given conditions occurs by diffusion. It is also clearly observed that the amount of particles sticking to or uptaken by the HeLa cells in cell culture medium without serum is only slightly lower than predicted, whereas the amount was clearly lower for particles in cell medium with 10% serum. These results indicate that if the hydrodynamic particle size and the density of the particles as a function of incubation time are known, prediction of the transported amount in cell culture with a fixed cell should be possible. It also appears that for protein adsorption, the adhesion of NPs and their uptake is much lower, or in other words, the sticking coefficient for particles at the cell membrane must be much smaller than for particles without adsorbed proteins. Interestingly, the difference between the expected amount calculated with the particle size measured before adding the particles to the cell culture medium, which is the typical procedure for toxicity tests, is 10–20% of the expected value for cell culture medium with serum and 30–40% for serum-free cell culture medium. This result implies that in typical *in vitro* toxicity studies, the real amount of NPs that is in contact with cells is difficult to predict and will be overestimated if the particles are mainly in the diffusion regime or underestimated if the particles are in the sedimentation regime. The reason for this over-/underestimation is that with protein adsorption and agglomeration, the size is increasing, which leads to a decrease in the transported amount for diffusion-controlled particle transport and vice versa in the sedimentation regime.

16.7 Conclusions

Most of the large number of product applications using particles with a diameter <100 nm are mainly based on well-known materials such as silica, titanium dioxide, or silver. Products that use NPs because their properties differ from those of bulk materials are rare and are mostly still in the development phase. One possible reason for this rarity is that the processing, including the synthesis, remains a very challenging task for research and development as well as for translation to the industrial production scale. The particle–particle interaction, controlled by the forces between the particles, can be calculated on the basis of the well-established DLVO theory if the correct material properties and ion distribution around the nanosized particles are known. However, these properties are mostly not known, and bulk values are used instead. This substitution leads to the calculation of incorrect interaction potentials and eventually to false colloidal stability predictions. Fortunately, it appears that NPs are colloidally much more stable than predicted using standard DLVO theory. For applications in medicine or for investigations of toxicity, the coating of the particles is much more important than the core material in terms of the behavior of the particles. The well-known corona approach, adsorption of proteins at the NP surface, is very helpful in developing a framework for understanding NPs in biological environments. Adsorption is a dynamic process, which means it is very difficult to link the actual protein composition and conformation with the behavior of the particles in the body. Additionally, the large number of different proteins observed on the surface of the particles indicates that an inhomogeneous coating exists. Therefore, each particle could have a different composition and thus exhibit different behavior *in vivo*. A positive effect is that in 1 pg of NPs, the typical number of particles per cell is 1 million particles, which means we normally observe the average behavior of the particles. The combination of protein corona, modified attraction, and repulsion forces and the high number of species in suspension defines the agglomeration behavior. In complex media, such as cell culture medium with serum or in blood, the interaction of platelets and other blood cells with the NPs also affects the colloidal stability. The complex behavior remains unpredictable; however, it is evident that in addition to van der Waals, electrostatic forces and the surface charge (ζ-potential) affect the protein corona composition and therefore the colloidal stability. The difficulty is even greater because as mentioned earlier, protein adsorption is a dynamic process and the loosely bound highly abounded proteins could be exchanged over time with strongly binding low abounded proteins (the Vroman effect). It is evident that all these parameters have an effect on the transport behavior. However, it is astonishing that most of the *in vitro* experiments that have quantitatively investigated the effect of NPs on cells have not considered the transport behavior of NPs in complex suspensions. As discussed, while the theoretical background and mathematical solutions for ideal situations are available, only in the last 2 to 3 years has the discussion started on how to apply this knowledge to real *in vitro* experiments. The first results indicate that the transport could be even slower than predicted by the theory, which suggests that more detailed research is necessary to understand the diffusion, sedimentation, and, most important, the adhesion of NPs on cell membrane surfaces.

Acknowledgments

LM, MGB, US, and VB thank the European Union Seventh Framework Programme NMP-2008-4.0-1, grant agreement no. 228929, for financial support. The results described in this chapter were elaborated in projects granted by the Swiss National Science Foundation (grant nos. SNF 200020-109490 and SNF 205321-111908) and the European Union Seventh Framework Programme NMP-2008-4.0-1, grant agreement no. 228929.

References

Barnard A.S. 2011. Mapping the photocatalytic activity or potential free radical toxicity of nanoscale titania. *Energy Environ Sci* 4: 439–443.

Barnard A.S. and Xu H. 2008. An environmentally sensitive phase map of titania nanocrystals. *ACS Nano* 2: 2237.

Bell G.M., Levine S. and McCartney L.N. 1970. Approximate methods of determining the double-layer free energy of interaction between two charged colloidal spheres. *J Colloid Interface Sci* 33: 335–359.

Brune H., Ernst H., Grunwald A., Grünwald W., Hofmann H., Janich P., Krug H., Mayor M., Rathgeber W., Schmid G., Simon U., Vogel V. and Wyrwa D. 2006. Nanotechnology: Assessment and perspective. In: *Wissenschaftsethik und Technikfolgen-Beurteilung*, Hrsg. von C. F. Gethmann, vol. 27. Berlin: Springer, ISBN: 354032819.

Chen K.L. and Elimelech M. 2008. Interaction of fullerene (C-60) nanoparticles with humic acid and alginate coated silica surfaces: Measurements, mechanisms, and environmental implications. *Environ Sci Technol* 42: 7607–7614.

Cho E.C., Zhang Q. and Xia Y. 2011. The effect of sedimentation and diffusion on cellular uptake of gold nanoparticles. *Nat Nanotechnol* 6: 385–391.

Dahirel V. and Jardat M. 2010. Effective interactions between charged nanoparticles in water: What is left from the DLVO theory? *Curr Opin Colloid Interface Sci* 15: 2–7.

Dobrovolskaia M.A., Aggarwal P., Hall J.B. and McNeil S.E. 2008. Preclinical studies to understand nanoparticle interaction with the immune system and its potential effects on nanoparticle biodistribution. *Mol Pharm* 5: 487–495.

Ehrenberg M.S., Friedman A.E., Finkelstein J.N., Oberdörster G. and McGrath J.L. 2009. The influence of protein adsorption on nanoparticle association with cultured endothelial cells. *Biomaterials* 30: 603–610.

Fatisson J., Domingos R.F., Wilkinson K.J. and Tufenkji N. 2009. Deposition of TiO_2 nanoparticles onto silica measured using a quartz crystal microbalance with dissipation monitoring. *Langmuir* 25(11): 6062–6069.

Faure B., Salazar-Alvarez G. and Bergström L. 2011. Hamaker constants of iron oxide nanoparticles. *Langmuir* 27: 8659–8664.

French R.H. 2000. Origins and applications of London dispersion forces and Hamaker constants in ceramics. *J Am Ceram Soc* 83: 2117–2146.

He Y.T., Wan J. and Tokunaga T. 2008. Kinetic stability of hematite nanoparticles: The effect of particle sizes. *J Nanopart Res* 10: 321–332.

Hinderliter P.M., Minard K.R., Orr G., Chrisler W.B., Thrall B.T., Pounds J.G. and Teeguarden J.G. 2010. ISDD: A computational model of particle sedimentation, diffusion and target cell dosimetry for *in vitro* toxicity studies. *Part Fibre Toxicol* 7: 36.

Hogg R., Healy T.W. and Fuerstenau D.W. 1966. Mutual coagulation of colloidal dispersions. *Trans Faraday Soc* 62: 1638.

Hough D.B. and White L.R. 1980. The calculation of Hamaker constants from Liftshitz theory with applications to wetting phenomena. *Adv Colloid Interface Sci* 14: 3–41. Available at http://europa.eu/rapid/pressReleasesAction.do?reference=IP/11/1202&format=HTML&aged=0&language=DE&guiLanguage=en (accessed August 2012).

Huwiler C., Kunzler T.P., Textor M., Voros J. and Spencer N.D. 2007. Functionalizable nanomorphology gradients via colloidal self-assembly. *Langmuir* 23: 5929–5935.

Jansch M., Stumpf P., Graf C., Rühl E. and Müller R.H. 2012. Adsorption kinetics of plasma proteins on ultrasmall superparamagnetic iron, oxide (USPIO) nanoparticles. *Int J Pharm* 428: 125–133.

Ji Z., Jin X., George S., Xia T., Meng H., Wang X., Suarez E., Zhang H., Hoek E.M.V., Godwin H., Nel A.E. and Zink J.I. 2010. Dispersion and stability optimization of TiO_2 nanoparticles in cell culture media. *Environ Sci Technol* 44: 7309–7314.

Kim H.-K., Sofo J.O., Velegol D., Cole M.W. and Lucas A.A. 2007. van der Waals dispersion forces between dielectric nanoclusters. *Langmuir* 2: 1735–1740.

Lifshitz E.M. 1956. The theory of molecular attractive forces between solids. *Sov Phys* 2: 73.

Lundqvist M., Stigler J., Elia G., Lynch I., Cedervall T. and Dawson K.A. 2008. Nanoparticle size and surface properties determine the protein corona with possible implications for biological impacts. *Proc Natl Acad Sci U S A* 105: 14265–14270.

Mason M. and Weave H. 1924. The settling of small particles in a fluid. *Phys Rev* 23: 412–426.

Mériguet G., Jardat M. and Turq P. 2004. Structural properties of charge-stabilized ferrofluids under a magnetic field: A Brownian dynamics study. *J Chem Phys* 121: 6078.

Monopoli M.P., Walczyk D., Campbell A., Elia G., Lynch I., Bombelli F.B. and Dawson K.A. 2011. Physical–chemical aspects of protein corona: Relevance to *in vitro* and *in vivo* biological impacts of nanoparticles. *J Am Chem Soc* 133: 2525–2534.

Nel A.E., Mädler L., Velegol S., Xia T., Hoek E.M.V., Somasundaran P., Klaessig F., Castranova V. and Thompson M. 2009. Understanding biophysicochemical interactions at the nano-biointerface. *Nat Mater* 8: 543–557.

Neuberger T., Schöpf B., Hofmann M., Hofmann H. and von Rechenberg B. 2005. Superparamagnetic nanoparticles for biomedical applications: Possibilities and limitations of a new drug delivery system. *J Magn Magn Mater* 293: 483–496.

Oberdörster G. 2010. Safety assessment for nanotechnology and nanomedicine: Concepts of nanotoxicology. *J Intern Med* 267: 89–105.

Parsegian V.A. and Ninham B.W. 1969. Application of the Lifshitz theory to the calculation of van der Waals forces across thin lipid films. *Nature* 224: 1197–1198.

Petri-Fink A. and Hofmann H. 2007. Superparamagnetic iron oxide nanoparticles (SPIONs): From synthesis to *in vivo* studies—A summary of the synthesis, characterization, *in vitro* and *in vivo* investigations of SPIONs with particular focus on surface and colloidal properties. *IEEE Trans Nanobiosci* 6: 289–295.

Rivera-Gil P., Jimenez de Aberasturi D., Wulf V., Pelaz B., del Pino P., Zhao Y., de la Fuente J., Ruiz de Larramendi I., Rojo T., Liang X.-J. and Parak W.J. 2013. The challenge to relate the physicochemical properties of colloidal nanoparticles to their cytotoxicity. *Acc Chem Res* 46: 743–749.

Shang W., Nuffer J.H., Muniz-Papandrea V., Colon W., Siegel R.W. and Dordick S. 2009. Cytochrome *c* on silica nanoparticles: Influence of nanoparticle size on protein structure, stability, and activity. *Small* 4: 470–476.

Smoluchowski M. 1917. Attempt for a mathematical theory of kinetic coagulation of colloid solutions. *Z. Phys. Chemie* 92: 129–168.

Soos M., Lattuada M. and Sefcik J. 2009. Interpretation of light scattering and turbidity measurements in aggregated systems: Effect of intra-cluster multiple-light scattering. *J Phys Chem B* 113: 14962–14970.

Sugio S., Kashima A., Mochizuki S., Noda M. and Kobayashi K. 1999. Crystal structure of human serum albumin at 2.5 Å resolution. *Protein Eng* 12: 439–446.

Tepper T., Ross C.A. and Dionne G.F. 2004. Microstructure and optical properties of pulsed-laser-deposited iron oxide films. *IEEE Trans Magn* 40: 1685–1690.

Thomas D.N., Judd S.J. and Fawcett N. 1999. Flocculation modelling: A review. *Water Res* 33: 1579–1592.

Vervey J. and Overbeek J.T.G. 1948. *Theory of the Stability of Lyophobic Colloids*. Amsterdam: Elsevier.

Vincent B. 1973. The van der Waals attraction between colloid particles having adsorbed layers. II. Calculation of interaction curves. *J Colloid Interface Sci* 42: 270–285.

Wittmaack K. 2011. Excessive delivery of nanostructured matter to submersed cells caused by rapid gravitational settling. *ACS Nano* 5: 3766–3778.

Wongsagonsup R., Shobsngob S., Oonkhanond B. and Varavinit S. 2005. Zeta potential analysis for the determination of protein content in rice flour. *Starch* 57: 25–31.

Xin-Rui X., Monteiro-Riviere N.A. and Riviere J.E. 2010. An index for characterization of nanomaterials in biological systems. *Nat Nanotechnol* 5: 671–676.

Zhang F., Skoda M.W.A., Jacobs R.M.J., Martin R.A., Martin C.M. and Schreiber F. 2007. Protein interactions studied by SAXS: Effect of ionic strength and protein concentration for BSA in aqueous solutions. *J Phys Chem B* 111: 251.

17

Dosimetry for In Vitro *Nanotoxicology: Too Complicated to Consider, Too Important to Ignore*

Joel M. Cohen and Philip Demokritou

CONTENTS

17.1 Introduction

The widespread manufacture and use of engineered nanomaterials (ENMs) in consumer products have led to increasing concerns of environmental and occupational exposures (Aitken et al. 2006; Bello et al. 2012). The unique physicochemical properties exhibited by ENMs that are distinct from those of their micron-sized counterparts (reactive surface area, surface energy, mobility, quantum size effects, etc.) endow them with exceptional performance in consumer products, and may also be responsible for unique biological effects that can render them unsafe for humans and for the environment (Nel et al. 2006, 2009; Oberdorster 2007).

The most common exposure pathways for ENMs include ingestion (pharmaceutical products, food), dermal contact (occupational exposure, cosmetics), injection (nanomedicines and drug delivery mechanisms), and inhalation (occupational and consumer exposure). These varied exposure pathways each present unique challenges to biological systems and may result in diverse toxicological outcomes. There is growing evidence that suggests exposure to ENMs may cause a variety of pulmonary and cardiovascular effects (Brain 2009; Mills et al. 2009; Choi et al. 2010; Demokritou et al. 2012a; Sotiriou et al. 2012). Although preliminary evidence demonstrates the potential for ENMs to cause adverse biological effects, the underlying toxicity mechanisms are not currently well understood (Oberdorster 2007; Nel et al. 2009), and major knowledge gaps along the exposure–disease continuum still exist.

Owing to the high cost and laborious nature of *in vivo* toxicity studies, most nanotoxicology efforts have focused on *in vitro* methods (Balbus et al. 2007; Lai 2012). The development of *in vitro* bioassays is imperative to keep apace of the rapidly growing number and variety of ENMs entering the consumer market (Nel et al. 2006; Krewski et al. 2010), as global ENM production rates are expected to increase by >20 times within the next 15 years (The Royal Society and the Royal Academy of Engineering 2004). Therefore, the development of cheap, accurate, and reproducible *in vitro* screening assays is recognized by all stakeholders as a major goal for nanoenvironmental health and safety research in the next decade (National Science and Technology Council Committee on Technology 2011).

A number of studies have endeavored to develop efficient and inexpensive screening tools to correlate mechanisms of biological activity and toxicity with ENM characteristics such as size, shape, and surface area (Balbus et al. 2007; Shaw et al. 2008; National Science and Technology Council Committee on Technology 2011; Puzyn et al. 2011; Rallo et al. 2011; Lai 2012). High-throughput *in vitro* toxicity assays have also recently been employed to assess multiple toxicity endpoints, in multiple cell lines, of libraries of ENMs over a range of exposure times and concentrations (Shaw et al. 2008; George et al. 2011; Zhang et al. 2012). To date, however, the results of *in vitro* assays have too often conflicted with those of animal studies (Rivera Gil et al. 2010; Han et al. 2012; Demokritou et al. 2012a; Lai 2012), and so have not earned widespread acceptance. Much of this disparity may be accounted for by complex interactions between multiple cell types and organismic processes present in whole animal exposure studies that are absent from most simplified *in vitro* models. Various groups have addressed these issues by developing increasingly sophisticated *in vitro* models that introduce multiple cell lines into coculture systems for evaluation of nanomaterial processing and toxicity (Rothen-Rutishauser et al. 2005; Lehmann et al. 2011; Muller et al. 2011), although widespread adoption of such sophisticated systems by nanotoxicologists has been slow.

Another major obstacle to the development of cost-effective and reliable *in vitro* toxicological screening methods for ENMs is the need for accurate dosimetry (Rivera Gil et al. 2010; Gangwal et al. 2011; Han et al. 2012; Lai 2012; Oberdorster 2012). Conflicting *in vitro* toxicological data have been reported in the literature, and these disparities likely result from a lack of harmonized, or standardized, *in vitro* method and proper ENM characterization, which can lead to erroneous reporting of exposure conditions and effective doses delivered to cells (Teeguarden et al. 2007; Hinderliter et al. 2010; Cohen et al. 2013; Deloid et al. 2014).

In a typical *in vitro* cytotoxicity study, ENMs are suspended in liquid medium for application to cells. Comparisons of biological response to ENM exposure usually relies on administered dose metrics based on ENM properties as measured in the dry powder form (e.g., mass, surface area, or particle number per volume), without taking into account particle–particle and particle–medium interactions (Oberdorster et al. 1994; Oberdorster and Oberdorster 2005; Wittmaack 2007; Jiang 2008; Rushton et al. 2010). Many studies have shown that ENMs form large agglomerates in liquid medium (Schulze et al. 2008; Verma and Stellacci 2010; Cohen et al. 2013; Deloid et al. 2014), which can alter the total number of free particles in suspension, as well as the total surface area available for biointeractions before they even reach the cells cultured *in vitro* (Figure 17.1a). Furthermore, these newly formed agglomerates have a different density compared with that of the raw material (Demokritou et al. 2012b; Deloid et al. 2014). More important, the aforementioned agglomeration state and the effective density of the ENMs have a major impact on their fate and transport in an *in vitro* system, which determines the effective dose delivered to cells in culture (Figure 17.1b) (Limbach 2005; Teeguarden et al. 2007; Lison et al. 2008; Hinderliter et al. 2010; Cho et al. 2011).

It is apparent that the lack of dosimetric considerations in most *in vitro* nanotoxicology studies introduces a significant bias and hinders our ability to develop accurate and cost-effective screening strategies of ENMs. One way of eliminating biases introduced from using a liquid suspension approach to deliver ENMs to cells is to develop more physiologically relevant inhalation exposure models for delivering aerosolized ENMs to cells cultured at an air–liquid interface (Savi et al. 2008; Lenz et al. 2009; Muller et al. 2011). However, commonly used aerosol generators used to disperse nanopowders for exposure to cells (e.g., nebulizers, fluidized beds, and other venturi aspirator–type systems) are unable to consistently produce realistic nanosized aerosol distributions, and often lack the ability to accurately control and change exposure concentrations for dose–response investigations (Limbach 2005; Fischer and Chan 2007; Schmoll 2009; Demokritou et al. 2012a).

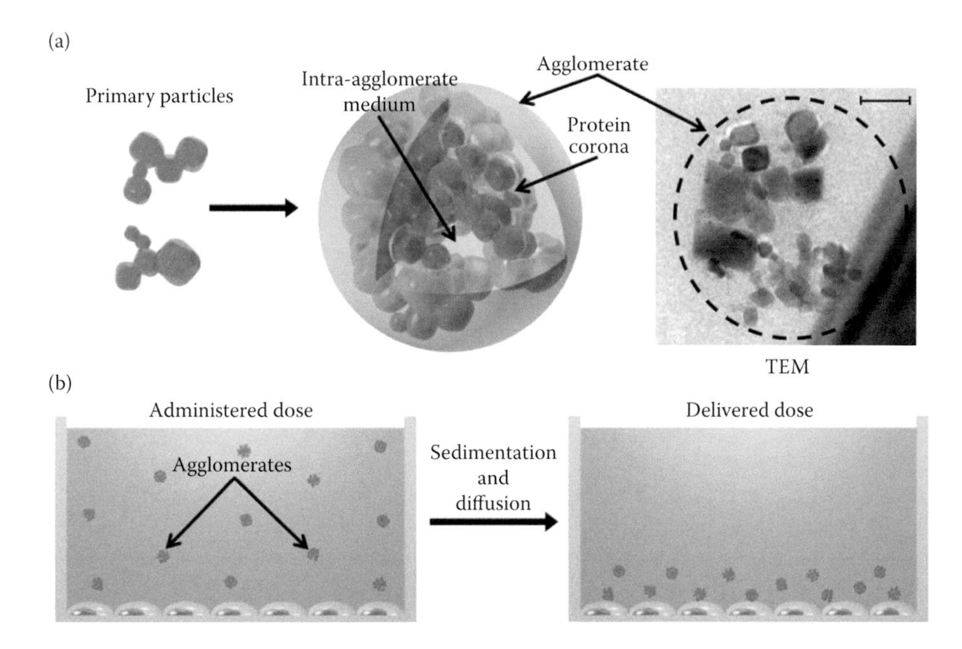

FIGURE 17.1
Agglomeration of ENMs and transport in culture. (a) ENM primary particles suspended in cell culture medium exist as agglomerates consisting of multiple primary particles, which may be enveloped by a corona of proteins from the medium, and liquid medium and medium components may be trapped between primary particles (intra-agglomerate medium). (b) ENM agglomerates within suspensions applied to cells settle toward the cells over time as a result of mass transport (sedimentation and diffusion). The initial administered dose is the concentration of ENM in the initially homogeneous suspension. As transport progresses, agglomerates are concentrated near or deposited onto the cells. The mass of ENM deposited per area is the delivered dose. (Adapted from Deloid, G. et al., *Nat Commun*, 5, 2014, doi:10.1038/ncomms4514.)

Furthermore, selection of exposure concentrations for *in vitro* study often lacks scientific justification, with doses often chosen to be very high and well above physiologically relevant levels (Oberdorster and Oberdorster 2005). It is equally important to validate *in vitro* assays with *in vivo* animal data by matching the effective target doses investigated (Teeguarden et al. 2007; Cohen Hubal 2009; Gangwal et al. 2011; Demokritou et al. 2012a). In some instances, nanotoxicologists have used particle concentrations for their *in vitro* investigations at levels that cause "overload" *in vivo*, where pulmonary clearance becomes severely impaired (Morrow 1988; Cohen Hubal 2009; Warheit et al. 2009; Gangwal et al. 2011; Demokritou et al. 2012a). Therefore, careful consideration must be given toward selecting relevant doses for *in vitro* study, and identifying equivalent effective exposures between these biological systems.

This chapter presents a method for *in vitro* nanotoxicology (be it monoculture or a more sophisticated coculture system) designed to improve the accuracy and relevance of reported dosimetry and exposure conditions. The proposed multistep method consists of (i) standardization of ENM liquid suspension preparation; (ii) careful characterization of ENM transformations in physiological medium; and (iii) dosimetric considerations in terms of the delivered to cell dose, and equivalent effective doses for comparison between *in vitro* and *in vivo* data.

17.2 Preparation of ENM Suspension in Liquid for *In Vitro* Studies

Figure 17.2 illustrates an optimized ENM sample preparation protocol for *in vitro* nantoxicology studies, which consists of the following critical steps.

17.2.1 Sonication

Standardizing sample preparation protocols is a critical step toward improving the accuracy and validity of reported dosimetry for *in vitro* nanotoxicology studies. Currently, methods used to disperse nanoparticles (NPs) in culture medium for application to cells differ widely between laboratories (Roco 2011). When ENMs are not already fully dispersed in liquid medium (such as ENMs generated by wet methods that are stored and stabilized in liquid suspension), the source ENM in dry powder form is dispersed into the test medium at desired mass concentrations (Murdock et al. 2008; Schulze et al. 2008; Jiang 2009). To optimize ENM dispersibility in suspension, ultrasonic energy is used in a process known as sonication, or the propagation of sound waves into liquid medium leading to alternating high-pressure (compression) and low-pressure (rarefaction) cycles (Taurozzi et al. 2011). During the low-pressure cycle, high-intensity ultrasonic waves create small vacuum bubbles or voids in the liquid. When these bubbles attain a critical volume, they collapse violently during a high-pressure cycle. This process of bubble formation, growth, and collapse is known as cavitation, and creates microjet streams that force ENM particles to collide with each other at velocities of up to 1000 km h^{-1} (Hielscher 2005). Interparticle collisions or the collapse of cavitation bubbles formed on a particle surface can lead to particle size reduction and reduced agglomeration. Sonication has proven more energy efficient and can achieve a higher degree of deagglomeration, at constant specific energy, than other conventional dispersion techniques (Hielscher 2005; Mandzy 2005; Taurozzi 2010). A number of factors influence the efficacy of sonication at reducing agglomeration

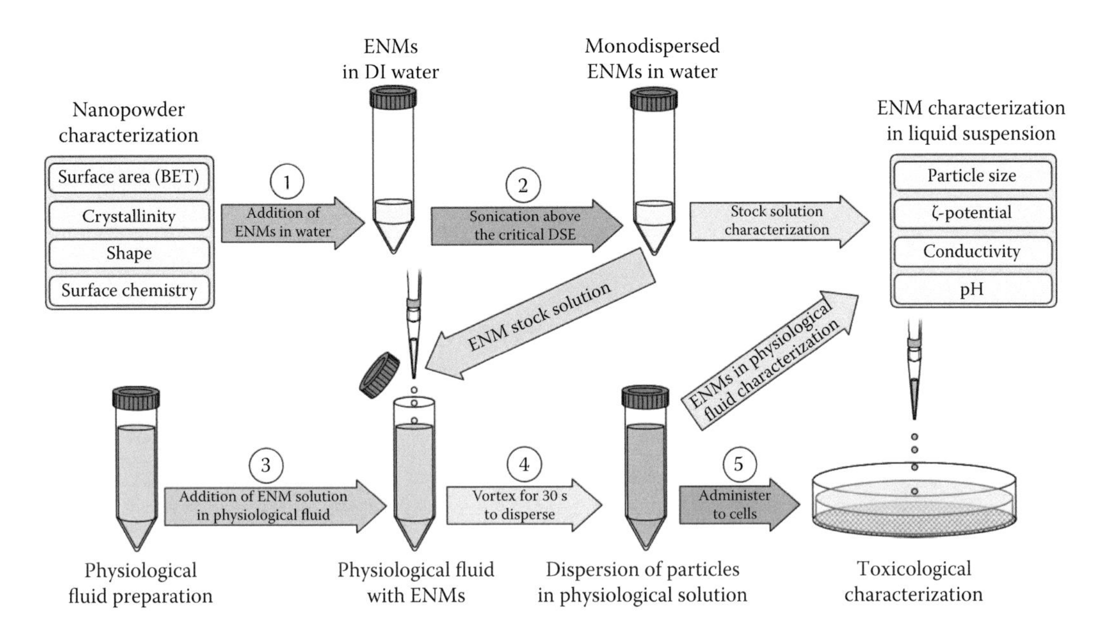

FIGURE 17.2
Optimized ENM sample preparation protocol for *in vitro* nantoxicology study. The proposed standardized protocol consists of first suspending ENMs in deionized water and sonicating above the previously determined critical energy required to achieve monodisperse and stable agglomerates (DSE_{cr}). The stock solution may then be diluted in cell culture medium at concentrations desired for toxicological characterization. To ensure accurate dosimetry reporting, the final solution should be characterized for hydrodynamic diameter by dynamic light scattering or another feasible method, and effective density by volumetric centrifugation or other acceptable method. (Previously published in Cohen, J. D. et al., *Nanotoxicology*, 7, 417–31, 2013.)

and optimizing dispersion stability over time, such as sonication energy (measured in watts), the duration of sonication, and the volume of dispersion undergoing sonication. For laboratories performing *in vitro* toxicity assessments of ENM dispersions, inconsistencies across any of these three areas may substantially increase variability in dispersion characteristics, with subsequent effects on *in vitro* dosimetry and observed cellular toxicity (Roebben et al. 2011).

17.2.2 Potential Pitfalls

Sonication can be performed by either immersing an ultrasound probe into the dispersion (direct sonication), or by immersing a sample container into a liquid bath that is propagating ultrasonic waves (indirect sonication). Both methods are widely used across the literature with little attention to their relative efficacy at reducing agglomeration. Recent studies investigating the dispersibility of TiO_2 and Ag NPs in various media suggest that the reported energy delivered by these different systems (direct sonication vs. indirect) may not accurately reflect the actual delivered sonication energy (DSE) to which ENM dispersions are subjected (Murdock et al. 2008; Jiang 2009; Taurozzi et al. 2011; Cohen et al. 2013). Improved reliability and reproducibility of dispersion protocols between laboratories therefore requires standardized and accurate reporting of sonication energy delivered to a sample for a given sonicator. In addition to standardized reporting of sonication energy, the selection of sonication medium is of critical importance. Sonolysis of medium components can lead to the generation of reactive oxygen species (ROS), artificially increasing the likelihood of cellular toxicity (Mason and Peters 2003). Additionally, sonication of typical cell culture medium containing serum components and other organics can degrade medium components (Lorimer et al. 1995; Naddeo et al. 2007), and can denature proteins or alter their physicochemistry, thereby altering the surface coating, agglomeration state, and surface reactivity of the ENMs in suspension (Wang et al. 2009; Taurozzi et al. 2011). Considering the potential pitfalls described previously, it is critical that *in vitro* nanotoxicologists establish and adopt a standardized sample preparation protocol, designed to improve reproducibility between different laboratories, to minimize undesired chemical reactions due to sonication (ROS generation, protein denaturing, etc.), and to achieve monodisperse suspensions of the smallest possible ENM agglomerates in solution that are stable over time.

17.2.3 Calorimetric Calibration of Sonication Equipment

A standardized sample preparation protocol previously published by our group, and based on those established by the Organization for Economic Cooperation and Development (OECD) and the National Institute of Standards and Technology (OECD 2010b; Taurozzi et al. 2011; Cohen et al. 2013) is presented in Figure 17.2. Calorimetric calibration of sonication equipment is recommended before sample preparation to ensure accurate application and reporting of DSE, and thereby to improve the reproducibility of ENM dispersions. This method involves measuring the temperature change over time for a sample of deionized water during sonication at a specific power setting. From these data, one can determine the effective delivered power for the given setting by the following equation:

$$P = \frac{dT}{dt} M C_p \tag{17.1}$$

where P is the delivered acoustic power (W), T is temperature (K), t is time (s), M is the mass of liquid (g), and C_p is the specific heat of the liquid (J g^{-1} K^{-1}). Sonication efficacy is also influenced by the duration of sonication and the volume of dispersion undergoing sonication. Therefore, we recommend reporting the DSE to a sample of given volume in units of W*s ml^{-1}, or J ml^{-1}. By determining the calorimetric curves for a given power setting, users can report and reproduce the effective power applied to a sonicated suspension in a manner that is easily transferrable and independent of the specific sonicating instrument (direct vs. indirect sonication, make and model of sonicator, etc.).

17.2.4 Sonication Media

To minimize undesired chemical reactions, degradation of medium components, or generation of ROS during sonication, we recommend selecting deionized water (DI H_2O) (or other appropriate pH-adjusted aqueous solutions) as a sonication medium. ENMs fully dispersed in DI H_2O can then be diluted to the desired concentrations in media appropriate for cell culture (Figure 17.2).

17.2.5 Critical Delivered Sonication Energy

Once the sonication equipment is sufficiently calibrated and appropriate sonication medium is selected, it is important to determine for each sample material the critical energy required to achieve monodisperse solutions at the lowest agglomeration state, or the critical DSE (DSE_{cr}). Many groups seeking to optimize NP agglomeration in suspension have found that increasing ultrasound energy and duration of sonication decreases agglomeration up to a certain point, above which no further particle size reduction is observed even with increasing sonication (Bihari et al. 2008; Jiang 2009; Cohen et al. 2013). An example protocol for determining the material-specific DSE_{cr} is detailed in recent publication (Cohen et al. 2013). In brief, ENMs are dispersed at 1 mg ml^{-1} in 3 ml solute in 15 ml conical polyethylene tubes using a Branson Sonifier S-450A (Branson Ultrasonic, Danbury, CT, USA), calibrated by the calorimetric calibration method whereby the power delivered to the sample was determined to be 1.75 W, fitted with a 3-in cup horn (maximum power output of 400 W at 60 Hz, continuous mode, output level 3) in which tubes were immersed so that the sample and cup water menisci were aligned. Suspensions are then characterized for hydrodynamic diameter (nm), polydispersity, ζ-potential (mV), and specific conductance (mS cm^{-1}) by dynamic light scattering (DLS) using a Zetasizer Nano-ZS (Malvern Instruments, Worcestershire, UK). Plots of hydrodynamic diameter as a function of DSE exhibiting asymptotic deagglomeration trends are derived for each ENM and used to determine the material-specific DSE_{cr}, corresponding to the lowest agglomeration state. From each curve, an ENM-specific DSE_{cr} is estimated by determining the DSE value at which the dispersed ENMs were within 10% of their observed minimum hydrodynamic diameter as measured by DLS (Figure 17.3). To determine the stability of the sonicated suspension over time, hydrodynamic diameter measurements should be repeated for several hours following sonication, up to the duration of a typical toxicology exposure time (usually up to 72 h). It is important to note that this exercise reveals how different materials may exhibit different DSE_{cr}, where some require relatively extensive sonication to achieve monodisperse suspensions of small agglomerates (Figure 17.3).

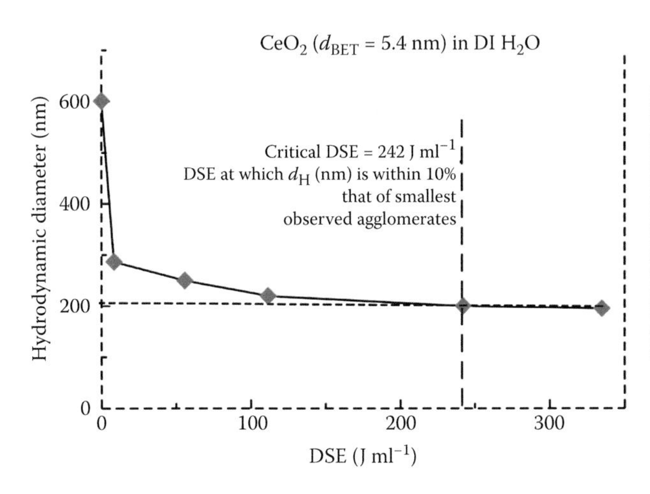

Material	d_{BET} (nm)	DSE_{cr} (J/ml^{-1})
VENGES SiO_2	11.9	161
VENGES SiO_2	18.6	161
VENGES Fe_2O_3	27.6	242
VENGES CeO_2	5.4	242
VENGES CeO_2	27.9	242
VENGES CeO_2	71.3	242
VENGES Ag	41.8	161
VENGES ZnO	26.7	242

FIGURE 17.3
Determination of critical delivered sonication energy (DSE_{cr}). Asymptotic deagglomeration curve associated with increasing sonication energy is presented for a representative material (CeO_2 d_{BET} = 5.4 nm). DSE_{cr} is defined as the DSE value at which the dispersed ENM agglomerates are within 10% of their lowest observed minimum hydrodynamic diameter as measured by DLS. Table of DSE_{cr} for representative ENMs is included to demonstrate how different materials may be characterized by different DSE_{cr}. d_{BET}, primary particle diameter determined from specific surface area measured by nitrogen adsorption/Brunauer–Emmett–Teller (BET) method (nm).

17.2.6 Dispersion in Physiological Medium and Final Characterization

The stock solution of ENM suspended in DI H_2O, sonicated at or above the DSE_{cr}, may then be diluted in physiological media used for cell culture at concentrations desired for toxicological evaluation. These suspensions of ENMs in physiological media must also be characterized for hydrodynamic diameter, polydispersity, ζ-potential, and specific conductance by the best available methods, be it DLS or analytical ultracentrifugation (AUC), as the agglomeration state may change once ENMs are introduced into new complex media (as described in more detail in Section 17.3). The stability of suspensions in physiological media should be characterized, and hydrodynamic diameter measurements should be repeated for several hours following sonication, up to the duration of a typical toxicology exposure time (usually up to 72 h).

17.3 ENM Transformations in Physiological Media

ENMs suspended in complex physiological media may flocculate, agglomerate, dissolve, and interact with serum components, which can alter their biological properties as well as determine their mass transport and delivered to cell rate in an *in vitro* system. (Schulze et al. 2008; Jones and Grainger 2009; Fadeel 2010; Verma and Stellacci 2010; Cohen et al. 2013). These interactions depend largely on the dispersion protocol (as described previously); particle characteristics such as primary particle size, shape, chemical composition, and surface chemistry (Murdock et al. 2008; Jiang 2009; Ji et al. 2010; Zook et al. 2010); and medium properties such as ionic strength, specific conductance, pH, and protein content (Laxen 1977; Bihari et al. 2008; Murdock et al. 2008; Elzey 2009; Zook et al. 2010; Wiogo et al. 2011). Agglomeration influences dosimetry by altering the total number of free particles in suspension, by altering the total surface area available for interaction with cells *in vitro*, as well as the mass transport of ENMs that directly influences the delivery of particles to cells. NPs suspended in physiological media can form either nonagglomerating units such as spheres or rods, or fractal and porous agglomerates consisting of multiple primary ENM particles with trapped suspension fluid and associated proteins, such as that presented in the electron microscopy image in Figure 17.1a. The degree and fractal nature of ENM agglomeration can influence the amount of fluid trapped within the agglomerate's pores, subsequently influencing the effective density of the agglomerate unit (Figure 17.1a), which, as we will soon show, can sometimes lie closer to the density of the medium than that of the original material.

17.3.1 Agglomerate Diameter

Agglomerate size distributions can be measured by a variety of methods, including AUC, DLS, hydrodynamic chromatography, NP tracking analysis, laser diffraction spectrometry, and x-ray disc centrifugation (Wohlleben 2012; Taurozzi et al. 2013). Table 17.1 presents hydrodynamic diameters for a variety of flame pyrolysis–generated metal oxides currently under investigation by the OECD (2010a) suspended in various physiological media commonly used in toxicity studies. These data highlight the material and medium-specific effects of agglomerate size, where materials such as CeO_2 require the presence of stabilizing serum proteins, either fetal bovine serum (FBS) or bovine serum albumin (BSA), to retain average

TABLE 17.1

Properties of ENM Dispersions in Various Physiological Media

Material	Media	d_H (nm)
VENGES SiO_2 (d_{BET} = 11.9 nm)	DI H_2O	124 ± 7.19
	RPMI	115 ± 2.40
	RPMI/10% FBS	204 ± 13.5
	RPMI/1% BSA	144 ± 5.59
VENGES CeO_2 (d_{BET} = 13 nm)	DI H_2O	136 ± 0.404
	RPMI	1678 ± 260
	RPMI/10% FBS	235 ± 5.92
	RPMI/1% BSA	369 ± 6.46
VENGES Fe_2O_3 (d_{BET} = 4.8 nm)	DI H_2O	151 ± 9.14
	RPMI	4333 ± 400
	RPMI/10% FBS	226 ± 4.20
	RPMI/1% BSA	937 ± 104

Note: d_{BET}, primary particle diameter determined from specific surface area measured by nitrogen adsorption/Brunauer–Emmett–Teller (BET) method (nm); d_H, hydrodynamic diameter measured by DLS (nm). Errors (±) indicate standard deviations based on three measurements.

agglomerate diameters on the nanoscale (for CeO_2 d_{BET} = 13 nm, d_H = 235 nm, 369 nm, and 1678 nm suspended in RPMI/10% FBS, RPMI/1% BSA, and RPMI alone, respectively). In contrast, SiO_2 retains a consistent nanosize diameter even in the absence of stabilizing serum proteins (for SiO_2 d_{BET} = 11.9 nm, d_H = 204 nm, 144 nm, and 115 nm suspended in RPMI/10% FBS, RPMI/1% BSA, and RPMI alone, respectively). Such medium- and ENM-specific interactions have implications for dosimetry, specifically the particle number and surface area available for cellular interactions, as well as the mass transport of agglomerates to cells in culture.

17.3.2 Effective Density

Determination of agglomerate density in suspension (or effective density) has presented a significantly greater challenge, and is often overlooked by nanotoxicologists in spite of the large implications it holds for particle delivery to cells and dosimetry. The Sterling equation, based on a theoretical fractal-based model of agglomerate density, has been used to obtain rough estimates for the effective density of agglomerates suspended in various liquid media (Sterling et al. 2005; Cohen et al. 2013), although the practical validity of this method remains unproved. Alternatively, the sedimentation coefficient of a suspended ENM can be measured directly by AUC (Carney et al. 2011; Zook et al. 2011; Wohlleben 2012). Although the combination of DLS and AUC results can provide highly accurate measurements of ENM effective density, AUC requires relatively expensive equipment that is not standard in most nanotoxicology laboratories, and is relatively limited in terms of throughput, which can make characterization of a growing number of commercially available ENMs by AUC relatively costly and time consuming. Multiple samples can be run by the LUMiFuge (LUM, Boulder, CO); however, this instrument has only rudimentary optics and is not ideal for sizing. The more popular analytical centrifuges by CPS and Brookhaven, although more modestly priced, can only run one sample at a time. Importantly, most nanotoxicology laboratories do not currently have analytical centrifuges.

The Harvard volumetric centrifugation method recently developed can be used to measure effective density of ENMs in liquids. This relatively fast and easy-to-use approach provides an estimate of the agglomeration state and the amount of fluid trapped within the agglomerate's pores (Demokritou et al. 2012b; Deloid et al. 2014) (Figure 17.1a). Following this method, the effective densities of ENMs suspended in physiological fluids can be accurately estimated in high throughput using a standard bench-top centrifuge (such as the Beckman Allegra series in which up to 148 samples can be run simultaneously), and relatively inexpensive packed cell volume (PCV) tubes (Demokritou et al. 2012b; Deloid et al. 2014). The PCV tube consists of a wide-bore sample loading upper chamber that tapers to a volumetric pellet-capturing capillary (see Figure 17.4).

Low-speed centrifugation of ENM suspensions at known mass concentrations in the PCV tubes forces ENM agglomerates into the pellet-capturing graduated capillary, which can then be measured for volume. Because the pellet contains a known volume of raw ENM (determined from the known mass concentration and volume of the ENM suspension), we can assume that the pellet volume consists of medium trapped within and between ENM particles or agglomerates as well as stacked ENM agglomerates. Therefore, the effective density of ENM agglomerates in their suspended form can be calculated as a volume-weighted average of the raw ENM and medium densities (Demokritou et al. 2012b; Deloid et al. 2014).

While the pellet volume is approximate to the true volume of agglomerates in suspension, agglomerate stacking in the capillary tube is not perfect, and the medium can remain interspersed between the stacked agglomerates (inter-agglomerate medium, Figure 17.3). Thus, the true volume of the agglomerates in suspension comprise only a fraction of the measured pellet, defined as the stacking factor (SF). The remainder of the pellet volume, 1 – SF, represents the medium interspersed between stacked agglomerates. The value of SF depends on the efficiency of agglomerate stacking, and much work has been done to characterize theoretical SFs for different scenarios. For example, a theoretical maximum

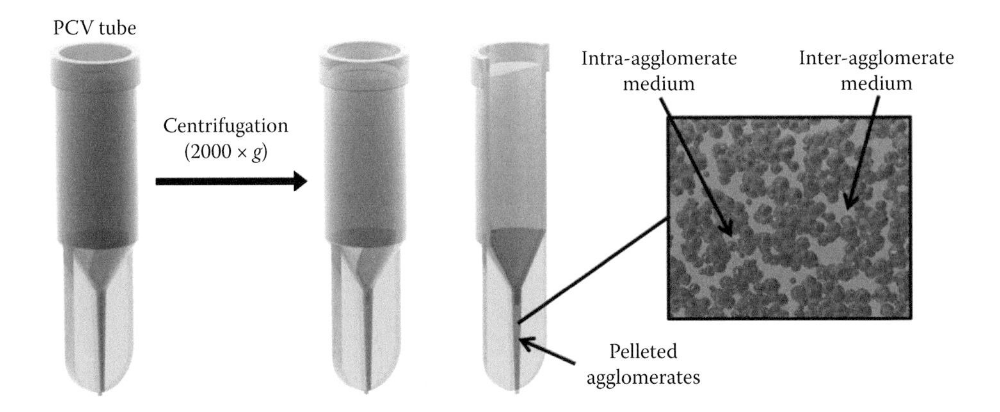

FIGURE 17.4
PCV tubes for volumetric centrifugation. In volumetric centrifugation, a sample of ENM suspension is centrifuged in a PCV tube to produce a pellet, the volume of which can be used to estimate the effective density of the ENM in suspension. The pellets contain both packed agglomerates and the medium remaining between them (inter-agglomerate medium). We refer to fraction of the pellet volume occupied by agglomerates as the SF. (Adapted from Deloid, G. et al., *Nat Commun*, 5, 2014, doi:10.1038/ncomms4514.)

SF of 0.74 was estimated for ordered stacking of uniform spheres (Gauss 1831). In the case of random close packing of irregular spheres, a theoretical maximum of 0.634 was estimated (Song et al. 2008). SF values for nonagglomerating gold nanospheres, as well as roughly spherical agglomerating nanosized metal oxides, were estimated by comparing the effective density measured by AUC with that measured by the volumetric centrifugation method. In both cases, the estimated SFs were extremely close to the theoretical values (Gauss 1831; Song et al. 2008; Demokritou et al. 2012b; Deloid et al. 2014), enabling use of the estimated SF values to calculate the effective densities for a variety of flame-generated metal oxide materials with fractal morphology by our volumetric centrifugation method and the following equation:

$$\rho_{EV} = \rho_{medium} + \left[\left(\frac{M_{ENM} - M_{ENMsol}}{V_{pellet} SF} \right) \left(1 - \frac{\rho_{medium}}{\rho_{ENM}} \right) \right] \quad (17.2)$$

where ρ_{EV} is effective density (g cm^{-3}), ρ_{medium} is density of the dispersion medium (g cm^{-3}), M_{ENM} is the total mass of ENM in suspension (g), M_{ENMsol} is the mass of ions associated with dissolution of soluble particles such as ZnO or CuO (g), V_{pellet} is the measured volume of the pellet (cm^3), and SF is the stacking factor (unitless). Table 17.2 presents the material density and measured effective density for a variety of flame pyrolysis–generated metal oxides currently under investigation by the OECD (2010a), as well as gold nanospheres developed for biomedical applications.

All ENMs were dispersed in cell culture medium typically used for *in vitro* nanotoxicity studies (RPMI supplemented with 10% FBS, RPMI/10% FBS). Gold nanospheres exhibit an effective density only slightly less than the density of elemental gold (17.18 vs. 19.3 g cm^{-3}), consistent with minimal agglomeration. On the other hand, the flame-generated ENMs exhibit effective density values significantly lower than that of their material density, and closer to the density of the dispersion medium (RPMI/10% FBS, 1.0084 g cm^{-3}). Section 17.4 will describe in greater detail the impact of agglomerate diameter and effective density on *in vitro* dosimetry.

TABLE 17.2

ENM Properties and Effective Densities by Volumetric Centrifugation

Material	d_{BET} (nm)	d_H (nm)	ρ_{ENM} (g cm^{-3})	ρ_{EV} (g cm^{-3})
VENGES SiO_2	18.6	135.5 ± 9.53	2.648	1.131 ± 0.001
VENGES Fe_2O_3	27.6	380 ± 3.60	5.242	1.571 ± 0.033
VENGES CeO_2	5.4	179 ± 3.76	7.215	1.492 ± 0.007
VENGES CeO_2	27.9	181 ± 29.8	7.215	1.650 ± 0.008
VENGES CeO_2	71.3	131 ± 5.17	7.215	2.421 ± 0.041
Au nanospheres	NA[a]	42.2 ± 24.7	19.3	17.18 ± 0.35

Note: d_{BET}, primary particle diameter determined from specific surface area measured by nitrogen adsorption/Brunauer–Emmett–Teller (BET) method (nm); d_H, hydrodynamic diameter determined by DLS (nm); ρ_{ENM}, raw ENM material density; ρ_{EV}, effective density estimated by volumetric centrifugation. Errors (±) indicate standard deviations based on three measurements.

[a] Gold nanospheres were generated in suspension, and were therefore not characterized by BET or XRD.

17.4 Determining the Delivered to Cell Dose

Accurate dosimetry requires characterization of agglomerate properties in liquid suspension, particularly their effective diameter and density, which, along with the density and viscosity of the suspending fluid, determine mass transport phenomena (Teeguarden et al. 2007; Hinderliter et al. 2010; Cohen et al. 2013; Deloid et al. 2014). In a typical temperature- and humidity-controlled nanotoxicity experiment, delivery of ENMs in liquid suspension to cells in culture is determined by two fundamental transport mechanisms—diffusion and sedimentation. Diffusion is the spontaneous, passive movement of particles in response to a gradient—from areas of high chemical potential to areas of low chemical potential. Rates of diffusional transport are a function of particle size and the viscosity of the medium, and can be estimated by the Stokes–Einstein equation:

$$D = \frac{k_B T}{3\pi\eta d} \tag{17.3}$$

where D is the diffusion coefficient ($m^2 s^{-1}$), k_B is the Boltzmann constant ($kg\ m^2 S^2 K^{-1}$), T is the absolute temperature (K), η is the medium's dynamic viscosity ($kg\ m^{-1} s^{-1}$), and d is the particle diameter (m) in suspension. The time required to diffuse a given distance in one dimension can be calculated from the equation

$$t = \frac{\langle \bar{r}^2 \rangle}{2D} \tag{17.4}$$

where the numerator is defined as the root mean squared distance, or the distance that the average particle of a given size will travel by diffusion (Einstein 1905; Teeguarden et al. 2007). On the other hand, a particle sediments at a rate determined by the balance of opposing forces acting on the particle: the acceleration force (e.g., gravitational or centrifugal), the counterbuoyant force caused by displacement of medium by the particle, and the frictional or drag force. The net effect of these forces is dependent on particle size, shape, and density, and can be calculated from the following equation (assuming spherical particles):

$$v_s = \frac{g(\rho_E - \rho_{medium})d_2}{18\eta} \tag{17.5}$$

where v_s is the settling velocity ($m\ s^{-1}$), g is acceleration due to gravity ($m\ s^{-2}$), ρ_E is the effective density ($kg\ m^{-3}$), and ρ_{medium} is the medium's density ($kg\ m^{-3}$).

On the basis of the two equations listed above for sedimentation and diffusion, fate and transport models can be used to accurately estimate particle deposition over time in an *in vitro* system. For example, the recently developed one-dimensional ISDD (*in vitro* sedimentation, diffusion, and dosimetry) model (Hinderliter et al. 2010) enables for a given ENM–medium suspension estimation of the fraction of administered particles that would deposit on cells as a function of time. ISDD is readily available from the authors, is simple to use, and allows nanotoxicologists to improve on typically reported administered *in vitro* dose metrics. However, ISDD is a one-dimensional model that assumes homogenous conditions within the cell culture well, assumes an agglomerate density on the basis of unvalidated theoretical estimations for fractal materials (Sterling et al. 2005; Hinderliter et al. 2010; Deloid et al. 2014), and does not account for dynamic changes of agglomerate diameter and effective density over time associated with dissolution of soluble and partially soluble particles. As an alternative, multidimensional computational fluid dynamics models are currently under development by our group and others. These more sophisticated models allow for consideration of spatial variability of particle concentration over time, and consideration of the impact of dynamic changes to agglomerate diameter and effective density over time on particle deposition *in vitro*.

It is necessary to note that for partially soluble materials, changes in agglomerate diameter and effective density that result from mass loss due to dissolution must be resolved over time and addressed to accurately estimate the delivered dose. Thus, for greatest accuracy, mass loss due to dissolution, d_H, and ρ_{EV} should be measured over the time of exposure. These time-resolved values should then be used by transport simulation models to accurately estimate the delivered dose. It is worth noting that the dissolution of 24 metal oxide ENMs in cell culture medium following a 24-h incubation has been previously reported (Zhang et al. 2012), and only a few materials exhibit ≥10% dissolution (ZnO, CuO, and WO_3). Considering that mass loss due to dissolution will likely result in a net decrease in effective density, dosimetry calculations for soluble materials based on effective density measured immediately after sample preparation provide an overestimate of particle deposition over time. In addition, for completion of dosimetry calculations, both the particulate and soluble components must be correctly identified and considered separately, as previously described (Teeguarden et al. 2007).

17.5 Relevant *In Vitro* Dosimetry Functions

Recently, the authors introduced a new tool for nanotoxicologists: relevant *in vitro* dose (RID) functions (Cohen et al. 2013, 2014b) RID functions are simple mathematical equations that provide an easy way to estimate the delivered to cell dose metrics for well-characterized ENM dispersions without the use of sophisticated fate and transport numerical algorithms.

Great emphasis has been placed on determining the most appropriate dose metric for *in vitro* toxicity, be it mass, particle number, surface area, etc. (Lison et al. 2008; Rushton et al. 2010; Wittmaack 2011). As described previously, ENMs in suspension can form large agglomerates close to 10 times their primary particle size, thereby altering the total number of free particles in suspension as well as the total surface area available for biointeractions before they even reach the cells cultured *in vitro*. Therefore, considerations of particle number and surface area dose metrics should take into account ENM transformations in liquid as described in the following.

For an ENM suspension of known mass concentration, γ ($\mu g\ ml^{-1}$), the total mass dose, M (μg), can be calculated as

$$M = \gamma \times V \tag{17.6}$$

where V is the volume of exposure medium (ml) applied directly to the cells in culture. The total particle number dose, N (#), can be calculated from the total mass, M, hydrodynamic radius, r_H (cm, determined by DLS for ENMs in suspension), and agglomerate effective density, ρ_E (g cm^{-3}), as

$$N = \frac{M}{\left(\frac{4}{3}\pi r_H^3\right) \times \rho_E} \tag{17.7}$$

The total surface area dose, SA (cm^2), can then be calculated as

$$\text{SA} = \left(4\pi r_H^2\right) \times N \tag{17.8}$$

It is worth noting that cellular response to a biologically active material reflects the quantity of the substance actually coming into contact with the cells. For example, Wittmaack recently reported significant correlation between the *in vitro* toxicity of SiO_2 NPs and the areal density of NP mass delivered to cells over the exposure duration (Wittmaack 2011). Cellular toxicity *in vitro* is therefore more accurately represented in relation to the delivered dose for any selected dose metric, rather than the typically reported administered mass concentration of ENMs in suspension.

Using material–medium-specific parameters for agglomerate hydrodynamic diameter and effective density as inputs to any fate and transport algorithms (e.g., ISDD), the fraction of administered particles that would deposit onto cells as a function of time, $f(t)$, can be calculated. The estimated $f(t)$ function can then be fitted to a Gompertz sigmoidal function as previously described by the authors (Cohen et al. 2013) as follows:

$$f_D(t) = 1 - e^{-\alpha t} \tag{17.9}$$

where α (h^{-1}) is the material–medium-specific deposition fraction constant. Figure 17.5 provides an example of the estimated deposition for a representative material (CeO_2) dispersed in RPMI/10% FBS in a 96-well plate over time.

The RID functions $(\text{RID})_f$ for delivered mass, particle number, and surface area for a given material–medium system can then be derived by combining Equation 17.9 with Equations 17.6 through 17.8, as follows.

For delivered to cell mass (RID_M, μg):

$$\text{RID}_M = (1 - e^{-\alpha t}) \times M \tag{17.10}$$

For delivered to cell particle number (RID_N, number of particles):

$$\text{RID}_N = (1 - e^{-\alpha t}) \times N \tag{17.11}$$

For total delivered to cell surface area (RID_{SA}, cm^2):

$$\text{RID}_{SA} = (1 - e^{-\alpha t}) \times \text{SA} \tag{17.12}$$

While validated computational models such as the ISDD are readily available to nanotoxicologists for performing accurate estimates of delivered dose, the RID functions provide

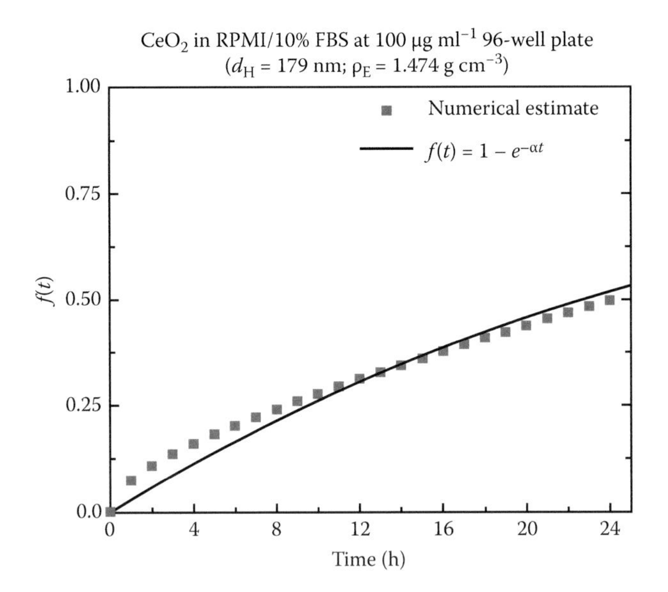

FIGURE 17.5
Estimated *in vitro* dosimetry based on integrated approach. Estimated deposition of CeO_2, indicated here by $f(t)$, the fraction of the administered dose (total mass of ENM in suspension) delivered to cells in a 96-well plate over time. Particle deposition based on measured values for agglomerate diameter and effective density measured for dispersion in typical cell culture medium used for nanotoxicology study (RPMI/10% FBS). Solid line indicates sigmoidal function curve fit to estimated deposition data ($f(t) = 1 - e^{-\alpha t}$).

an easy way to estimate the delivered to cell dose metrics for well-characterized ENM dispersions without the use of sophisticated fate and transport numerical algorithms. The only unknown parameters required are the material–medium-specific deposition fraction constant (α), agglomerate size (r_H), and agglomerate effective density (ρ_E), as measured for various mass concentrations and well plate geometries (96-well, 384-well, etc.).

In summary, the aforementioned method for preparing stable ENM dispersions and characterizing agglomerate diameter and effective density can be followed for any ENM, and the specific RID function can be derived for any ENM, medium, and well plate configuration. The authors have recently derived and reported the RID functions for 24 metal and metal oxide ENMs in various configurations, which can be used by nanotoxicologists to take into consideration dosimetry for future *in vitro* studies (Cohen et al. 2014b).

17.6 Validation of the Integrated Method for Estimating Dosimetry *In Vitro*

The ISDD dosimetry model has been previously validated for nonagglomerating, nonindustrially relevant fluorescently labeled polystyrene beads of various agglomerate diameter (Hinderliter et al. 2010; Ahmad Khanbeigi et al. 2012), as well as for superparamagnetic iron oxide particles (Hinderliter et al. 2010). However, these validation experiments relied on unvalidated theoretical estimations for effective density, which can vary greatly from

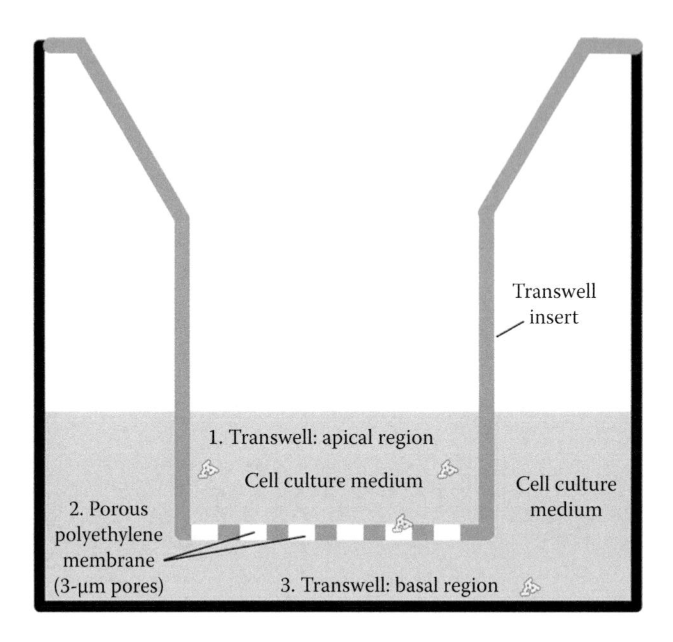

FIGURE 17.6
Experimental setup for dosimetry method validation experiments. (Adapted from Deloid, G. et al., *Nat Commun*, 5, 2014, doi:10.1038/ncomms4514.)

the true agglomerate effective density (Deloid et al. 2014) and therefore lead to miscalculation of sedimentation speed.

To validate our proposed integrated method for estimating dosimetry *in vitro*, which includes accurate assessment of agglomerate effective density, suspensions of industry-relevant flame-generated ENMs were neutron activated and applied to transwell insert membranes with 3-μm pores (Figure 17.6) (Cohen et al. 2014a).

Neutron activation of ENMs has been used routinely in our laboratory for biokinetic studies of ENMs, and is a well-established method for accurately tracking the gamma-emitting isotopes ^{141}Ce and ^{65}Zn, with high sensitivity and correlation with total particle mass (He et al. 2010). Following 2-, 4-, and 24-h incubations, we measured the delivered dose by gamma spectroscopy, defined as the sum of particles that deposited on or passed through the membrane. For all materials, <4% of the administered dose remained stuck to the membrane, suggesting easy passage of ENMs through the 3-μm pores (data not shown). Additionally, the delivered dose was estimated using the ISDD model, and measured values for agglomerate diameter and effective density and inputs. As shown in Figure 17.7, there is close agreement between the measured delivered dose and the estimated delivered dose for all materials at each time point. These data provide a clear validation of the proposed dosimetric approach for ENMs.

17.7 Effect of Agglomerate Diameter and Effective Density on Dosimetry

The effect of agglomerate diameter on mass transport in an *in vitro* system has recently been evaluated, demonstrating the importance of proper dispersion characterization.

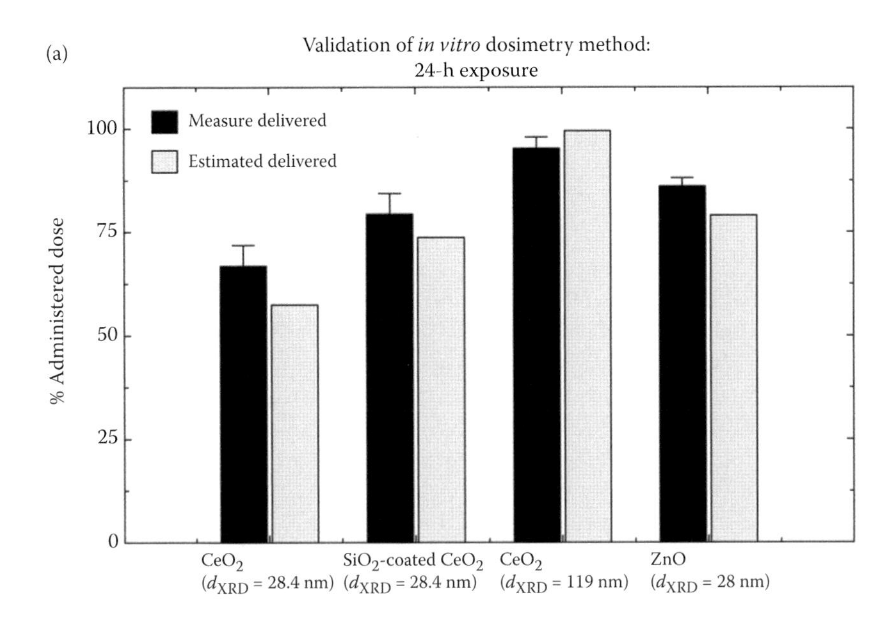

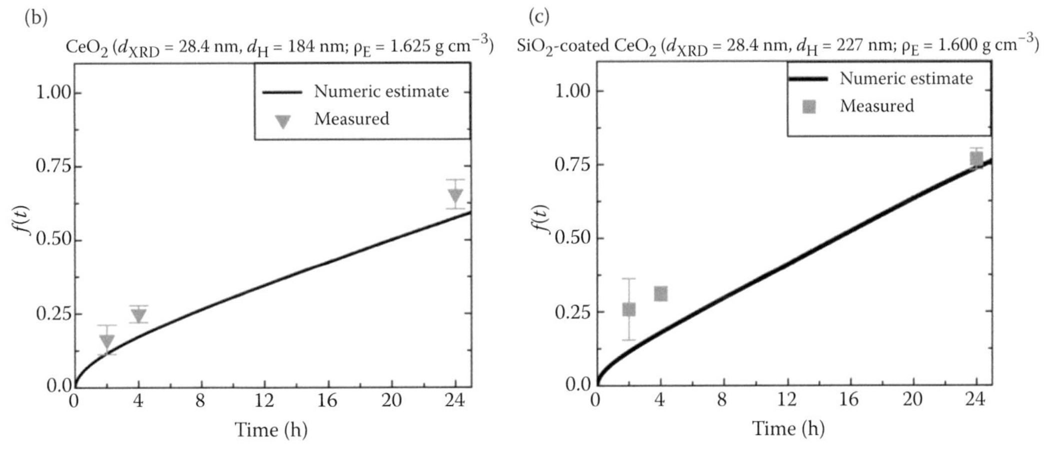

FIGURE 17.7
Validation of integrated *in vitro* dosimetry approach. (a) Validation for five neutron-activated ENMs suspended in DMEM after 24-h exposure. (b) Validation for CeO_2 (d_{XRD} = 28.4 nm) and (c) SiO_2-coated CeO_2 (d_{XRD} = 28.4 nm) following 2-, 4-, and 24-h exposures. All experiments were done in triplicate; error bars represent standard deviation.

Studies investigating cellular interactions with and uptake of nonagglomerating gold nanospheres report that for particles of hydrodynamic diameter 15 nm, diffusion is the prevailing means for transporting NPs to cells in culture, while for particles of hydrodynamic diameter 100 nm sedimentation is the prevailing means for transport (Cho et al. 2011). The sedimentation and diffusion transport rates for gold nanospheres of known density (19.3 g cm^{-3}), suspended in cell culture medium typically used for *in vitro* nanotoxicity experiments (RPMI/10% FBS: viscosity = 0.00089 Pa s; density = 1.00 g cm^{-3}), at 37°C, were calculated using Equations 17.3 and 17.5 for a range of primary particle diameters.

The time required to travel 1 mm (approximately half the medium column height used in a typical cell culture experiment) by sedimentation was estimated by simply dividing the transport rate by the distance, and was estimated for diffusion via Equation 17.4. Figure 17.8 demonstrates a clear size dependency for which transport mechanism drives deposition *in vitro*, where ENMs ≤40 nm in diameter are primarily driven by diffusion and ENMs >40 nm are primarily driven by sedimentation. A 1-nm gold nanosphere is estimated to travel 1 mm by diffusion in 16 min versus approximately 1,500,000 min by sedimentation. In contrast, a 1000-nm gold nanosphere is estimated to travel 1 mm by diffusion in approximately 16,000 min, versus only approximately 1 min by sedimentation.

Most nonsoluble flame-generated metal oxide NPs, such as those currently under investigation by the OECD (2010a), agglomerate in liquid suspension in the size range of ~100–300 nm, and thus transport is expected to be dominated by sedimentation (Bihari et al. 2008; Murdock et al. 2008; Deloid et al. 2014).

Dosimetric modeling also requires accurate assessment of effective density for agglomerates in liquid suspension. Figure 17.9 presents the estimated deposition rates (using the ISDD model) for a representative material, VENGES SiO_2 (d_{BET} = 18.6 nm), estimated using either the material density or the volumetric centrifugation–derived effective density as input.

After 24 h, an estimated 80% of the administered dose will be delivered to cells in a standard 96-well plate based on the material density input, while only 20% is estimated on the basis of the more accurate effective density input. Clearly, agglomerate density greatly influences sedimentation and mass transport in an *in vitro* system. It is also evident from these data that effective doses can vary greatly from the typically reported administered dose, highlighting the necessity for accurate characterization of ENM effective density in suspension.

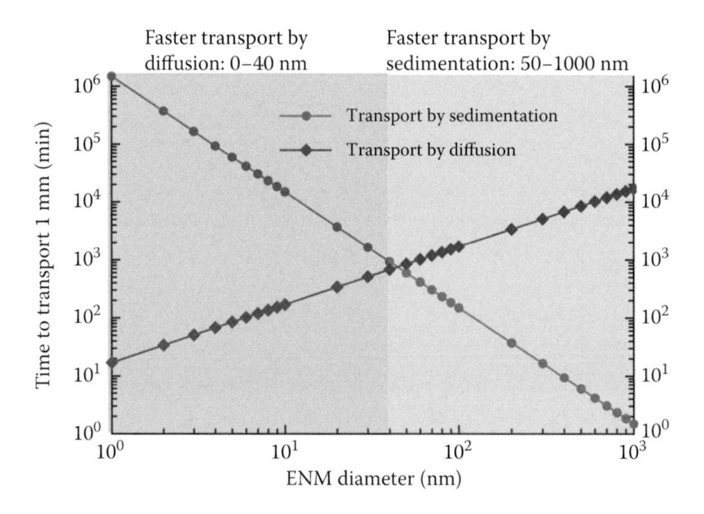

FIGURE 17.8
Impact of ENM diameter on mass transport *in vitro*. The theoretical time to transport nonagglomerating gold nanospheres of known density to 1 mm (approximately half the medium column height used in a typical cell culture experiment) is plotted against primary particle diameter, demonstrating a clear size dependency for which transport mechanism drives deposition *in vitro*. ENMs ≤40 nm in diameter are primarily driven by diffusion, and ENMs >40 nm are primarily driven by sedimentation. The sedimentation and diffusion transport rates were estimated for particles ranging from 1 to 1000 nm in diameter using Equations 17.3 and 17.4, and the following inputs: material effective density, 19.3 g cm^{-3}; media viscosity, 0.00089 Pa s; media density, 1.00 g cm^{-3}; temperature, 37°C. The time required to travel 1 mm by sedimentation was estimated by dividing the transport rate by the distance, and was estimated for diffusion via Equation 17.4.

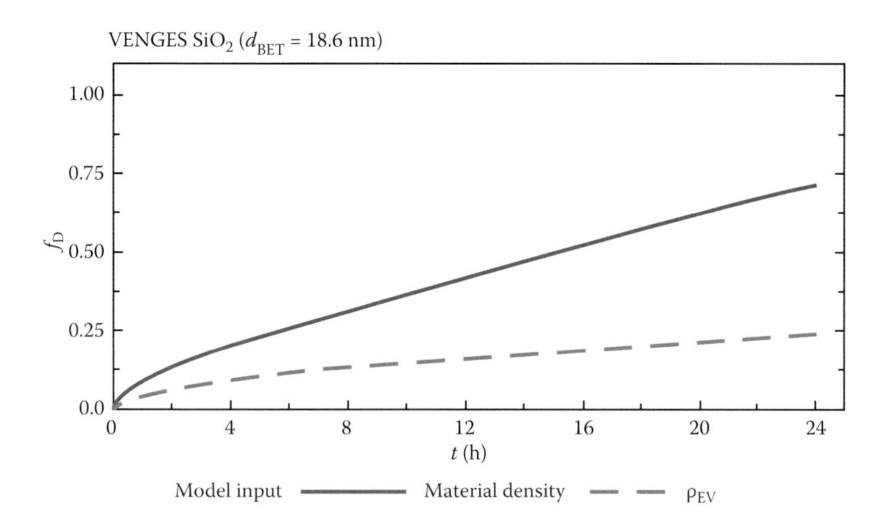

FIGURE 17.9
Role of effective density in the rate of ENM deposition. Modeling of transport over time demonstrates the important role of ENM density in determining the rate of ENM deposition, indicated here by f_D, the fraction of the administered dose (total mass of ENM in suspension) deposited at a given time. Assuming a density equal to that of the raw material (solid line) results in an overestimation of deposition rate. Deposition rate curves are shown for VENGES SiO_2 (d_{BET} = 18.6 nm) suspended in cell culture media typically used for *in vitro* nanotoxicology studies (RPMI/10% FBS). The solid and dashed line represent deposition based on material density, and effective density determined by volumetric centrifugation, respectively. (Adapted from Deloid, G. et al., *Nat Commun*, 5, 2014, doi:10.1038/ncomms4514.)

Another useful application of dosimetric modeling is to estimate the time required for delivery of the entire administered dose of ENM in suspension. Solving Equation 17.9 for time t at which the fraction $f_D(t)$ of administered particles is delivered yields

$$t = -\frac{\ln(1 - f_D(t))}{\alpha} \tag{17.13}$$

Equation 17.7 can then be used to calculate the time required for delivery of 90% of the administered dose, t_{90}, for a given fully characterized ENM–medium combination using the specific deposition fraction constant, α, and an $f_D(t)$ value of 0.90. Figure 17.10 presents estimated t_{90} values for a large panel of metal oxide NPs, along with nonagglomerating gold nanospheres, and further demonstrates the importance of effective density, along with agglomerate diameter, in determining the time required for ENM delivery.

In general, ENMs having relatively greater values for both effective density and agglomerate diameter are estimated to deposit more rapidly than those with smaller values for both properties (e.g., CuO, d_H = 310 nm, ρ_{EV} = 2.214 g cm^{-3}, t_{90} = 16 h vs. VENGES SiO_2, d_H = 135 nm, ρ_{EV} = 1.131 g cm^{-3}, t_{90} = 171 h). These results demonstrate that for liquid-based *in vitro* systems, the dose rates and target cell doses can vary significantly depending on material and medium properties, and emphasize the import of characterizing ENM suspensions for agglomerate diameter and effective density. Since cellular response to a biologically active substance delivered by mass transport over time should reflect the quantity of the substance actually coming in contact with cells, nanotoxicity *in vitro* should be better

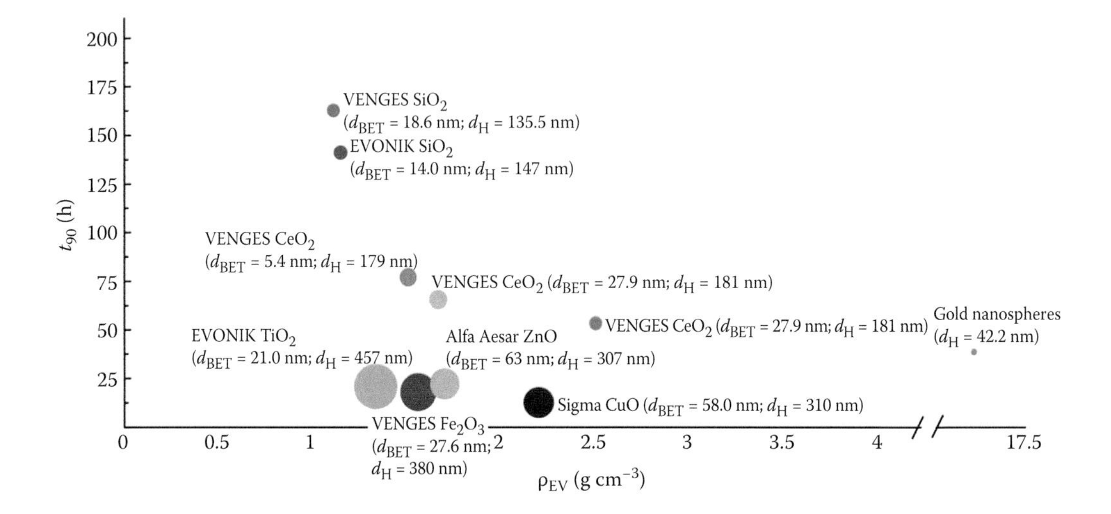

FIGURE 17.10
Roles of agglomerate density and diameter in time to deposit 90% of suspended ENM. Density and particle size both play important roles in determining the rate of ENM deposition. The combined effects of these two parameters result in a wide range, among the ENMs investigated, in the rate of deposition and consequently the time required for 90% of the ENM in suspension to be deposited, indicated here by t_{90}. Effective density (ρ_{EV}) is represented along the x-axis. Bubble diameters are relatively scaled to the agglomerate or hydrodynamic diameter (d_H) of the specific ENM represented. The primary particle diameter (d_{BET}) is indicated, along with d_H, for each ENM in the text corresponding to its bubble in the graph. (Adapted from Deloid, G. et al., *Nat Commun*, 5, 2014, doi:10.1038/ncomms4514.)

represented in relation to the mean delivered dose based on accurate modeling of mass transport, rather than to the typically reported initial administered mass concentration of ENM in suspension.

It is necessary to note that for partially soluble materials (ZnO, CuO), the dosimetry data presented in Figure 17.10 are limited by the fact that effective density measurements were obtained immediately following dispersion, and do not reflect dissolution during dispersion or the dynamic nature of agglomerate size and density due to dissolution over time. Considering that mass loss due to dissolution will likely result in a net decrease in effective density, the dosimetry calculations for soluble materials based on effective density measured immediately after sample preparation provide an overestimate of particle deposition over time. For simplicity, dissolution effects were ignored here, resulting in an overestimate of effective density and delivered to cell dose.

17.8 Identifying Physiologically Relevant Exposure Levels for Comparison between *In Vitro* and *In Vivo* Bioassays

Thoughtful selection of concentrations used for *in vitro* nanotoxicity studies is critical for accurate comparison of *in vitro* results with those used in *in vivo* studies (Cohen Hubal 2009; Gangwal et al. 2011; Demokritou et al. 2012a). Therefore, careful consideration must be given to identifying equivalent effective doses between these two biological systems. As more and improved exposure assessments for aerosolized ENMs in occupational and other settings become available, the measured aerosol concentrations and particle size distributions can be used to estimate deposited and retained mass doses of ENMs in the pulmonary alveolar region of humans and animals. Currently, a number of numerical models are available such as the widely used and well-accepted multiple-path particle dosimetry model (MPPD) (Asgharian 2001; Asgharian and Price 2007; US Environmental Protection Agency 2009), which can be employed for this purpose. A recent publication by Gangwal et al. used this approach to determine human lung surface mass concentrations for a given ENM exposure scenario, which was subsequently converted to testing solution mass concentrations for an *in vitro* screening based on the reported cell culture well-bottom surface area (Gangwal et al. 2011). The reported equivalent *in vitro* dose for a peak lung surface concentration following 24-h exposure to 1 mg m^{-3} aerosol concentration of TiO_2 NPs ranging in aerodynamic diameter from 5 to 100 nm, estimated using the method described previously, was 0.240 μg ml^{-1} (assuming a standard 96-well cell culture plate) (Gangwal et al. 2011). While this value represents a peak exposure value for humans in an occupational setting, it is on the low end of doses typically tested *in vitro*, (Gangwal et al. 2011; George et al. 2011; Kroll et al. 2011; Cohen et al. 2013) suggesting that future studies should aim to characterize toxicity at doses equal to and below this value. Importantly, this approach may only be appropriate for short-term or daily human exposures. Equating lifetime doses accumulated in the alveolar region following long-term chronic inhalation exposure with doses delivered all at once as a bolus in a short-term *in vitro* system ignores the differences in dose rate (which may span many orders of magnitude) and can be highly misleading. For example, Gangwal et al. report that a 45-year accumulated surface area dose equates to *in vitro* concentrations of 50–69 μg ml^{-1}, extremely high doses that should only be considered as the high-end limit of an *in vitro* investigation using a wide range of doses (Oberdorster 2012).

Another limitation to the above approach is the assumption that all the applied ENMs in suspension are delivered to cells at the bottom of the well over the course of the exposure, which, as detailed in Section 17.7 on accurate dosimetry, is rarely the case. A similar method proposed by Demokritou et al. for comparing the *in vivo* toxicity of carefully controlled aerosolized ENM exposures with *in vitro* results incorporates characterization of ENM transformations in liquid suspension and their influence on particle kinetics and mass transport using the ISDD model (Demokritou et al. 2012a). Similar to Gangwal's approach, the MPPD model was used to estimate the lung surface mass concentrations associated with the rodent *in vivo* exposures. The estimated *in vitro* equivalent dose, which represents the delivered to cell dose, was then converted to an administered mass concentration using output from the ISDD model. In this instance, the toxicity results between the *in vivo* and *in vitro* studies were incompatible even at equivalent doses. While more work is still needed to improve the sophistication and sensitivity of these methods, as well as the application of increasingly sophisticated *in vitro* models attempting to mimic the biological complexity of a whole-animal exposure scenario, such efforts toward improving the accuracy and relevance of *in vitro* dosimetry are laudable.

17.9 Summary and Conclusions

Improving the accuracy and relevance of *in vitro* dosimetry will have positive impacts in several major nanotoxicology research areas. Cellular toxicity studies will be able to properly isolate and evaluate the effect of fundamental particle characteristics (size, surface area, particle number, surface chemistry, etc.) on cytotoxicity and uptake. Several groups have endeavored to associate ENM toxicity *in vitro* with chemical reactivity due to increased surface area per mass (Jiang 2008; Rushton et al. 2010; Cho et al. 2012; Donaldson et al. 2012). With regard to cellular uptake studies, the recent literature suggests the rate of delivery of ENMs to cells may be a rate-limiting factor for cellular internalization (Limbach 2005). Furthermore, our group recently reported that particle translocation across cellular monolayers *in vitro* may also be limited by the time required to deliver particles to cells (Cohen et al. 2014a). Additionally, particle size, agglomeration state, and concentration-dependent differences in macrophage and epithelial cell uptake of NPs have been reported (Chithrani et al. 2006; Xia et al. 2008; Tedja et al. 2012). While these studies are extremely valuable to the nanotoxicology community, it is not clear whether the results reflect fundamental differences in chemical reactivity, rates of cellular uptake, or rather ENM-specific differences in mass transport and particle delivery to cells. Characterization of ENM transformations in liquid suspension and estimation of particle mobility *in vitro* will help address these limitations.

Further benefits of improved *in vitro* dosimetry include the possibility for results of cytotoxicity studies from various laboratories to have comparable data that can be consolidated into large databases, with the goal of comprehensive toxicological characterization (Shaw et al. 2008; Puzyn et al. 2011). Dose–response characterization for ENMs *in vitro* may be compared with effective doses and measured biological responses *in vivo* (Demokritou et al. 2012a). Additionally, these results can be compared with physiologically relevant exposures experienced by humans in occupational settings (Gangwal et al. 2011). The Nanotechnology Working Group of the US National Institutes of Health National Cancer Informatics Program has begun developing a database repository in which researchers can organize, analyze, and compare nanotoxicity data from human, *in vivo*, and *in vitro* studies, referred to as "ISA-TAB-Nano" (National Cancer Institute). For such an endeavor to succeed in consolidating nano–biointeraction results from various laboratories, a standard of practice must be set for accurate and relevant reporting and selection of *in vitro* dosimetry.

Accurate reporting of dosimetry *in vitro* requires acknowledgment that mass transport (sedimentation and/or diffusion) of ENMs to cells in a typical culture system proceeds at a rate governed by the mass transport properties of the physical particulate forms in which ENMs exist in suspension. Improved control and characterization of the factors influencing mass transport of ENMs *in vitro* can be achieved by (i) standardizing sample preparation protocols for suspending ENMs in liquid; (ii) careful characterization of ENM transformations in physiological media; and (iii) computational modeling of the impact these transformations have on ENM delivery to cells in culture, and estimation of ENM delivery to cells over time. Once particokinetics in an *in vitro* system have been considered, equivalent doses can be estimated and compared with physiologically relevant *in vivo* animal and human exposures. Adoption of such a method will be a major step toward the development of cheap, accurate, and reproducible *in vitro* screening assays, and a major advancement for nano–environmental health and safety research.

References

Ahmad Khanbeigi, R.; Kumar, A.; Sadouki, F.; Lorenz, C.; Forbes, B.; Dailey, L. A.; Collins, H. 2012. The delivered dose: Applying particokinetics to *in vitro* investigations of nanoparticle internalization by macrophages. *J Control Release* 162, 259–66.

Aitken, R. J.; Chaudhry, M. Q.; Boxall, A. B.; Hull, M. 2006. Manufacture and use of nanomaterials: Current status in the UK and global trends. *Occup Med* 56, 300–6.

Asgharian, B. 2001. Particle deposition in a multiple-path model of the human lung. *Aerosol Sci Technol* 34, 332–9.

Asgharian, B.; Price, O. T. 2007. Deposition of ultrafine (nano) particles in the human lung. *Inhal Toxicol* 19, 1045–54.

Balbus, J. M.; Maynard, A. D.; Colvin, V. L.; Castranova, V.; Daston, G. P.; Denison, R. A. et al. 2007. Meeting report: Hazard assessment for nanoparticles—Report from an interdisciplinary workshop. *Environ Health Perspect* 115, 1654–9.

Bello, D.; Martin, J.; Santeufemio, C.; Sun, Q.; Lee Bunker, K.; Shafer, M.; Demokritou, P. 2012. Physicochemical and morphological characterisation of nanoparticles from photocopiers: Implications for environmental health. *Nanotoxicology* 7, 989–1003.

Bihari, P.; Vippola, M.; Schultes, S.; Praetner, M.; Khandoga, A. G.; Reichel, C. A.; Coester, C.; Tuomi, T.; Rehberg, M.; Krombach, F. 2008. Optimized dispersion of nanoparticles for biological *in vitro* and *in vivo* studies. *Part Fibre Toxicol* 5, 14. doi: 10.1186/1743-8977-5-14.

Brain, J. 2009. Biologic responses to nanomaterials depend on exposure, clearance, and material characteristics. *Nanotoxicology* 3, 175–80.

Carney, R. P.; Kim, J. Y.; Qian, H.; Jin, R.; Mehenni, H.; Stellacci, F.; Bakr, O. M. 2011. Determination of nanoparticle size distribution together with density or molecular weight by 2D analytical ultracentrifugation. *Nat Commun* 2, 335. doi: 10.1038/ncomms1338.

Chithrani, B. D.; Ghazani, A. A.; Chan, W. C. 2006. Determining the size and shape dependence of gold nanoparticle uptake into mammalian cells. *Nano Lett* 6, 662–8.

Cho, E. C.; Zhang, Q.; Xia, Y. 2011. The effect of sedimentation and diffusion on cellular uptake of gold nanoparticles. *Nat Nanotechnol* 6, 385–91.

Cho, W. S.; Duffin, R.; Thielbeer, F.; Bradley, M.; Megson, I. L.; Macnee, W.; Poland, C. A.; Tran, C. L.; Donaldson, K. 2012. Zeta potential and solubility to toxic ions as mechanisms of lung inflammation caused by metal/metal oxide nanoparticles. *Toxicol Sci* 126, 469–77.

Choi, H. S.; Ashitate, Y.; Lee, J. H.; Kim, S. H.; Matsui, A.; Insin, N.; Bawendi, M. G.; Semmler-Behnke, M.; Frangioni, J. V.; Tsuda, A. 2010. Rapid translocation of nanoparticles from the lung airspaces to the body. *Nat Biotechnol* 28, 1300–3.

Cohen, J.; Deloid, G.; Pyrgiotakis, G.; Demokritou, P. 2013. Interactions of engineered nanomaterials in physiological media and implications for *in vitro* dosimetry. *Nanotoxicology* 7, 417–31.

Cohen, J. M.; Derk, R.; Wang, L.; Godleski, J.; Kobzik, L.; Brain, J.; Demokritou, P. 2014. Tracking translocation of industrially relevant engineered nanomaterials (ENMs) across alveolar epithelial monolayers *in vitro*. *Nanotoxicology* 8, 216–25. doi: 10.3109/17435390.2013.879612.

Cohen, J. M.; Teeguarden, J. G.; Demokritou, P. 2014. An integrated approach for the *in vitro* dosimetry of engineered nanomaterials. *Part Fibre Toxicol* 11, 20.

Cohen Hubal, E. 2009. Biologically relevant exposure science for 21st century toxicity testing. *Toxicol Sci* 111, 226–32.

Deloid, G.; Cohen, J.; Darrah, T.; Derk, R.; Wang, L.; Pyrgiotakis, G.; Wohlleben, W.; Demokritou, P. 2014. Estimating effective density of engineered nanomaterials for *in vitro* dosimetry. *Nat Commun* 5, 3514. doi: 10.1038/ncomms4514.

Demokritou, P.; Cohen, J.; DeLoid, G. 2012a. Novel methods of measuring effective density of nanoparticles in fluids. U.S. Patent application 61/661,895.

Demokritou, P.; Gass, S.; Pyrgiotakis, G.; Cohen, J. M.; Goldsmith, W.; McKinney, W. et al. 2012b. An *in vivo* and *in vitro* toxicological characterization of realistic nanoscale CeO_2 inhalation exposures. *Nanotoxicology* 7, 1338–50.

Donaldson, K.; Schinwald, A.; Murphy, F.; Cho, W. S.; Duffin, R.; Tran, L.; Poland, C. 2012. The biologically effective dose in inhalation nanotoxicology. *Acc Chem Res* 46, 723–32.

Einstein, A. 1905. Uber die von der molekularkinestischen Theorie der Warme geforderte Bewegung von in ruhenden Flussigkeiten suspendierten Teilchen. *Ann Phys* 322, 549–60.

Elzey, S. 2009. Agglomeration, isolation and dissolution of commercially manufactured silver nanoparticles in aqueous environments. *J Nanopart Res* 12, 1945–58.

Fadeel, B. 2010. Better safe than sorry: Understanding the toxicological properties of inorganic nanoparticles manufactured for biomedical applications. *Adv Drug Deliv Rev* 8, 362–74.

Fischer, H. C.; Chan, W. C. 2007. Nanotoxicity: The growing need for *in vivo* study. *Curr Opin Biotechnol* 18, 565–71.

Gangwal, S.; Brown, J. S.; Wang, A.; Houck, K. A.; Dix, D. J.; Kavlock, R. J.; Hubal, E. A. 2011. Informing selection of nanomaterial concentrations for ToxCast *in vitro* testing based on occupational exposure potential. *Environ Health Perspect* 119, 1539–46.

Gauss, C. F. 1831. Besprechung des Buchs von L. A. Seeber: Intersuchungen über die Eigenschaften der positiven ternären quadratischen Formen usw. *GöttingscheGelehrteAnzeigen* 2, 9.

George, S.; Xia, T.; Rallo, R.; Zhao, Y.; Ji, Z.; Lin, S. et al. 2011. Use of a high-throughput screening approach coupled with *in vivo* zebrafish embryo screening to develop hazard ranking for engineered nanomaterials. *ACS Nano* 5, 1805–17.

Han, X.; Corson, N.; Wade-Mercer, P.; Gelein, R.; Jiang, J.; Sahu, M.; Biswas, P.; Finkelstein, J. N.; Elder, A.; Oberdorster, G. 2012. Assessing the relevance of *in vitro* studies in nanotoxicology by examining correlations between *in vitro* and *in vivo* data. *Toxicology* 297, 1–9.

He, X.; Zhang, H.; Ma, Y.; Bai, W.; Zhang, Z.; Lu, K.; Ding, Y.; Zhao, Y.; Chai, Z. 2010. Lung deposition and extrapulmonary translocation of nano-ceria after intratracheal instillation. *Nanotechnology* 21, 285103. doi: 1088/0957-4484/21/28/285103.

Hielscher, T. 2005. Ultrasonic production of nano-size dispersions and emulsions. In: *ENS'05*. Accessed at http://www.euronanoforum2011.eu/wp-content/uploads/2011/10/enf2011_production-nanostructures_hielscher_fin.pdf.

Hinderliter, P. M.; Minard, K. R.; Orr, G.; Chrisler, W. B.; Thrall, B. D.; Pounds, J. G.; Teeguarden, J. G. 2010. ISDD: A computational model of particle sedimentation, diffusion and target cell dosimetry for *in vitro* toxicity studies. *Part Fibre Toxicol* 7, 36. doi: 10.1186/1743-8977-7-36.

Ji, Z.; Jin, X.; George, S.; Xia, T.; Meng, H.; Wang, X. et al. 2010. Dispersion and stability optimization of TiO_2 nanoparticles in cell culture media. *Environ Sci Technol* 44, 7309–14.

Jiang, J. 2008. Does nanoparticle activity depend upon size and crystal phase. *Nanotoxicology* 2, 33–42.

Jiang, J. 2009. Characterization of size, surface charge, and agglomeration state of nanoparticle dispersions for toxicological studies. *J Nanopart Res* 11, 77–89.

Jones, C. F.; Grainger, D. W. 2009. *In vitro* assessments of nanomaterial toxicity. *Adv Drug Deliv Rev* 61, 438–56.

Krewski, D.; Acosta, D., Jr.; Andersen, M.; Anderson, H.; Bailar, J. C., III; Boekelheide, K. et al. 2010. Toxicity testing in the 21st century: A vision and a strategy. *J Toxicol Environ Health B Crit Rev* 13, 51–138.

Kroll, A.; Dierker, C.; Rommel, C.; Hahn, D.; Wohlleben, W.; Schulze-Isfort, C.; Gobbert, C.; Voetz, M.; Hardinghaus, F.; Schnekenburger, J. 2011. Cytotoxicity screening of 23 engineered nanomaterials using a test matrix of ten cell lines and three different assays. *Part Fibre Toxicol* 8, 9. doi: 10.1186/1743-8977-8-9.

Lai, D. Y. 2012. Toward toxicity testing of nanomaterials in the 21st century: A paradigm for moving forward. *Wiley Interdiscip Rev Nanomed Nanobiotechnol* 4, 1–15.

Laxen, D. P. H. 1977. A specific conductance method for quality control in water analysis. *Water Res* 11, 91–4.

Lehmann, A. D.; Daum, N.; Bur, M.; Lehr, C. M.; Gehr, P.; Rothen-Rutishauser, B. M. 2011. An *in vitro* triple cell co-culture model with primary cells mimicking the human alveolar epithelial barrier. *Eur J Pharm Biopharm* 77, 398–406.

Lenz, A. G.; Karg, E.; Lentner, B.; Dittrich, V.; Brandenberger, C.; Rothen-Rutishauser, B.; Schulz, H.; Ferron, G. A.; Schmid, O. 2009. A dose-controlled system for air–liquid interface cell exposure and application to zinc oxide nanoparticles. *Part Fibre Toxicol* 6, 32. doi: 10.1186/1743-8977-6-32.

Limbach, L. 2005. Oxide nanoparticle uptake in human lung fibroblasts: Effects of particle size, agglomeration, and diffusion at low concentrations. *Environ Sci Technol* 39, 9370–6.

Lison, D.; Thomassen, L. C.; Rabolli, V.; Gonzalez, L.; Napierska, D.; Seo, J. W.; Kirsch-Volders, M.; Hoet, P.; Kirschhock, C. E.; Martens, J. A. 2008. Nominal and effective dosimetry of silica nanoparticles in cytotoxicity assays. *Toxicol Sci* 104, 155–62.

Lorimer, J. P.; Mason, T. J.; Cuthbert, T. C.; Brookfield, E. A. 1995. Effect of ultrasound on the degradation of aqueous native dextran. *Ultrason Sonochem* 2, 55–9.

Mandzy, N. 2005. Breakage of TiO_2 agglomerates in electrostatically stabilized aqueous dispersions. *Powder Technol* 160, 121–6.

Mason, T.J.; Peters, D. 2003. *Practical Sonochemistry: Power Ultrasound Uses and Applications*. Chichester, UK: Woodhead Publishing.

Mills, N. L.; Donaldson, K.; Hadoke, P. W.; Boon, N. A.; MacNee, W.; Cassee, F. R.; Sandstrom, T.; Blomberg, A.; Newby, D. E. 2009. Adverse cardiovascular effects of air pollution. *Nat Clin Pract Cardiovasc Med* 6, 36–44.

Morrow, P. E. 1988. Possible mechanisms to explain dust overloading of the lungs. *Fundam Appl Toxicol* 10, 369–84.

Muller, L.; Gasser, M.; Raemy, D. O.; Herzog, F.; Brandenberger, C.; Schmid, O.; Gehr, P.; Rothen-Rutishauser, B.; Clift, M. J. D. 2011. Realistic exposure methods for investigating the interaction of nanoparticles with the lung at the air–liquid interface *in vitro*. *Insciences* 1, 30–64.

Murdock, R. C.; Braydich-Stolle, L.; Schrand, A. M.; Schlager, J. J.; Hussain, S. M. 2008. Characterization of nanomaterial dispersion in solution prior to *in vitro* exposure using dynamic light scattering technique. *Toxicol Sci* 101, 239–53.

Naddeo, V.; Belgiorno, V.; Napoli, R. M. A. 2007. Behaviour of natural organic matter during ultrasonic irradiation. *Desalination* 210, 175–82.

National Cancer Institute. *Nanoinformatics*. Accessed at http://nanoinformatics.org/2012/webinar/isa-tab-nano.

National Science and Technology Council Committee on Technology, Subcommittee on Nanoscale Science, Engineering and Technology. 2011. *National Nanotechnology Initiative Strategic Plan*. National Science and Technology Council, Ed.: Washington, DC. Accessed at http://www.nano.gov/sites/default/files/pub_resource/2011_strategic_plan.pdf.

Nel, A. E.; Madler, L.; Velegol, D.; Xia, T.; Hoek, E. M.; Somasundaran, P.; Klaessig, F.; Castranova, V.; Thompson, M. 2009. Understanding biophysicochemical interactions at the nano–bio interface. *Nat Mater* 8, 543–57.

Nel, A.; Xia, T.; Madler, L.; Li, N. 2006. Toxic potential of materials at the nanolevel. *Science* 311, 622–7.

Oberdorster, E.; Oberdorster, J. 2005. Nanotoxicology: An emerging discipline evolving from studies of ultrafine particles. *Environ Health Perspect* 113, 823–39.

Oberdorster, G. 2007. Toxicology of nanoparticles: A historical perspective. *Nanotoxicology* 1, 2–25.

Oberdorster, G. 2012. Nanotoxicology: *In vitro–in vivo* dosimetry. *Environ Health Perspect* 120, A13; author reply A13.

Oberdorster, G.; Ferin, J.; Lehnert, B. E. 1994. Correlation between particle size, *in vivo* particle persistence, and lung injury. *Environ Health Perspect* 102 Suppl 5, 173–9.

OECD. 2010a. *List of Manufactured Nanomaterials and List of Endpoints for Phase One of the OECD Testing Programme*. Paris: OECD. Accessed at http://search.oecd.org/officialdocuments/displaydocumentpdf/?doclanguage=en&cote=env/jm/mono(2008)13.

OECD. 2010b. *Preliminary Guidance Notes on Sample Preparation and Dosimetry for the Safety Testing of Manufactured Nanomaterials*. Paris: OECD. Accessed at http://search.oecd.org/officialdocuments/displaydocumentpdf/?cote=env/jm/mono(2012)40&doclanguage=en.

Puzyn, T.; Rasulev, B.; Gajewicz, A.; Hu, X.; Dasari, T. P.; Michalkova, A.; Hwang, H. M.; Toropov, A.; Leszczynska, D.; Leszczynski, J. 2011. Using nano-QSAR to predict the cytotoxicity of metal oxide nanoparticles. *Nat Nanotechnol* 6, 175–8.

Rallo, R.; France, B.; Liu, R.; Nair, S.; George, S.; Damoiseaux, R.; Giralt, F.; Nel, A.; Bradley, K.; Cohen, Y. 2011. Self-organizing map analysis of toxicity-related cell signaling pathways for metal and metal oxide nanoparticles. *Environ Sci Technol* 15, 1695–702.

Rivera Gil, P.; Oberdorster, G.; Elder, A.; Puntes, V.; Parak, W. J. 2010. Correlating physico–chemical with toxicological properties of nanoparticles: The present and the future. *ACS Nano* 4, 5527–31.

Roco, M. C. 2011. Nanotechnology research directions for societal needs in 2020. *J Nanopart Res* 13, 897–919.

Roebben, G.; Ramirez-Garcia, S; Hackley, V. A.; Roesslein, M.; Klaessig, F.; Kestens, V. et al. 2011. Interlaboratory comparison of size and surface charge measurements on nanoparticles prior to biological impact assessment. *J Nanopart Res* 13, 2675–87.
Rothen-Rutishauser, B. M.; Kiama, S. G.; Gehr, P. 2005. A three-dimensional cellular model of the human respiratory tract to study the interaction with particles. *Am J Respir Cell Mol Biol* 32, 281–9.
Rushton, E. K.; Jiang, J.; Leonard, S. S.; Eberly, S.; Castranova, V.; Biswas, P. et al. 2010. Concept of assessing nanoparticle hazards considering nanoparticle dosemetric and chemical/biological response metrics. *J Toxicol Environ Health A* 73, 445–61.
Savi, M.; Kalberer, M.; Lang, D.; Ryser, M.; Fierz, M.; Gaschen, A.; Ricka, J.; Geiser, M. 2008. A novel exposure system for the efficient and controlled deposition of aerosol particles onto cell cultures. *Environ Sci Technol* 42, 5667–74.
Schmoll, L. H. 2009. Nanoparticle aerosol generation methods from bulk powders for inhalation exposure studies. *Nanotoxicology* 3, 265–75.
Schulze, C.; Kroll, A.; Lehr, C. M.; Schafer, U.; Becker, K.; Schnekenburger, J.; Schulze-Isfort, C.; Landsiedel, R.; Wohllebene, W. 2008. Not ready to use—Overcoming pitfalls when dispersing nanoparticles in physiological media. *Nanotoxicology* 2, 51–61.
Shaw, S. Y.; Westly, E. C.; Pittet, M. J.; Subramanian, A.; Schreiber, S. L.; Weissleder, R. 2008. Perturbational profiling of nanomaterial biologic activity. *Proc Natl Acad Sci U S A* 105, 7387–92.
Song, C.; Wang, P.; Makse, H. A. 2008. A phase diagram for jammed matter. *Nature* 453, 629–32.
Sotiriou, G. A.; Diaz, E.; Long, M. S.; Godleski, J.; Brain, J.; Pratsinis, S. E.; Demokritou, P. 2012. A novel platform for pulmonary and cardiovascular toxicological characterization of inhaled engineered nanomaterials. *Nanotoxicology* 6, 680–90.
Sterling, M. C., Jr.; Bonner, J. S.; Ernest, A. N.; Page, C. A.; Autenrieth, R. L. 2005. Application of fractal flocculation and vertical transport model to aquatic sol–sediment systems. *Water Res* 39, 1818–30.
Taurozzi, J. 2010. Protocol for preparation of nanoparticle dispersions from powdered material using ultrasonic disruption. In: *CEINT/NIST*. Accessed at http://nvlpubs.nist.gov/nistpubs/special publications/nist.sp.1200-2.pdf.
Taurozzi, J. S.; Hackley, V. A.; Wiesner, M. R. 2011. Ultrasonic dispersion of nanoparticles for environmental, health and safety assessment—Issues and recommendations. *Nanotoxicology* 5, 711–29.
Taurozzi, J. S.; Hackley, V. A.; Wiesner, M. R. 2013. A standardised approach for the dispersion of titanium dioxide nanoparticles in biological media. *Nanotoxicology*7, 389–401.
Tedja, R.; Lim, M.; Amal, R.; Marquis, C. 2012. Effects of serum adsorption on cellular uptake profile and consequent impact of titanium dioxide nanoparticles on human lung cell lines. *ACS Nano* 6, 4083–93.
Teeguarden, J. G.; Hinderliter, P. M.; Orr, G.; Thrall, B. D.; Pounds, J. G. 2007. Particokinetics *in vitro*: Dosimetry considerations for *in vitro* nanoparticle toxicity assessments. *Toxicol Sci* 95, 300–12.
The Royal Society and the Royal Academy of Engineering. 2004. *Nanoscience and Nanotechnologies: Opportunities and Uncertainties*. Cardiff, UK: Clyvedon Press. Accessed at http://www.nanotec.org.uk/report/Nano%20report%202004%20fin.pdf.
US Environmental Protection Agency. 2009. *Integrated Science Assessment for Particulate Matter (Final Report)*. US Environmental Protection Agency: Washington, DC. Accessed at http://www.epa.gov/ncea/pdfs/partmatt/Dec2009/PM_ISA_full.pdf.
Verma, A.; Stellacci, F. 2010. Effect of surface properties on nanoparticle–cell interactions. *Small* 6, 12–21.
Wang, J.; Wang, Y.; Gao, J.; Hu, P.; Guan, H.; Zhang, L.; Xu, R.; Chen, X.; Zhang, X. 2009. Investigation on damage of BSA molecules under irradiation of low frequency ultrasound in the presence of FeIII–tartrate complexes. *Ultrason Sonochem* 16, 41–9.
Warheit, D. B.; Sayes, C. M.; Reed, K. L. 2009. Nanoscale and fine zinc oxide particles: Can *in vitro* assays accurately forecast lung hazards following inhalation exposures? *Environ Sci Technol* 43, 7939–45.

Wiogo, H. T.; Lim, M.; Bulmus, V.; Yun, J.; Amal, R. 2011. Stabilization of magnetic iron oxide nanoparticles in biological media by fetal bovine serum (FBS). *Langmuir* 27, 843–50.
Wittmaack, K. 2007. In search of the most relevant parameter for quantifying lung inflammatory response to nanoparticle exposure: Particle number, surface area, or what? *Environ Health Perspect* 115, 187–94.
Wittmaack, K. 2011. Novel dose metric for apparent cytotoxicity effects generated by *in vitro* cell exposure to silica nanoparticles. *Chem Res Toxicol* 24, 150–8.
Wohlleben, W. 2012. Validity range of centrifuges for the regulation of nanomaterials: From classification to as-tested coronas. *J Nanopart Res* 14, 1300. doi: 10.1007/s11051-012-1300-z.
Xia, T.; Kovochich, M.; Liong, M.; Zink, J. I.; Nel, A. E. 2008. Cationic polystyrene nanosphere toxicity depends on cell-specific endocytic and mitochondrial injury pathways. *ACS Nano* 2, 85–96.
Zhang, H.; Ji, Z.; Xia, T.; Meng, H.; Low-Kam, C.; Liu, R. et al. 2012. Use of metal oxide nanoparticle band gap to develop a predictive paradigm for oxidative stress and acute pulmonary inflammation. *ACS Nano* 6, 4349–68.
Zook, J. M.; Maccuspie, R. I.; Locascio, L. E.; Halter, M. D.; Elliott, J. T. 2010. Stable nanoparticle aggregates/agglomerates of different sizes and the effect of their size on hemolytic cytotoxicity. *Nanotoxicology* 5, 517–30.
Zook, J. M.; Rastogi, V.; Maccuspie, R. I.; Keene, A. M.; Fagan, J. 2011. Measuring agglomerate size distribution and dependence of localized surface plasmon resonance absorbance on gold nanoparticle agglomerate size using analytical ultracentrifugation. *ACS Nano* 5, 8070–9.

18

Potential for Nose-to-Brain Delivery of Drugs

Lisbeth Illum

CONTENTS

18.1 Introduction

To treat serious diseases of the central nervous system (CNS) such as Parkinson's, meningitis, and Alzheimer's, it is necessary for the drug to cross the endothelial membrane, in the form of the blood–brain barrier (BBB), separating the systemic circulation from the brain extracellular fluid (BECF). However, owing to the impervious nature of the BBB, which comprises endothelial cells joined by tight junctions, only small and sufficiently lipophilic drugs can cross this barrier, resulting in hydrophilic drugs and large molecular weight drugs, such as peptides, proteins, and antibodies, only reaching negligible levels in the CNS after parenteral application. Hence, it has not been possible to fully exploit many of these therapeutic agents for the treatment of CNS-related diseases.

It has been well documented, mostly in animal models such as rats, but also in sheep and monkeys, that it is possible to transport drugs and other materials (transmucosally) to the CNS directly from the nasal cavity via the olfactory epithelium or via the trigeminal nerve endings and nerves residing in the respiratory epithelium in the nasal cavity (Faber 1937; Cauna et al. 1969; Gopinath et al. 1978; Shipley 1985; Chou and Donovan 1998; Illum 2000; Thorne et al. 2004, 2008; Dhuria et al. 2009; Johnson et al. 2010; Lockhead and

Thorne 2012; Schiöth et al. 2012). For obvious reasons, it has been more difficult to show direct pharmacokinetic evidence in man that drugs can be delivered to a higher degree into the CNS after nasal than after parenteral administration, and only few articles are published to this extent (Born et al. 2002; Merkus 2003; Merkus et al. 2003; Illum 2004). Hence, Born et al. (2002) investigated the nasal administration of three peptides, melanocortin, vasopressin, and insulin, in 36 female and male volunteers and measured cerebral spinal fluid (CSF) and plasma levels as compared with placebo. Although the interpretation of the data from this study should be done with caution, owing to the lack of a parenteral control, it was reported that for melanocortin and insulin, increased levels in the CSF were detected without an increase in plasma levels as compared with placebo, indicating a direct transport from the nasal cavity to the CNS. In comparison, it was found that the plasma levels of vasopressin were increased after nasal application, which was indicative of the transport to the CNS being due to nasal absorption to the systemic circulation, followed by transport over the BBB. As discussed by Illum (2004), Merkus et al. (2003) suggested that no direct transport from the nose to the brain was shown (in terms of CSF levels) in their study, investigating the nasal application of hydroxycobalamin and the more lipophilic drug melatonin in 28 postoperative neurology (ICU) patients. However, when recalculating the CSF ratios for hydroxycobalamin (which gives an indication of whether direct transport from nose to brain has occurred or not) on the basis of individual patient area under the concentration curve (AUC) values rather than the mean AUC values, and also changing zero CSF values to the values for the limit of detection, the results showed a CSF ratio of 1.61, indicating a distinct direct transport from the nose to the brain. This matter is still under discussion between research groups, with diverging beliefs, and more studies may need to be carried out to convince the pharmaceutical society about the merit of nose-to-brain delivery.

However, in general, the amount of drug actually reaching the CNS via the nasal cavity is low (normally well below 1% of the administered dose) and may not result in a therapeutically sufficient amount reaching the CNS. Although it was reported in a review by Lockhead and Thorne (2012) that a pharmacokinetic study (Fliedner et al. 2006) of intranasal [^{125}I]leptin showed that 81.5% was delivered intact to the CNS within 30 min, this is misleading since in reality the "bioavailability" of the drug in the CNS was not disclosed, and the 81.5% related to the percentage of the amount of drug that was intact after having reached the CNS. Hence, it is important to develop nose-to-brain drug delivery systems that enable the enhanced delivery of the drug to the CNS. Such systems may include bioadhesives and absorption enhancers and micro- and nanoparticulate delivery systems. Furthermore, it is also important to consider that although targeting to the olfactory region in the rat nose, where this region constitutes about 50% of the nasal surface area, is easily achievable, it is somewhat more difficult reaching the olfactory region in man where this region constitutes only about 2.5% of the surface area and is placed high up in the nasal cavity above the superior turbinates. It has been demonstrated that the anatomical configurations in the nasal cavity affect the airflow so that only between 15% and 20% of the airflow entering into the nasal cavity reach the olfactory region (Keyhani et al. 1997). Hence, specifically designed nasal delivery devices may be needed to optimize the targeting of the delivery system in this area. Less is known about the relative importance of the trigeminal nerve system in comparison with the olfactory region for transport of drugs from the nose to the brain, and how this can be optimized without increasing the systemic absorption of the drugs.

The use of nanoparticles (NPs) as "nanomedicines," especially for the enhanced delivery of drugs to the body, has been extensively studied in the last decade (Donovan and Huang 1998; Vila et al. 2004; Illum 2007; Haskell and Constantinides 2012). However, it is still unclear whether NPs are of a real benefit for nasal absorption of drugs as compared with the use of a simpler delivery system. Hence, in a previous review, Illum (2007) discussed the scientific evidence for the efficacy of NPs in nasal drug delivery and concluded that NP systems (such as those made from absorption enhancing polymers, e.g., chitosan), did not result in an improved drug efficacy over and above the efficiency found when using solution or powder control systems of the same polymer. It was also noted that in many articles, a proper solution control of the absorption-enhancing polymer was not included, and hence it was difficult to judge the real effect of the NP formulation (Illum 2007; Al-Ghananeem et al. 2010). A range of articles evaluating the benefit of NPs after nasal administration have since appeared in the literature; however, their absolute benefit is still not fully substantiated, except potentially as vaccine delivery systems (Köping-Höggård et al. 2005; Slütter et al. 2010).

In line with this interest, evaluation of the potential use of nanomedicines for delivery of drugs directly from the nasal cavity to the CNS via the olfactory region or the trigeminal nerves in the nasal cavity has also been reported by a range of laboratories (Elder et al. 2006; Illum 2007; Mistry et al. 2009a,b; Liu et al. 2012; Lucchini et al. 2012). The present chapter will discuss the scientific evidence and the potential for the enhanced delivery of drugs from the nasal cavity to the brain using nanoparticulate systems. The chapter will also give some background information in terms of routes of transport from the nasal cavity to the CNS and discuss factors affecting the efficacy of the nanoparticulate systems.

18.2 Nasal Anatomy and Physiology

To more fully understand the potential of an efficient transport of NPs from the nose to the brain, it is important to start this chapter with a description of the most relevant morphological structures and the physiological factors of the nasal cavity most important for this transport function. Additional information can be found in several reviews of this topic (e.g., Mygind 1978; Hilger 1989; Illum 2000, 2004; Mistry et al. 2009a).

18.2.1 Nasal Cavity

The total surface area of the human nasal cavity is about 150 cm^2 for men and slightly smaller for women. The main features of the human nasal cavity are outlined in Figure 18.1. The nasal cavity is divided into identical halves by the nasal septum, and each of these sections can be subdivided into the nasal vestibule, the respiratory region, and the olfactory region. The two cavities open anteriorly to the outer nose by the narrow nasal apertures or the nasal valve with an opening diameter of about 0.3 cm^2. Posteriorly, the nasal cavity leads into the nasopharynx, where the nasal lymphoid tissue is mainly situated, for example, in Waldeyer's ring. In comparison, the surface area of the nasal cavity in rats (most often used as a model for nose-to-brain studies) is about 13 cm^2 (at 16 weeks), of which about 50% constitute the olfactory region situated in the posterior part of the nasal cavity (Figure 18.2, Table 18.1). From Table 18.1, it can be seen that the differences in morphometry of the nasal cavities of rat and man are significant and should be considered when evaluating data.

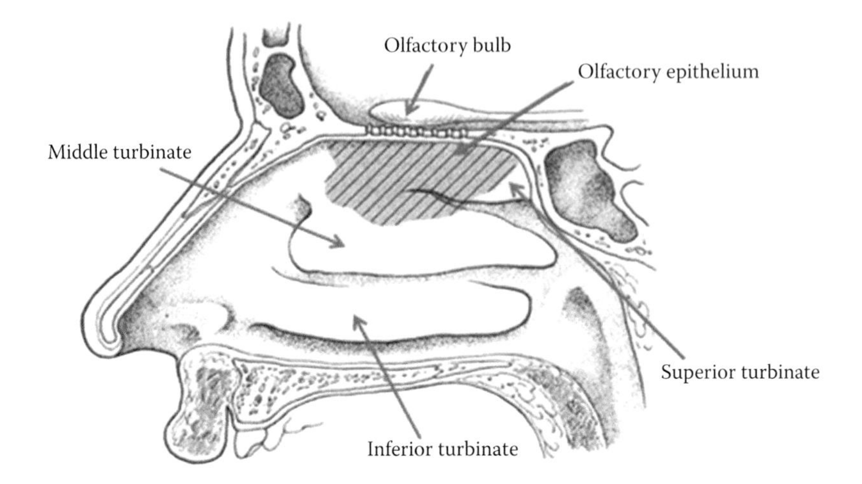

FIGURE 18.1
Human nasal cavity showing the inferior, middle, and superior turbinates and the olfactory region placed high in the nasal cavity. (Adapted from Mygind N and Dahl R, *Adv. Drug Del. Rev.*, 29, 3, 1998.)

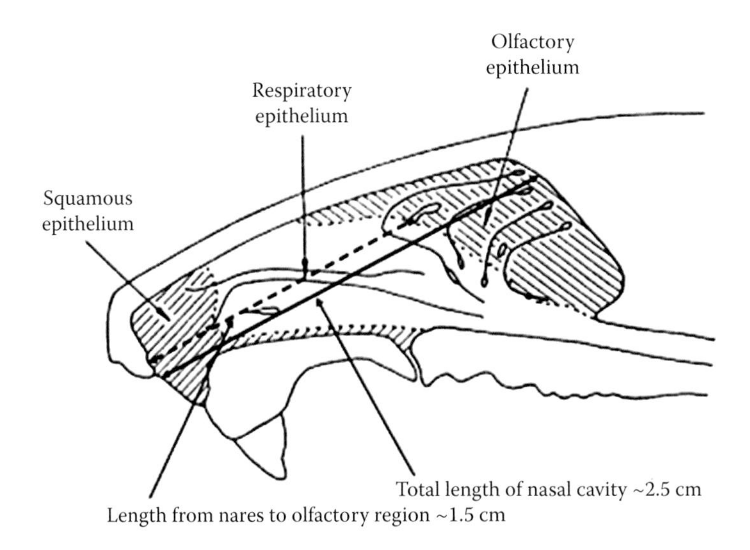

FIGURE 18.2
Rat nasal cavity indicating the position of the squamous, respiratory, and olfactory epithelia. The total length of the nasal cavity and the distance from the naris to the olfactory region are also indicated on the drawing. (From Charlton ST et al., *Int. J. Pharm.*, 338, 94, 2007.)

The anterior part of the nasal cavity (the nasal vestibule, ~0.6 cm^2) is covered with stratified squamous epithelium and in this sense resembles skin. In the roof of the nasal cavity, partly on the nasal septum and partly on the superior and middle conchae (turbinates), one finds the olfactory epithelium, which constitutes about 1.25%–3% of the surface area of the nasal cavity. It is possible (as suggested by N. Jones, personal communication) that the olfactory sensory neurons can reach further into the nasal cavity.

TABLE 18.1
Comparative Nasal Cavity Characteristics of Man, Monkey, and Rat

Species	Nasal Cavity Length (cm)	Nasal Surface Area (cm^2)	Nasal Volume (cm^3)	Surface Area (kg/ body weight)	Surface Area/ Volume	Surface Area Olfactory Region (cm^2)	Turbinate Complexity
Rat (0.25 kg)	2.5	13.4	0.26	53.6	51.5	6.7 (50%)	Complex scroll
Monkey (7 kg)	5.3	61.6	8.0	8.8	7.7	8.0 (12.9%)	Simple scroll
Man (70 kg)	8.0	150.0	15.0	2.2	10.0	3.1 (1.7%)[a]	Simple scroll

Source: Harkema JR, *Environ. Health Perspect.*, 85, 231, 1990; Jafek BW, Olfactory mucosal biopsy and related histology. In: Seiden AM et al. (ed.), *Taste and Smell Disorders*. Thieme, New York, pp. 107–127, 1997; Gross EA et al., *J. Anat.*, 135, 83, 1982.

[a] The estimate of the size of the olfactory epithelium in man vary between references but is generally thought to be between 1.25% and 3% of the nasal surface area.

18.2.2 Nasal Respiratory Epithelium

From the nasal vestibule, the epithelium gradually changes posteriorly into the respiratory epithelium, which comprises pseudostratified columnar epithelium covered in microvilli (Figure 18.3). Owing to the large surface area created by the microvilli, this epithelium is the major site for systemic absorption of drugs administered into the nasal cavity. The

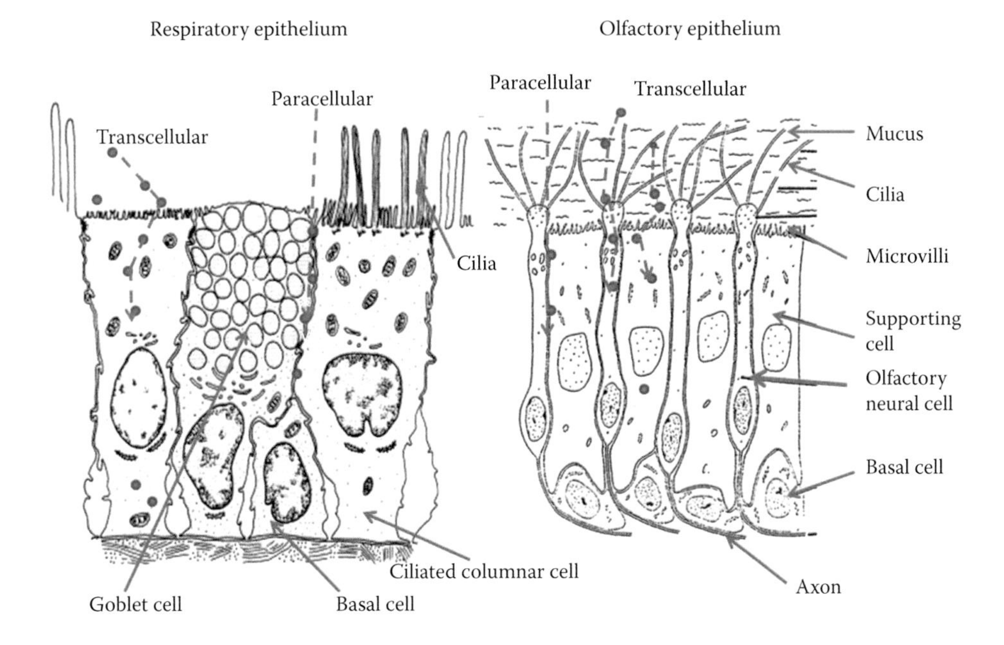

FIGURE 18.3
Respiratory and olfactory epithelia. The respiratory epithelium comprises ciliated or nonciliated pseudostratified columnar epithelial cells, goblet cells providing mucus, and basal cells. The olfactory epithelium comprises olfactory neural cells with long immovable cilia, supporting cells, and basal cells that are able to replace the neural cells. The schematic drawing also shows the paracellular (through the tight junctions) and the transcellular routes of transepithelial passage of NPs. (Adapted from Mygind N and Dahl R, *Adv. Drug Del. Rev.*, 29, 3, 1998; Gold GH, *Ann. Rev. Phys.*, 61, 857, 1999.)

respiratory epithelium is composed of four major cell types, namely the ciliated and the nonciliated columnar cells, the goblet cells, and the basal cells. The mobile cilia project 2–4 μm from the surface of the cells. The respiratory epithelium is coved by mucus (mainly derived from the goblet cells) and comprises two layers, a viscous gel layer (the mucus blanket, 2–4 μm thickness) that floats on the serous fluid layer (the sol layer, 3–5 μm thickness) surrounding the cilia. The mucus blanket is moved mainly posteriorly through the nasal cavity by the mucociliary clearance mechanism of the cilia, covering about 15%–20% of the columnar cells and beating with a frequency of 1000 strokes per minute. Materials including NPs applied to the nasal mucosa, with no particular bioadhesive or interactive function, will hence be transported posteriorly with a speed of 5 mm/min.

18.2.3 Nasal Olfactory Epithelium

The olfactory epithelium is located within recess of the skull, below the cribriform plate of the ethmoid bone, which separates the nasal cavity from the brain (Figure 18.1). The epithelium is situated approximately 7 cm from the nostril, at the top of the nasal cavity, partly lying on the septum and partly on the superior conchae. The olfactory epithelium is not easily accessible by physical means, since any instrument will have to pass through a 1.5-mm crevasse between the closely opposed nasal conchae and septum structures, and is also above the normal airflow path. Hence, odorants normally reach the sensitive receptors by diffusion.

The olfactory epithelium (Figure 18.3) comprises a modified respiratory epithelium, consisting of olfactory sensory neurons, sustentacular (supporting) cells, and basal cells. The supporting cells ensheath the receptor neurons, providing mechanical support and also physiological support, in terms of provision of the necessary extracellular potassium levels needed for neuronal activity. The basal cells can differentiate into neuronal receptor cells and replace these as needed every 40 days. The lamina propria underneath contains nerve fascicles and Bowman's glands that provide the mucus for the epithelium.

The olfactory receptor cells, which are interspaced between the sustentacular cells, are bipolar neurons with a round cell body (Figures 18.3 and 18.4). They originate in the olfactory bulb and terminate at the apical surface of the olfactory epithelium as small knob-like swellings. From these extend numerous, up to 200-μm-long, nonmotile cilia. From the knob, the neurons taper into a nonmyelinated axon that bundle together with other axons in the lamina propria. Here, they are ensheathed by glial cells (Schwann cells), cross via small holes in the cribriform plate, and enter the cranial cavity and synapse in the olfactory bulb. From there, the projections go to the amygdala, the prepyriform cortex, the anterior olfactory nucleus, and the entorhinal cortex, as well as the hippocampus, hypothalamus, and thalamus.

The olfactory epithelium is covered by a thick layer of mucus (mainly excreted by the Bowman's glands), but since the cilia are nonmotile the mucus is not cleared from the region by a mucociliary clearance mechanism. However, the mucus will gradually clear at overproduction and also due to the upright position of a normal human.

18.2.4 Nasal Trigeminal Nerve System

The largest of the cranial nerves is the trigeminal nerve, which provides the sensory perception and participates in motor functions in the nasal cavity (Figure 18.5). The nerve has (on each side of the pons) three major branches, the ophthalmic, the maxillary, and the mandibular nerves, which converge at the trigeminal ganglion, located within Meckel's

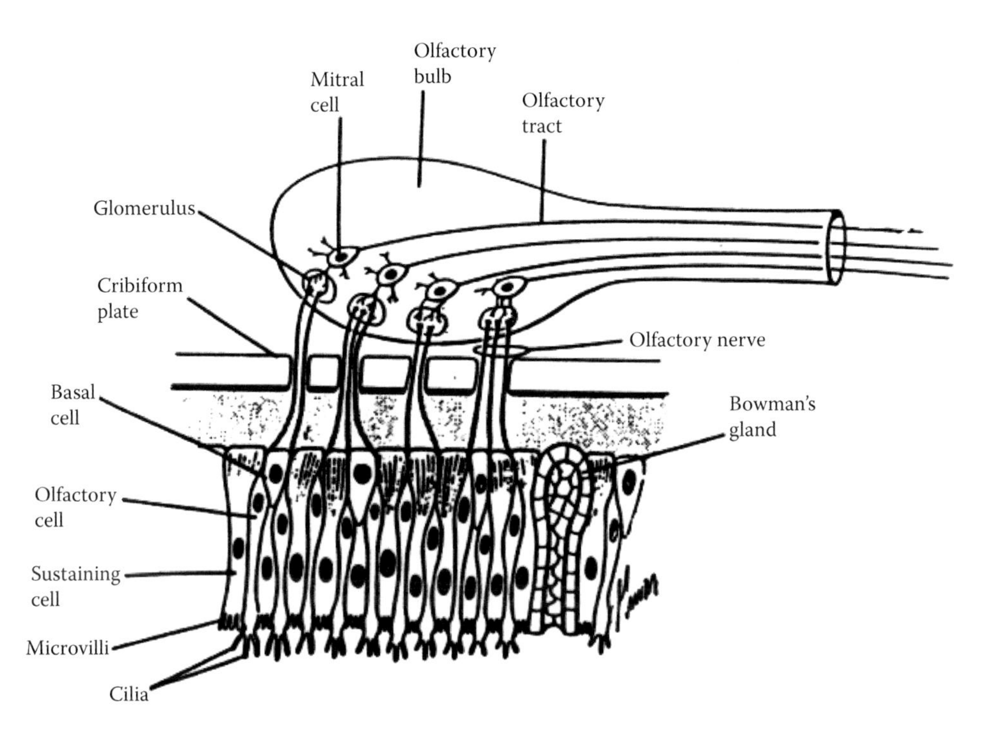

FIGURE 18.4
Schematic drawing illustrating the olfactory epithelium and the olfactory bulb showing the synapses of neural axons in the olfactory bulb and the continuation of the olfactory nerves through the olfactory tract. The axons are shown to pass through the cribriform plate in axon bundles. (Adapted from Illum L, *Eur. J. Pharm. Sci.*, 11, 1, 2000.)

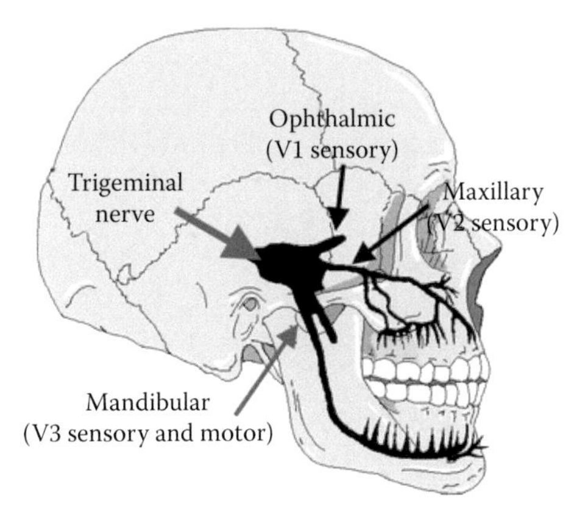

FIGURE 18.5
Schematic drawing showing the three major branches of the trigeminal nerve, the ophthalmic, mandibular, and maxillary nerves, of which the maxillary nerve provides sensory functions to the nasal mucosa.

cave. The trigeminal ganglion contains cell bodies of incoming sensory nerve fibers, and from here a single large sensory root enters the brain stem at the level of the pons. The trigeminal nerve passes out of the cranial cavity through openings at the base of the cranium. The nerve is composed of axons leading to and from several trigeminal nuclei in the midbrain and hindbrain. The bundles of its axons located within the brain are referred to as trigeminal nerve fibers, tracts, and roots.

The ophthalmic and the maxillary nerves only provide sensory functions: the ophthalmic nerve from the scalp, forehead, upper eyelid, the conjunctiva and the cornea of the eye, the nose, the nasal mucosa, the frontal sinuses, and part of the meninges; the maxillary nerve from the lower eyelid and cheek, the nares and upper lip, the upper teeth and gums, the nasal mucosa, the palate and roof of the pharynx, the maxillary, ethmoid, and sphenoid sinuses, and part of the meninges. The mandibular nerve, on the other hand, also participates in motor functions, and carries sensory information from the lower lip, the lower teeth and gums, the chin and jaw, parts of the external ear, and parts of the meninges.

Whereas in mammals the olfactory receptors are found in discrete locations (olfactory region) within the nasal cavity, the trigeminal nerve endings are distributed throughout the nasal cavity and function, in part, as a detection system for irritants and potentially noxious chemicals. These nasal trigeminal chemoreceptors are believed to be intraepithelial free nerve endings arising from Aδ and C fibers of the nasopalatine and ethmoid branches of the trigeminal nerve (Silver 1992). The anterior nasal mucosa (as shown in rat) is innervated by the ethmoid nerve that bifurcates repeatedly, sending nonmyelinated and thinly myelinated fibers with diameters of 0.2–1.5 μm throughout the respiratory system. Similarly, major clusters of nasal trigeminal neurons have been identified in the rat in the near-posterior, central–lateral region, and in an anterior region on both sides relative to the midline of the ganglion (Rothermel et al. 2011).

The morphology of the intraepithelial corpuscular nerve endings in the respiratory mucosa of the dog was investigated by Yamamoto et al. (1998). Complex corpuscular endings (300–500 μm long, 100–250 μm wide) displaying bulbous, laminar, and varicose expansions were found on the dorsal elevated part of the nasal septum and on the nasal concha, showing either single or branched endings. The corpuscular endings were located within the respiratory epithelium, near the lumen of the nasal cavity. Such terminal regions were covered by the cytoplasmic process of ciliated cells and were never exposed to the lumen of the nasal cavity. Similarly, it was found in humans that the nasal respiratory epithelium was supplied by nonmyelinated nerves that approached the mucosa in fascicles comprising up to 200 axons and devoid of perineurial sheaths (Cauna et al. 1969). The fascicles produced one type of receptor organ, a simple terminal arborization either in the cell free zone of the lamina propria or in the spaces between the epithelial cells next to the basement membrane.

Hence, neurons from the ophthalmic and maxillary nerve branches of the trigeminal nerve pass directly though the nasal mucosa, and have a direct transsynaptic connection with the CNS beginning at the entry through the pons and then through the rest of the hindbrain. A portion of the trigeminal nerve also passes through the cribriform plate and hence is connected to the forebrain. Although it is apparent from the description above that the nerve endings are not directly exposed to the lumen of the nasal cavity, the trigeminal nerve system is considered important for the nose-to-brain delivery of drugs, since these nerve endings can be reached by transcellular and extracellular routes through the respiratory epithelium (Lochhead and Thorne 2012).

18.2.5 Epithelial Cell Barrier

The epithelial cells of the respiratory and olfactory epithelia are closely connected by intercellular junctions, narrow belt-like structures that totally surround the cells. These junctional complexes comprise three regions, the zona occludens (the junction nearest the apical surface) also called the "tight junction," the zona adherens, and the macula adherence (nearest the basal surface), and form a dynamic regulatable semipermeable diffusion barrier between the epithelial cells (Madara 2000). The diameter of the tight junctions in the nasal cavity is in the order of 3.9–8.4 Å (Hayashi et al. 1985), and transport across the tight junction is considered limited for molecules with a hydrodynamic radius of >3.9 Å and negligible for molecules with a radius >15 Å (Stevenson et al. 1988). It is difficult to relate these sizes to exact molecular weights because the size of biological molecules will be determined by the physicochemical environment and potentially also secondary and tertiary structures of the molecules. As an example, Shorten et al. (2007) has reported that the size of an insulin monomer molecule (MW 5808 Da) is in the order of 26.8 Å in diameter and peroxidase (MW 44,000 Da) has a structural diameter of 40 Å. The use of absorption enhancers in the nasal cavity can open the tight junctions; however, it is unlikely that such tight junction permeation enhancers would be able to increase the diameter more than 10–15 times. It was reported by Costantino et al. (2007) that the maximal diameter reached when using absorption enhancers in the nasal tissue was in the order of 15 nm in diameter.

18.3 Transport Routes and Mechanisms

Following nasal administration, it has been shown that a drug can reach the brain via three main routes: (i) absorption through the nasal respiratory epithelium into the systemic circulation followed by transport across the BBB ("systemic pathway"), (ii) direct transport paracellularly or transcellularly via the olfactory epithelium ("olfactory epithelial pathway") or the olfactory neurons ("olfactory neural pathway") connecting the nasal passages into (mostly) the rostral brain area, and (iii) transport via the trigeminal nerves ("trigeminal pathway") connecting to mainly the caudal brain area (Illum 2000; Thorne et al. 2004).

The different pathways that a drug or other materials can follow when transported via the olfactory region to the CNS has been thoroughly described by different authors (Dahlin 2000; Illum 2000; Thorne and Frey 2001; Thorne et al. 2004), and hence only the minimal necessary details will be given here. The systemic pathway is only relevant for small lipophilic drugs that can be readily absorbed across the nasal epithelium into the systemic circulation and then due to their lipophilicity cross the BBB into the brain tissue.

Using the olfactory neural pathway, the drug or particulate material can be internalized into the primary neurons of the olfactory epithelium by mechanisms of endocytosis or pinocytosis, and transported by intracellular axonal transport along the neuron, and via nerve bundles, transverse the cribriform plate and reach directly into the highly organized olfactory bulb, located in the forebrain on the other side of the cribriform plate (Figure 18.4). From the first-order synapse in the olfactory bulb, several dendrites are emitted further into the CNS, and material can hence be carried further along. The existence and use of the olfactory neural pathway has been described both for drugs, enzymes, and for particles by several authors, for example, IGF-1 (Thorne et al. 2004), wheat germ agglutinin (WGA) horseradish peroxidase (Itaya 1987; Thorne et al. 1995), gold particles (De Lorenzo 1960; Gopinath et al. 1978), WGA-conjugated polyethylene glycol–poly-L-lactic acid (PEG–PLA) NPs (Liu et al. 2012), manganese oxide (Elder et al. 2006; Lucchini et al. 2012), and aluminum lactate (Perl and Good 1987). It has been shown that the olfactory neural pathway is slow, with a transport time of up to 24 h for the drug to reach the CNS (Kristensson and Olsson 1971).

In comparison with the olfactory neural pathway, the olfactory epithelial pathway is able to transport drugs or other material much faster, in terms of a few minutes, from the nasal cavity to the CSF and the brain tissue as shown, for example, for dihydroergotamine (Wang et al. 1998), lidocaine (Chou and Donovan 1998), cocaine (Chow et al. 1999), and cefalexin (Sakane et al. 1991). However, only a few articles have shown the transport of NPs via the olfactory epithelial pathway (Garzotto and De Marchis 2010; Liu et al. 2012), and then in general the NPs in question were of a size about or below 20 nm, enabling them to penetrate the tight junctions as discussed previously.

The extracellular olfactory epithelial pathway transports material through tight junctions between the olfactory neural cells and supporting cells into the CSF and potentially further into the brain tissue, and relies on a direct connection between the submucosa and the subarachnoid extensions (the perineural space surrounding the olfactory nerves), as the nerve bundles penetrate the cribriform space. The way materials enter into the perineural space is not totally clear but is either through a loosely adherent epithelium surrounding the nerve or if the epithelium is closely adherent, then through the tight junctions (Jackson et al. 1979). It should be noted that it is also possible for drugs and particulate materials to pass through the olfactory supporting epithelial cells by a transcellular route, and to reach the perineural space surrounding the axons, and then reach the CSF, olfactory bulb, and further into the brain.

Since most studies on nose-to-brain delivery are short term, that is, <4 h duration, it is not always possible to deduce from such published studies whether for some drugs and materials both the neural pathway and the olfactory epithelial pathway are exploited.

Although the trigeminal nerve endings are not exposed in the nasal cavity, drugs and other materials should be able to reach the nerve endings and the perineural space surrounding the nerves dispersed in the epithelium. It is not totally known at which site in the nasal cavity materials enter the nerve; however, it is likely both from the ophthalmic and maxillary nerves. The trigeminal nerve pathway transport drugs and other materials to the brainstem beginning at the entry through the pons and further through the rest of the hindbrain. Since a portion of the trigeminal nerve passes through the cribriform plate, the trigeminal nerve pathway can also contribute to the concentration of drug and other materials in the forebrain. There are, as far as known, only very few published articles on transport of NPs via the trigeminal nerve pathway; from data in these articles, it is likely that this trigeminal nerve pathway is feasible in line with transport of NPs through the olfactory nerves.

It is possible that all three transport mechanisms, the olfactory epithelial pathway, the olfactory neural pathway, and the trigeminal nerve pathway, could be exploited simultaneously or selectively after application of drugs and other materials in the nasal cavity.

18.4 Uptake of NPs in Epithelial or Neural Cells

As discussed previously, NPs can either pass epithelial membranes transcellularly or paracellularly. For paracellular transport, the NPs will need to be about ≤20 nm to have the ability to pass through the epithelial tight junctions. In general, NPs used for drug transport are larger, in the order of 50–200 nm, and would be expected to be taken up into the cell by processes of endocytosis, that is, macropinocytosis (either clathrin or caveolin mediated, or clathrin or caveolin independent), or by phagocytosis. In macropinocytosis, sealing of the aperture into discrete vacuoles forms the macropinosomes (0.5–5 μm), which effectively take up considerable volumes of suspended NPs into the cell. Receptor-mediated endocytosis is a process where material is initiating the endocytic process by stimulation/complementing a receptor on the cell membrane. This process can be either clathrin dependent or independent. In the former, clathrin, endogenously expressed in the cell as a heavy and a light chain, fuse and form a triskelion structure that produces a clathrin coat (Edeling et al. 2006). The clathrin coat gives rise to coated pit regions on the cell surface during the initial stages of the endocytosis process, and invagination of these pits results in clathrin-coated vesicles that are taken up into the cell cytosol. These vesicles are <150 nm in diameter. Phagocytosis is an uptake of exogenous material by specialized phagocytic cells, e.g., macrophages, where large patches of membrane are internalized with the exogenous material and the material transported into lysosomes for destruction or antigen presentation. Since phagocytes are not common in the nasal cavity, this type of uptake mechanism does not contribute significantly to cellular uptake of NPs in the nasal cavity. However, M-like cells (part of the nasal-associated lymphoid tissue [NALT]) are present in the nasal tissue and have been shown to take up particles and transport them further into the tissue, the lymph nodes, and the blood vessels (Des Rieux et al. 2005).

The cellular internalization of NPs is thought to be dependent on particle diameter (Rejman et al. 2004), NP concentration (Jones 2008), and surface charge and other surface characteristics (Harush-Frenkel et al. 2008).

Alternatively, as discussed previously, the NPs are thought to be able to be transported from the nasal cavity via the neurons, either the olfactory neurons or the trigeminal neurons. It is thought that the initial process is an endocytotic or pinocytotic process whereby the particle is taken into the cell and the further transport is possibly by a mechanism related to microtubules (De Lorenzo 1960). However, the exact mechanism of this uptake and the further transport process has not been described in detail (see Section 18.5).

18.5 Transport of Viral and Elemental NPs from Nose to Brain

Very early studies investigated the transport of 30-nm poliovirus from the nasal cavity into the brain (Howe and Bodian 1940; Bodian and Howe 1941). They showed that these virus particles were transported via the olfactory nerves into the olfactory bulb, with a transport rate of 2.4 mm/h. A range of studies has thus far examined the direct transport of elemental NPs from the nasal cavity to the CNS. Early studies made use of basic particulate materials such as colloidal gold (De Lorenzo 1960; Czerniawska 1970; Gopinath et al. 1978), carbon-13 (Oberdörster et al. 2004), silica (Hunter et al. 1998), iridium-192 (Semmler et al. 2004), manganese oxide (Elder et al. 2006), iron (II) oxide (Wang et al. 2007), titanium dioxide (Wang et al. 2008), and quantum dots (Garzotto and De Marchis 2010).

Hence, De Lorenzo (1960) administered 50-nm gold particles to the nasal cavity of squirrel monkeys and visualized the particles inside the olfactory neurons using TEM. After 30–60 min, the particles had traveled to the olfactory bulb and the rate of transport was calculated to 2.5 mm/h. This was confirmed in a study in rabbits where colloidal gold was injected in the olfactory submucosa and seen to reach the CSF within 1–2 h (Czerniawska 1970). It was later shown by Gopinath et al. (1978) that not only did colloidal gold particles enter and were transported by the neural cells, they were also taken up by the supporting cells but not found in tight junctions of the epithelial cells.

Hunter et al. (1998) exposed rats to inhalation of fresh silica (~1000 nm) for 6 months. Following exposure, the trigeminal ganglia was removed, and it was shown by means of substance P (SP) immunocytochemistry and preprotachykinin (PPT) autoradiography that the SP immunoreactivity and PPT mRNA expression in trigeminal ganglion neurons were significantly increased after the silica inhalation. This suggested that silica was taken up by the trigeminal nerves and activated the sensory pathways. Oberdörster et al. (2004) later exposed rats twice to insoluble ^{13}C NPs (~35 nm) in an inhalation chamber for 6 h and found an increase in ^{13}C in the olfactory bulb between days 1 and 7 (0.35–0.43 μg/g). However, the authors could not locate any particles deeper in the brain (i.e., in the cerebellum and cerebrum), which suggested that the NPs did not move deeper into the brain. It should be noted that from the article, it can be calculated that the total amount of particles transported from the nasal cavity to the olfactory bulb in 6 days was in the order of 3%. Similar studies were carried out by Feikert et al. (2004), who found a 3.5-fold increase in manganese oxide (30 nm) in the olfactory bulb after nasal application.

Elder et al. (2006) showed after 6 and 12 days that rats inhaling poorly soluble salts of manganese oxide (30 nm diameter), for 6 h a day, 5 days a week, contained manganese particles in the olfactory bulb and also in deeper brain structures such as the cortex and cerebellum. The particles were detected by means of graphite absorption spectroscopy; however, unfortunately, no images of the particles in the brain were shown in the article, and hence it is possible that the particles could have been converted to soluble salts and transported in solution.

Wang et al. (2007) administered 40 mg/kg iron oxide NPs (~280 nm) nasally to mice and found that these particles had penetrated as far as to the hippocampus. Fourteen days after the exposure, the olfactory bulb and the brain stem contained a larger amount of NPs than of the control. Unfortunately, no intravenous control was added to the study, and hence it cannot be excluded that the particles may have reached the brain via the systemic pathway, although this is unlikely. Later, similar studies were carried out in mice with titanium dioxide NPs of about 70 nm in diameter (Wang et al. 2008a). It was found that these particles likely passed to the hippocampus via the olfactory neurons.

Garzotto and De Marchis (2010) using quantum dots found that the olfactory neural pathway was not preferentially involved in the uptake of these fluorescent NPs (size 2–10 nm), hence suggesting that the particles could also cross the olfactory epithelium through extracellular pathways (i.e., the olfactory epithelial pathway). Quantum dots are tiny particles, or "NPs," of a semiconductor material, traditionally chalcogenides (selenides or sulfides) of metals like cadmium or zinc (e.g., CdSe or ZnS), which range from 2 to 10 nm in diameter (about the width of 50 atoms).

18.6 Transport of Drugs with Nanoparticulate Systems from Nose to Brain

The ability of nasally applied NPs to appear in the bloodstream in animal models after crossing the nasal mucosa has previously been reviewed by Illum (2007) and will only briefly be mentioned here. Also, *in vitro* studies in cell cultures have evaluated the potential for

NPs to cross the olfactory epithelium, and these studies also includes an evaluation of the effect of surface characteristics of the NPs. Furthermore, a range of studies have been carried out with different types of nanosystems for improved delivery of drugs from the nose to the brain. These studies will be discussed here.

18.6.1 Transport of NPs

18.6.1.1 Cell Cultures and Respiratory Epithelium

A study was carried out by Huang and Donovan (1996) evaluating the transport of fluorescently labeled polystyrene particles across excised nasal tissue. The particles were of sizes between 10 and 500 nm, and with different surface groups giving negative and positive and more or less hydrophobic surface characteristics. It was found that the smallest particles (10 nm, carboxylated, hydrophilic) were transported via the tight junctions, whereas the other larger particles independent of charge and hydrophilicity crossed the membrane transcellularly. The amount of particles transported was for all particles very low (between 0.011% and 0.022%). Des Rieux et al. (2005) later investigated the transport of carboxylated and aminated polystyrene particles of 200 and 500 nm in diameter across monolayers of Caco-2 cells and cocultures of Caco-2 cells and human Raji B lymphocytes comprising M-like cells. In general, low amounts of particles (about 100 NPs per insert for Caco-2 cells and 10,000 NPs per insert for the human Raji B lymphocytes during 90 min) were transported across both of the cell layers and were dependent on the size, the surface characteristics, and the type of cell monolayer used. The smaller the particle, the higher the transport; carboxylated particles were transported to a lesser degree than aminated particles and due to the presence of the M-cells the transport was significantly higher for the cocultures than for the simple Caco-2 cell monolayers.

As opposed to the studies by Des Rieux et al. (2005), Ma and Lim (2003) found that chitosan NPs (500 nm in diameter), although they were internalized into the Caco-2 cell monolayers, were not transported across the monolayer within the time frame of the study (4 h). In support of this result, when insulin was encapsulated into the NPs, no insulin was found to be transported across the monolayer, although the insulin was detected in the cell layer. A similar study was performed by Sadeghi et al. (2008), who found that insulin encapsulated in chitosan or derivatized NPs (200 nm in diameter) was transported to a lower degree across Caco-2 cell monolayers as compared with the insulin administered with the chitosan or derivatized chitosan in solution.

Behrens et al. (2002) investigated the uptake of polystyrene, chitosan, and PLA–PEG NPs (200–300 nm in diameter) by Caco-2 cell monolayers and the mucus-producing HT29-MTX cell line. In Caco-2 cells, the hydrophobic polystryrene particles showed the highest uptake, followed by the chitosan NPs and then the hydrophilic and sterically stabilized PLA–PEG NPs, which showed only negligible association with the cells. However, for the HT29-MTX cells, the chitosan NPs showed the highest uptake followed by the polystyrene NPs. Again the uptake of the PLA–PEG NPs was negligible. These results are in line with the studies of Brooking et al. (2001) discussed in the following, where PEGylation decreased the transport of particles across the nasal mucosa, and work by Hillary and Florence (1996). It is suggested that the low uptake of PEGylated NPs is due to the surface-exposed PEG, preventing interaction with the cell surface and potentially also the entanglement between mucus and the PEG chains. It has been shown that the ability for PEGylated particles to penetrate mucus was highly dependent on the density and chain length of the PEG on the surface of the NPs (Wang et al. 2008b).

In an *in vivo* study, Almeida et al. (1993) found that 0.96% of carboxylated polystyrene particles of 510 nm in diameter and fluorescently labeled, administered to the nasal cavity of rabbits, was able to cross the nasal mucosa and reach the systemic circulation. Similarly for 830-nm particles, 1.0%–2.9% was found to cross. No discussion was forthcoming as to how these particles were able to cross the mucosa. It was later found by Brooking et al. (2001) in the rat model that sulfated polystyrene particles of diameters from 20 to 1000 nm were transported across the nasal mucosa to the bloodstream in rats, with the transport decreasing with increasing diameter of the particles. For the smallest 20-nm particle, the transport during 3 h was 3.25% and for the 1000 nm particles 1.25%, with the numbers for the 100 and 500 nm being 2.0% and 1.25%, respectively. Owing to the hydrophobicity of the particles, they most likely were transported via the transcellular route. It was also suggested that the particles could have been taken up by the antigen sampling M-like cells in the nasal cavity and in this way transported into blood. Negligible amounts of the 100-nm particles were found in the brain homogenate 3 h after application. It was further shown in the same study that coating of the 100-nm particles with poly-L-lysine, poloxamine 908, and chitosan resulted in no significant change in transport, a significant decrease in transport (from 2.00% to 0.75%), and a significant increase in transport (from 2.0% to 3.25%), respectively. In support of these results, in a later study by Amidi et al. (2006), TMC–chitosan NPs loaded with FITC–albumin and administered nasally to rats were found to be taken up by the epithelial cells and trafficked into the cytoplasm throughout the epithelial cell layer, as shown by confocal microscopy.

The above studies were all carried out either with cell culture models, excised tissue, or in live animals, and focused on the evaluation of models of the respiratory epithelium or of the respiratory epithelium itself. It is important to consider in a simple model whether NPs are able to be transported through epithelial cell membranes and hence potentially further into the tissue, although it is evident that there will be a difference in results found in cell cultures to results found in live animals, where in the latter uptake could to some degree be attributed to interaction of the NPs with the NALT (Liu et al. 2012).

18.6.1.2 Olfactory Epithelium and Trigeminal Nerves

Mistry et al. (2008, 2009) evaluated the transport of 20, 100, and 200 nm polystyrene NPs in *in vitro* studies using excised olfactory epithelial tissue using a Franz diffusion cell and *in vivo* in a murine animal model. Hence, polystyrene particles were either used noncoated or surface coated with chitosan at pH 6.0 or pH 4.5, and Polysorbate 80. No transport across the olfactory epithelium was detected during the length of the study (90 min) for noncoated NPs of any size. However, it was found that 8%–12% of the NPs were either associated with the epithelial surface or taken up into the cells. No significant size effect was found for this interaction. When the NPs were coated with chitosan, the interaction with the epithelium or uptake into the same increased significantly to about 40% of the NPs at the pH of 4.5 where the chitosan is highly protonated. However, no NPs were detected in the receiver chamber of the Franz diffusion cell. It was shown using confocal microscopy that the chitosan-coated NPs were highly associated with the olfactory mucosa, that is, the mucus and the sialic acid residues on the epithelial surface, but also that some NPs were transported into the epithelium with a larger part of the 100-nm chitosan-coated NPs entering into the epithelium than that of the 200-nm NPs.

The *in vivo* studies in mice evaluated the uptake of 100- and 200-nm polystyrene particles, noncoated or coated with chitosan or Polysorbate 80 into the olfactory epithelium and in the olfactory bulb after nasal administration (Mistry et al. 2009b). A total of 4.55×10^{10} NPs in a 15 μl test formulation were administered nasally every 24 h for 3 days. All NPs were found to some degree to be transported across the olfactory membrane into the tissue, using a transcellular pathway, entering the apical cells by a transcytotic mechanism. None of the NPs showed preference for axonal transport over and above transcellular transport in olfactory supporting cells, and hence NPs were found both in the olfactory neural cells and in the sustentacular cells. Coating of the NPs caused adherence to the extracellular mucus present on the olfactory epithelium and the respiratory epithelium and significantly reduced the transport of particles into the olfactory epithelium. Coating with Polysorbate 80 had no significant effect on the degree of transport. It was also shown that a greater number of 100-nm NPs were taken into the cells as compared with the 200-nm NPs both for Polysorbate 80–coated and noncoated NPs. No NPs of any kind used in the study were found in the olfactory bulb at the time of sacrifice at day 4 after the first nasal dose.

In an earlier study, Hunter et al. (1998) administered rhodamine-labeled latex NPs (20–200 nm) in one nostril of rats and after 7, 10, and 14 days removed both trigeminal ganglia. It was demonstrated that the NPs translocated to the ganglia via the ophthalmic and maxillary branches of the trigeminal nerve. Of interest was also that the NPs were localized only in the epithelial layer of the nasal mucosa and did not enter the lamina propria. Hence, the transport of the NPs was confined to trigeminal nerve fibers present in the epithelial cell layer.

In a very recent study, Musumeci et al. (2014) investigated the uptake of different types of rhodamine-loaded NPs (poly-D,L-lactide-co-glycolide [PLGA], PLA, and chitosan) and correlated the uptake into olfactory ensheathing cells using confocal microscopy. They found that the uptake was time dependent and influenced by the charge of the NPs, with PLGA NP showing a higher increase in uptake compared with PLA NP and chitosan NP 1 h after administration.

Gao et al. (2006) investigated the effect of conjugating the lectin WGA to the surface of PLA–PEG NPs (70–80 nm in diameter) on the transport of these NPs from the nasal cavity to the CNS in rats, by incorporating the fluorescent marker 6-coumarin in the NPs. WGA binds specifically to *N*-acetylglucosamine and sialic acid and shows a greater binding affinity to the olfactory than to the respiratory epithelium. The brain uptake (olfactory bulb, olfactory tract, cerebrum, and cerebellum) in terms of $AUC_{0-24\ h}$ for the WGA–NPs were found to be about 2-fold higher than that found for the nonfunctionalized NPs. It was later reported by Shen et al. (2011) that the density of the WGA on the surface of the PLA–PEG particles is important for both cellular uptake by receptor-mediated endocytosis and cytotoxicity to cells. The same group later investigated the nose-to-brain transport pathway for the WGA–NPs labeled with ^{125}I and containing 6-coumarin (Liu et al. 2012). They found that the WGA–NPs were taken up by the transcellular route in the olfactory epithelium, both by the olfactory neural cells and by the sustentacular cells. The transport was very fast with the first NPs found in the olfactory bulb after 5 min and the concentration increasing with time. It was shown that NPs were transported both inside the axons and also by moving in the surrounding continuous perineural channels generated by the ensheathing cells enveloping the axons, which could explain the rapid transport time. In addition, WGA–PLA–PEG NPs were also detected in the cervical spinal cord and medulla, demonstrating transport by the trigeminal nerves. No NPs were found in the CSF.

Gao et al. (2007a) also investigated the nose-to-brain delivery in rats of PLA–PEG NPs (~111 nm) with a surface-conjugated targeting ligand in the form of the lectin *Ulex europeus* agglutinin I (UEA I). This lectin specifically binds to L-fucose, which is mainly located in the olfactory epithelium. The presence of the particles was qualified by the incorporation of 6-coumarin in the NPs. *In vivo* results in rats showed that the lectin-coated NPs were transported to different brain tissues to a higher extent ($AUC_{0-24\,h}$, 1.7 times larger) than was the case for the unmodified NPs. It was also found that the UEA I–NPs had a higher brain targeting efficiency (expressed as DTE) as compared with the WGA–NPs discussed previously. The same group also investigated the use of the lectin odorranalectin (OL) for conjugation onto the PLA–PEG NPs and the effect on nose-to-brain transport in rats (Wen et al. 2011a). The OL is much smaller than other lectins with less immunogenicity. The brain-targeting index was found to double for the OL–PLA–PEG NPs as compared with the PLA–PEG NPs. Similarly, the group also evaluated the use of low molecular weight protamine for conjugation to PLA–PEG NPs for improved transport to the brain from the nasal cavity in rats (Xia et al. 2011). They found that this cell-penetrating peptide enabled an increased transport (2–3-fold) of the functionalized as compared with the nonfunctionalized NPs. A recent study by the same group evaluated *Solanum tuberosum* lectin-conjugated PLGA NPs and found a similar improved brain uptake after administration to the nasal cavity of rats as compared with nonconjugated PLGA NPs (Chen et al. 2012).

The studies discussed here have demonstrated that NPs can be translocated from the nasal cavity to the brain tissues using transport routes such as the olfactory neural pathway, the olfactory epithelial pathway, and also the trigeminal pathway, and that the amount of NPs targeted to the brain can be increased by functionalizing the NPs with lectins that have specific receptors in the olfactory region. It should also be noted that the studies discussed here have indicated that the most common means of uptake of the NPs into the epithelial tissue is by transcellular uptake. Only NPs <20 nm in diameter could be considered small enough to pass through the tight junctions. Only few studies give information concerning the percentage of the dose of NPs reaching the brain. It should be noted that the surface characteristics of the NPs are important for the transport from the nose to the brain. As discussed previously, a chitosan coating appeared to halt the transport of particles owing to interaction with the mucus, whereas a surface labeling with targeting ligands highly improved the transport.

18.6.2 Transport of Drugs in NPs

Attempts to improve the transport of drugs from the nasal cavity to the CNS have involved a range of delivery systems that can be considered a particulate carrier system, for example, micellar nanocarriers, nanoemulsions, solid lipid NPs, and a range a of different NPs made from materials such as chitosan, PLA, and PLGA.

Jain et al. (2010) encapsulated zolmitriptan in micellar nanocarriers produced from a mixture of Transcutol P®, benzyl alcohol, Pluronic® F127, PEG-400, and vitamin E–D-α-tocopheryl PEG succinate. The NP size was 24 nm, and for the rat studies the zolmitriptan was labeled with ^{99m}Tc. The authors found that after nasal administration, the encapsulated zolmitriptan accumulated in the brain at a 5-fold higher amount (5%–6% of dose administered) than was seen for the zolmitriptan nasal solution and the IV injection (both 1% of administered dose). It was also found that the amount of drug was highest in the olfactory bulb and then decreased through the frontal cortex, olfactory tubercle, hippocampus, and diencephalon, whereas the amount in the midbrain, cerebellum, and pons increased to the level of the frontal cortex. This indicates that the micellar carrier and/or the drug was transported via the olfactory epithelium and also via the trigeminal nerve.

Abdelbary et al. (2013) produced nanomicelles from a mixture of Pluronic L121 and P123, and loaded these with olanzapine using a thin-film hydration method. The olanzapine–nanomicelles were found to a provide significantly higher brain/blood ratio, drug-targeting index and drug-targeting efficiency, and direct nose-to-brain transport percentage than a simple olanzapine solution after nasal administration to rats.

Zhang et al. (2004) investigated the effect of encapsulating nimodipine in microemulsions (26 nm in diameter) on the brain targeting of the drug after nasal administration in rats compared with an IV injection of the drug. The data showed, not surprisingly, that the uptake into the brain tissue (especially the olfactory bulb, but also in the cerebrum, cerebellum, and CSF) was significantly higher after nasal than after parenteral administration. It is unfortunate that no intranasal control (e.g., in the form of a simple drug solution) was used, to be able to judge the effect of the (particulate) emulsion system.

Kumar et al. (2008a) produced a nanoemulsion (risperidone [RSP]) and a mucoadhesive nanoemulsion (RMNE), both containing RSP. The RSP was mixed with chitosan to produce the bioadhesive coating. The emulsion droplets were about 16 nm in diameter. The radiolabeled (^{99m}Tc) formulations were administered nasally to rats, and the biodistribution was compared with an IV injection of RSP. The study showed that RMNE resulted in the highest levels of drug/label in the brain as demonstrated by the higher ratio between blood/brain counts, the DTE%, and the drug targeting percentage (DTP). There was, however, no significant difference between the IV injection and the RSP formulation. Gamma scintigraphy studies supported this view. The same group produced similar nanoemulsions for olanzapine, performed identical studies, and confirmed the increased brain uptake for the mucoadhesive emulsion system (Kumar et al. 2008b).

It was shown by Zhao et al. (2014) that gelatin nanostructured lipid carriers loaded with fibroblast growth factor (bFGF) (143 nm) enhanced the levels of the drug in the olfactory bulb and in the striatum as compared with normal gelatin NPs (91 nm) and a solution of bFGF. Furthermore, the lipid nanostructures also showed obvious therapeutic effects on hemiparkinsonian rats.

Gamma scintigraphy images in rabbits after intranasal administration of solid lipid NPs of ondansetron HCl (320–498 nm), as carried out by Joshi et al. (2012), showed rapid localization of the drug in the brain. It was not clear whether the drug was still encapsulated in the NPs.

The nasal administration in mice of maltodextrin NPs (~60 nm) (Biovector) in conjunction with [^{3}H]morphine resulted in an increased (doubled) morphine antinociception in mice as compared with a nasal administration of [^{3}H]morphine solution (Betbeder et al. 2000). Since a subcutaneous dose of morphine produced higher blood levels than for the nasal doses, but lower analgesia, the increase in analgesic effect could not be due to transport via the BBB. Furthermore, there were no significant difference between the plasma levels for the nasal solution and the NP formulation. The authors suggested that the nasal application of the NP formulation had resulted in direct nose-to-brain transport. It was interesting to note that the morphine did not (at least *in vitro*) interact with the NPs, although if the NPs and the morphine were administered separated by 2 min, the NPs resulted in no increased effect.

Another NP system used for transport of RSP from the nasal cavity to the CNS was solid lipid NPs (RSLNs) made by a solvent diffusion–solvent evaporation method using Compritol 888 ATO as the lipid and Pluronic F127 as the surfactant (Patel et al. 2011). The particles (~150 nm in diameter) with encapsulated RSP were administered nasally to mice, and the pharmacodynamic and pharmacokinetic data were evaluated and compared with the data from IV injections of RSP and RSLNs. It was shown that the brain/blood ratio for the RSLNs was 1.36 as compared with 0.17 and 0.78 for the RSP and the RSLNs injected IV, indicating a clear direct nose-to-brain transport of the drug after nasal administration. These results were consistent with the higher hind-limb retraction time found for the nasal application. It should be noted that the gamma scintigraphy images shown in the article for RSLNs given nasally and intravenously cannot be compared because of the activity (color) in the brain area being relative to the much stronger signal in the liver for the parenteral administration. Hence, no conclusions can be drawn from these data. Also, unfortunately, there was no nasal control solution included in the studies and, therefore, the true effect of the RSLNs cannot be deduced from the study.

A study by Ruan et al. (2011) encapsulated neurotoxin-1 (NT) in ethyl acetate NPs (particle size not given) coated with Polysorbate 80 and administered these to the nasal cavity of mice, using a nasal solution of NT and an IV injection of NT as controls. It was found that as compared with the brain concentration of nasal control solution (6.26 ng/ml), the brain concentration after application of the NPs was 3-fold higher (18.23 ng/ml). At the same time, the concentration after the IV injection was 8.56 ng/ml. Although the brain concentrations of the drug were similar for the IV and nasal controls, the nasal NT solution was not able to fully inhibit constriction in acetic acid–induced writhing reactions, which were surprising since at the same time the concentration of drug in the blood was much higher for the nasal control compared with the nasal NP formulation. The inhibition after NP administration matched the effect of the IV injection of NT.

Zhang et al. (2006) evaluated the encapsulation of nimodipine in MPEG–PLA NPs (~76 nm in diameter) and the resultant biodistribution after nasal administration in rats as compared with the nasal administration of a nimodipine solution formulation and an IV injection of the drug. They found that compared with the controls, the NPs significantly increased the levels of drug in the CSF, whereas for the other brain tissues investigated, the levels were higher for the nasal solution formulation or no difference in levels was seen. It would be expected that nimodipine would be readily absorbed across the nasal membrane into the systemic circulation and also be able to cross the BBB. This is confirmed by the high bioavailability of 67% for the drug (into the systemic circulation) after nasal administration. Hence, the solution formulation would feed to the brain both by direct transfer from the nasal cavity and by the systemic route.

Wang et al. (2008a) prepared chitosan NPs containing estradiol (E_2) (~270 nm in diameter) and investigated the blood and CSF levels after nasal administration to rats. The CSF levels achieved after nasal administration were about double of those achieved after IV administration of the same NPs. The drug-targeting index was 3.2, indicating a direct transport from the nose to the brain. The article contained no nasal solution control; however, comparing data from a previous article (Wang et al. 2006), where E_2 was administered nasally as an inclusion complex in methylated β-cyclodextrin, it can be concluded that the chitosan NPs enabled a higher CSF AUC than was obtained for the inclusion complex.

Similarly, Al-Ghananeem et al. (2010) used chitosan NPs for encapsulation of didanosine to increase the brain-targeting efficiency and administered these nasally to rats, using a drug solution given nasally and by the IV route as controls. They found that the brain/plasma, olfactory bulb/plasma, and the CSF/plasma ratios were significantly higher for the two nasal formulations as compared with the IV formulation, and also that the NP formulation was superior in efficacy compared with the simple nasal solution.

A later article by Fazil et al. (2012) used rhodamine encapsulated in chitosan NPs (165 nm) as a marker to estimate the distribution of the NPs in the brain after nasal administration to rats as compared with a rhodamine solution and rhodamine NPs given by the IV route. A higher intensity of rhodamine when administered nasally in NPs was seen in the brain by means of confocal microscopy as compared with the controls. Furthermore, when rivastigmine was administered in the 966 ng/ml chitosan NPs, the brain concentrations of the drug was significantly higher (than for the intranasal [508 ng/ml] and IV [387 ng/ml] controls). This was accompanied by a higher DTE% and a higher DTP%. The same group, later in the same way, investigated bromocriptine-loaded chitosan NPs and found a higher uptake of the labeled drug in the brain of mice for the NP formulation as compared with the controls, which was supported by gamma scintigraphy studies (Md et al. 2013). Furthermore, a reversal in catalepsy and akinesia behavior was especially pronounced in haloperidol-treated mice after the mice received bromocriptine-loaded chitosan NPs.

An improved nasal transport and brain uptake of tizanidine HCl encapsulated in thiolated chitosan NPs (~270 nm) after nasal administration to mice was recently reported by Patel et al. (2012). Particles made from different molecular weights of chitosan were used both thiolated and nonthiolated. The brain/blood ratios for the different formulations were shown to be highest for the thiolated medium molecular weight chitosan NPs. This was supported by the highest drug targeting index (DTI)% and DTP% values and a high antinociceptive effect. Unfortunately, the quality of the printed article in the journal was poor; some figures were missing and it was not possible to read the numbers off the graph.

Kumar et al. (2013) later evaluated the brain uptake of fluorescent labeled leucine–enkephalin (Leu–Enk), a neurotransmitter or neuromodulator in pain transmission, encapsulated in trimethyl chitosan NPs (443 nm) by means of fluorescence microscopy of sectioned mice brains. It was found that the NPs improved the uptake of the labeled drug into the brain as compared with a labeled solution, and that this enhancement also resulted in an improvement of the antinociceptive effect of Leu–Enk as shown by a hot plate and acetic acid–induced writhing assay.

Recently, Seju et al. (2011) demonstrated that the encapsulation of olanzapine in PLGA NPs (~90 nm) increased the efficiency of nose-to-brain delivery (6–10-fold) as compared with nasal delivery of a drug solution and an IV injection.

The group from Fudan University, Shanghai, had, as discussed above, previously evaluated the effect of conjugating PLA–PEG NPs with WGA for improved transport of the particles to the brain (Gao et al. 2006), and in a later article presented the data obtained in terms of drug biodistribution, brain uptake, and neuroprotective effect, after incorporating vasoactive intestinal peptide (VIP), a neuroprotective peptide, into the NPs (100–120 nm in diameter) and administering these nasally to mice (Gao et al. 2007b). As measured by the $AUC_{1-12\ h}$ of VIP in the brain, the nonfunctionalized and WGA-functionalized NPs enhanced the brain AUC as compared with a nasal solution control by 3.5–4.7- and 5.6–7.7-fold, respectively. This was supported by the pharmacodynamic studies in which a dose of 25 and 12.5 μg/kg VIP given in the nonfunctionalized and in WGA-functionalized NPs, respectively, resulted in the same improvements in spatial memory in ethylcholine aziridium–treated rats. It was found by histology that the WGA-functionalized NPs accumulated to a higher degree in the olfactory epithelium than the respiratory epithelium, and that these particles as compared with the nonfunctionalized NPs were observed to a higher degree to be internalized in the tissue. It was later shown by the same group that ^{125}I-labeled WGA–PEG–PLA NPs were transported into the brain via both the olfactory and the trigeminal nerve pathways (Liu et al. 2012). The authors made an attempt to quantify the uptake into the brain of the VIP by measuring not only radioactivity counts but also detecting the presence of intact VIP. The bioavailability of the VIP administered in the WGA-functionalized NPs was given as about 28% ID g/h for a combined olfactory bulb and tract cerebrum and cerebellum, calculated by dividing the radioactivity count per gram tissue by the administration dose per gram body weight of the animal. It is questionable that this is the correct way of presenting the data since the body weight of a rat vastly overwhelms the weight of the brain and hence calculates an artificially higher bioavailability.

Similarly, the OL-conjugated NPs (~120 nm in diameter) described previously by the same group (Wen et al. 2011a) were used for incorporation of urocortin peptide, and the system was evaluated for improved brain targeting (in mice) and therapeutic effect on hemiparkinsonian rats following nasal administration in comparison with nonconjugated NPs (Wen et al. 2011b). The biodistribution of the nanoparticulate systems were monitored using an *in vivo* imaging system and for these studies, the NPs contained a fluorescent marker, DiR. It was evident from the study that the OL–PLA–PEG particles compared with the nonfunctionalized NPs, to a significant higher degree, accumulated in the brain region. This was supported by the improved pharmacodynamic effect. There was no attempt to quantify the uptake in the CNS. The same group later used lactoferrin-modified PEG-co-PCL NPs (Lf-NP) for brain delivery of NAP peptide after nasal administration to rats, and found that the neuroprotective and memory improvement effect of the drug in a Morris water maze experiment when encapsulated in the Lf-NP was even lower than when encapsulated into nonmodified NPs (Liu et al. 2013).

It is evident from the above studies that it is possible by means of nanoparticulate carrier systems to enhance the direct transport of drugs and/or NPs from the nasal cavity to the CNS. However, it should be noted that only in a few quoted references have the authors made an attempt to quantify the bioavailability of the transported drug in terms of the percentage of administered dose (Gao et al. 2007b; Jain et al. 2010; Md et al. 2013). For example, it can be calculated from the AUC values given in Md et al. (2013), by estimating the brain weight of a 20-g mouse to 0.5 g and the blood weight to 1.4 g, that the bioavailability of the drug was 4.3% for the NP formulation and 1.8% for the solution formulation. However, these numbers are based on measurement of the ^{99m}Tc label, not on the drug. Another problem with some of the earlier studies discussed above is that the drug molecule is not detected by analytical procedures but rather is radiolabeled and the radioactivity in the brain tissue counted (see, e.g., Kumar et al. 2008b; Jain et al. 2010). This leaves the problem of a potential separation of the label from the drug. Furthermore, for some of the studies, a control in terms of a drug solution has not been included, which makes it difficult to judge completely the added effect of the nanoparticulate carrier system. Hence, it is not always possible from these studies to conclude whether the systems would have a real potential for efficient therapeutic exploitation in the clinic.

18.7 Conclusion

This chapter set out to discuss the scientific evidence and the potential for enhanced delivery of drugs directly from the nasal cavity to the brain using nanoparticulate carrier systems, thereby bypassing the BBB and enabling the therapeutic exploitation of drugs that do not easily cross the BBB from the systemic circulation. Furthermore, the chapter also set out to give some up-to-date background information in terms of routes of transport available for drugs and NPs from the nasal cavity to the CNS and discuss factors affecting the efficacy of the nanoparticulate systems. It has been suggested that a drug and a nanoparticulate system can reach the brain from the nasal cavity using the systemic pathway, or through the olfactory region using the olfactory epithelial pathway or the olfactory neural pathway and via the trigeminal nerves using the trigeminal pathway. Evidence has been given here that nanoparticulate materials, both in the form of viral, elemental, and polymer composites, after administration to the nasal cavity of animal models, are able to reach the olfactory bulb and often much deeper brain areas such as the cortex, cerebrum, and cerebellum, showing involvement of the olfactory pathway and caudal areas of the brain such as the cervical spinal cord, the medulla, and the pons, showing involvement of the trigeminal pathway. It was also evident from the articles discussed here that the degree of transport of the NPs could be enhanced by modifying the surface characteristics by adsorption of bioadhesive and/or absorption-enhancing materials such as chitosan or by functionalizing the NPs by lectins with specific receptors in the olfactory region. Furthermore, it was shown here that when drugs were encapsulated in the polymer NPs, especially when the NPs were functionalized, the transport of the drug (albeit often not directly analyzed as the drug but as a radiolabel) to the brain was also enhanced together with the pharmacological effect of the drug. Also of special interest is that there is some evidence from articles published from the Fudan University in Shanghai that not only drugs but also NPs can be transported through the trigeminal nerves to the brain. However, it is not clear how large a portion of the NPs are transported through the trigeminal pathway and which through the olfactory pathway.

The studies described here have all been carried out in animal models, mostly rats but also in mice. It is a distinct feature of these animals that 50% of the nasal cavity is covered by olfactory epithelium and that this region is easily reached when administering the nanoparticulate formulations. Hence, data obtained in these models may not easily translate to results in humans. Importantly, the olfactory region in humans only constitute 1.25%–3.00% of the epithelial surface and it is not easily accessible being situated high up and behind the upper conchae in the nasal cavity. However, if the importance of the trigeminal pathway in animal models translates to humans, then this may compensate for the smaller and difficult-to-reach olfactory region, and may suggest that formulations given to the nasal cavity for delivery to the brain should both be deposited in the respiratory and in the olfactory region. Thus far, such studies have not been carried out in humans neither for drugs nor for nanoparticulates.

It is not very clear from the data available how large a portion of the NPs (and drugs) administered to the nasal cavity actually reach the brain. Oberdörster's work indicates that a maximum of 3% of carbon NPs of 35 nm applied for 6 days reached the brain; other publications quote the amount of drug administered in the NPs reaching the brain as 5%–6% of drug administered. However, the latter was measured as the amount of radioactivity and not directly by measuring the drug, and it is not certain that the drug was still associated with the NPs. Hence, more studies need to be carried out to evaluate whether such nanoparticulate systems are able to sufficiently enhance the transport. Also, again, it should be noted that these studies are performed in animals and likely would show different (lower) results in humans.

Finally, it is important to consider the potential toxicity of such nanoparticulate systems especially for chronic indications. One has to ask the question whether it is safe to continuously accumulate solid albeit biodegradable NPs into the brain. Hence, detailed toxicological studies would need to be undertaken before one could consider therapeutic use.

References

Abdelbary G. A. and Tadros M. I., 2013. Brain targeting of olanzapine via intranasal delivery of core–shell difunctional block copolymer mixed micellar carriers: *In vitro* characterization, *ex vivo* estimation of nasal toxicity and *in vivo* biodistribution studies. *Int. J. Pharm.* 452: 300–310.

Al-Ghananeem A. M., Saeed H., Florence R., Yokel R. A. and Malkawi A. H., 2010. Intranasal drug delivery of didanosine loaded chitosan nanoparticles for brain targeting; An attractive route against infections caused by aids viruses. *J. Drug Target.* 18: 381–388.

Almeida A. J., Alpar H. O. and Brown M. R. W., 1993. Immune response to nasal delivery of antigenically intact tetanus toxoid associated with poly(L-lactic acid) microspheres in rats, rabbits and guinea pigs. *J. Pharm. Pharmacol.* 45: 198–203.

Amidi M., Romeijn S. G., Borchard G., Junginger H. E., Hennink W. E. and Jiskoot W., 2006. Preparation and characterization of protein-loaded *N*-trimethyl chitosan nanoparticles as nasal delivery system. *J. Control. Rel.* 111: 107–116.

Behrens I., Vila Pena A. I., Alonso M. J. and Kissel T., 2002. Comparative uptake studies of bioadhesive nanoparticles in human intestinal cell lines and rats: The effect of mucus on particle adsorption and transport. *Pharm. Res.* 19: 1185–1193.

Betbeder D., Sperandio S., Latapie J.-P., de Nadai J., Etienne A., Zajac J.-M. and Frances B., 2000. Biovector™ nanoparticles improve antinociceptive efficacy of nasal morphine. *Pharm. Res.* 17: 743–748.

Bodian D. and Howe H. A., 1941. The rate of progression of poliomyelitis virus in nerves. *Bull. Johns Hopkins Hosp.* 69: 79–85.

Born J., Lange T., Kern W., McGregor G. P., Bickel U. and Fehm H. L., 2002. Sniffing neuropeptides: A transnasal approach to the human brain. *Nat. Neurosci.* 5: 514–516.

Brooking J., Davis S. S. and Illum L., 2001. Transport of nanoparticles across the rat nasal mucosa. *J. Drug Target.* 9: 267–279.

Cauna N., Hinderer K. H. and Wentges R. T., 1969. Sensory receptor organs of the human nasal respiratory mucosa. *Am. J. Anat.* 124: 187–209.

Charlton S. T., Davis S. S. and Illum L., 2007. Nasal administration of an angiotension antagonist in the rat model: Effect of bioadhesive formulations on the distribution of drugs to the systemic and central nervous systems. *Int. J. Pharm.* 338: 94–103.

Chen J., Zhang C., Liu Q., Shao X., Feng C., Shen Y., Zhang Q. and Jiang X., 2012. *Solanum tuberosum* lectin-conjugated PLGA nanoparticles for nose-to-brain delivery: *In vivo* and *in vitro* evaluations. *J. Drug Target.* 20: 174–184.

Chou K. J. and Donovan M. D., 1998. Distribution of antihistamine into the CSF following intranasal delivery. *Biopharm. Drug Dispos.* 18: 335–346.

Chow H. H. S., Chen Z. and Matsuura G. T., 1999. Direct transport of cocaine from the nasal cavity to the brain following intranasal cocaine administration in rats. *J. Pharm. Sci.* 88: 754–758.

Costantino H. R., Illum L., Brandt G., Johnson P. and Quay S. C., 2007. Intranasal delivery: Physiochemical and therapeutic aspects. *Int. J. Pharm.* 337: 1–24.

Czerniawska A., 1970. Experimental investigations on the penetration of ^{198}Au from nasal mucous membrane into cerebrospinal fluid. *Acta Otolaryngol.* 70: 58–61.

Dahlin M., 2000. Nasal administration of compounds active in the central nervous system. Dissertation, Uppsala University, Uppsala, Sweden.

De Lorenzo A., 1960. Electron microscopy of the olfactory and gustatory pathways. *Ann. Otol. Rhinol. Laryngol.* 68: 410–420.

Des Rieux A., Ragnarsson E. G. E., Gullberg E., Preat V., Schneider Y.-J. and Artursson P., 2005. Transport of nanoparticles across an *in vitro* model of the human intestinal follicle associated epithelium. *Eur. J. Pharm. Sci.* 25: 455–465.

Dhuria S. V., Hanson L. R., Frey W. H. 2nd, 2009. Intranasal drug targeting of hypocretin-1 (orexin-A) to the central nervous system. *J. Pharm. Sci.* 98(7): 2501–2515.

Donovan M. D. and Huang Y., 1998. Large molecule and particulate uptake in the nasal cavity: The effect of size on nasal absorption. *Adv. Drug Deliv. Rev.* 29: 147–155.

Edeling M. A., Smith C. and Owen D., 2006. Life of a clathrin coat: Insights from clathrin and AP structures. *Nat. Rev. Mol. Cell. Biol.* 7: 32–44.

Elder A., Gelein R., Silva V., Feikert T., Opanashuk L., Carter J., Potter R., Maynard A., Ito Y., Finkelstein J. and Oberdörster G., 2006. Translocation of inhaled ultrafine manganese oxide particles to the central nervous system. *Environ. Health Perspect.* 114: 1172–1178.

Faber W. M., 1937. The nasal mucosa and the subarachnoid space. *Am. J. Anat.* 62: 121–148.

Fazil M., Md, S., Haque S., Kumar M., Baboota S., Sahni J. K. and Ali J., 2012. Development and evaluation of rivastigmine loaded chitosan nanoparticles for brain targeting. *Eur. J. Pharm. Sci.* 47: 6–15.

Feikert T., Mercer P., Corson N., Gelein R., Opanashuk L. and Elder A., 2004. Inhaled solid ultrafine particles (UFP) are efficiently translocated via neuronal naso-olfactory pathways. *Toxicologist* 78 (Suppl 1): 435–436.

Fliedner S., Schulz C. and Lehnert H., 2006. Brain uptake of intranasally applied radioiodinated leptin in Wistar rats. *Endocrinology* 147: 2088–2094.

Gao X., Chen J., Tao W., Zhu J., Zhang Q., Chen H. and Jiang X., 2007a. UEA I-bearing nanoparticles for brain delivery following intranasal administration. *Int. J. Pharm.* 340: 207–215.

Gao X., Tao W., Lu W., Zhang Q., Zhang Y., Jiang X. and Fu S., 2006. Lectin-conjugated PEG–PLA nanoparticles: Preparation and brain delivery after intranasal administration. *Biomaterials* 27: 3482–3490.

Gao X., Wu B., Zhang Q., Chen J., Zhu J., Zhang W., Rong Z., Chen H. and Jiang X., 2007b. Brain delivery of vasoactive intestinal peptide enhanced with the nanoparticles conjugated with wheat germ agglutinin following intranasal administration. *J. Control. Rel.* 121: 156–167.

Garzotto D. and De Marchis S., 2010. Quantum dot distribution in the olfactory epithelium after nasal delivery. *AIP Conf. Proc.* 1275: 118–123.

Gold G. H., 1999. Controversial issues in vertebrate olfactory transduction. *Ann. Rev. Phys.* 61: 857–871.

Gopinath P. G., Gopinath G. and Kumar T. C. A., 1978. Target site of intranasally sprayed substances and their transport across the nasal mucosa: A new insight into the intranasal route of drug delivery. *Curr. Ther. Res.* 23: 596–607.

Gross E. A., Swenberg J. A., Fields S. and Popp J. A., 1982. Comparative morphometry of the nasal cavity in rats and mice. *J. Anat.* 135: 83–88.

Harkema J. R., 1990. Comparative pathology of the nasal mucosa in laboratory animals exposed to inhaled irritants. *Environ. Health Perspect.* 85: 231–238.

Harush-Frenkel O., Rozentur E., Benita S. and Altschuler Y., 2008. Surface charge of nanoparticles determines their endocytotic and transcytotic pathway in polarized MDCK cells. *Biomacromolecules* 9: 435–443.

Haskell R. and Constantinides P. P., 2012. Perspectives in pharmaceutical nanotechnology. *AAPS Newsmag.* 15: 16–23.

Hayashi M., Hirasawa T., Muraoka T., Shiga M. and Awaza S., 1985. Comparison of water influx and sieving coefficient in rat jejunal, rectal and nasal absorption of antipyrine. *Chem. Pharm. Bull.* 33: 2149–2152.

Hilger P. A., 1989. Applied anatomy and physiology of the nose. In: *Otolaryngology. A Textbook of Ear, Nose and Throat Diseases*. W.B. Saunders, Philadelphia, PA, pp. 177–195.

Hillary A. M. and Florence A. T., 1996. The effect of absorbed poloxamer 188 and 407 surfactants on the intestinal uptake of 60 nm polystyrene particles after oral administration in the rats. *Int. J. Pharm.* 132: 123–130.

Howe H. A. and Bodian D., 1940. Portals of entry of poliomyelitis virus in the chimpanzee. *Proc. Soc. Exp. Biol. Med.* 43: 718–721.

Huang Y. and Donovan M. D., 1996. Microsphere transport pathways in the rabbit nasal mucosa. *Int. J. Pharm. Adv.* 1: 298–309.

Hunter D. D., Castranova V., Stanley C. and Dey R. D., 1998. Effects of silica exposure on substance P immunoreactivity and preprotachykinin mRNA expression in trigeminal sensory neurons in Fischer 344 rats. *J. Toxicol. Environ. Health A* 53: 593–605.
Illum L., 2000. Transport of drugs from the nasal cavity to the central nervous system. *Eur. J. Pharm. Sci.* 11: 1–18.
Illum L., 2004. Is nose-to-brain transport of drugs in man a reality? *J. Pharm. Pharmacol.* 56: 3–17.
Illum L., 2007. Nanoparticulate systems for nasal delivery of drugs: A real improvement over simple systems? *J. Pharm. Sci.* 96: 473–483.
Itaya S. K., 1987. Anterograde transsynaptic transport of WGA–HRP in rat olfactory pathways. *Brain Res.* 409: 205–214.
Jackson R. T., Tigges J and Arnold W, 1979. Subarachnoid space of CNS, nasal mucosa and lymphatic system. *Arch. Otolaryngol.* 105: 180–184.
Jafek B. W., 1997. Olfactory mucosal biopsy and related histology. In: Seiden A. M. (ed.) *Taste and Smell Disorders*. Thieme, New York, pp. 107–127.
Jain R., Nabar S., Dandekar P. and Patravale V., 2010. Micellar nanocarriers: Potential nose-to-brain delivery of zolmitriptan as novel migraine therapy. *Pharm. Res.* 27: 655–664.
Johnson N. J., Hanson L. R. and Frey II, W. H., 2010. Trigeminal pathways deliver a low molecular weight drug from the nose to the brain and orofacial structures. *Mol. Pharm.* 7: 884–893.
Jones A. T., 2008. Gateways and tools for drug delivery: Endocytic pathways and the cellular dynamics of cell penetrating peptides. *Int. J. Pharm.* 354: 34–38.
Joshi A. S., Patel H. S., Belgamwar V. S., Agrawal A. and Tekade A. R., 2012. Solid lipid nanoparticles of ondansetron HCl for intranasal delivery: Development, optimization and evaluation. *J. Mater. Sci. Mater. Med.* 23: 2163–2175.
Keyhani K., Scherer P. W. and Mozell M. M., 1997. A numerical model of nasal odorant transport for the analysis of human olfaction. *J. Theor. Biol.* 186: 279–301.
Köping-Höggård M., Sánchez A. and Alonso M. J., 2005. Nanoparticles as carriers for nasal vaccine delivery. *Expert Rev. Vaccines* 4: 185–196.
Kristensson K. and Olsson Y., 1971. Uptake of exogenous proteins in mouse olfactory cells. *Acta Neuropathol.* 19: 145–154.
Kumar M., Misra A., Babbar A. K., Mishra A. K., Mishra P. and Pathak K., 2008a. Intranasal nanoemulsion based brain targeting drug delivery system of risperidone. *Int. J. Pharm.* 358: 285–291.
Kumar M., Misra A., Mishra A. K., Mishra P. and Pathak K., 2008b. Mucoadhesive nanoemulsion-based intranasal drug delivery system of olanzapine for brain targeting. *J. Drug Target.* 16: 806–814.
Kumar M., Pandey R. S., Patra K. C., Jain S. K., Soni M. L., Dangi J. S. and Madan J., 2013. Evaluation of neuropeptide loaded trimethyl chitosan nanoparticles for nose to brain delivery. *Int. J. Biol. Macromol.* 61: 189–195.
Liu Q., Shen Y., Chen J., Gao X., Feng C., Wang L., Zhang Q. and Jiang X., 2012. Nose-to-brain transport pathways of wheat germ agglutinin conjugated PEG-PLA nanoparticles. *Pharm. Res.* 29: 546–558.
Liu Z., Jiang M., Kang T., Miao D., Gu G., Song Q., Yao L., Hu Q., Tu Y., Pang Z., Chen H., Jiang X., Gao X. and Chen J., 2013. Lactoferrin-modified PEG-co-PCL nanoparticles for enhanced brain delivery of NAP peptide following intranasal administration. *Biomaterials* 34: 3870–3881.
Lochhead J. J. and Thorne R. G., 2012. Intranasal delivery of biologics to the central nervous system. *Adv. Drug Deliv. Rev.* 64: 614–628.
Lucchini R. G., Dorman D. C., Elder A. and Veronesi B., 2012. Neurological impacts from inhalation of pollutants and the nose–brain connection. *Neurotoxicology* 33: 838–841.
Ma Z. and Lim L.-Y., 2003. Uptake of chitosan and associated insulin in Caco-2 cell monolayers: A comparison between chitosan molecules and chitosan nanoparticles. *Pharm. Res.* 20: 1812–1819.
Madara J. L., 2000. Modulation of tight junctional permeability. *Adv. Drug Deliv. Rev.* 41: 251–253.

Md S., Khan R. A., Mustafa G., Chuttani K., Baboota S., Sahni J. K. and Ali J., 2013. Bromocriptine loaded chitosan nanoparticles intended for direct nose to brain delivery: Pharmacodynamic, pharmacokinetic and scintigraphy study in mice model. *Eur. J. Pharm. Sci.* 48: 393–405.
Merkus P., 2003. Transport of non-peptide drugs from the nose to the CSF. In: *Proceedings from Nasal Drug Delivery Meeting, Management Forum*, London, 24–25th March.
Merkus P., Guchelaar H.-J., Bosch A. and Merkus F. W. H. M., 2003. Direct access to human brain after intranasal administration? *Neurology* 60: 1669–1671.
Mistry A., Stolnik S. and Illum L., 2009a. Nanoparticles for direct nose-to-brain delivery of drugs. *Int. J. Pharm.* 379: 146–157.
Mistry A., Zoffmann Glud S., Kjems J., Randel J., Howard K. A., Stolnik S. and Illum L., 2009b. Effect of physicochemical properties on intranasal nanoparticle transit into murine olfactory epithelium. *J. Drug Target.* 17: 543–552.
Musumeci T., Pellitteri R., Spatuzza M. and Puglisi G., 2014. Nose-to-brain delivery: Evaluation of polymeric nanoparticles on olfactory ensheathing cells uptake. *J. Pharm. Sci.* 103: 628–635. doi:101002/jps.23836 (Epub ahead of print).
Mygind N., 1978. *Nasal Allergy*. Blackwell Science, Oxford, UK.
Mygind N. and Dahl R., 1998. Anatomy, physiology and function of the nasal cavities in health and disease. *Adv. Drug Deliv. Rev.* 29: 3–12.
Oberdörster G., Sharp Z., Atudorei V., Elder A., Gelein R., Kreyling W. and Cox C., 2004. Translocation of inhaled ultrafine particles to the brain. *Inhal. Toxicol.* 16: 437–445.
Patel D., Naik S. and Misra A., 2012. Improved transnasal transport and brain uptake of tizanidine HCl-loaded thiolated chitosan nanoparticles for alleviation of pain. *J. Pharm. Sci.* 101: 690–706.
Patel S., Chavhan S., Soni H., Babbar A. K., Mathur R., Mishra A. K. and Sawant K., 2011. Brain targeting of risperidone-loaded solid lipid nanoparticles by intranasal route. *J. Drug Target.* 19: 468–474.
Perl D. P. and Good P. F., 1987. The association of aluminum Alzheimer's disease, and neurofibrillary tangles. *J. Neural Transm. Suppl.* 24: 205–211.
Rejman J., Oberle V., Zuhorn I. S. and Hoekstra D., 2004. Size dependent internalisation of particles via the pathway of clathrin- and caveolae-mediated endocytosis. *Biochem. J.* 377: 159–169.
Rothermel M., Ng B. S. W., Grabska-Barwinska A., Hatt H. and Jancke D., 2011. Nasal chemosensory-stimulation evoked activity patters in the rat trigeminal ganglion visualised by *in vivo* voltage-sensitive dye imaging. *PLoS One* 6:e26158. doi:10.1371/journal.pone.0026158.
Ruan Y., Yao L., Zhang B., Zhang S. and Guo J., 2011. Antinoceptive properties of nasal delivery of neurotoxin-loaded nanoparticles coated with polysorbate 80. *Peptides* 32: 1526–1529.
Sadeghi A. M. M., Dorkoosh F. A., Avadi M. R., Weinhold M., Bayat A., Delie F., Gurny R., Larijani B., Rafiee-Tehrani M. and Junginger H. E., 2008. Permeation enhancer effect of chitosan and chitosan derivatives: Comparison of formulations as soluble polymers and nanoparticulate systems on insulin absorption in Caco-2 cells. *Eur. J. Pharm. Biopharm.* 70: 270–278.
Sakane T., Akizuki M., Yoshida M., Yamashita S., Nadai T., Hashida M. and Sezaki H., 1991. Transport of cephalexin to the cerebrospinal fluid directly from the nasal cavity. *J. Pharm. Pharmacol.* 43(6): 449–451.
Schiöth H. B., Craft S., Brooks S. J., Frey II, W. H. and Benedict C., 2012. Brain insulin signalling and Alzheimer's disease: Current evidence and future directions. *Mol. Neurobiol.* 46: 4–10.
Seju U., Kumar A. and Sawant K. K., 2011. Development and evaluation of olanzapine–loaded PLGA nanoparticles for nose-to-brain delivery: *In vitro* and *in vivo* studies. *Acta Biomater.* 7: 4169–4176.
Semmler M., Seitz J., Erbe F., Mayer P., Heyder J., Oberdorster G. and Kreyling W. G., 2004. Long-term clearance kinetics of inhaled ultrafine insoluble iridium particles from the rat lung, including transient translocation into secondary organs. *Inhal. Toxicol.* 16: 453–459.
Shen Y., Chen J., Liu Q., Feng C., Gao X., Wang L., Zhang Q. and Jiang X., 2011. Effect of wheat germ agglutinin density on cellular uptake and toxicity of wheat germ agglutinin conjugated PEG–PLA nanoparticles in Calu-3 cells. *Int. J. Pharm.* 413: 184–193.

Shipley M. T., 1985. Transport of molecules from nose to brain: Transneural anterograde and retrograde labeling in the rat olfactory system by wheat germ agglutinin–horseradish peroxidase applied to the nasal epithelium. *Brain Res. Bull.* 15: 129–142.

Shorten P. R., McMahon C. D. and Soboleva T. K., 2007. Insulin transport within skeletal muscle transverse tubule networks. *Biophys. J.* 93: 3001–3007.

Silver W. L., 1992. Neural and pharmacological basis for nasal irritation. *Ann. N. Y. Acad. Sci.* 641: 152–163.

Slütter B., Bal S., Keijzer C., Mallants R., Hagenaars N., Que I., Kaijzel E., van Eden W., Augustijns P., Löwik C., Bouwstra J., Broere F. and Jiskoot W., 2010. Nasal vaccination with N-trimethyl chitosan and PLGA based nanoparticles: Nanoparticle characteristics determine quality and strength of the antibody response in mice against the encapsulated antigen. *Vaccine* 28: 6282–6291.

Stevenson B. R., Anderson J. M. and Bullivant S., 1988. The epithelial tight junction: Structure, function and preliminary biochemical characterisation. *Mol. Cell Biochem.* 83: 129–145.

Thorne R. G., Emory C. R., Ala T. A. and Frey II, W. H., 1995. Quantitative analysis of the olfactory pathway for drug delivery to the brain. *Brain Res.* 692: 278–282.

Thorne R. G. and Frey W. H., 2001. Delivery of neurotropic factors to the central nervous system. *Clin. Pharmacokinet.* 40: 907–946.

Thorne R. G., Hanson L. R., Ross T. M., Tung D. and Frey II, W. H., 2008. Delivery of interferon-b to the monkey nervous system following intranasal administration. *Neuroscience* 152: 785–797.

Thorne R. G., Pronk G. J., Padmanabhan V. and Frey W. H., 2004. Delivery of insulin-like growth factor-I to the rat brain and spinal cord along olfactory and trigeminal pathways following intranasal administration. *Neuroscience* 127: 481–496.

Vila A., Sanchez A., James K., Behrens I., Kissel T., Jato J. L. V. and Alonso M. J., 2004. Low molecular weight chitosan nanoparticles as new carriers for nasal vaccine delivery in mice. *Eur. J. Pharm. Biopharm.* 57: 123–131.

Wang B., Feng W. Y., Wang M., Shi J. W., Zhang F., Ouyang H., Zhao Y. L., Chai Z. F., Huang Y. Y., Wang H. F. and Wang J., 2007. Transport of intranasally instilled fine Fe2)3 particles into the brain: Micro-distribution, chemical states and histopathological observation. *Biol. Trace Elem. Res.* 118: 233–243.

Wang J., Liu Y., Jiao F., Lao F., Li W., Gu Y., Li Y., Ge C., Zhou G., Li B., Zhao Y., Chai Z. and Chen C., 2008a. Time-dependent translocation and potential impairment on central nervous system by intranasally instilled TiO(2) nanoparticles. *Toxicology* 254: 82–90.

Wang X., Chi N. and Tang X., 2008. Preparation of estradiol chitosan nanoparticles for improving nasal absorption and brain targeting. *Eur. J. Pharm. Biopharm.* 70: 735–740.

Wang X., He H., Leng W. and Tang X., 2006. Evaluation of brain targeting for the nasal delivery of estradiol by the microdialysis method. *Int. J. Pharm.* 317: 40–46.

Wang Y., Aun R. and Tse F. L. S., 1998. Brain uptake of dihydroergotamine after intravenous and nasal administration in the rat. *Biopharm. Drug Dispos.* 19: 571–575.

Wang Y.-Y., Lai S. K., Suk J. S., Pace A., Cone R. and Hanes J., 2008b. Addressing the PEG mucoadhesivity paradox to engineer nanoparticles that 'slip' through the human mucus barrier. *Angew. Chem. Int. Ed. Engl.* 47: 9726–9729.

Wen Z., Yan Z., He R., Pang Z., Guo L., Qian Y., Jiang X. and Fang L., 2011a. Brain targeting and toxicity study of odorranalectin-conjugated nanoparticles following intranasal administration. *Drug Deliv.* 18: 556–561.

Wen Z., Yan Z., Hu K., Pang Z., Cheng X., Guo L., Zhang Q., Jiang X., Fang L. and Lai R., 2011b. Odorranalectin-conjugated nanoparticles: Preparation, brain delivery and pharmacodynamic study on Parkinson's disease following intranasal administration. *J. Control. Rel.* 151: 131–138.

Xia H., Gao X., Gu G., Liu Z., Zeng N., Hu Q., Song Q., Yao L., Pang Z., Jiang X., Chen J. and Chen H., 2011. Low molecular weight protamine-functionalised nanoparticles for drug delivery to the brain after intranasal administration. *Biomaterials* 32: 9888–9898.

Yamamoto Y., Kondo A., Atoji Y., Tsubone H. and Suzuki Y., 1998. Morphology of intraepithelial corpuscular nerve endings in the nasal respiratory mucosa of the dog. *J. Anat.* 193: 581–586.

Zhang Q., Jiang X., Jiang W., Lu W., Su L. and Shi Z., 2004. Preparation of nimodipine–loaded micro-emulsion for intranasal delivery and evaluation on the targeting efficiency to the brain. *Int. J. Pharm.* 275: 85–96.

Zhang Q.-Z., Zha L.-S., Zhang Y., Jiang W.-M., Lu W., Shi Z.-Q., Jiang X.-G. and Fu S.-K., 2006. The brain targeting efficiency following nasally applied MPEG-PLA nanoparticles in rats. *J. Drug Target.* 14: 281–290.

Zhao Y. Z., Li X., Lu C. T., Lin M., Chen L. J., Xiang Q., Zhang M., Jin R. R., Jiang X., Shen X. T., Li X. K. and Cai J., 2014. Gelatin nanostructured lipid carriers-mediated intranasal delivery of basic fibroblast growth factor enhances functional recovery in hemiparkinsonian rats. *Nanomedicine.* 10(4): 755–764.

19

The Developing Lungs

Akira Tsuda and Frank S. Henry

CONTENTS

19.1 Introduction

The effects of nanoparticles (NPs) on lung tissues, regardless of whether or not they are beneficial or detrimental, are likely to be amplified in the developing lungs. For instance, it is well known that children are at higher risk of exposure when subjected to particulate air pollution than the adult population. An increasing number of epidemiological studies also indicate that environmental air pollutants (e.g., tobacco smoke, nitrogen dioxide, acid vapor, ozone, particles) impair the developing pulmonary system of children (e.g., Heinrich et al. 2002; Finkelstein and Johnston 2004; Rauh et al. 2004; Schwartz 2004; Perera et al. 2005; Wang and Pinkerton 2007). Both prenatal and postnatal exposures can be associated with negative pulmonary health outcomes. Smoking during pregnancy, for example, may increase the susceptibility of infants to the development of a number of pediatric lung disorders, including bronchopulmonary dysplasia (BPD) (Schuller et al. 2000; Ueda et al. 2006; Cao et al. 2009). In particular, the epidemiological data strongly suggest the adverse health effects of particulate air pollution on lung growth and function in children (e.g., Bates 1995; Pope 2000; Gauderman et al. 2000, 2002, 2004; Gilliland et al. 2000; Avol et al. 2001; Horak et al. 2002; Mathieu-Nolf 2002; Kim 2004; Sram et al. 2005). Similarly, the health effects in the developing lungs associated with therapeutic drug delivery may also be amplified. Here, we first review the normal process of postnatal lung development, followed by a description of particle exposure and biological responses.

19.2 Postnatal Lung Development

19.2.1 Anatomy

19.2.1.1 Structural Alveolation and Septal Tissue Development

During the process of lung development from birth to adulthood in humans (and similarly, in rodents, which are often used as a surrogate animal model), the lungs undergo dramatic structural changes, not only in size but also in acinar architecture (Figure 19.1). At birth, the acinus consists of wide and smooth-walled saccular airspaces (Figure 19.1, left). Rapid structural alveolation takes place in the first few years after birth (equivalent to the first few weeks in rats; Figure 19.1, middle), followed by a slow and gradual increase in alveolar size until later childhood (~90 days [adult] in rats; Figure 19.1, right). Detailed analyses of the postnatal changes

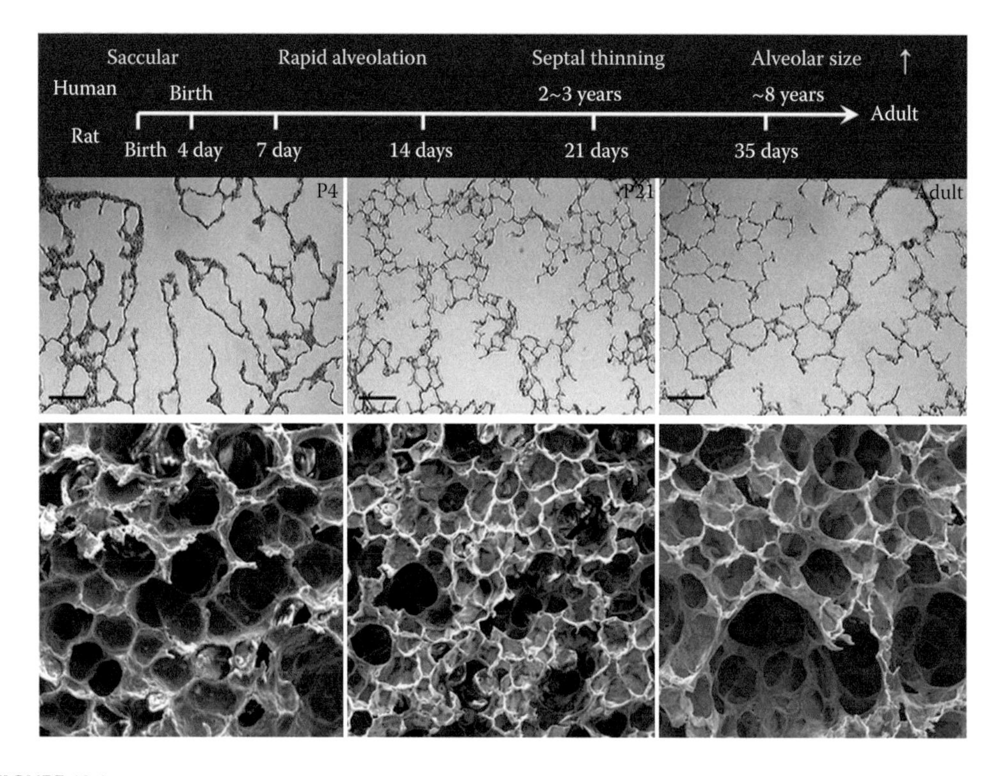

FIGURE 19.1
Postnatal lung development. From left: day 4, day 21, adult rats. Top: Hematoxylin and eosin (H&E) staining. Bar, 100 μm. Bottom: SEM. Because of obvious ethical constraints, a surrogate of the human infants is necessary for any experimental studies. The rat lung is the most widely investigated laboratory model for human pulmonary development because the remodeling process associated with development and resulting structural change is essentially the same in humans and rodents, only with differing time scales (Zeltner and Burri 1987; Zeltner et al. 1987), and the morphometry of postnatal rat lung maturation has been well characterized (for review, see Burri et al. 1974; Burri 2006). Rats possess saccular lungs as neonates, and undergo an expansionary phase from birth to postnatal day 4 (Burri 1974) before progressing to the alveolar stage (Zoetis and Hurtt 2003). Bulk alveolation occurs during postnatal days 4–13 (Vidic and Burri 1983). Interalveolar septa are formed in concert with the folding up of one of the two capillary layers. Thinning of the septal interstitium and microvascular maturation produces a miniature version of adult lung morphology after ~3 weeks in rats. Lung maturation then proceeds predominantly through lung growth. It is also important to note that molecular probes and reagents are also available for rats. (Adapted from Semmler-Behnke M et al. *Proc. Natl. Acad. Sci. U. S. A.*, 109, 5092, 2012.)

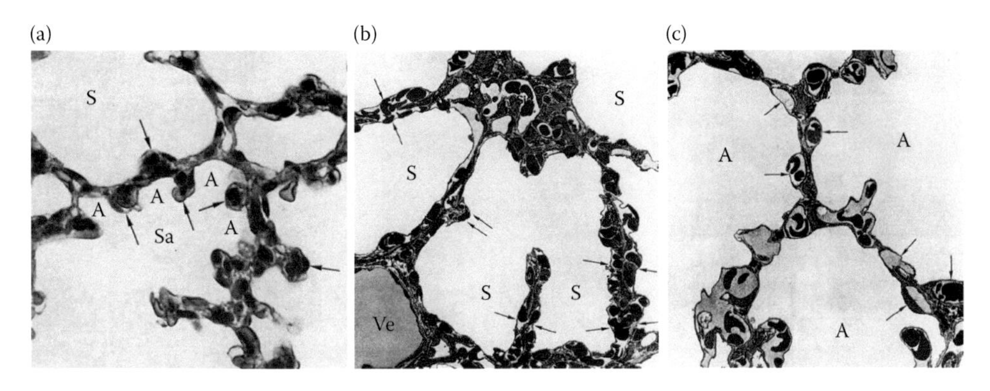

FIGURE 19.2
(a) Formation of new septa, day 7 rat (from Figure 3b in Burri 1974). (b) Double-layer capillaries, day 1 rat. (c) Single-layer capillaries, day 21. S: saccules; A: alveoli; Sa: alveolar sacs; Ve: blood vessel. (Adapted from Figures 3b, 5, and 11, respectively, of Burri PH. *Anat. Rec.*, 180, 77, 1974. With permission.)

of parenchymal ultrastructure, which have been performed in the ultrastructure of rats (e.g., Burri 1974, 2006), show that the rapid structural alveolation in early postnatal development is achieved by the formation of new septa (denoted secondary septa) from shallow indentations (denoted primary septa) lining the saccules and transitory ducts present at birth (Figure 19.2a). When bulk alveolation wanes, lung development enters the stage of microvasculature remodeling and septal thinning (several months to 2–3 years in humans) (Burri 2006). The key feature of this stage is the restructuring of double-layer septal capillaries (Figure 19.2b) to more matured single capillaries (Figure 19.2c). Detailed morphometric studies in rats (Burri 2006) show that because of this dramatic microvascular restructuring, the interstitial tissue volume significantly decreases. After completion of microvascular maturation and septal thinning, the shape of the acinus is essentially the same as that of the adult lung morphology (after ~3 years in humans; Burri 2006). The lung volume also rapidly increases during this period, by about 13 times from birth to age 6 in children, primarily by the process of terminal buds developing into new ducts and associated alveoli, and by three times from age 6 to adulthood, mainly due to the increase in the size of the alveoli. The major structural changes in the rapidly growing lungs are also characterized by lengthening of the acinar and conducting airways.

19.2.2 Breathing Patterns

19.2.2.1 Changes in Breathing Patterns during Postnatal Lung Development

The breathing pattern (tidal volume [VT], breathing frequency [f]) is one of the major parameters that determine particle deposition in the lung, and it is different between children and adults. While for the fully developed adult animals, it is generally known that metabolic rate can scale as 2/3 the power of body weight (BW) (Heusner 1982), the exponent value of the allometric relation between minute ventilation (MV; a product of VT and f) and BW is likely to be different in the developing lungs. To demonstrate this, we measured the tidal volume (VT), breathing frequency (f), and minute ventilation (MV) in awake, spontaneously breathing infant rats at various ages during development, and found that these breathing parameters scale allometrically with body weight (BW) to exponents of 1.06, −0.12 (note that Fleming et al. [2011] reported a sharp decrease in breathing frequency in human infants <2 years of age; this is consistent with our finding of negative exponent value), and 0.91, respectively (Figure 19.3). It is interesting to note that the

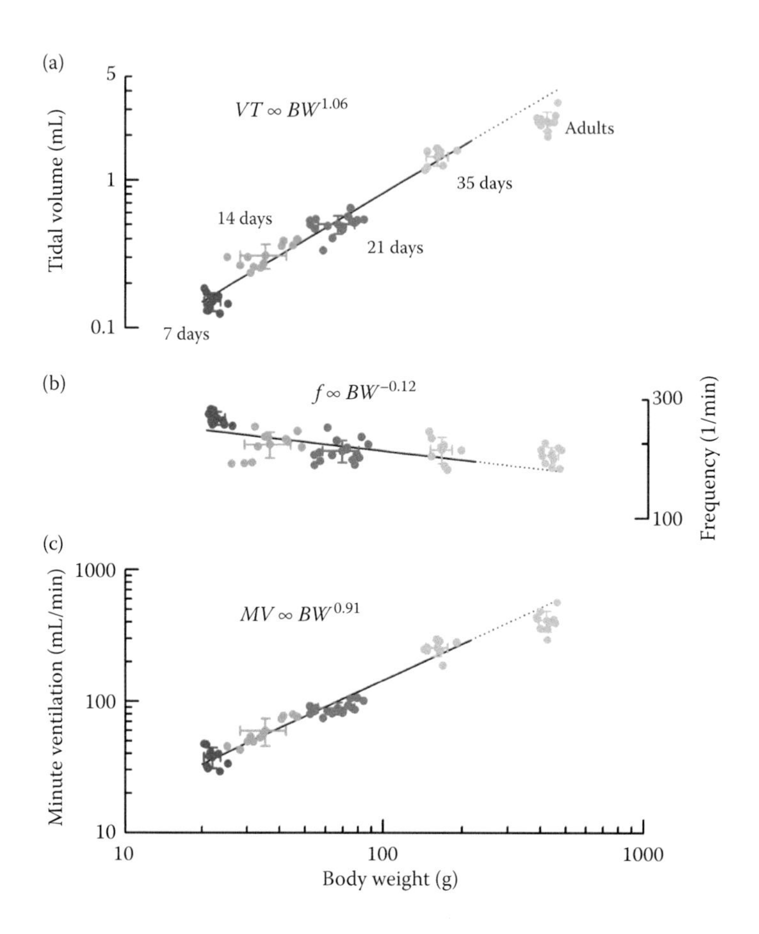

FIGURE 19.3
(a) Tidal volume (VT), (b) breathing frequency (f), and (c) minute ventilation (MV) vs. body weight (BW) in developing rats. The data from the animals during development (7–35 days; $n \geq 10$ each; body weights from ~20 to 200 g) appear to follow the allometric scaling of ventilation with body mass. For minute ventilation (c), for instance, the intraspecies power law equation is given by $MV = 2.14 \times BW^{0.91}$ ($r^2 = 0.93$) in the case of the developing rats. Slight deviation of the data of the adult animals (90 days) from the developmental scaling suggests that different biological features are coming into play for the variation in minute ventilation with body mass in the fully developed animal. (Adapted from Semmler-Behnke M et al. *Proc. Natl. Acad. Sci. U. S. A.*, 109, 5092, 2012.)

exponent value for MV (= $VT \times f$) in the developing lungs (0.91) is different from that in fully developed adult lungs (2/3). The fact that the intraspecies allometric exponent value for MV, which is a surrogate for metabolism, is significantly higher in the developing lungs than in fully developed adult lungs may be interpreted as a consequence of the increased metabolic need that is directly associated with growth. Another noteworthy fact is that the exponent value (0.91) of the allometric relation between MV and BW is statistically not exactly 1, suggesting that MV (e.g., infant ventilation) cannot simply be normalized by BW (discussed in more detail on page 332). In addition, it is interesting to note that, recently, Bonafide et al. (2013) reported that the respiratory rates of up to 40% of hospitalized children are outside of the normal textbook reference range (Ralston et al. 2006; Hartman and Cheifetz 2011; Kirk 2011; Lennox 2011). Bonafide et al. concluded that a high proportion of vital signs (heart and respiratory rates) among hospitalized children would be considered out of range according to existing reference ranges and pediatric early warning score parameters (Akre et al. 2010; Parshuram et al. 2011).

19.2.3 Gene Expression

19.2.3.1 Gene Expression Profiles during Normal Lung Development

Genome-wide gene expression profiling performed for the entire process of normal murine lung development (Mariani et al. 2002; Kho et al. 2009; http://www.ncbi.nlm.nih.gov/geo/query/acc.cgi?acc=GSE10889, homogenized whole lungs) showed complex patterns of gene expression associated with various steps of the morphogenetic processes, beginning at embryonic day 12 (E12) and continuing to adulthood in mice. The complexity of the developmental process is illustrated by changes in the expression of genes coding for some important structural proteins, such as tropoelastin, and interstitial and basement membrane collagens. The expression of tropoelastin shows two peaks: the first on embryonic day 18 (E18), corresponding to the formation of vascular structures of the lungs, and the second on postnatal days 10–14 (P10–14), at the time of alveolation, corresponding to the formation of alveolar interstitium. Similarly, interstitial collagen genes showed high levels of expression at an early embryonic stage (E12), when the lung structures were first established, and again peaked on P7, at the time of the formation of alveolar septa. Basement membrane collagen genes, on the other hand, increased expression late in embryonic development (E18), and then on P7, with the establishment of alveolar vascularization. These data provide guidance to the understanding of the sequential expression of gene clusters at major steps of lung morphogenesis in general.

19.2.4 Immunology

Birth is a major transition from the sterile intrauterine environment to the outside world full of microbes. Within days, the newborn skin and intestine is colonized by commensal bacteria, and immediately after birth the respiratory tract is exposed to inhaled bacteria, viruses, and fungi. In response to all these stimuli, the neonatal immune system undergoes major changes after birth, but still differs in many aspects from the adult one. In general, neonatal immune responses are biased against pro-inflammatory cytokine production (Levy 2007). That means that interleukin-1β (IL-1 β) and tumor necrosis factor (TNF) secretion is diminished in infants; however, IL-6, IL-10, and IL-23 responses of neonatal antigen-presenting cells exceed those of the adults. In animal models in mice and rats, the expression of innate immune receptors, Toll-like receptor 2 (TLR2) and TLR4, in the lungs is low early in life and increases after birth. In rats, the lung TLR2 and TLR4 expression does not reach adult levels till 4 weeks of age. Consistent with the gradual upregulation of TLRs, antimicrobial β-defensin responses of the respiratory epithelium to endotoxin, a TLR4 ligand, increase with age in humans, as well as in animals. Innate immune responses in the respiratory tract may also affect pre- and postnatal lung development. IL-6, a cytokine induced in the fetal environment by chorioamnionitis, a bacterial infection of the placenta and the amniotic fluid, enhances fetal lung branching in a lung explant model (Nogueira-Silva et al. 2006). Moreover, a respiratory pathogen (*Ureaplasma urealyticum*) that is often found in the lungs of prematurely born babies with BPD is a TLR agonist inducing TNF responses in neonatal monocytes. Infants infected with respiratory syncytial virus during the first few years of life have an increased risk of respiratory pathology and abnormal lung function later (Halfhide and Smyth 2008). In mouse experiments, exposure of fetal lungs to TLR ligands (e.g., lipopolysaccharide) inhibited structural lung development (Prince et al. 2005). Activated macrophages with enhanced nuclear factor-κB-mediated signaling and IL-1 secretion were responsible for dysregulated lung morphogenesis manifested by thickened lung interstitium and reduced airway branching in this model (Blackwell et al. 2011). These observations, taken together, suggest that innate immune responses during pre- and postnatal lung development may have a long-lasting, and not yet fully understood, effect on lung structure and function.

19.3 Acinar Fluid Mechanics and NP Deposition in the Developing Acinus

As was described in Chapter 2 (Deposition), a large portion of inhaled NP deposits in the pulmonary acinus by a combination of convective (airflow) and diffusive (NPs' intrinsic motion) transport. An NP travels with its surrounding airflow for a long distance (from airway opening to the gas-exchange region of the lung), and on arriving at close to the alveolar surface it deviates from convective airflow streamline by diffusion and deposits on the alveolar wall surface. Whereas the diffusion process governs the final short distance of the NP's journey, alveolar convective airflow patterns make a major contribution to NP transport, and are critically important for NP deposition. Therefore, it is important to know the characteristics of alveolar airflow in the developing lungs, and whether they are very different from those in the fully developed lungs.

Taking into account the two major findings, that (i) the geometry of alveolated airway walls and their tidal motion become ultimately very important in determining the airflow patterns (e.g., Tsuda et al. 1995, 2002; Henry et al. 2002; Karl et al. 2004) because airflow momentum in the acinus is largely governed by the interaction of pressure drops and viscous forces; and (ii) dramatic structural changes occur during postnatal lung development (e.g., Burri 1974; Burri et al. 1974; Dickie et al. 2007; Semmler-Behnke et al. 2012), acinar fluid mechanics and subsequent deposition of inhaled particles within the developing acinus must be strongly age dependent.

To examine the effects of postnatal structural changes (depth and size of alveoli) and breathing patterns (which conditions alveolar wall motion) on acinar airflow patterns, we conducted computational fluid dynamics analyses (Semmler-Behnke et al. 2012). We were specifically interested in the presence of rotational flow inside the alveolar cavity because it critically conditions the existence of chaos in the alveolus (the onset of chaotic flows; Tsuda et al. 1995). As the alveolar cavity's shape changes from shallow newborn saccular alveoli to more deep mature fully formed alveoli during postnatal lung development (Figure 19.4, upper and middle), the flow must also change from smooth flow without rotation (thus, reversible) (Figure 19.4, bottom left) to largely rotational, thus potentially irreversible flow (Tsuda et al. 1995, 2011) (Figure 19.4, bottom right). We hypothesized that there must be a critical developmental stage at which the lengthening secondary septa become large enough to make the alveolar cavity sufficiently deep (Figure 19.1) to cause the cavity flow to rotate (Figure 19.4, bottom center). At that stage of development, the characteristics of the acinar flow should qualitatively change (Tsuda et al. 2011), triggering chaotic dynamics (see particle irreversibility in Figure 19.5). We predicted that this dramatic change in the fluid dynamics would lead to significant mixing (Tsuda et al. 2002, 2011; Semmler-Behnke et al. 2012) of the particles in the inhaled tidal air with the residual gas and, consequently, would result in increased deposition of particles on the alveolar walls in the developing lungs (Semmler-Behnke et al. 2012).

To test these predictions, which were based on computational fluid dynamics analysis, we performed *in vivo* exposure experiments with rats of five different age groups (7, 14, 21,

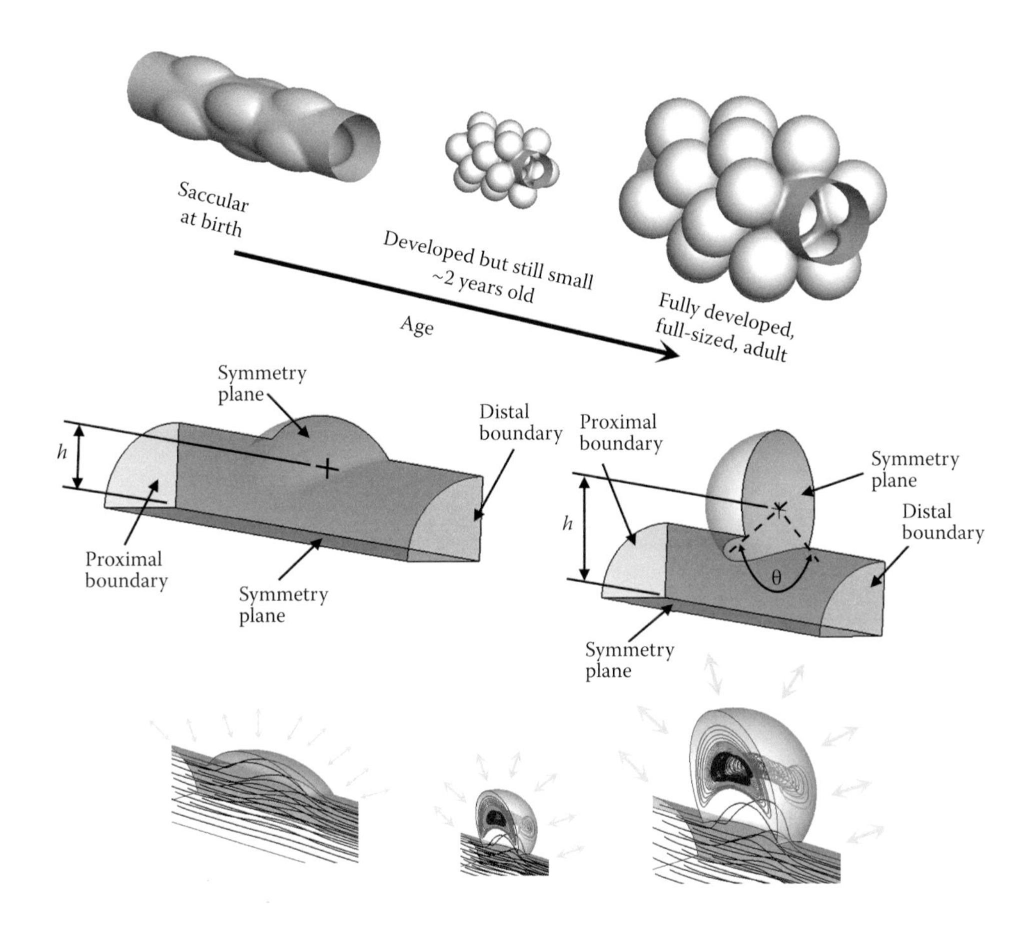

FIGURE 19.4
Computational simulation of airflow patterns in acinar airway geometries at three different stages of human lung development. Middle: Computational models of the developing lung. The newborn saccular duct was modeled as a cylindrical central channel (center duct) and an ellipsoid (alveolus). The alveolated duct was modeled as a cylindrical central channel (center duct) and a truncated sphere (alveolus). h: distance between the center of the alveolus and the center line of the duct; θ: the alveolar opening angle. See details in Semmler-Behnke et al. (2012). Upper: Acinar airway geometric models at three different stages of human lung development: a saccular airway at birth (left), an alveolated but small airway at 2 years old (center), and a fully developed adult airway (right). Bottom: Airflow patterns corresponding to the three developmental stages: a simple flow pattern with no recirculation (left), a funnel-like spiral rotational flow in the fully developed duct (center), and a similar pattern in the adult case (right). (Adapted from Semmler-Behnke M et al. *Proc. Natl. Acad. Sci. U. S. A.*, 109, 5092, 2012.)

35, and 90 days old, $n = 16$ each). These different age groups represented the various stages of postnatal acinar structural development (Burri 1985). We exposed each age group (nose-only exposure) to insoluble, radioactively labeled iridium NPs of 20 or 80 nm. The NPs were generated by a spark generator, neutralized, humidified, and continuously monitored (Semmler-Behnke et al. 2012). After 1-h exposure, the animals were killed either immediately ($n = 8$ for each age group) or at 24 h ($n = 8$ for each age group). From the groups scarified immediately after exposure, the total deposition fraction was calculated as a ratio of the measured radioactivity in the body to the amount of radioactivity inhaled as aerosols (Kreyling et al. 2002; Semmler et al. 2004). From the groups scarified at 24 h after exposure,

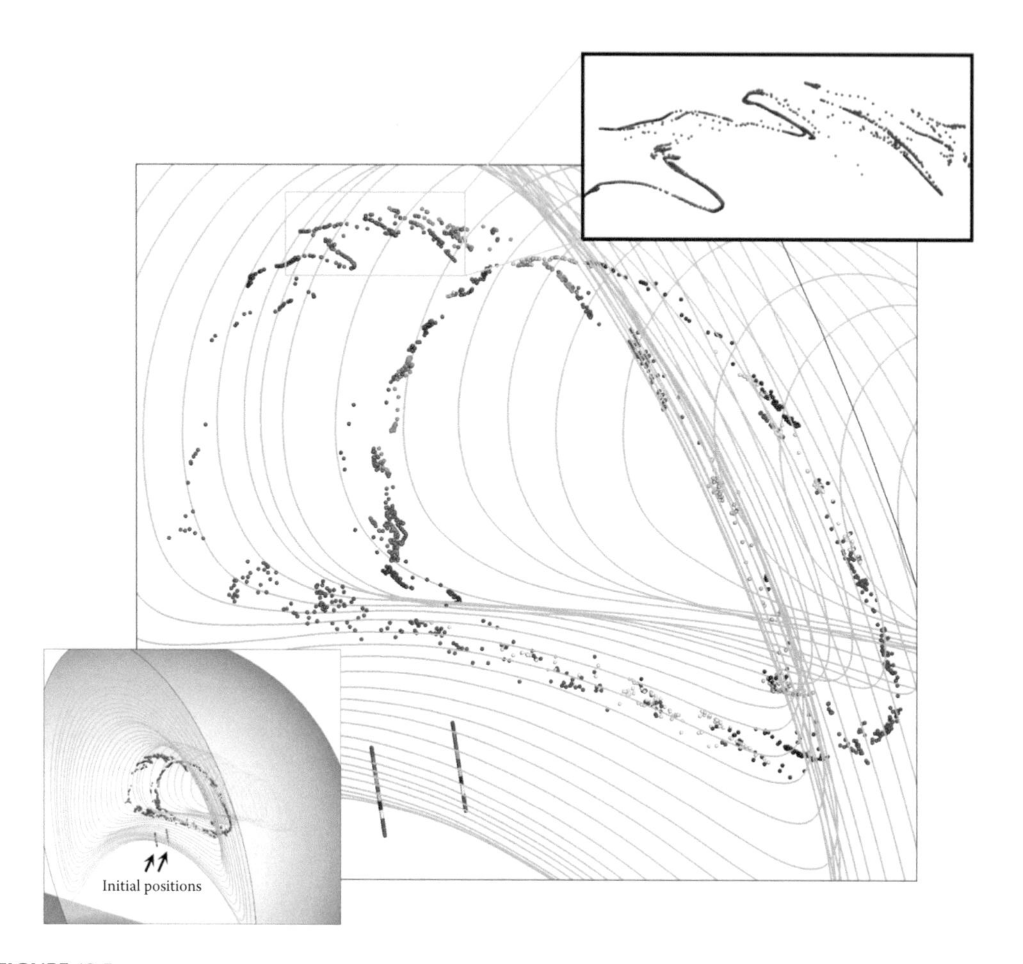

FIGURE 19.5
Different views of the simulated position of 2000 fluid particles at the end of one cycle in the 2-year-old alveolus. Panels show that two short lines, each of 1000 fluid particles, placed near the center of the alveolus (shown as vertical lines) at the beginning of inspiration ($t = 0$) spread out and form fractured ring-like structures inside the alveolus by the end one cycle. (Inset, top right) Stretch-and-fold flow pattern, a hallmark of chaotic flow. (Adapted from Semmler-Behnke M et al. *Proc. Natl. Acad. Sci. U. S. A.*, 109, 5092, 2012.)

the regional deposition (lung parenchyma vs. other sites) was estimated (see Semmler-Behnke et al. 2012, for more details).

Total deposition strongly depended on the age of the rats. In the youngest age group tested (7 days), the deposition was the lowest. The acini at this age group have few relatively shallow alveoli; thus, alveolar airflows must be simpler and reversible according to our computational model (Figure 19.4). In addition to the absence of chaotic mixing due to a lack of recirculation flow, diffusional deposition must be less effective because of the relatively large airspaces in the airways of saccular geometry. In the age group of 14 days, the deposition rapidly increased, and it reached peak values in the age group of 21 days (Figure 19.6). As described in Section 19.2, structural alveolation is just completed by this age; thus, the shape of the acini of this age group is similar to that of the adult animal. A central channel flow, passing by a sufficiently deep alveolus, causes fluid in an alveolus to rotate (Figures 19.4 and 19.5); a rotational alveolar flow induces chaos in each alveolus, and, consequently, chaotic mixing increases particle deposition. In rats older than 21 days, the

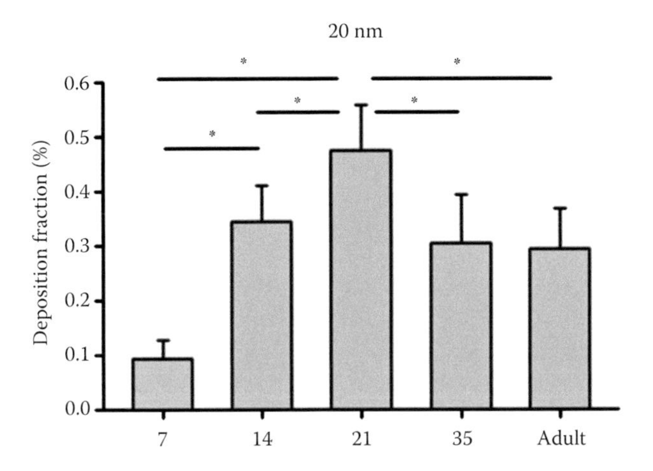

FIGURE 19.6
Deposition fraction versus rat age for inhaled 20-nm particles. A dramatic acinar structural remodeling takes place postnatally in the developing infant lungs; deposition of NPs appears to be significantly influenced by the process of structural alveolation. NP deposition was initially very low (in 7-day-old rats) but later it significantly increased and peaked at the end of bulk alveolation stage of postnatal lung development (21 day-old rats). n = 16 for age group; $p < .05$; error bars, ±SD. (Adapted from Semmler-Behnke M et al. *Proc. Natl. Acad. Sci. U. S. A.*, 109, 5092, 2012.)

shape of the acini does not change since structural alveolation has been already completed by 21 days; however, the size of the acini increases. Therefore, while alveolar airflows are likely rotational and chaotic, alveolar airspaces of the rats older than 21 days, through which particles need to be transported for deposition, become large relative to the diffusional distance. As a consequence, deposition in rats older than 21 days decreases relative to the value observed at 21 days (Figure 19.6).

As experimental observations of particle deposition (Figure 19.6) in rats of various ages were consistent with the predictions of our computational studies (Figures 19.4 and 19.5), our findings strongly support the idea that age-dependent changes in the alveolar duct structure and breathing patterns, and the resulting changes in acinar fluid mechanics, indeed play a critical role in determining the fate of inhaled particles. We note, although data are not shown here (interested readers may see Semmler-Behnke et al. 2012), that the following two results support our conclusion mentioned above: (i) a similar age dependency was found with 80-nm NPs although the deposition was generally lower with 80-nm NPs compared with 20-nm NPs owing to their relatively lower intrinsic diffusivity; and (ii) a similar age dependency was found in acinar deposition of both 20- and 80-nm NPs (i.e., regional deposition [parenchyma vs. other sites] was estimated from measurements made at 24 h after exposure).

The significance of our findings in terms of inhalation therapy for infants is discussed as follows. In current neonatal care, drug dose is simply estimated by scaling the dose (National Institutes of Health Guidelines, 1997) to the infant's body mass from that of an adult, ("simple scaling" implies $MV \propto BW^{1.0}$; i.e., an allometric exponent of 1.0). This classical approach, however, has the following problems.

Particle deposition (D) is given by the following time integral:

$$D = \int DF \cdot C \cdot MV \; \mathrm{d}t$$

where DF is deposition fraction and C is particle concentration. Thus, the total dose D of particles deposited per body weight (BW) can be expressed as

$$\frac{D}{BW} = \frac{\int DF \cdot C \cdot MV \, \mathrm{d}t}{BW}$$

Although the current estimation, which is calculated by using the simple scaling ($MV \propto BW^{1.0}$), is based on the idea that the dose normalized by body weight can be independent from breathing patterns $\left(D/BW = \int DF \cdot C \, \mathrm{d}t \right)$, this is not the case in reality. Since the minute ventilation (i.e., metabolism) is indeed dependent on body weight ($MV \propto BW^{0.91}$, shown in Figure 19.3), the dose (D) per body weight (BW) is given by

$$\frac{D}{BW} \left[\int DF \cdot C \ \mathrm{d}t \right] BW^{-0.09}$$

The fact that the exponent of BW on the right-hand side is not zero demonstrates that drug dose cannot be simply estimated as proportional to body mass (although its dependency on body mass is relatively weak). Furthermore, and more importantly, a major difference in drug dose (D) from the conventional estimation is in reality due to how one considers the deposition fraction (DF). While in the current approach, DF is considered to remain constant, in reality, it does not. In fact, during postnatal lung development, DF changes substantially throughout the acinar structure remodeling phase (Figure 19.6); it rapidly increases nearly 5-fold during the active alveolation stage and decreases after this period. This substantial change in DF during various stages of postnatal lung development has a direct influence on particle deposition, thus affecting the delivered drug dose (D). Comprehensive knowledge of the amount of particles deposited in the acinus is fundamental for inhalation therapy, especially because drugs administered into the lung periphery, where there is no fast clearance mechanism, remain there for extended periods (Semmler et al. 2004; Semmler-Behnke et al. 2007). This long biological half-life offers some advantages for therapeutic drug delivery. For instance, it prolongs the time available for a drug to be released, and it increases the chance that the particle crosses the air–blood barrier.

The fact that acinar deposition is strongly age dependent can be explained as a consequence of the combined effects of the changes in geometric characteristics of the developing acinus, changes in breathing characteristics, and chaotic fluid dynamical phenomena. Deposition peaks in 21-day-old rats, when the acinus has just completed the bulk structural alveolation stage. At this time, the shape of each alveolus is already similar to the deep, matured alveoli; however, the size is still small. Finally, it is important to note that the end of bulk alveolation in rats occurs at 21 days of age, which corresponds to ~2 years old in humans, suggesting that human infants of ~2 years old are, in general, at greater risk than newborns or older children up to adulthood. In summary, as infant lungs are not miniature versions of adult lungs, inhalation therapy dose estimations should not simply be based on scaling an infant's body to that of an adult.

We note that while there is the large difference in size between the human and the rat lung, it is known that particle deposition in these two species is similar (McMahon et al. 1977; Schulz and Muhle 2000). The similarity can be explained theoretically by the law of dynamic similitude; that is, two flows of vastly different sizes can be shown to be similar if certain nondimensional parameters are the same in both cases. In the case of rat and human lungs, the pertinent parameter is the airflow Reynolds number, and this parameter varies over essentially the same range in both respiratory tracts. Also, while there is a significant difference in the branching pattern of the conducting airways of the two species, the rat acinus is morphologically very similar to that of the human, and it is in this region that NPs mainly deposit (Tsuda et al. 2002; Semmler-Behnke et al. 2012).

19.4 Deposition Distribution along the Acinus of the Developing Lungs

Inhaled NPs predominantly deposit in the pulmonary acinus, and the experiments described in Section 19.3 clearly show that the deposition of inhaled NPs within the developing acinus is strongly age dependent.

As we have described in Chapter 2, in adult subjects, there is a significant heterogeneity in the distribution of inflammatory lesions within the acinus after inhalation of pollutant particulates; the proximal region of the acini is the most frequent site of inflammation (e.g., Pinkerton et al. 2000; Saldiva et al. 2002). Determination of whether the distribution of deposited particles within the acinus varies with age (postnatal developmental stage), and whether the site of inflammation or particle-induced damage in the acinus directly corresponds to high local particle dose are important subjects to address because the findings of such investigation would likely provide us with important clues to understand the link between (local) dose and biological responses in the developing lungs.

Because the deposition process is strongly dependent on structure, and the lung structure is changing during postnatal development, deposition distribution of inhaled particles is likely to be strongly age dependent. We speculate that in the newborn acinus, because the acinar airways are wide, relatively smooth, and saccular (Figure 19.7, top left), the inhaled air is expected to enter the acinar airways along the centerline, owing to the viscous nature of air, like a narrow tongue (Figure 19.7, bottom left). Hence, the residual air forms a barrier between the particle-laden air and the acinar surface. As the airways are relatively smooth, we expect that the incoming air travels to the periphery relatively undisturbed. This leads to the prediction that particle deposition in the newborn acinus occurs at a relatively low rate but uniformly over the entire acinar tree.

On the other hand, as the acinus develops, secondary septa grow and alveoli are formed (Figure 19.7, top right). The *QA*/*QD* ratio (i.e., *QA* is the rate of flow entering the alveolus and *QD* is the ductal flow passing by the alveolar opening) starts to play an important role in determining alveolar flow. At some point in development when the alveoli become sufficiently deep, the alveoli in the entrance region of the acinus, in particular, start to exhibit secondary rotation flow (Tsuda et al. 1995, 2008; Henry et al. 2009, 2012; Figure 19.7, bottom right). This rotation flow would enhance particle deposition in the entrance region of the acinus. By contrast, alveolar flow in the distal alveoli is largely radial without rotation due to a large *QA*/*QD* ratio, resulting in reduced deposition. This leads to the hypothesis that as the acinus becomes alveolated, the distribution of particle deposition becomes heterogeneous with preferentially higher deposition in the entrance region of the acinus.

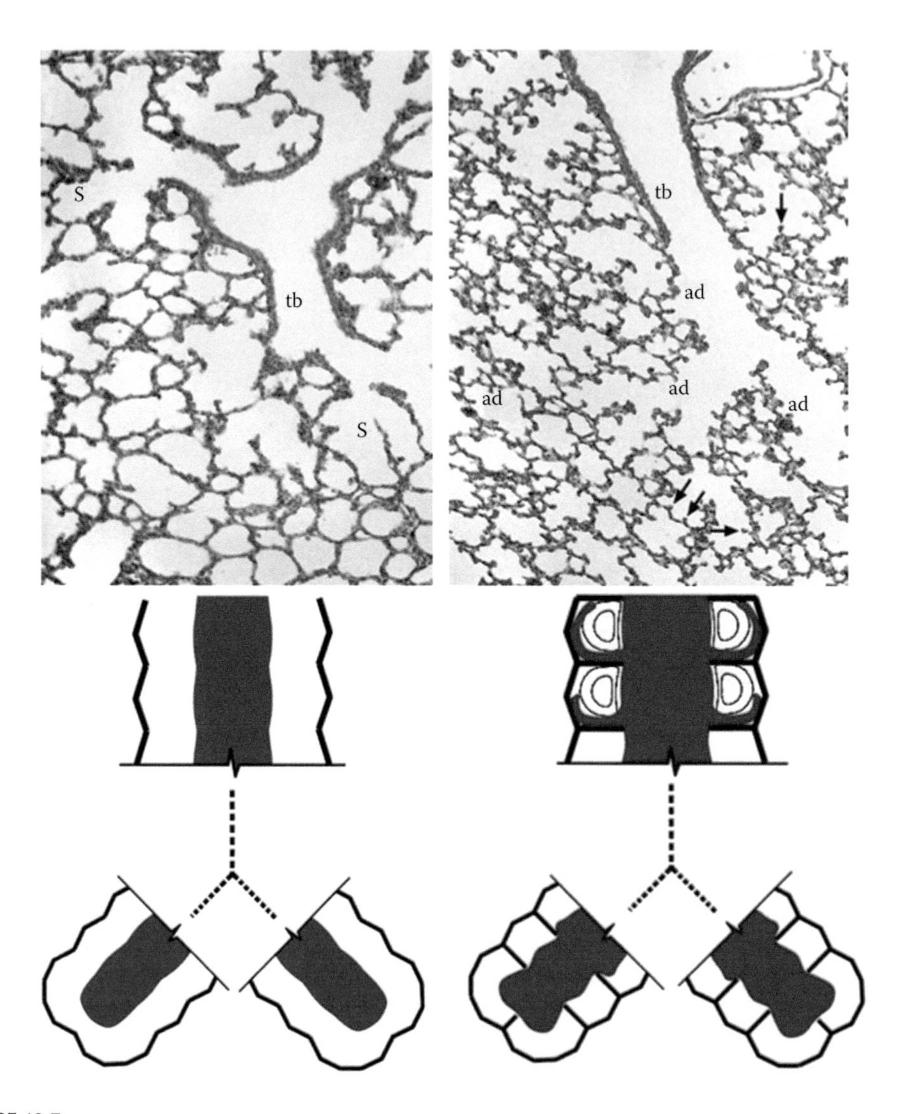

FIGURE 19.7
Top: Acinar structure of the immature lung of a 1-day old rat (left) and the lung of a 21-day-old rat whose alveolar shape is nearly fully developed but small in size (right). tb: terminal bronchioles; S: saccules; ad: alveolar ducts; arrows: secondary septa. (Adapted from Burri PH. Structural aspects of prenatal and postnatal development and growth of the lung. In: McDonald JA, ed. *Lung Growth and Development*. New York: Marcel Dekker; pp. 1–35, 1997. With permission.) Bottom: Schematics of our hypothesis. Particle-laden inhaled airflow is shown in gray (see text).

19.4.1 Future Directions

To elucidate the basic physics determining the site-specific deposition distribution of NPs within the acinus at various stages of development, massive numerical simulations are necessary. Such simulations also require detailed anatomical data for different stages of lung development. Detailed imaging data resulting from our recent synchrotron-based, high-resolution, acinar imaging studies (e.g., Figure 2.6 in Chapter 2; Mund et al. 2008; Schittny et al. 2008; Tsuda et al. 2008; Henry et al. 2012) can be used for this purpose. From such data, we can extract and collect anatomically accurate structural data to build representative geometric models of the acinus at various developmental stages.

19.5 Effects of Inhaled NPs in the Developing Lungs

Prenatal and/or postnatal exposure to inhaled environmental pollutants, such as tobacco smoke and ozone, is known to cause morphological alterations in the developing lung (e.g., Stocks and Dzateux 2003; Hoo et al. 2004; Plopper 2004; Tran et al. 2004; Rouse et al. 2007; Wang and Pinkerton 2007; Joad et al. 2009). Although the reduced cell proliferation in response to inhaled particles in neonate rats was not accompanied by overt histological damage within the short time points examined by Pinkerton et al. (2004), morphological change may be detectable after longer periods (Health Review Committee 2008). Mauad et al. (2008) showed that pre- and postnatal exposure of mice to urban $PM_{2.5}$ (which include NPs) resulted in smaller surface-to-volume ratios compared with nonexposed animals.

In one study (Cormack et al. 2010), we exposed immature rats of various developmental stages to nanosized or microsized CuO particles. The data revealed that the influx of neutrophils into the lung, which indicates inflammatory reaction (Figure 19.8), and gene expression profile patterns of lung tissues were age and particle size dependent (data not shown; see Cormack et al. 2010). These results suggest that acute exposure of postnatally developing lungs, which have rapidly growing and differentiating alveolar structure, to nanosized or microsized particles induces different biological responses.

Inhaled NPs may be used therapeutically to treat the developing lungs. For instance, one of the most devastating infant respiratory diseases that are often encountered in neonatal care clinics is BPD, which is associated with mechanical ventilation of premature infants. The typical morphological features of BPD are large alveoli, which result in a reduction in the overall surface available for gas exchange (Baraldi and Filippone 2007), thereby causing impaired lung function. Inhaled therapeutics for the treatment of lung function impairments (e.g., BPD) is particularly attractive for infants because significant barriers to therapeutic efficacy (poor gastrointestinal absorption and proteolytic activity; Patton 1996) and first-pass metabolism in the liver (Labiris and Dolovich 2003) could be avoided. In addition, local delivery of drugs in the lungs is noninvasive (Sung et al. 2007). However, the challenge ahead is that there are no NP deposition models specifically made for the

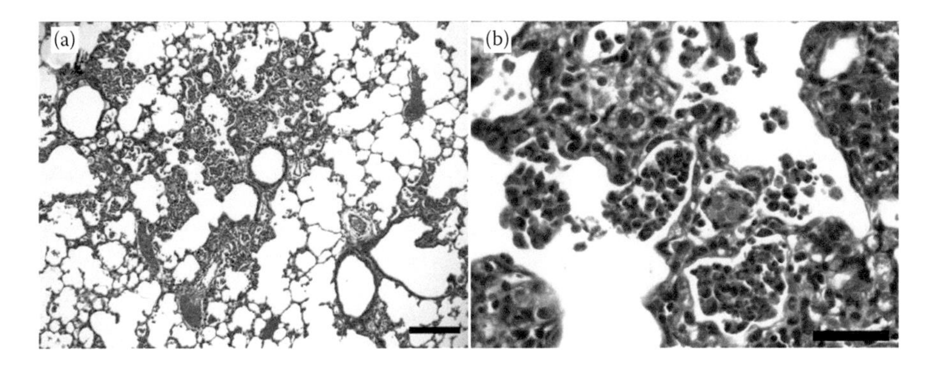

FIGURE 19.8
H&E staining of 14-day-old rat lung exposed to nano-CuO particles. (a) Area view. Severe inflammation is often observed at the junction of the terminal bronchioles and alveolar duct. The alveoli are filled with neutrophils (bar, 500 μm). (b) Fragmented nuclei show that many neutrophils in the alveoli are apoptotic (bar, 100 μm). (From Cormack M et al. Age-dependent changes in gene expression profiles of postnatally developing rat lungs exposed to nano-size and micro-size CuO particles. AAAR, San Diego, 2010 [abstract].)

developing lungs, except ours (Semmler-Behnke et al. 2012), that consider the structural change of developing airways (even nondiseased, normal lungs). A combination of morphological studies of the developing (both normal and diseased) lungs with state-of-the-art computational fluid mechanics analyses would be helpful to shed more light on this important subject.

Acknowledgments

We thank the Dana Farber/Harvard Cancer Center Rodent Histopathology Core as well as Dr. Renee Dickie's help. This work was supported in part by research grants from the National Institutes of Health (HL054885, HL070542, HL074022, HL094567, and ES000002).

References

Akre M, Finkelstein M, Erickson M, Liu M, Vanderbilt L, Billman G. 2010. Sensitivity of the pediatric early warning score to identify patient deterioration. *Pediatrics* 125(4): e763–e769. Available at www.pediatrics.org/cgi/content/full/125/4/e763.

Avol EL, Gauderman WJ, Tan SM, London SJ, Peters JM. 2001. Respiratory effects of relocating to areas of differing air pollution levels. *Am J Respir Crit Care Med* 164(11):2067–2072.

Baraldi E, Filippone M. 2007. Chronic lung disease after premature birth. *N Engl J Med* 357(19):1946–1955.

Bates DV. 1995. The effects of air pollution on children. *Environ Health Perspect* 103(Suppl 6):49–53.

Blackwell TS, Hipps AN, Yamamoto Y, Han W, Barham WJ, Ostrowski MC, Yull FE, Prince LS. 2011. NF-κB signaling in fetal lung macrophages disrupts airway morphogenesis. *J Immunol* 187:2740–2747.

Bonafide CP, Brady PW, Keren R, Conway PH, Marsolo K, Daymont C. 2013. Development of heart and respiratory rate percentile curves for hospitalized children. *Pediatrics* 131:e1150–e1157.

Burri PH. 1974. The postnatal growth of the rat lung. 3. Morphology. *Anat Rec* 180(1):77–98.

Burri PH. 1985. Development and growth of the human lung. In *Handbook of Physiology. Section 3: The Respiratory System*, vol. 1, Fishman AP, Fisher AB, eds. William & Wilkins, Baltimore, pp. 1–46.

Burri PH. 1997. Structural aspects of prenatal and postnatal development and growth of the lung. In *Lung Growth and Development*, McDonald JA, ed. Marcel Dekker, New York, pp. 1–35.

Burri PH. 2006. Structural aspects of postnatal lung development—Alveolar formation and growth. *Biol Neonate* 89(4):313–322.

Burri PH, Dbaly J, Weibel ER. 1974. The postnatal growth of the rat lung. I. Morphometry. *Anat Rec* 178(4):711–730.

Cao L, Wang J, Tseu I, Luo D, Post M. 2009. Maternal exposure to endotoxin delays alveolarization during postnatal rat lung development. *Am J Physiol Lung Cell Mol Physiol* 296(5):L726–L737.

Cormack M, Lin M, Fedulov A, Mentzer SJ, Tsuda A. 2010. Age-dependent changes in gene expression profiles of postnatally developing rat lungs exposed to nano-size and micro-size CuO particles (abstract). AAAR, San Diego.

Dickie R, Wang YT, Butler JP, Szymanska K, Schulz H, Tsuda A. 2007. Spatial and temporal distribution of smooth muscle actin in the alveolar interstitium of the postnatal rat lung. *Am J Respir Crit Care Med* 175:A88.

Finkelstein JN, Johnston CJ. 2004. Enhanced sensitivity of the postnatal lung to environmental insults and oxidant stress. *Pediatrics* 113(4 Suppl):1092–1096.

Fleming S, Thompson M, Stevens R, Heneghan C, Pluddemann A, Maconochie I, Tarassenko L, Mant D. 2011. Normal ranges of heart rate and respiratory rate in children from birth to 18 years of age: A systematic review of observational studies. *Lancet* 377:1011–1018.

Gauderman WJ, Avol E, Gilliland F, Vora H, Thomas D, Berhane K, McConnell R, Kuenzli N, Lurmann F, Rappaport E, Margolis H, Bates D, Peters J. 2004. The effect of air pollution on lung development from 10 to 18 years of age. *N Engl J Med* 351(11):1057–1067.

Gauderman WJ, Gilliland GF, Vora H, Avol E, Stram D, McConnell R, Thomas D, Lurmann F, Margolis HG, Rappaport EB, Berhane K, Peters JM. 2002. Association between air pollution and lung function growth in southern California children: Results from a second cohort. *Am J Respir Crit Care Med* 166(1):76–84.

Gauderman WJ, McConnell R, Gilliland F, London SJ, Thomas D, Avol EL, Vora H, Berhane K, Rappaport EB, Lurmann F, Marigolis HG, Peters J. 2000. Association between air pollution and lung function growth in southern California children. *Am J Respir Crit Care Med* 162:1383–1390.

Gilliland FD, Berhane K, McConnell R, Gauderman WJ, Vora H, Rappaport EB, Avol E, Peters JM. 2000. Maternal smoking during pregnancy, environmental tobacco smoke exposure and childhood lung function. *Thorax* 55:271–276.

Halfhide C, Smyth RL. 2008. Innate immune response and bronchiolitis and preschool recurrent wheeze. *Paediatr Respir Rev* 4:251–262.

Hartman ME, Cheifetz IM. 2011. Pediatric emergencies and resuscitation. In *Nelson Textbook of Pediatrics*, 19th ed., Kliegman RM, Stanton BF, St Geme JW III, Schor NF, Behrman RE, eds. Elsevier Saunders, Philadelphia, p. 280.

Health Review Committee. Critique of Pinkerton KE, Zhou Y, Zhong C, Smith KR, Teague SV, Kennedy IM, Ménache MG. 2008. Mechanisms of particulate matter toxicity in neonatal and young adult rat lungs. *HEI Research Report* 135. Health Effects Institute, Boston.

Heinrich J, Hoelscher B, Frye C, Meyer I, Pitz M, Cyrys J, Wjst M, Neas L, Wichmann HE. 2002. Improved air quality in reunified Germany and decreases in respiratory symptoms. *Epidemiology* 13(4):394–401.

Henry FS, Butler JP, Tsuda A. 2002. Kinematically irreversible flow and aerosol transport in the pulmonary acinus: A departure from classical dispersive transport. *J Appl Physiol* 92:835–845.

Henry FS, Haber S, Haberthür D, Filipovic N, Milasinovic D, Schittny JC, Tsuda A. 2012. Fluid mechanics of the gas-exchange region of the lung. *ASME J Biomech Eng* 134:121001-1–121001-11.

Henry FS, Laine-Pearson FE, Tsuda A. 2009. Hamiltonian chaos in a model alveolus. *ASME J Biomech Eng* 131(1):011006.

Heusner AA. 1982. Energy metabolism and body size. I. Is the 0.75 mass exponent of Kleiber's equation a statistical artifact? *Respir Physiol* 48:1–12.

Hoo AF, Stocks J, Lum S, Wade AM, Castle RA, Costeloe KL, Dezateux C. 2004. Development of lung function in early life: Influence of birth weight in infants of nonsmokers. *Am J Respir Crit Care Med* 170(5):527–533.

Horak F, Studnicka M, Gartner C, Spengler JD, Tauber E, Urbanek R, Veiter A, Frischer T. 2002. Particulate matter and lung function growth in children: A 3-yr follow-up study in Austrian schoolchildren. *Eur Respir J* 19(5):838–845.

Joad JP, Kott KS, Bric JM, Peake JL, Pinkerton KE. 2009. Effect of perinatal secondhand tobacco smoke exposure on *in vivo* and intrinsic airway structure/function in non-human primates. *Toxicol Appl Pharmacol* 234(3):339–344.

Karl A, Henry FS, Tsuda A. 2004. Low Reynolds number viscous flow in an alveolated duct. *ASME J Biomech Eng* 126(1):13–19.

Kho AT, Bhattacharya S, Mecham BH, Hong J, Kohane IS, Mariani TJ. 2009. Expression profiles of the mouse lung identify a molecular signature of time-to-birth. *Am J Respir Cell Mol Biol* 40(1):47–57.

Kim JJ. 2004. Ambient air pollution: Health hazards to children. *Pediatrics* 114(6):1699–1707.

Kirk A. 2011. Pulmonology In *The Harriet Lane Handbook*, 19th ed., Tschudy MM, Arcara KM, eds. Elsevier Mosby, Philadelphia, p. 585.

Kreyling WG, Semmler M, Erbe F, Mayer P, Takenaka S, Schulz H, Oberdörster G, Ziesenis A. 2002. Translocation of ultrafine insoluble iridium particles from lung epithelium to extrapulmonary organs is size dependent but very low. *J Toxicol Environ Health* 65:1513–1530.

Labiris NR, Dolovich MB. 2003. Pulmonary drug delivery. Part I: Physiological factors affecting therapeutic effectiveness of aerosolized medications. *Br J Clin Pharmacol* 56:588–599.

Lennox EG. 2011. Cardiology. In *The Harriet Lane Handbook*, 19th ed., Tschudy MM, Arcara KM, eds. Elsevier Mosby, Philadelphia, p. 170.

Levy O. 2007. Innate immunity of the newborn: Basic mechanisms and clinical correlates. *Nat Rev Immunol* 7:379–390.

Mariani TJ, Reed JJ, Shapiro SD. 2002. Expression profiling of the developing mouse lung Insights into the establishment of the extracellular matrix. *Am J Respir Cell Mol Biol* 26:541–548.

Mathieu-Nolf M. 2002. Poisons in the air: A cause of chronic disease in children. *J Toxicol Clin Toxicol* 40(4):483–491.

Mauad T, Rivero DH, de Oliveira RC, Lichtenfels AJ, Guimarães ET, de Andre PA, Kasahara DI, Bueno HM, Saldiva PH. 2008. Chronic exposure to ambient levels of urban particles affects mouse lung development. *Am J Respir Crit Care Med* 178(7):721–728.

McMahon TA, Brain JD, LeMott S. 1977. Species differences in aerosol deposition. In *Inhaled Particles, IV*, Walton WH, ed. Pergamon, Oxford, UK, pp. 23–33.

Mund SI, Stampanoni M, Schittny JC. 2008. Developmental alveolarization of the mouse lung. *Dev Dyn* 237(8):2108–2116.

National Institutes of Health Guidelines for the Diagnosis and Management of Asthma. 1997. NIH Publ No. 97-40511995. National Institutes of Health, Bethesda, MD.

Nogueira-Silva C, Santos M, Baptista MJ, Moura RS, Correia-Pinto J. 2006. IL-6 is constitutively expressed during lung morphogenesis and enhances fetal lung explant branching. *Pediatr Res* 60:530–536.

Parshuram CS, Duncan HP, Joffe AR, Farrell CA, Lacroix JR, Middaugh KL, Hutchison JS, Wensley D, Blanchard N, Beyene J, Parkin PC. 2011. Multicentre validation of the bedside paediatric early warning system score: A severity of illness score to detect evolving critical illness in hospitalised children. *Crit Care* 15(4):R184.

Patton JS. 1996. Mechanisms of macromolecule absorption by the lungs. *Adv Drug Deliv Rev* 19:3–36.

Perera FP, Rauh V, Whyatt RM, Tang D, Tsai WY, Bernert JT, Tu YH, Andrews H, Barr DB, Camann DE, Diaz D, Dietrich J, Reyes A, Kinney PL. 2005. A summary of recent findings on birth outcomes and developmental effects of prenatal ETS, PAH, and pesticide exposures. *Neurotoxicology* 26(4):573–587.

Pinkerton KE, Green FH, Saiki C, Vallyathan V, Plopper CG, Gopal V, Hung D, Bahne EB, Lin SS, Ménache MG, Schenker MB. 2000. Distribution of particulate matter and tissue remodeling in the human lung. *Environ Health Perspect* 108(11):1063–1069.

Pinkerton KE, Zhou YM, Teague SV, Peake JL, Walther RC, Kennedy IM, Leppert VJ, Aust AE. 2004. Reduced lung cell proliferation following short-term exposure to ultrafine soot and iron particles in neonatal rats: Key to impaired lung growth? *Inhal Toxicol* 16(Suppl 1):73–81.

Plopper C. 2004. Stunting of conducting airways begins early during postnatal development in infant rhesus monkeys exposed to ozone and/or allergen. *Am J Respir Crit Care Med* 169(7):A697.

Pope CA. 2000. Epidemiology of fine particulate air pollution and human health: Biologic mechanisms and who's at risk? *Environ Health Perspect* 108(Suppl 4):713–723.

Prince LS, Dieperink HI, Okoh VO, Fierro-Perez GA, Lallone RL. 2005. Toll-like receptor signaling inhibits structural development of the distal fetal mouse lung. *Dev Dyn* 233:553–561.

Ralston M, Hazinski MF, Zaritsky AL, Schexnayder SM, Kleinman ME, eds. 2006. Pediatric assessment. In *Pediatric Advanced Life Support Provider Manual*. American Heart Association, Dallas, pp. 9–16.

Rauh VA, Whyatt RM, Garfinkel R, Andrews H, Hoepner L, Reyes A, Diaz D, Camann D, Perera FP. 2004. Developmental effects of exposure to environmental tobacco smoke and material hardship among inner-city children. *Neurotoxicol Teratol* 26(3):373–385.

Rouse RL, Boudreaux MJ, Penn AL. 2007. *In utero* environmental tobacco smoke exposure alters gene expression in lungs of adult BALB/c mice. *Environ Health Perspect* 115(12):1757–1766.

Saldiva PH, Clarke RW, Coull BA, Stearns RC, Lawrence J, Murthy GG, Diaz E, Koutrakis P, Suh H, Tsuda A, Godleski JJ. 2002. Lung inflammation induced by concentrated ambient air particles is related to particle composition. *Am J Respir Crit Care Med* 165(12):1610–1617.

Schittny JC, Mund SI, Stampanoni M. 2008. Evidence and structural mechanism for late lung alveolarization. *Am J Physiol Lung Cell Mol Physiol* 294(2): L246–L254.

Schuller HM, Jull BA, Sheppard BJ, Plummer HK. 2000. Interaction of tobacco-specific toxicants with the neuronal alpha(7) nicotinic acetylcholine receptor and its associated mitogenic signal transduction pathway: Potential role in lung carcinogenesis and pediatric lung disorders. *Eur J Pharmacol* 30:393(1–3):265–277.

Schulz H, Muhle H. Respiration. 2000. In *Handbook of Experimental Animals: The Laboratory Rat*, Krinke GJ, ed. Academic, London, pp. 323–344.

Schwartz J. 2004. Air pollution and children's health. *Pediatrics* 113(4 Suppl):1037–1043.

Semmler M, Seitz J, Erbe F, Mayer P, Heyder J, Oberdorster G, Kreyling WG. 2004. Long-term clearance kinetics of inhaled ultrafine insoluble iridium particles from the rat lung, including transient translocation into secondary organs. *Inhal Toxicol* 16:453–459.

Semmler-Behnke M, Kreyling WG, Schulz H, Takenaka S, Butler JP, Henry FS, Tsuda A. 2012. Nanoparticle delivery in infant lungs. *Proc Natl Acad Sci U S A* 109(13):5092–5097.

Semmler-Behnke M, Takenaka S, Fertsch S, Wenk A, Seitz J, Mayer P, Oberdörster G, Kreyling WG. 2007. Efficient elimination of inhaled nanoparticles from the alveolar region: Evidence for interstitial uptake and subsequent reentrainment onto airways epithelium. *Environ Health Perspect* 115:728–733.

Sram RJ, Binkova B, Dejmek J, Bobak M. 2005. Ambient air pollution and pregnancy outcomes: A review of the literature. *Environ Health Perspect* 113(4):375–382.

Stocks J, Dzateux C. 2003. The effect of parental smoking on lung function and development during infancy. *Respirology* 8(3):266–285.

Sung JC, Pulliam BL, Edwards DA. 2007. Nanoparticles for drug delivery to the lungs. *Trends Biotechnol* 25(12):563–570.

Tran MU, Weir AJ, Fanucchi MV, Murphy AE, Van Winkle LS, Evans MJ, Smiley-Jewell SM, Miller L, Schelegle ES, Gershwin LJ, Hyde DM, Plopper CG. 2004. Smooth muscle development during postnatal growth of distal bronchioles in infant rhesus monkeys. *J Appl Physiol* 97(6):2364–2371.

Tsuda A, Filipovic N, Haberthür D, Dickie R, Stampanoni M, Matsui Y, Schittny JC. 2008. The finite element 3D reconstruction of the pulmonary acinus imaged by synchrotron X-ray tomography. *J Appl Physiol* 105:964–976.

Tsuda A, Henry FS, Butler JP. 1995. Chaotic mixing of alveolated duct flow in rhythmically expanding pulmonary acinus. *J Appl Physiol* 79:1055–1063.

Tsuda A, Laine-Pearson FE, Hydon PE. 2011. Why chaotic mixing of particles is inevitable in the deep lung. *J Theor Biol* 286:57–66.

Tsuda A, Rogers RA, Hydon PE, Butler JP. 2002. Chaotic mixing deep in the lung. *Proc Natl Acad Sci U S A* 99:10173–10178.

Ueda K, Cho K, Matsuda T, Okajima S, Uchida M, Kobayashi Y, Minakami H, Kobayashi K. 2006. A rat model for arrest of alveolarization induced by antenatal endotoxin administration. *Pediatr Res* 59(3):396–400.

Vidic B, Burri PH. 1983. Morphometric analysis of the remodeling of the rat pulmonary epithelium during early postnatal development. *Anat Rec* 207(2):317–324.

Wang L, Pinkerton KE. 2007. Air pollutant effects on fetal and early postnatal development. *Birth Defects Res C Embryo Today* 81(3):144–154.

Zeltner TB, Burri PH. 1987. The postnatal development and growth of the human lung. II. Morphology. *Respir Physiol* 67(3):269–282.

Zeltner TB, Caduff JH, Gehr P, Pfenninger J, Burri PH. 1987. The postnatal development and growth of the human lung. I. Morphometry. *Respir Physiol* 67(3):247–267.

Zoetis T, Hurtt ME. 2003. Species comparison of lung development. *Birth Defects Res B Dev Reprod Toxicol* 68(2):121–124.

20

Nanotoxicology

Dominique Balharry, Eva Gubbins, Helinor Johnston, Ali Kermanizadeh, and Vicki Stone

CONTENTS

20.1 Introduction

As the production and use of nanomaterials (NMs) in medicine and many other applications develops, so the need to understand the potential risks posed by NMs to human health (and the environment) increases (Aitken et al. 2006). At the nanoscale (1–100 nm), materials exhibit properties that are different to larger or bulk materials. These new properties are exploited by researchers and industry to generate new products; however, the same properties can also influence how the NM behaves in biological systems, including affecting toxicity. Nanotoxicology is a relatively new field of research that aims to assess the human and environmental hazard of nanomaterials. In recent years, this new discipline has seen a rapid expansion in the number of studies concerned with assessing the safety of engineered NMs (Figure 20.1).

20.2 Ultrafine Particle Toxicology

Particle toxicology has a long history dating back to the middle ages where writings observing lung health in miners were first recorded by Paracelsus and Agricola (Gochfeld 2005). The study of the toxicity of NMs has evolved from the knowledge gained from particulate

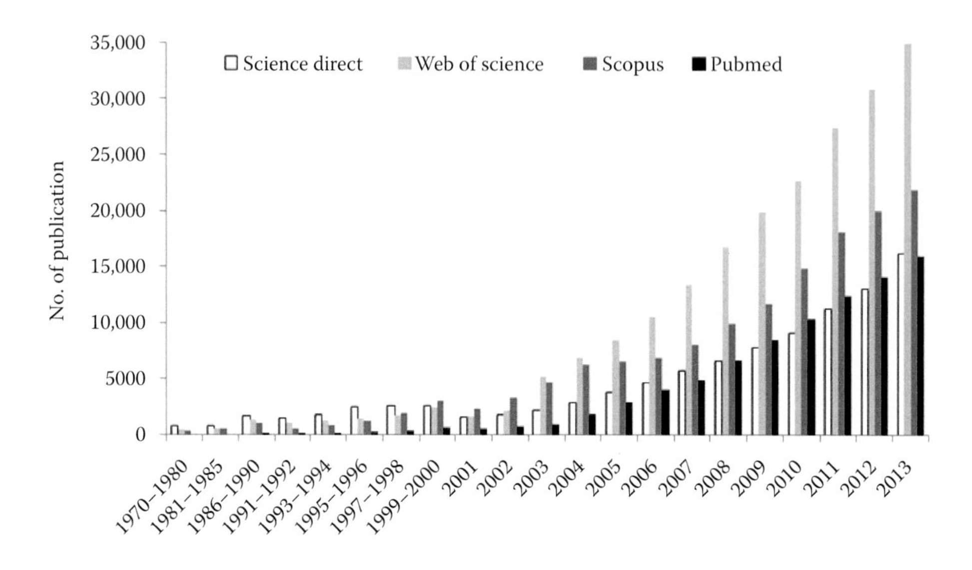

FIGURE 20.1
The number of articles identified per year from 1970 to the end of 2013, using a web-based search of online publications databases (NCBI PubMed, Science Direct, Web of Science and Scopus). Searches were carried out using the Boolean search terms "nanotoxicology OR nanotoxicity OR NPs OR ultrafine."

toxicology, and from research into the adverse health effects of air pollution, in particular focusing on respirable and ultrafine particles (Oberdorster et al. 2005a). Ultrafine particles are defined as particles <100 nm in diameter, and are therefore nanosized; however, they are not manufactured intentionally and in general are not of a uniform size or composition. It was well known by the 1900s that air pollution could be associated with adverse effects (Wilson and Spengler 1996), and by the 1990s epidemiology evidence had identified that ambient air particle exposure is associated with increased disease and mortality (Dockery et al. 1993).

Today there is considerable evidence that for respirable air pollution, particulate matter with dimensions of 2.5 μm or less (PM2.5), ultrafine particles are the most toxic component on a per mass basis. An increase in particulate matter, including the PM2.5 and ultrafine fractions, has been found to correlate with morbidity and mortality resulting from cardiovascular and respiratory distress in humans (Peters et al. 2001; Pope et al. 2002, 2004; Kappos et al. 2004). This has led to the "ultrafine hypothesis," which states that very small particles can be associated with distinct biological effects (Seaton et al. 1995). There are various toxicological mechanisms by which particles exert their effect, which will depend on exposure dose, particle type (physicochemical characteristics), the initial deposition site, and translocation (Ayres et al. 2006).

It has been proposed that the ultrafine paradigm of particle toxicology is also applicable to NMs, indicating that nanotoxicology as a discipline essentially follows on from the well-established particle toxicology field. A more thorough description of the history of nanotoxicology related to ultrafine particle toxicology has been provided by Oberdorster et al. (2007).

The research first conducted on ultrafine particles identified particle-induced reactive oxygen species (ROS) and oxidative stress as playing a key role in the mechanisms of toxicity (Stone et al. 1998) and pro-inflammatory signaling (Stone et al. 2000; Brown et al. 2004). Nel et al. (2006) describe the hierarchical oxidative stress hypothesis, where a lower level of oxidative stress is associated with the induction of antioxidant and detoxification enzymes. At higher levels of oxidative stress, activation of pro-inflammatory mediators can lead to inflammation and cytotoxicity, and prolonged or excessive ROS generation can lead to cell death. It is widely accepted that ROS can damage virtually any biomolecules, and oxidative stress is a significant event in the mechanism for inflammation-based chronic degenerative diseases (Galli et al. 2005), such as atherosclerosis, pulmonary fibrosis, cancer, neurodegenerative diseases, and aging (Thannickal and Fanburgh 2000). Therefore, the potential for ultrafine particles to induce oxidative stress and inflammation is a significant mechanism by which these particles can enhance existing disease or induce new disease.

Increased understanding of the toxicity of nanoscale particulates, derived from the ultrafine studies described above, prompted concern about the potential adverse health effects of other nanoscale materials. This was considered of particular importance given the predicted increase in production and use of NMs, leading to an increase in human and environmental exposure (Royal Society/Royal Academy of Engineering 2004; Royal Commission on Environmental Pollution 2008). A pressing need still exists to conduct the in-depth risk assessments required to inform the regulatory decision-making process on the safety of NMs, which is why nanotoxicology continues to develop and gain momentum.

20.3 Current Status of Nanotoxicology

Since the publication of reports identifying the need to better understand the hazard of nanomaterials (e.g., Royal Society/Royal Academy of Engineering 2004; Royal Commission on Environmental Pollution 2008), there has been a steady increase in funding and therefore research output in this area (Figure 20.1). Much of the work continues to focus on the lung as a route of exposure and as a target (Figure 20.2). In comparison, little work has been conducted in relation to the hazard of ingested nanomaterials, despite their use in food, food packaging, and health remedies (Figure 20.2). There are also relatively few published studies on dermal exposure and effects (Figure 20.2); however, this may, in part, be due to the current generally accepted understanding that intact, healthy skin forms a good and effective barrier to NMs (e.g., Landsiedel et al. 2012). However, studies on compromised skin with reduced barrier function (e.g., sunburned, dermatitis) are lacking.

Inhalation is a major route of exposure to NMs and, as stated in the preceding paragraph, a large proportion of nanotoxicology has been carried out following pulmonary exposure (Figure 20.2). The pulmonary bias also stems from the natural progression of nanotoxicology from ultrafine and air pollution particle toxicology. The lung not only has physical defense mechanisms that can filter and eliminate particles (i.e., nasal hairs and the mucociliary escalator) but also specialized cells such as type I and type II epithelial cells and macrophages (interstitial, intravascular and alveolar macrophages). Type I cells contain numerous small pinocytotic vesicles that are involved in the turnover of pulmonary surfactant and the removal of small particles from the alveolar surfaces, while type II cells mediate immune responses and act as type I progenitor cells to repair damaged epithelium. As discussed in Chapter 6, macrophages play a key role in particulate clearance from the lung; however, Renwick et al. (2001) showed a significant reduction in the ability of macrophages (J774.2) to phagocytose particles (fine TiO_2, ultrafine TiO_2, carbon black [CB], and ultrafine CB) and attributed this slowed clearance, in part, to a particle-mediated impairment of macrophage phagocytosis. Such an effect may lead to persistence,

Key: n = no. of publications (NCBI Pubmed search, October 2012)

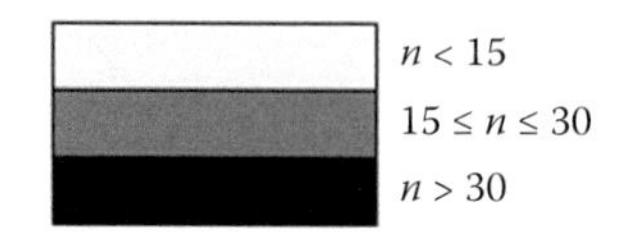

(a) Pulmonary (local)

	Biological Impact					
PC Properties	Cytotox	Inflam	Ox Stress	Fibrosis	Genotox	Carcinog
Size	58	156	56	17	6	7
SA	10	33	13	2	2	3
Charge	1	3	1	1	0	0
Aspect ratio	1	3	0	0	0	2
Solubility	4	6	4	1	3	1
Crystallinity	0	0	0	0	0	0
Composition	3	9	5	0	0	0

(b) Pulmonary (systemic)

	Biological Impact						
Target	Biokinetics	Cytotox	Inflam	Ox Stress	Fibrosis	Genotox	Carcinog
Lung	106	91	275	95	23	17	17
Liver	37	8	13	7	1	1	1
Spleen/immune	26	4	14	4	1	0	1
CNS	22	4	13	7	1	1	1
GI tract	2	1	2	0	0	0	0
Kidney	18	7	4	2	1	0	1
CV	32	15	51	25	1	1	1
Repro/dev	2	0	2	0	1	1	1
Pleura (retention)	0	0	9	3	0	1	0

Acute → Chronic

(c) Ingestion (local)

	Biological Impact					
PC Properties	Cytotoxicity	Inflam	Ox Stress	Fibrosis	Genotox	Carcinog
Size	3	2	2	0	1	3
SA	1	0	1	0	0	0
Charge	0	0	0	0	0	0
Aspect ratio	0	0	0	0	0	0
Solubility	0	0	0	0	0	0
Crystallinity	0	0	0	0	0	0
Composition	0	0	0	0	0	0

FIGURE 20.2
"Heat maps" showing the number of publications identified from nanospecific key word searches on National Center for Biotechnology Information's PubMed Central database. (a) Pulmonary exposure, local effects (mechanistic studies); (b) pulmonary exposure, systemic effects; (c) ingestion exposure, local effects. Cytotox = cytotoxicity; Inflam = inflammation; Ox stress = oxidative stress; Genotox = genotoxicity; Carcinog = carcinogenicity.

(Continued)

(d) Ingestion (systemic)

	Biological Impact						
Target	Biokinetics	Cytotox	Inflam	Ox Stress	Fibrosis	Genotox	Carcinog
Lung	4	3	1	0	0	0	2
Liver	5	2	0	1	0	1	0
Spleen/immune	1	0	1	1	0	0	0
CNS	3	0	0	1	0	0	0
GI tract	16	7	2	4	0	2	3
Kidney	3	2	0	0	0	1	0
CV	5	0	0	1	0	1	1
Repro/dev	0	1	0	2	0	0	0
Pleura (retention)	1	2	0	1	0	1	0

Acute → Chronic

(e) Dermal (local)

	Biological Impact					
PC Properties	Cytotox	Inflam	Ox Stress	Fibrosis	Genotox	Carcinog
Size	23	16	10	6	5	2
SA	1	2	1	0	0	1
Charge	0	0	0	0	0	0
Aspect ratio	0	0	0	0	0	0
Solubility	1	1	0	0	1	1
Crystallinity	0	0	0	0	0	0
Composition	1	1	1	0	0	0

(f) Dermal (systemic)

	Biological Impact						
Target	Biokinetics	Cytotox	Inflam	Ox Stress	Fibrosis	Genotox	Carcinog
Lung	1	0	1	0	0	0	0
Liver	5	1	1	1	2	1	1
Spleen/immune	4	1	2	0	0	0	1
CNS	2	1	0	1	2	0	0
GI tract	1	0	1	0	0	0	0
Kidney	3	0	0	0	1	0	0
CV	4	0	1	0	0	0	0
Repro/dev	0	0	0	0	0	0	0
Pleura (retention)	0	0	0	0	0	0	0

Acute → Chronic

FIGURE 20.2 (CONTINUED)
"Heat maps" showing the number of publications identified from nanospecific key word searches on National Center for Biotechnology Information's PubMed Central database. (d) Ingestion exposure, systemic effects; (e) dermal exposure, local effects; (f) dermal exposure, systemic effects. Cytotox = cytotoxicity; Inflam = inflammation; Ox stress = oxidative stress; Genotox = genotoxicity; Carcinog = carcinogenicity. The full method and detailed description is given in the "Identification of Knowledge Gaps and Strategic Priorities for Human and Environmental Hazard, Exposure and Risk Assessment of Engineered Nanomaterials" document carried out for the ITS-NANO project (http://www.nano.hw.ac..uk/research-projects/itsnano.html).

translocation to other body compartments, interaction with other cell types, and induction of a biological response. Compounding this is the ability of transition metals (e.g., zinc and iron) to potentiate the toxic effects of NMs (Wilson et al. 2007).

NMs that are effectively cleared from the lungs may potentially exert toxic effects in the gut, or may translocate via the circulatory system to a different tissue or organ. An ever-increasing number of *in vitro* studies find that NMs of many types are taken up by a wide variety of mammalian cell types (e.g., hepatocytes, macrophages, fibroblasts, epithelial cells, and red blood cells) (Johnston et al. 2010; Sohaebuddin et al. 2010; Mu et al. 2012; Wang et al. 2012). *In vivo* studies show evidence that owing to their nanoscale size, NMs can cross barriers from lungs and the gastrointestinal tract, into the bloodstream, and migrate to other parts of the body such as the liver (Jani et al. 1989; Oberdorster et al. 2002, 2005a, 2007). A further understanding of how NMs are taken up into the body, how they are distributed, and how they are cleared is essential, as this research will maximize the potential of engineered and medical NMs. A key role for nanotoxicology is to identify and investigate potential systemic toxicity resulting from translocation of NMs from their portal of entry into various other organs. If a potential target organ can be identified, the mechanism by which the material can produce its effect at this secondary site must be considered. The rate and frequency of exposure, as well as the absorption, distribution, metabolism, and elimination, which influence the fate of NMs in the body, are crucial factors when considering the toxicity of NMs. Many *in vivo* studies have focused on pulmonary toxicity (e.g., Jani et al. 1989; Brown et al. 2007; Poland et al. 2008; Brain et al. 2009; Murphy et al. 2012); however, NMs such as those designed for drug delivery (e.g., iron oxide NMs used in magnetic hyperthermia) can also be deliberately introduced into the body by injection. In such cases, NMs are generally developed to have specific coatings to increase blood circulation time, reduce adsorption of protein, and target NMs to specific cell types (Walczyk et al. 2010).

The systemic distribution of NMs to secondary target organs following various routes of exposure has been investigated in a number of studies. Biodistribution studies have shown that following exposure by various routes, NMs will translocate and accumulate in various organs, and that the pattern of distribution tends to be size and charge dependent (Semmler-Behnke et al. 2008; Hirn et al. 2011; Schleh et al. 2012). Following exposure, NMs can be detected in many organs of the body, including the lung, kidney, heart, brain, and spleen, with the greatest accumulation tending to be in the liver (e.g., Oberdorster et al. 2002; De Jong et al. 2008; Sonavane et al. 2008; Kreyling et al. 2009; Sadauskas et al. 2009; Almeida et al. 2011; Schleh et al. 2012). It is largely unknown whether there is a toxic effect at these distal sites; however, tissues, cells, and organs have limited and well-defined responses to exposure to any form of toxicant or substance. To predict, or measure, these responses in nanotoxicology, it is important to consider areas such as (i) the association of NMs with cells, how they interact with cell membranes at the bio–nano interface, and their uptake into cells; (ii) direct toxicological effects such as the production of free radicals and oxidative stress; (iii) indirect effects such as the ability of ROS to induce inflammation; and (iv) damage to DNA or genotoxic potential, as some low soluble fibers/particles such as asbestos and crystalline silica are known carcinogens (Mossman et al. 1983, 1998).

NMs have been shown to penetrate cell membranes and transport into cells such as endothelial, pulmonary, macrophage, and neuronal cells (Zhao et al. 2011). Mechanisms of membrane penetration by NMs are not fully elucidated, as the process varies depending on the NM and cell type. NMs may cross the cellular membrane lipid bilayer by active uptake mechanisms (e.g., phagocytosis, pinocytosis, and clathrin- and caveolin-mediated endocytosis) or by passive penetration (Kunzmann et al. 2011; Wang et al. 2012). If NMs enter the cell by endocytosis, they are enclosed by endocytic vesicles and are not directly in contact with the cytosol of the cell; however, NMs internalized by membrane penetration enter the cytosol directly. As targeted entry of NMs into cells is an important feature of drug delivery, this is the subject of much ongoing research, and many cellular uptake pathways have been described for particle uptake (as well as drugs, bacteria, and viruses) (Doherty and McMahon 2009; Hubbell and Chilkoti 2012; Johnston et al. 2012). This is described in greater detail in Sections II and III. A study investigating a range of cell types, including HeLa (cervical cancer), A549 (lung carcinoma), and 1321N1 (brain astrocytoma) cells, concluded that no one mechanism of transport was responsible for uptake and suggested that even for one type of NM, the uptake mechanism could be different depending on the cell type (dos Santos et al. 2011a). As discussed in Chapters 3, 4, and 16, an emerging area of investigation is the concept of the NM "corona" or the ability of NMs to absorb biomolecules onto their surface (particularly proteins, therefore often termed the protein corona). This is central in shaping the surface properties, charge, aggregation behavior, and size of NMs within a biological system, and has been found to affect the amount of uptake into cells (Chithrani et al. 2006). Hypothetically, this could also affect the interaction of NMs with cell surface receptor proteins. An interaction such as this may lead to denaturation or malformation of these cell membrane receptors, leading to the induction of a biological response causing changes in the signaling pathway of target cells. However, this hypothesis could potentially lead away from the current paradigm of nanotoxicity, which is the role of oxidative stress–induced toxicity, unless ROS play a role in protein corona denaturation or alteration.

Oxidative stress, as described for ultrafine particles, remains a leading paradigm for NM toxicity (Donaldson and Borm 2007), resulting from highly reactive free radicals that react with cellular molecules (lipids, nucleic acids, etc.) leading to oxidation. There is evidence from many published studies (Duffin et al. 2002, 2007; Donaldson and Stone, 2003; Oberdorster et al. 2005a) that NMs have the ability to induce oxidative stress. NMs may cause oxidative stress either by cellular-mediated ROS production or by particle-mediated cellular ROS and reactive nitrogen species production (Duffin et al. 2007). A variety of nanomaterials, including CB (Stone et al. 1998), carbon nanotubes (CNTs) (Rothen-Rutishauser et al. 2010), fullerenes (Hotze et al. 2008), polystyrene beads (Brown et al. 2001), and metal oxides (Han et al. 2012) have been demonstrated to generate ROS in a cell-free system. Techniques used to estimate these measurements include fluorescent dyes (e.g., dichlorofluorescin; Wilson et al. 2002; Foucaud et al. 2007; Rothen-Rutishauser et al. 2010) and electron paramagnetic resonance (Hotze et al. 2008). In addition, these studies have often been linked to the ability of NPs to then produce ROS in cells (Wilson et al. 2002) or oxidative stress in cells/tissues (Stone et al. 1998; Kermanizadeh et al. 2013). Increased oxidative stress depletes cellular antioxidants and leads to the activation of transcription factors such as nuclear factor-κB, which activate the expression of pro-inflammatory cytokines and chemokines causing inflammation (e.g., Brown et al. 2004). Again, much research has focused on the role of inflammation in the mechanism of NM-induced toxicity or disease (Figure 20.2).

Oxidative stress results from, or is enhanced by, transition metal–derived ROS. The mechanism of toxic action of other particles such as asbestos has been shown to involve the ability of iron present in the fibers to form hydroxyl radicals due to iron-catalyzed reactions (Fenton chemistry). These hydroxyl radicals can go on to cause lipid peroxidation and DNA strand breaks. In addition, metal ions have been shown to enhance the ability of CB NPs to induce pro-inflammatory responses by macrophages *in vitro* (Wilson et al.

2002, 2007) and in the lung (Wilson et al. 2002), suggesting that it is potentially important to characterize and assess metal contamination of NMs.

In addition to promoting inflammation, NM-derived ROS and oxidative stress can lead to activation of apoptotic pathways, necrosis, fibrosis, genotoxicity, and cancer. Many of the nanotoxicology studies investigating genotoxicity have focused on DNA damage mediated by ROS. There are diverse cellular responses to genotoxicity, and these responses can vary depending on the different mechanisms of action. DNA damage stimulated by ROS (generated following NM exposure) is thought to be driven by the physicochemical properties attributed to NMs (e.g., surface properties, size, shape, crystallinity, and solubility). Other factors, all of which may respond differently at the nanoscale versus bulk, such as the cellular uptake, interaction with cell processes, and the presence of contaminants on the NM surface may also influence genotoxicity (Schins 2002). There are a number of standard toxicology assays for measuring genotoxicity resulting from direct oxidative interference with DNA (Lorge et al. 2007; Landsiedel et al. 2009). Most of these assays appear to be transferable to nanotoxicology studies and have successfully been used to assess the genotoxic impacts of NMs (e.g., Shinohara et al. 2009; Doak et al. 2012; Kermanizadeh et al. 2012).

Particle size can influence the fate and behavior of NMs in the body, resulting in altered responses to the exposure (Borm et al. 2006). For example, much research in the area of inhaled particles, which have been obtained predominantly from rodent models, finds that the mechanism of inhaled particle deposition is to a great extent dependent on the aerodynamic diameter of the inhaled particles (Geiser and Kreyling 2010). When particles are inhaled, the size will determine not only the mechanism of deposition but also where in the lung it is deposited. Particles in the micrometer range will deposit preferentially in the upper airways, or mouth and throat, with only a fraction of the inhaled amount reaching deep into the lung, whereas smaller particles tend to deposit throughout the lung, and with greater efficiency in the alveolar region (Geiser and Kreyling 2010). Oberdorster et al. (1994, 2005a) have reported that the situation for NMs is a little more complicated, and the deposition behavior in the respiratory system is not linear with size, with NPs depositing efficiently throughout the respiratory tract, with peaks in both the nasopharyngeal–laryngeal and alveolar regions. In general, NMs in the size range 10–50 nm are deposited mainly in the alveoli, while smaller and larger particles are more efficiently deposited in the higher regions. NMs will not necessarily remain at their sites of deposition in the respiratory tract; instead, they can undergo numerous transport processes, including clearance from the lungs (Geiser and Kreyling 2010; Kreyling et al. 2013). A study by Möller et al. (2008) on the retention times of NMs in the airways also points to a size relationship. For example, a human inhalation study of 100-nm carbon particles found that 25% of the particles were cleared after 1 day (due to mucociliary action), and 75% of the particles persisted in the airways for more than 48 h. It has been proposed that the increased NM retention time (compared with microparticles) is due to the particles no longer being accessible for clearance. This could be because of deeper penetration of the NM through the mucosal layer (so mucociliary clearance is no longer possible), or that they were deposited in areas with a reduced lung lining layer (Geiser and Kreyling 2010).

Data from a pulmonary toxicity study in rats show that inhalation exposure to NMs (TiO_2, polystyrene beads, and CB) results in enhanced toxicity responses when compared with microsized particles with a comparable chemical composition. For example, in the case of TiO_2 at high dose concentrations, the ultrafine TiO_2 caused inflammation and at low doses resulted in pro-inflammatory effects (Donaldson et al. 2001; Oberdorster 2001), suggesting that the NMs may have greater biological activity than larger particles. *In vitro* experiments have also demonstrated size-specific effects; for example, in rat alveolar macrophages and human blood monocyte cells, 14-nm CB NPs were shown to induce calcium signaling, including a calcium influx (Stone et al. 2000). This activated transcription factors and stimulated significant production of the pro-inflammatory cytokine tumor necrosis factor (TNF)-α, when compared with treatment with 260-nm CB (Brown et al. 2004). Similar pro-inflammatory responses to NPs have been observed in other cell types, such as alveolar epithelial cells (A549s), which have also been shown to induce increased interleukin-8 expression in comparison with bulk equivalents (e.g., Brown et al. 2001). It is proposed that, in most cases, NPs will elicit a greater inflammatory response, in various cell types, than the equivalent microparticle of the same chemical composition (Oberdorster et al. 2005a). This apparent increase in biological reactivity may, in part, be due to increased uptake of NPs into cells. It has been suggested that, in general, particles >100 nm in size will less readily enter cells (e.g., Chithrani and Chan, 2007; Clift et al. 2008), and that those <40 nm may have greater access to the nucleus and the mitochondria (Brain et al. 2009). However, the specific mechanisms of interaction and uptake of NMs into various cell types is still under investigation, and it is known that specific cells may respond differently to materials of various sizes (dos Santos et al. 2011b). For example, while it has been shown that smaller particles may enter macrophages more rapidly and to a greater extent than larger particles (e.g., Clift et al. 2008; dos Santos et al. 2011b), it has also been demonstrated that some particles <200 nm can avoid phagocytic uptake (e.g., Geiser et al. 2008; Li et al. 2013). These differences may be related to the specific mechanism of uptake, as it has been suggested that although phagocytosis is the major pathway for macrophage internalization of particles, smaller particles are primarily processed by endocytotic pathways (e.g., Pauluhn, 2009; Lansiedel et al. 2012; Herd et al. 2013).

As the size of NMs decreases, so the surface area per unit mass increases and forces such as gravity become insignificant compared with those of surface forces like van der Waals, electrostatic forces, and chemical bonding (Hosokawa et al. 2012). As the surface area of a particle increases, a greater proportion of its atoms or molecules are exposed at the surface rather than contained within the interior of the material. An increased surface-to-bulk ratio has been linked to greater biological reactivity of NMs (Duffin et al. 2002, 2007), providing a relatively larger surface area for interaction with cells and molecules. A larger surface area of course provides more potential for molecules to adsorb (Johnston et al. 2012) and relates to the corona hypothesis (Lundqvist et al. 2011; Tenzer et al. 2011), which is discussed in the last paragraph of this section (20.3). The positive effects of a large surface-to-bulk ratio is being exploited in nanomedicine in creating novel ways of enabling drug delivery. However, it is vital to consider any possible negative toxic effects (side effects), as the interaction of NMs with subcellular structures may be different from that of the bulk material. It is therefore important when considering dose and effect, that properties such as size and surface area are investigated (Duffin et al. 2002, 2007; Oberdorster et al. 2007).

Shape is another important factor to consider in NM toxicity, as evidence from investigating materials such as asbestos has shown that shape can drive the biological response (Mossman et al. 1998; Donaldson and Borm 2000). Fiber-like NMs such as CNTs have been investigated for shape-related toxic effects. For example, Brown et al. (2007) demonstrated the connection between morphology and cellular response, finding that a monocyte cell line exposed to long, straight multiwalled CNTs produced significantly more frustrated phagocytosis associated with TNF-α and ROS, compared with highly curved materials. Poland et al. (2008) showed a similar result *in vivo*, where long straight nanotubes injected into the abdominal cavity of mice produced a similar result to that of long fiber amosite asbestos, causing an inflammatory response associated with thickening of the mesothelium. This was in contrast to the entangled nanotubes, which did not produce a significant inflammatory response. A study on titanium dioxide NMs of different shapes concluded that alveolar macrophages had more difficulty in processing longer fibers than the shorter fibers, and concluded that toxicity was altered by the shape of the NM (Hamilton et al. 2009). In recent years, much research has been carried out investigating whether fibers such as CNTs translocate in a similar way to biopersistent asbestos owing to their fiber-like shape. There is growing evidence that suggests CNT are able to translocate; for example, multiwalled CNTs *in vivo* (inhalation or aspiration) translocate to subpleural tissue, where they cause fibrosis (Ryman-Rasmussen et al. 2009; Porter et al. 2010). While many studies on CNT have focused on shape, other features of their physicochemical characteristics (e.g., hydrophobicity, functionalization, and metal content) are also likely to influence their toxicity.

The charge of NMs is another aspect that needs consideration. The charge of NMs, measured as the zeta potential (ZP), is a measure of the electrostatic forces and will vary depending on the material of interest. Any surface coatings or adsorbed material will affect this ZP measurement, and therefore a single ZP measurement without regard to the biological environment (pH adsorbed molecules, etc.) will not be relevant. As the cell wall has a negative surface charge (Silva et al. 1987), it has been proposed that positively charged particles would be attracted to the cell, which may potentially lead to greater toxicity. Indeed, many *in vivo* and *in vitro* studies have found that cationic particles are more cytotoxic (Nel et al. 2009); for example, Goodman et al. (2004) found cells exposed to gold NMs were more likely to induce hemolysis and platelet aggregation than anionic particles (Goodman et al. 2004). However, there is conflicting evidence, and the effect of surface charge on toxicity has been reported for both anionic and cationic particles (Lockman et al. 2004). This may be due, at least in part, to the surface charge affecting what molecules (such as proteins and lipids) adhere to the surface of the NP, rather than direct associations between the charged surface of the NP and the target tissue/cell.

This ability of NMs to bind biological molecules such as proteins and lipids, once in the body, is an important factor in nanotoxicity. After particles and NMs enter the body, they immediately come into contact with biological fluids, the content of which is dependent on the route of entry (Cedervall et al. 2007; Lynch and Dawson 2008; Brain et al. 2009; Monopoli et al. 2011). It is well known that early biological responses to foreign objects in the body are influenced by their attached protein (Hara et al. 2006), and the ability to control what binds to the surface of an NM leads to the potential for numerous applications, for example, in nanobiotechnology (Salata 2004; Parashar et al. 2008). As discussed in Chapter 16, much research has been carried out on these protein–surface interactions in the biomedical industry; however, to date, there is limited understanding of how these interactions may influence NM fate and behavior within the body, and hence toxicity.

20.4 Designing a Nanotoxicology Study

The increased research output in relation to nanotoxicology has provided information not only on the mechanisms of biological interaction but also on how study design can influence results and result interpretation (Stone et al. 2009). This has led to an improved understanding of the many parameters that should be considered when designing a nanotoxicology study (Johnston et al. 2013). Such parameters include

1. Physicochemical characterization of the NM
2. Suitable methods of NM dispersion
3. Identification of relevant NM doses
4. Study of appropriate *in vivo* models, targets, and endpoints
5. Development of improved *in vitro* alternatives
6. Design of hazard studies to allow hazard data to be used by industry and regulators for risk assessment purposes

There are a number of efforts focusing on the development and potential standardization of each of these stages in order to progress the field, which requires comparable output from different sources.

20.4.1 Physicochemical Characterization of Nanomaterials

There are multiple reasons for assessing the physicochemical characteristics of NMs. First, characterization allows confirmation or identification of the NM being studied. Second, it is not possible to consider the biological reactivity of all NMs as one class or group (Hussain et al. 2005; Soto et al. 2005), as different NMs of the same material may have diverse physicochemical characteristics according to their source and how they have been made. This may lead to differences in their toxicity, as the biological reactivity of NMs will vary according to their physicochemical properties (Aitken et al. 2009). There are a number of physicochemical properties that have been studied, including particle size, size distribution, specific surface area, surface structure, molecular and crystal structure, chemical composition (including surface coating), physical form, particle density, solubility, colloidal stability, bulk density, agglomeration state, porosity, and surface charge (Oberdorster et al. 2005b; Medley et al. 2007). As mentioned previously, owing to the influence of size, when NMs are composed of the same elements as their bulk counterparts they behave differently. Improved understanding of how the physicochemical characteristics influence toxicity will allow better prediction in future of NM toxicity, possibly through the use of *in silico* approaches. However, until researchers have a better understanding of how each individual characteristic influences toxicity, and the how the combined effects of different characteristics modify this response, it remains necessary to assess a wide array of physicochemical characteristics (Fubini et al. 2010; Bouwmeester et al. 2011; Bohnsack et al. 2012). Ideally, an extensive characterization of NMs would be performed before (pristine and dispersed), during (at several time points), and at the end of all hazard studies. Unfortunately, this is not realistic due to financial and time constraints, as well as due to limited access to advanced specialist technical expertise and instrumentation (Stone et al. 2010). In response to this, it has been suggested that routine characterization should be carried out for NM size, surface area, shape, crystal structure, aggregation/agglomeration, composition, surface chemistry, solubility, and charge (Warheit 2008). Methodological advances have improved the accuracy and reliability of characterization of NMs in complex medium and tissues; however, a long-term goal is to reduce the need for exhaustive characterization, with specific shortlists of characterization requirements that will vary depending on the requirements and research questions of the study.

20.4.2 Suitable Methods of Nanomaterial Dispersion

The physicochemical properties of NMs can be significantly affected by the procedures used in their preparation for experimental studies. This can especially influence their surface properties and agglomeration status and hence their potential toxicity.

The methods required for dispersion of NMs vary according to the desired dispersion characteristics. For example, in some studies, monodispersed solutions may be required (e.g., for regulatory or risk assessment purposes where this may present a "worst-case" situation). However, to determine real-life hazards or investigate the mode of action of an NM, for the majority of nanotoxicology studies a dispersion that is closer to the real-life scenario may be more appropriate. A number of approaches can be taken for the dispersion of NMs, including the use of dispersants (such as proteins, detergents, and solvents) and mechanical or physical processes (such as sonication, and manipulation of ion strength or pH). The dispersion characteristics can alter or enhance the toxicity of NMs (Foucaud et al. 2007); therefore, it is important to take this into account and prevent miscalculation of NM toxicity in cases where real-life exposure is likely to present the NM in a different state (i.e., highly dispersed vs. highly agglomerated). The preparation of NMs for hazard studies is unlikely to be the same for all scenarios, and the aim of the study should influence the approach taken, ensuring the dispersion is physiologically or environmentally relevant. This flexible approach should consider a number of experimental features, such as the NM under investigation (e.g., its hydrophobicity, its production and use), the likely exposure route (whether the NM is likely to gain a protein coating, etc.), and the potential for translocation from the exposure site. All of these aspects should be taken into consideration when planning dispersion protocols, with the ideal dispersion medium replicating the physiological conditions as closely as possible. For example, the protein coating bound to the surface of a NM will be different if exposure occurs via inhalation, than if via injection, which will likely alter the toxicity and translocation potential of the NM. Dispersants and mechanical processes (such as sonication) can influence not only the dispersion but also the toxicity as measured by a biological assay; for example, it has been suggested that the toxicity associated with C_{60} derives from the presence of residual solvent or its derivatives (Isakovic et al. 2006).

20.4.3 Identification of Relevant Nanomaterial Doses

The dose–response paradigm is central to toxicology; however, in the case of nanotoxicology, there is currently some debate about the best dose metric for determining this relationship. In toxicology studies, the mass dose of particles is normally used to quantify exposure (e.g., μg/ml, mg/kg, μg/cm^3); however, in nanotoxicology, relating dose to particle size (e.g., Ferin et al. 1992; Brown et al. 2001; Gaiser et al. 2009), surface area (e.g., Oberdorster 2000; Duffin et al. 2007; Jacobsen et al. 2009), or number (Ferin et al. 1992) may be more useful. Parameters such as surface area are logical, as it is the surface that interacts with cells. To allow comparisons across studies, and facilitate standardization and harmonization of nanotoxicology, a dose metric that is able to account for the response of cells and organisms to NMs would be beneficial. In addition to the potential dose metrics previously mentioned (surface area and particle number), other novel metrics and models are currently emerging (Teeguarden et al. 2007; Hinderliter et al. 2010).

Conventional monitoring approaches are generally based on a mass or number, and there is currently a lack of suitable technologies available for other parameters such as surface area (Abbott and Maynard 2010). It is a pressing requirement to develop technologies and methods that can accurately identify NMs, monitor, quantify, and measure their concentrations (number and/or mass) and physicochemical properties at various stages of their life cycle (manufacture, use and disposal) and in various media (e.g., blood, cells, food, air, and water). Even when the technology becomes available, there will be an additional lag while scientists accumulate data to ascertain the suitability of those new dose metrics in predicting hazard or risk.

This technological bottleneck means that for consumer products and environmental NMs, there is currently a lack of exposure data available that would enable researchers to determine the real-life concentrations of NMs at the site of action. The lack of exposure data is less of a problem for medical NMs where exposure dose is better controlled. However, inhalation or ingestion of a known amount of NM into the body does not reflect how much is actually absorbed or translocated to target sites.

While there are many challenges to selecting the most appropriate NM dose/concentration for nanotoxicology studies, the key factors remain: (i) identifying a physiologically relevant dose utilizing known exposure information; and (ii) investigating the dose–response relationship to determine toxicity of NMs at a range of doses. The dose–response relationship cannot be viewed in isolation when considering the "risk" of a substance as the risk assessment paradigm must be considered, which states that risk reflects both hazard (in this case toxicity) and exposure. Thus, while the relationship between a range of doses and toxicity may not directly reflect the risk associated with a specific level of predicted exposure, it is expected that the range of doses will capture levels anticipated in occupational, environmental, and consumer settings. This technique has proved useful in identifying the NOAEL and LOAEL (no/lowest observable adverse effect level), which are key for benchmarking and regulatory purposes (Aschberger et al. 2010a,b; Christensen et al. 2010 2011; Shinohara et al. 2011).

20.4.4 Study of Appropriate *In Vivo* Models, Targets, and Endpoints

To assess the toxicological effect of NMs, knowledge of the route of entry and fate of the NM are required. NMs can enter the body via inhalation, dermal contact, ingestion, injection, or implantation. Each of these routes will influence the toxicological consequence of the NM. Exposure protocols can vary for each route of exposure such as inhalation studies (in comparison with intratracheal instillation) or ingestion (vs. gavage). The most physiologically relevant method is often not feasible because of cost and technological requirements. Such results are easier to interpret if data allowing comparison between different models of exposure exist; however, this is not always possible. It is therefore often necessary to understand the limitations of each model used, and to build these into any conclusions drawn. The same is also true for the animal model employed, as *in vivo* models are often restricted to healthy rodents. Data are therefore required to allow extrapolation from rodents to humans, as well as to extrapolate from healthy rats to diseased humans. There is a need to develop better disease models in the future, especially since NMs have the potential to activate processes (oxidative stress and inflammation) that exacerbate many of the high-prevalence diseases (e.g., cardiovascular disease and cancer). Many animal studies with NMs have thus far concentrated on short-term end points (Figure 20.2) and so there is also a need in future studies to consider the long-term impacts, as well as the effect of repeat dosing (Stone et al. 2014).

Following exposure to medical NMs, there will be a requirement to assess translocation and biodistribution to ensure that the NM reaches its proposed target, to assess potential side effects and to determine elimination from the body (metabolism and excretion). Some medical NMs are easily traced because of their physicochemical properties. For example, paramagnetic iron oxide can be imaged by magnetic resonance imaging (MRI; Choi et al. 2004; Oghabian et al. 2006; Mohammadi-Nejad et al. 2010), while gold nanoshells and nanorods can be imaged in the near infrared range (NIR; Hirsch et al. 2003; Pissuwan et al. 2008). Other types of NM may be much more difficult to trace and detect but some chemical detection techniques (e.g., inductively coupled plasma–mass spectrometry [ICP–MS]) can provide information about their biodistribution. Techniques such as NIR and MRI have the advantage of being noninvasive, and also have greater sensitivity, allowing identification of intact NMs, which is not possible using techniques such as ICP–MS. Further technological advances to improve NM detection and localization would be extremely beneficial to aid biodistribution/biokinetics research.

The fate of NMs in the body will depend not only on the physicochemical properties of the NMs but also on the forces acting at the interface between the NM and the cell (Dockery et al. 1993; Nel et al. 2009). Studying this interaction of NMs with cell membranes is a crucial step in understanding nanotoxicity, as this interface offers a first line of defense after NM exposure to the body, acting as a barrier protecting the tissues/cells/cell organelles. A review by Nel et al. (2009) lists "hydrodynamic, electrodynamic, electrostatic (or ionic), solvent and steric interactions, and polymer binding" as the main forces governing the NM–cell interactions. As these interactions will vary according to the physicochemical makeup of the NM, this begins to explain why, if we are to understand NM behavior at the cell interface, a good knowledge of the NM physicochemical makeup is desirable.

As described previously, the mechanisms of toxicity of ultrafine particles in the lung and NMs, in general, are largely dependent on stimulation of ROS and oxidative stress leading to inflammatory responses. Therefore, many of the endpoints currently investigated in nanotoxicology tend to focus in this area. There are a number of relatively common approaches that assess oxidative stress, inflammation, cell signaling pathways, and genotoxicity while new techniques such as genomics, proteomics, and metabolomics are now being applied in an effort to gain a better understanding of the mechanisms of action of NM toxicity (Park et al. 2011).

A number of the biological assays used for determining NM toxicity can be affected by NM interference. This can lead to false-negative or false-positive data (Worle-Knirsch et al. 2006), reinforcing the usefulness of carrying out at least two independent test systems for each endpoint assessed.

20.4.5 Development of Improved *In Vitro* Alternatives

Many of the studies carried out to investigate NM hazard, are currently *in vitro* rather than *in vivo* (Figure 20.2). In nanotoxicology, as with many fields of biological and medical research, there is a desire to reduce the amount of *in vivo* testing that is required. Reliable, predictive *in vitro* systems would be beneficial in terms of ethics, time, and costs for providing alternatives to animal testing. A key aspect of all toxicology studies (including nanotoxicology) is ensuring that the studies are relevant to realistic and likely situations, and that *in vitro* studies can be related to actual physiological responses *in vivo*.

Often, these studies include a single cell type in static conditions with short-term endpoints. There has been criticism of the relevance of such models, indicating that investment is needed in the development of more complex models that include multiple cell types from one organ to make a three-dimensional (3D) culture system (e.g., Rothen-Rutishauser et al. 2005, 2008; Lehmann et al. 2011), but also cells from different organs linked by a fluidics system (e.g., European Commission project InLiveTox). Recent data suggest that the simple single-cell type cultures using serum-dispersed NMs are quite good at predicting pro-inflammatory and oxidative stress responses of liver tissue to injected NMs. In contrast, *in vitro* comparisons with results for exposure via the lung were less promising, while results following gavage were very different. It has been suggested that current *in vitro* models cannot accurately replicate the *in vivo* responses, and are therefore not appropriate for predicting a physiological response to a toxic substance (Donaldson et al. 2009). The difference in responses from *in vivo* to *in vitro* test systems often stems from the interaction between various cell types and cell signaling cascades that occur *in vivo*. In single cell cultures, this cell signaling and interaction is missing; however, progress with multiple cell cultures, and differentiated 3D *in vitro* tissue models are beginning to overcome these issues (Duell et al. 2011; Huh et al. 2011). There remains, however, some challenges in correctly interpreting and extrapolating *in vitro* toxicity measurements to *in vivo* effects (Sayes et al. 2007). Despite this, *in vitro* studies have successfully been used in many cases to increase the understanding of the interactions and mechanisms of NM toxicity in humans (e.g., Donaldson et al. 2008; Jacobsen et al. 2009; Saber et al. 2012).

The relevance of dose or concentration *in vitro* is often hotly debated. Since it is often difficult to ascertain the concentration of NMs in key targets in the body of animal models or humans, it is therefore difficult to decide what doses are relevant *in vitro*.

20.4.6 Use of Hazard Data for Risk Assessment by Industry and Regulators

The risk assessment paradigm is based on the statement that risk = exposure × hazard, and this holds true for risk assessment of NMs (Scientific Committee on Emerging and Newly Identified Health Risks [SCENIHR] 2006, 2009; EFSA 2009). However, risk assessments for NMs have been hampered by the lack of exposure and hazard data available, which limits the applicability of classical regulatory approaches (Aschberger et al. 2010a,b; Christensen et al. 2010, 2011). To make use of existing risk assessment methods for regulatory decision making, they should be properly adapted to address the novelties of NMs, for example, to include extended datasets, assess various physicochemical properties determining toxicity, consider different metrics, develop standard testing protocols and detection techniques, and properly evaluate existing uncertainties due to lack of knowledge and methodological gaps. The design of nanotoxicology hazard tests should be informed by the risk assessment requirements, ideally utilizing standardized, validated methods and technologies. As previously discussed, these standardized protocols are not yet available for NMs; however progress is being made in this area. All available hazard evidence, in conjunction with the increasing amount of information on exposure routes and dose, combine to begin to fill the existing knowledge gaps in this area of nanosafety. This area of research is currently undergoing rapid expansion, and as understanding and knowledge grows, this will enable improved risk assessments allowing more robust conclusions concerning the risk of NMs.

20.5 Nanotoxicology in the Future

Much of the knowledge relating to the hazard of bulk (not nanosized) materials is not easily comparable to the potential hazard induced by NMs owing to the inherent novelty exhibited at the nanoscale. Not only their small size but also the impact this has on their surface characteristics can lead to vastly different effects *in vivo* when compared with their bulk counterparts. Because of this, nanotoxicology is a rapidly evolving field with a requirement for new, or significantly altered methods and technologies to ensure hazard studies are relevant for nanosized materials, and can meet the demand of the increasing number of NMs and their expanding range of uses. These new technologies include a requirement to improve detection of nanomaterials within complex media, including consumer products, the environment, and biological tissues. These techniques are essential to better understand and quantify exposure and dose.

Owing to the vast array of NMs produced and in development, as well as the wide array of products in which they are incorporated, testing of every individual NM in production is not a possibility considering the amount of experimental work, time, money, and animals that would be required. Instead, strategies are required that in the shorter term reduce the amount of testing necessary, possibly through the development of high-throughput screening. To identify relevant endpoints or biomarkers for such screens, an understanding of the mechanisms of NM-induced toxicity is needed. For example, the ability of NMs to generate oxidative stress or a pro-inflammatory marker could be used as such an indicator. Longer-term strategies aim to develop *in silico* approaches, such as quantitative nanostructure activity relationships that predict toxicity based on the physicochemical characteristics of nanomaterials (e.g., the European Commission–funded project ENPRA).

There are some ongoing efforts to coordinate and standardize research in the area of nanosafety, with several research projects currently supporting the development of nanotechnology in a safe and sustainable fashion (e.g., Nanosafety Vision 2015–2020, and the European Commission–funded FP7 projects MARINA, Nanovalid, etc.). In particular, the FP7 project, ITS-NANO (2012), has put forward an overarching strategy aimed at developing an intelligent testing strategy for exposure, hazard, and risk assessment for human and environmental exposure to engineered nanomaterials (Stone et al, 2014). In the context of ITS-NANO, intelligent testing strategies means the development of a strategy that will provide an efficient and highly effective means of assessing the risks of nanomaterials, while integrated testing strategies refers to the use of data from multiple sources and combining them into a single analysis to determine risk. ITS-NANO has identified and set a priority research agenda that aims to reduce the research gaps according to the need of stakeholders (industry, regulators) and at the same time to be economical and ethical (i.e., to adhere to the 3Rs principle). A framework has been proposed for grouping of NMs based on their chemical, physical, and biological properties and on their subsequent exposure routes and biological impacts to intelligently design next-generation nanosafety evaluation and risk assessment strategies.

References

Abbott LC, Maynard AD. 2010. Exposure assessment approaches for engineered nanomaterials. *Risk Anal* 30(11):1634–1644.

Aitken R, Borm P, Donaldson K, Ichihara G, Loft S, Marano F, Maynard A, Oberdoerster G, Stamm H, Stone V et al. 2009. Nanoparticles—One word: A multiplicity of different hazards. *Nanotoxicology* 3(4):263–264.

Aitken RJ, Chaudhry MQ, Boxall AB, Hull M. 2006. Manufacture and use of nanomaterials: Current status in the UK and global trends. *Occup Med (Lond)* 56(5):300–306.
Almeida JP, Chen AL, Foster A, Drezek R. 2011. *In vivo* biodistribution of nanoparticles. *Nanomedicine (Lond)* 6(5):815–835.
Aschberger K, Johnston HJ, Stone V, Aitken RJ, Hankin SM, Peters SA, Tran CL, Christensen FM. 2010a. Review of carbon nanotubes toxicity and exposure—Appraisal of human health risk assessment based on open literature. *Crit Rev Toxicol* 40(9):759–790.
Aschberger K, Johnston HJ, Stone V, Aitken RJ, Tran CL, Hankin SM, Peters SA, Christensen FM. 2010b. Review of fullerene toxicity and exposure—Appraisal of a human health risk assessment, based on open literature. *Regul Toxicol Pharmacol* 58(3):455–473.
Ayres J, Maynard R, Richards R (eds.). 2006. *Air Pollution and Health*: Imperial College Press, London.
Bohnsack JP, Assemi S, Miller JD, Furgeson DY. 2012. The primacy of physicochemical characterization of nanomaterials for reliable toxicity assessment: A review of the zebrafish nanotoxicology model. *Methods Mol Biol* 926:261–316.
Borm PJ, Robbins D, Haubold S, Kuhlbusch T, Fissan H, Donaldson K, Schins R, Stone V, Kreyling W, Lademann J et al. 2006. The potential risks of nanomaterials: A review carried out for ECETOC. *Part Fibre Toxicol* 3:11.
Bouwmeester H, Lynch I, Marvin HJ, Dawson KA, Berges M, Braguer D, Byrne HJ, Casey A, Chambers G, Clift MJ et al. 2011. Minimal analytical characterization of engineered nanomaterials needed for hazard assessment in biological matrices. *Nanotoxicology* 5(1):1–11.
Brain JD, Curran MA, Donaghey T, Molina RM. 2009. Biologic responses to nanomaterials depend on exposure, clearance, and material characteristics. *Nanotoxicology* 3(3):174–180.
Brown DM, Donaldson K, Borm PJ, Schins RP, Dehnhardt M, Gilmour P, Jimenez LA, Stone V. 2004. Calcium and ROS-mediated activation of transcription factors and TNF-alpha cytokine gene expression in macrophages exposed to ultrafine particles. *Am J Physiol Lung Cell Mol Physiol* 286(2):L344–353.
Brown DM, Kinloch IA, Bangert U, Windle AH, Walter DM, Walker GS, Scotchford CA, Donaldson K, Stone V. 2007. An *in vitro* study of the potential of carbon nanotubes and nanofibres to induce inflammatory mediators and frustrated phagocytosis. *Carbon* 45(9):1743–1756.
Brown DM, Wilson MR, MacNee W, Stone V, Donaldson K. 2001. Size-dependent proinflammatory effects of ultrafine polystyrene particles: A role for surface area and oxidative stress in the enhanced activity of ultrafines. *Toxicol Appl Pharmacol* 175(3):191–199.
Cedervall T, Lynch I, Lindman S, Berggard T, Thulin E, Nilsson H, Dawson KA, Linse S. 2007. Understanding the nanoparticle–protein corona using methods to quantify exchange rates and affinities of proteins for nanoparticles. *Proc Natl Acad Sci U S A* 104(7):2050–2055.
Chithrani BD, Chan WC. 2007. Elucidating the mechanism of cellular uptake and removal of protein-coated gold nanoparticles of different sizes and shapes. *Nano Lett* 7(6):1542–1550.
Chithrani BD, Ghazani AA, Chan WC. 2006. Determining the size and shape dependence of gold nanoparticle uptake into mammalian cells. *Nano Lett* 6(4):662–668.
Choi H, Choi SR, Zhou R, Kung HF, Chen IW. 2004. Iron oxide nanoparticles as magnetic resonance contrast agent for tumor imaging via folate receptor-targeted delivery. *Acad Radiol* 11(9):996–1004.
Christensen FM, Johnston HJ, Stone V, Aitken RJ, Hankin S, Peters S, Aschberger K. 2010. Nanosilver—Feasibility and challenges for human health risk assessment based on open literature. *Nanotoxicology* 4(3):284–295.
Christensen FM, Johnston HJ, Stone V, Aitken RJ, Hankin S, Peters S, Aschberger K. 2011. Nano-TiO(2)—Feasibility and challenges for human health risk assessment based on open literature. *Nanotoxicology* 5(2):110–124.
Clift MJ, Rothen-Rutishauser B, Brown DM, Duffin R, Donaldson K, Proudfoot L, Guy K, Stone V. 2008. The impact of different nanoparticle surface chemistry and size on uptake and toxicity in a murine macrophage cell line. *Toxicol Appl Pharmacol* 232(3):418–427.
De Jong WH, Hagens WI, Krystek P, Burger MC, Sips AJ, Geertsma RE. 2008. Particle size-dependent organ distribution of gold nanoparticles after intravenous administration. *Biomaterials* 29(12):1912–1919.

Doak SH, Manshian B, Jenkins GJ, Singh N. 2012. *In vitro* genotoxicity testing strategy for nanomaterials and the adaptation of current OECD guidelines. *Mutat Res* 745(1–2):104–111.
Dockery DW, Pope CA, 3rd, Xu X, Spengler JD, Ware JH, Fay ME, Ferris BG, Jr., Speizer FE. 1993. An association between air pollution and mortality in six U.S. cities. *N Engl J Med* 329(24):1753–1759.
Doherty GJ, McMahon HT. 2009. Mechanisms of endocytosis. *Annu Rev Biochem* 78:857–902.
Donaldson K, Borm P. 2000. Particle paradigms. *Inhal Toxicol* 12:1–6.
Donaldson K, Borm P (eds.). 2007. *Particle Toxicology*: Taylor & Francis Group, Bocca Raton, Florida.
Donaldson K, Borm PJ, Castranova V, Gulumian M. 2009. The limits of testing particle-mediated oxidative stress *in vitro* in predicting diverse pathologies; Relevance for testing of nanoparticles. *Part Fibre Toxicol* 6:13.
Donaldson K, Borm PJ, Oberdorster G, Pinkerton KE, Stone V, Tran CL. 2008. Concordance between *in vitro* and *in vivo* dosimetry in the proinflammatory effects of low-toxicity, low-solubility particles: The key role of the proximal alveolar region. *Inhal Toxicol* 20(1):53–62.
Donaldson K, Stone V. 2003. Current hypotheses on the mechanisms of toxicity of ultrafine particles. *Ann Ist Super Sanita* 39(3):405–410.
Donaldson K, Stone V, Clouter A, Renwick L, MacNee W. 2001. Ultrafine particles. *Occup Environ Med* 58(3):211–216, 199.
dos Santos T, Varela J, Lynch I, Salvati A, Dawson KA. 2011a. Effects of transport inhibitors on the cellular uptake of carboxylated polystyrene nanoparticles in different cell lines. *PLoS One* 6(9):e24438.
dos Santos, T, Varela, J, Lynch, I, Salvati, A, Dawson, KA. 2011b. Quantitative assessment of the comparative nanoparticle-uptake efficiency of a range of cell lines. *Small* 7:3341–3349.
Duell BL, Cripps AW, Schembri MA, Ulett GC. 2011. Epithelial cell coculture models for studying infectious diseases: Benefits and limitations. *J Biomed Biotechnol* 2011:852419.
Duffin R, Tran CL, Clouter A, Brown DM, Macnee W, Stone V, Donaldson K. 2002. The importance of surface area and specific reactivity in the acute pulmonary inflammatory response to particles. *Ann Occup Hyg* 46(Suppl 1):242–254.
Duffin R, Tran L, Brown D, Stone V, Donaldson K. 2007. Proinflammogenic effects of low-toxicity and metal nanoparticles *in vivo* and *in vitro*: Highlighting the role of particle surface area and surface reactivity. *Inhal Toxicol* 19(10):849–856.
EFSA. 2009. Opinion on the potential risks arising from nanoscience and nanotechnologies on food and feed safety. *EFSA J* 958:1–39.
Ferin J, Oberdorster G, Penney DP. 1992. Pulmonary retention of ultrafine and fine particles in rats. *Am J Respir Cell Mol Biol* 6(5):535–542.
Foucaud L, Wilson MR, Brown DM, Stone V. 2007. Measurement of reactive species production by nanoparticles prepared in biologically relevant media. *Toxicol Lett* 174(1–3):1–9.
Fubini B, Ghiazza M, Fenoglio I. 2010. Physico–chemical features of engineered nanoparticles relevant to their toxicity. *Nanotoxicology* 4:347–363.
Gaiser BK, Fernandes TF, Jepson M, Lead JR, Tyler CR, Stone V. 2009. Assessing exposure, uptake and toxicity of silver and cerium dioxide nanoparticles from contaminated environments. *Environ Health* 8 Suppl 1:S2.
Galli F, Piroddi M, Annetti C, Aisa C, Floridi E, Floridi A. 2005. Oxidative stress and reactive oxygen species. *Contrib Nephrol* 149:240–260.
Geiser, M, Casaulta, M, Kupferschmid, B, Schulz, H, Semmler-Behnke, M, Kreyling, W. 2008. The role of macrophages in the clearance of inhaled ultrafine titanium dioxide particles. *Am J Respir Cell Mol Biol* 38:371–376.
Geiser M, Kreyling WG. 2010. Deposition and biokinetics of inhaled nanoparticles. *Part Fibre Toxicol* 7:2.
Gochfeld M. 2005. Chronologic history of occupational medicine. *J Occup Environ Med* 47(2):96–114.
Goodman CM, McCusker CD, Yilmaz T, Rotello VM. 2004. Toxicity of gold nanoparticles functionalized with cationic and anionic side chains. *Bioconj Chem* 15(4):897–900.
Hamilton RF, Wu N, Porter D, Buford M, Wolfarth M, Holian A. 2009. Particle length-dependent titanium dioxide nanomaterials toxicity and bioactivity. *Part Fibre Toxicol* 6:35.

Han, X, Corson, N, Wade-Mercer, P, Gelein, R, Jiang, J, Sahu, M, Biswas, P, Finkelstein, JN, Elder, A, Oberdorster, G. 2012. Assessing the relevance of *in vitro* studies in nanotoxicology by examining correlations between *in vitro* and *in vivo* data. *Toxicology* 297:1–9.

Hara H, Nakamura M, Palmaz JC, Schwartz RS. 2006. Role of stent design and coatings on restenosis and thrombosis. *Adv Drug Deliv Rev* 58(3):377–386.

Herd, H, Daum, N, Jones, AT, Huwer, H, Ghandehari, H, Lehr, CM. 2013. Nanoparticle geometry and surface orientation influence mode of cellular uptake. *ACS Nano* 7:1961–1973.

Hinderliter PM, Minard KR, Orr G, Chrisler WB, Thrall BD, Pounds JG, Teeguarden JG. 2010. ISDD: A computational model of particle sedimentation, diffusion and target cell dosimetry for *in vitro* toxicity studies. *Part Fibre Toxicol* 7(1):36.

Hirn S, Semmler-Behnke M, Schleh C, Wenk A, Lipka J, Schaffler M, Takenaka S, Möller W, Schmid G, Simon U et al. 2011. Particle size-dependent and surface charge-dependent biodistribution of gold nanoparticles after intravenous administration. *Eur J Pharm Biopharm* 77(3):407–416.

Hirsch LR, Stafford RJ, Bankson JA, Sershen SR, Rivera B, Price RE, Hazle JD, Halas NJ, West JL. 2003. Nanoshell-mediated near-infrared thermal therapy of tumors under magnetic resonance guidance. *Proc Natl Acad Sci U S A* 100(23):13549–13554.

Hosokawa M, Nogi K, Naito M, Yokoyama T (eds.). 2012. *Nanoparticle Technology Handbook*, 2nd edn: Elsevier, Philadelphia.

Hotze EM, Labille J, Alvarez P, Wiesner MR. 2008. Mechanisms of photochemistry and reactive oxygen production by fullerene suspensions in water. *Environ Sci Technol* 42(11):4175–4180.

Hubbell JA, Chilkoti A. 2012. Chemistry. Nanomaterials for drug delivery. *Science* 337(6092):303–305.

Huh D, Hamilton GA, Ingber DE. 2011. From 3D cell culture to organs-on-chips. *Trends Cell Biol* 21(12):745–754.

Hussain SM, Hess KL, Gearhart JM, Geiss KT, Schlager JJ. 2005. *In vitro* toxicity of nanoparticles in BRL 3A rat liver cells. *Toxicol In Vitro* 19(7):975–983.

Isakovic A, Markovic Z, Nikolic N, Todorovic-Markovic B, Vranjes-Djuric S, Harhaji L, Raicevic N, Romcevic N, Vasiljevic-Radovic D, Dramicanin M et al. 2006. Inactivation of nanocrystalline C60 cytotoxicity by gamma-irradiation. *Biomaterials* 27(29):5049–5058.

Jacobsen NR, Möller P, Jensen KA, Vogel U, Ladefoged O, Loft S, Wallin H. 2009. Lung inflammation and genotoxicity following pulmonary exposure to nanoparticles in $ApoE^{-/-}$ mice. *Part Fibre Toxicol* 6:2.

Jani P, Halbert GW, Langridge J, Florence AT. 1989. The uptake and translocation of latex nanospheres and microspheres after oral administration to rats. *J Pharm Pharmacol* 41(12):809–812.

Johnston H, Brown D, Kermanizadeh A, Gubbins E, Stone V. 2012. Investigating the relationship between nanomaterial hazard and physicochemical properties: Informing the exploitation of nanomaterials within therapeutic and diagnostic applications. *J Control Release* 164(3):307–313.

Johnston H, Pojana G, Zuin S, Jacobsen NR, Möller P, Loft S, Semmler-Behnke M, McGuiness C, Balharry D, Marcomini A et al. 2013. Engineered nanomaterial risk. Lessons learnt from completed nanotoxicology studies: Potential solutions to current and future challenges. *Crit Rev Toxicol* 43(1):1–20.

Johnston HJ, Semmler-Behnke M, Brown DM, Kreyling W, Tran L, Stone V. 2010. Evaluating the uptake and intracellular fate of polystyrene nanoparticles by primary and hepatocyte cell lines *in vitro*. *Toxicol Appl Pharmacol* 242(1):66–78.

Kappos AD, Bruckmann P, Eikmann T, Englert N, Heinrich U, Hoppe P, Koch E, Krause GH, Kreyling WG, Rauchfuss K et al. 2004. Health effects of particles in ambient air. *Int J Hyg Environ Health* 207(4):399–407.

Kermanizadeh A, Gaiser BK, Hutchison GR, Stone V. 2012. An *in vitro* liver model—Assessing oxidative stress and genotoxicity following exposure of hepatocytes to a panel of engineered nanomaterials. *Part Fibre Toxicol* 9(1):28.

Kermanizadeh A, Vranic S, Boland S, Moreau K, Baeza-Squiban A, Gaiser BK, Andrzejczuk LA, Stone V. 2013. An *in vitro* assessment of panel of engineered nanomaterials using a human renal cell line: Cytotoxicity, pro-inflammatory response, oxidate stress and genotoxicity. *BMC Nephrol* 14(96): doi: 10.1186/1471-2369-14-96.

Kreyling WG, Semmler-Behnke M, Seitz J, Scymczak W, Wenk A, Mayer P, Takenaka S, Oberdorster G. 2009. Size dependence of the translocation of inhaled iridium and carbon nanoparticle aggregates from the lung of rats to the blood and secondary target organs. *Inhal Toxicol* 21 Suppl 1:55–60.

Kreyling WG, Semmler-Behnke M, Takenaka S, Möller W. 2013. Differences in the biokinetics of inhaled nano- versus micrometer-sized particles. *Acc Chem Res* 46(3):714–722.

Kunzmann A, Andersson B, Vogt C, Feliu N, Ye F, Gabrielsson S, Toprak MS, Buerki-Thurnherr T, Laurent S, Vahter M et al. 2011. Efficient internalization of silica-coated iron oxide nanoparticles of different sizes by primary human macrophages and dendritic cells. *Toxicol Appl Pharmacol* 253(2):81–93.

Landsiedel R, Fabian E, Ma-Hock L, van Ravenzwaay B, Wohlleben W, Wiench K, Oesch F. 2012. Toxico-/biokinetics of nanomaterials. *Arch Toxicol* 86(7):1021–1060.

Landsiedel R, Kapp MD, Schulz M, Wiench K, Oesch F. 2009. Genotoxicity investigations on nanomaterials: Methods, preparation and characterization of test material, potential artifacts and limitations—Many questions, some answers. *Mutat Res* 681(2–3):241–258.

Lehmann AD, Daum N, Bur M, Lehr CM, Gehr P, Rothen-Rutishauser BM. 2011. An *in vitro* triple cell co-culture model with primary cells mimicking the human alveolar epithelial barrier. *Eur J Pharm Biopharm* 77(3):398–406.

Li, F, Zhu, A, Song, X, Ji, L, Wang, J. 2013. The internalization of fluorescence-labeled PLA nanoparticles by macrophages. *Int J Pharm* 453:506–513.

Lockman PR, Koziara JM, Mumper RJ, Allen DD. 2004. Nanoparticle surface charges alter blood–brain barrier integrity and permeability. *J Drug Targeting* 12(9–10):635–641.

Lorge E, Gervais V, Becourt-Lhote N, Maisonneuve C, Delongeas JL, Claude N. 2007. Genetic toxicity assessment: Employing the best science for human safety evaluation part IV: A strategy in genotoxicity testing in drug development: Some examples. *Toxicol Sci* 98(1):39–42.

Lundqvist M, Stigler J, Cedervall T, Berggard T, Flanagan MB, Lynch I, Elia G, Dawson K. 2011. The evolution of the protein corona around nanoparticles: A test study. *ACS Nano* 5(9):7503–7509.

Lynch I, Dawson KA. 2008. Protein–nanoparticle interactions. *Nano Today* 3(1–2):40–47.

Medley T, Walsh S, Partnership E. 2007. *Nano Risk Framework*: Environmental Defense–DuPont Nano Partnership, available at http://www.nanoriskframework.com/files/2011/11/6496_Nano-Risk-Framework.pdf.

Mohammadi-Nejad AR, Hossein-Zadeh GA, Soltanian-Zadeh H. 2010. Quantitative evaluation of optimal imaging parameters for single-cell detection in MRI using simulation. *Magn Reson Imaging* 28(3):408–417.

Möller W, Felten K, Sommerer K, Scheuch G, Meyer G, Meyer P, Haussinger K, Kreyling WG. 2008. Deposition, retention, and translocation of ultrafine particles from the central airways and lung periphery. *Am J Respir Crit Care Med* 177(4):426–432.

Monopoli MP, Bombelli FB, Dawson KA. 2011. Nanobiotechnology: Nanoparticle coronas take shape. *Nat Nanotechnol* 6(1):11–12.

Mossman B, Light W, Wei E. 1983. Asbestos: Mechanisms of toxicity and carcinogenicity in the respiratory tract. *Annu Rev Pharmacol Toxicol* 23:595–615.

Mossman BT, Churg A. 1998. Mechanisms in the pathogenesis of asbestosis and silicosis. *Am J Respir Crit Care Med* 157(5 Pt 1):1666–1680.

Mu Q, Hondow NS, Krzeminski L, Brown AP, Jeuken LJ, Routledge MN. 2012. Mechanism of cellular uptake of genotoxic silica nanoparticles. *Part Fibre Toxicol* 9:29.

Murphy FA, Schinwald A, Poland CA, Donaldson K. 2012. The mechanism of pleural inflammation by long carbon nanotubes: Interaction of long fibres with macrophages stimulates them to amplify pro-inflammatory responses in mesothelial cells. *Part Fibre Toxicol* 9.

Nel A, Xia T, Madler L, Li N. 2006. Toxic potential of materials at the nanolevel. *Science* 311(5761):622–627.

Nel AE, Madler L, Velegol D, Xia T, Hoek EM, Somasundaran P, Klaessig F, Castranova V, Thompson M. 2009. Understanding biophysicochemical interactions at the nano–bio interface. *Nat Mater* 8(7):543–557.

Oberdorster G. 2000. Toxicology of ultrafine particles: *In vivo* studies. *Philos Trans R Soc Lond Ser A-Math Phys Eng Sci* 358(1775):2719–2739.

Oberdorster G. 2001. Pulmonary effects of inhaled ultrafine particles. *Int Arch Occup Environ Health* 74(1):1–8.

Oberdorster G, Ferin J, Lehnert BE. 1994. Correlation between particle size, *in vivo* particle persistence, and lung injury. *Environ Health Perspect* 102 Suppl 5:173–179.

Oberdorster G, Sharp Z, Atudorei V, Elder A, Gelein R, Lunts A, Kreyling W, Cox C. 2002. Extrapulmonary translocation of ultrafine carbon particles following whole-body inhalation exposure of rats. *J Toxicol Environ Health A* 65(20):1531–1543.

Oberdorster G, Oberdorster E, Oberdorster J. 2005a. Nanotoxicology: An emerging discipline evolving from studies of ultrafine particles. *Environ Health Perspect* 113(7):823–839.

Oberdorster G, Maynard A, Donaldson K, Castranova V, Fitzpatrick J, Ausman K, Carter J, Karn B, Kreyling W, Lai D et al. 2005b. Principles for characterizing the potential human health effects from exposure to nanomaterials: Elements of a screening strategy. *Part Fibre Toxicol* 2:8.

Oberdorster G, Stone V, Donaldson K. 2007. Toxicology of nanoparticles: A historical perspective. *Nanotoxicology* 1(1):2–25.

Oghabian MA, Guiti M, Haddad P, Gharehaghaji N, Saber R, Alam NR, Malekpour M, Rafie B. 2006. Detection sensitivity of MRI using ultra-small super paramagnetic iron oxide nano-particles (USPIO) in biological tissues. *Conf Proc IEEE Eng Med Biol Soc* 1:5625–5626.

Parashar UK, Saxena PS, Srivastava A. 2008. Role of nanomaterials in biotechnology. *Dig J Nanomater Biostruct* 3(2):81–87.

Park EJ, Choi K, Park K. 2011. Induction of inflammatory responses and gene expression by intratracheal instillation of silver nanoparticles in mice. *Arch Pharm Res* 34(2):299–307.

Pauluhn J. 2009. Pulmonary toxicity and fate of agglomerated 10 and 40 nm aluminum oxyhydroxides following 4-week inhalation exposure of rats: Toxic effects are determined by agglomerated, not primary particle size. *Toxicol Sci* 109(1):152–67.

Peters A, Dockery DW, Muller JE, Mittleman MA. 2001. Increased particulate air pollution and the triggering of myocardial infarction. *Circulation* 103(23):2810–2815.

Pissuwan D, Valenzuela S, Cortie MB. 2008. Prospects for gold nanorod particles in diagnostic and therapeutic applications. *Biotechnol Genet Eng Rev* 25:93–112.

Poland CA, Duffin R, Kinloch I, Maynard A, Wallace WA, Seaton A, Stone V, Brown S, Macnee W, Donaldson K. 2008. Carbon nanotubes introduced into the abdominal cavity of mice show asbestos-like pathogenicity in a pilot study. *Nat Nanotechnol* 3(7):423–428.

Pope CA, 3rd, Burnett RT, Thun MJ, Calle EE, Krewski D, Ito K, Thurston GD. 2002. Lung cancer, cardiopulmonary mortality, and long-term exposure to fine particulate air pollution. *JAMA* 287(9):1132–1141.

Pope CA, 3rd, Burnett RT, Thurston GD, Thun MJ, Calle EE, Krewski D, Godleski JJ. 2004. Cardiovascular mortality and long-term exposure to particulate air pollution: Epidemiological evidence of general pathophysiological pathways of disease. *Circulation* 109(1):71–77.

Porter DW, Hubbs AF, Mercer RR, Wu N, Wolfarth MG, Sriram K, Leonard S, Battelli L, Schwegler-Berry D, Friend S et al. 2010. Mouse pulmonary dose- and time course-responses induced by exposure to multi-walled carbon nanotubes. *Toxicology* 269(2–3):136–147.

Renwick LC, Donaldson K, Clouter A. 2001. Impairment of alveolar macrophage phagocytosis by ultrafine particles. *Toxicol Appl Pharmacol* 172(2):119–127.

Rothen-Rutishauser B, Brown DM, Piallier-Boyles M, Kinloch IA, Windle AH, Gehr P, Stone V. 2010. Relating the physicochemical characteristics and dispersion of multiwalled carbon nanotubes in different suspension media to their oxidative reactivity *in vitro* and inflammation *in vivo*. *Nanotoxicology* 4(3):331–342.

Rothen-Rutishauser B, Mueller L, Blank F, Brandenberger C, Muehlfeld C, Gehr P. 2008. A newly developed *in vitro* model of the human epithelial airway barrier to study the toxic potential of nanoparticles. *ALTEX* 25(3):191–196.

Rothen-Rutishauser BM, Kiama SG, Gehr P. 2005. A three-dimensional cellular model of the human respiratory tract to study the interaction with particles. *Am J Respir Cell Mol Biol* 32(4):281–289.

Royal Commission on Environmental Pollution (RCEP). 2008. *Novel Materials in the Environment: The Case of Nanotechnology*: RCEP, UK, London.

Royal Society/Royal Academy of Engineering. 2004. *Nanoscience and Nanotechnologies: Opportunities and Uncertainties*: Royal Society/Royal Academy of Engineering, London.

Ryman-Rasmussen JP, Cesta MF, Brody AR, Shipley-Phillips JK, Everitt JI, Tewksbury EW, Moss OR, Wong BA, Dodd DE, Andersen ME et al. 2009. Inhaled carbon nanotubes reach the subpleural tissue in mice. *Nat Nanotechnol* 4(11):747–751.

Saber AT, Jensen KA, Jacobsen NR, Birkedal R, Mikkelsen L, Möller P, Loft S, Wallin H, Vogel U. 2012. Inflammatory and genotoxic effects of nanoparticles designed for inclusion in paints and lacquers. *Nanotoxicology* 6(5):453–471.

Sadauskas E, Jacobsen NR, Danscher G, Stoltenberg M, Vogel U, Larsen A, Kreyling W, Wallin H. 2009. Biodistribution of gold nanoparticles in mouse lung following intratracheal instillation. *Chem Cent J* 3:16.

Salata O. 2004. Applications of nanoparticles in biology and medicine. *J Nanobiotechnol* 2(1):3.

Sayes CM, Reed KL, Warheit DB. 2007. Assessing toxicity of fine and nanoparticles: Comparing *in vitro* measurements to *in vivo* pulmonary toxicity profiles. *Toxicol Sci* 97(1):163–180.

SCENIHR. 2006. *Opinion on the Appropriateness of Existing Methodologies to Assess the Potential Risks Associated with Engineered and Adventitious Products of Nanotechnologies*: SCENIHR, Luxembourg, Luxembourg.

SCENIHR. 2009. *Risk Assessment of Products of Nanotechnologies*.

Schins RP. 2002. Mechanisms of genotoxicity of particles and fibers. *Inhal Toxicol* 14(1):57–78.

Schleh C, Semmler-Behnke M, Lipka J, Wenk A, Hirn S, Schaffler M, Schmid G, Simon U, Kreyling WG. 2012. Size and surface charge of gold nanoparticles determine absorption across intestinal barriers and accumulation in secondary target organs after oral administration. *Nanotoxicology* 6(1):36–46.

Seaton, A, MacNee, W, Donaldson, K, Godden, D. 1995. Particulate air pollution and acute health effects. *Lancet* 345:176–178.

Semmler-Behnke M, Kreyling WG, Lipka J, Fertsch S, Wenk A, Takenaka S, Schmid G, Brandau W. 2008. Biodistribution of 1.4- and 18-nm gold particles in rats. *Small* 4(12):2108–2111.

Shinohara N, Gamo M, Nakanishi J. 2011. Fullerene c60: Inhalation hazard assessment and derivation of a period-limited acceptable exposure level. *Toxicol Sci* 123(2):576–589.

Shinohara N, Matsumoto K, Endoh S, Maru J, Nakanishi J. 2009. *In vitro* and *in vivo* genotoxicity tests on fullerene C60 nanoparticles. *Toxicol Lett* 191(2–3):289–296.

Silva FC, Santos ABS, Decarvalho TMU, Desouza W. 1987. Surface-charge of resident, elicited, and activated mouse peritoneal macrophages. *J Leukoc Biol* 41(2):143–149.

Sohaebuddin SK, Thevenot PT, Baker D, Eaton JW, Tang L. 2010. Nanomaterial cytotoxicity is composition, size, and cell type dependent. *Part Fibre Toxicol* 7:22.

Sonavane G, Tomoda K, Makino K. 2008. Biodistribution of colloidal gold nanoparticles after intravenous administration: Effect of particle size. *Colloids Surf B Biointerfaces* 66(2):274–280.

Soto KF, Carrasco A, Powell TG, Garza KM, Murr LE. 2005. Comparative *in vitro* cytotoxicity assessment of some manufactured nanoparticulate materials characterized by transmission electron microscopy. *J Nanopart Res* 7(2–3):145–169.

Stone V, Johnston H, Schins RP. 2009. Development of *in vitro* systems for nanotoxicology: Methodological considerations. *Crit Rev Toxicol* 39(7):613–626.

Stone V, Nowack B, Baun A, van den Brink N, Kammer F, Dusinska M, Handy R, Hankin S, Hassellov M, Joner E et al. 2010. Nanomaterials for environmental studies: Classification, reference material issues, and strategies for physico–chemical characterisation. *Sci Total Environ* 408(7):1745–1754.

Stone V, Shaw J, Brown DM, Macnee W, Faux SP, Donaldson K. 1998. The role of oxidative stress in the prolonged inhibitory effect of ultrafine carbon black on epithelial cell function. *Toxicol In Vitro* 12(6):649–659.

Stone V, Tuinman M, Vamvakopoulos JE, Shaw J, Brown D, Petterson S, Faux SP, Borm P, MacNee W, Michaelangeli F et al. 2000. Increased calcium influx in a monocytic cell line on exposure to ultrafine carbon black. *Eur Respir J* 15(2):297–303.

Stone V, Pozzi-Mucelli S, Tran L, Aschberger K, Sabella S, Vogel U, Poland C, Balharry D, Fernandes T, Gottardo S et al. 2014. ITS-NANO–prioritising nanosafety research to develop a stakeholder driven intelligent testing strategy. *Part Fibre Toxicol* 11(9): doi: 10.1186/1743-8977-11-9.

Teeguarden JG, Hinderliter PM, Orr G, Thrall BD, Pounds JG. 2007. Particokinetics *in vitro*: Dosimetry considerations for *in vitro* nanoparticle toxicity assessments. *Toxicol Sci* 95(2):300–312.

Tenzer S, Docter D, Rosfa S, Wlodarski A, Kuharev J, Rekik A, Knauer SK, Bantz C, Nawroth T, Bier C et al. 2011. Nanoparticle size is a critical physicochemical determinant of the human blood plasma corona: A comprehensive quantitative proteomic analysis. *ACS Nano* 5(9):7155–7167.

Thannickal VJ, Fanburg BL. 2000. Reactive oxygen species in cell signaling. *Am J Physiol Lung Cell Mol Physiol* 279(6):L1005–1028.

Walczyk D, Bombelli FB, Monopoli MP, Lynch I, Dawson KA. 2010. What the cell "sees" in bionanoscience. *J Am Chem Soc* 132(16):5761–5768.

Wang T, Bai J, Jiang X, Nienhaus GU. 2012. Cellular uptake of nanoparticles by membrane penetration: A study combining confocal microscopy with FTIR spectroelectrochemistry. *ACS Nano* 6(2):1251–1259.

Warheit DB. 2008. How meaningful are the results of nanotoxicity studies in the absence of adequate material characterization? *Toxicol Sci* 101(2):183–185.

Wilson MR, Foucaud L, Barlow PG, Hutchison GR, Sales J, Simpson RJ, Stone V. 2007. Nanoparticle interactions with zinc and iron: Implications for toxicology and inflammation. *Toxicol Appl Pharmacol* 225(1):80–89.

Wilson MR, Lightbody JH, Donaldson K, Sales J, Stone V. 2002. Interactions between ultrafine particles and transition metals *in vivo* and *in vitro*. *Toxicol Appl Pharmacol* 184(3):172–179.

Wilson R, Spengler JD (eds.). 1996. *Particles in Our Air: Concentrations and Health Effects*: Harvard University Press, Harvard.

Worle-Knirsch JM, Pulskamp K, Krug HF. 2006. Oops they did it again! Carbon nanotubes hoax scientists in viability assays. *Nano Lett* 6(6):1261–1268.

Zhao F, Zhao Y, Liu Y, Chang X, Chen C. 2011. Cellular uptake, intracellular trafficking, and cytotoxicity of nanomaterials. *Small* 7(10):1322–1337.

21

Summary

Peter Gehr and Akira Tsuda

Nanoparticles in the Lung: Environmental Exposure and Drug Delivery—this book covers a challenging and important field of research activity: studies on the interaction of our organism with the environment. The lung is the main portal of entry of our organism for any type of particulate matter. Within the lung, it is the whole internal surface exposed to air with which particulate matter may interact. The book logically begins with an excursion on the deposition of particulate matter, in particular, its nano-fraction (Chapters 2 and 19). It is their physics and chemistry that already makes the deposition process of these minuscule particles a special one, quite different from that of larger, micron-sized particles. The book then continues reviewing the interaction of nanoparticles (NPs) with structures of the internal lung surface. It starts with the aqueous lining layer and the surfactant at the air–liquid interface, which is the first structure NPs encounter when deposited and continues with the subsequent processes affecting the structures. These processes determine the fate of the NPs in the pulmonary tissue, and they are governed by physics and chemistry.

The interaction of inhaled particles that deposit on the internal lung surface with surface structures, cells, and tissue largely depend on the size of the particles (Chapter 3). NPs behave differently from their larger counterparts, the micron particles. Of particular importance is the interaction of NPs with cells. In case of micron particles, that is, particles much larger than cell-surface molecules (which are nano-scale), the cells may just contact and react to a small surface area of the particle, but may not change their topology and their behavior (e.g., wrap around or engulf the entire particles). On the other hand, if the size of the inhaled particles is similar to those of the cell-surface molecules or proteins, the particle surface characteristics become critical and the cell may react totally differently to the nanosized particles than to the larger ones. What are the distinguishing features of nanosized particles? It is well documented that significant changes in physicochemical properties occur with decreasing particle size, especially as the size becomes nano-range (Chapter 16). The surface/volume ratio increases, causing an increased reactivity per mass (Oberdörster et al. 2005) and also affecting the thermodynamic properties of the particles (e.g., phase stability and equilibrium). As a result, the melting temperature, homogeneity of the particles, and their shape may also change. With decreasing particle size, the electronic band structure and optical properties also change (quantum effects). With decreasing numbers of atoms per particle, the interaction potential between the atoms decreases. The exchange energy—relative to the thermal energy—determines the ferroelectric properties of the particles. One of the important biological consequences of the reduced particle size (i.e., increased surface/volume ratio, thus increased surface reactivity per mass) is that the NP surface is coated by a variety of biomolecules; NPs tend to reduce their surface energy via adsorption of biomolecules (e.g., proteins) from their surroundings (e.g., Monopoli et al. 2011). By adsorbing biomolecules on the surface, NPs not only reduce their surface energy but they also dramatically change their subsequent interactions with cells, thus consequently changing their fate in the body. A full description of protein corona formed on NPs, especially with surfactant proteins, is given in Chapter 4, and several other chapters (Chapters 3, 5, 8–10, 16, 17, and 20) also discuss this important concept.

When the size of particles is reduced into the nano-range, a passive transmembrane behavior may also occur (demonstrated by the following experimental data: after inhalation exposure, non-membrane-bound NPs were found within red blood cells, which lack an endocytic machinery; Geiser et al. 2005; Lesniak et al. 2005; Rothen-Rutishauser et al. 2006; Mu et al. 2012). Thus, NPs—unlike larger particles (i.e., micron-size particles)—are able to traverse across the cell membrane and get into cells in a fashion not dependent on the well-known endocytic pathways (reviewed by Conner and Schmid 2003; Wang et al. 2012), wherein internalized particles are found exclusively within membrane-bound vesicles (Chithrani et al. 2006). This transmembrane pathway (Geiser et al. 2005; Rothen-Rutishauser et al. 2007; Wang et al. 2012) unique to NPs may open a new avenue to deliver therapeutic NPs directly to the cytosol (nanomedicine). These are exciting directions of future research (see Chapters 3, 6 through 8, 13 through 15).

Subsequent to entering is the intracellular trafficking of the NPs. The cytoskeleton may or may not be involved, probably depending on whether the NPs are membrane bound or not (Jiang et al. 2010; Shang et al. 2014). It could support a directed movement of NPs to certain locations, such as lysosomes, mitochondria, endoplasmic reticulum, Golgi apparatus, and the nucleus. Whereas it is not yet known of all the organelles what effect their interaction with NPs could have, it has been described that NPs in mitochondria could interfere with the respiratory chain, causing cytotoxic and genotoxic effects (Costa et al. 2010; Teodoro et al. 2011; Chichova et al. 2014). Few publications have described NPs entering the nucleus and interacting with the DNA, which could cause genotoxic effects (Magdolenova et al. 2013; Rim et al. 2013).

An interesting and well-described process is the translocation of NPs that had been deposited in the alveolar area to the capillary blood and to the secondary organs by the blood circulation (Chapters 10 and 12). NPs have been found in literally all organs including the brain (Kreyling et al. 2002). What this translocation really means is still unknown. It is not clear how and why NPs would leave the bloodstream in a certain organ and not in another one. Other than about the blood circulation, little is known about the NPs in the lymphatic system (Choi et al. 2010; Chapter 11).

It is important to realize that NPs behave like a double-edged sword. It is a concern that NPs could cause toxicity since the reduction of particle size into the nano-size range is accompanied with many new physical and chemical phenomena as well as interaction mechanisms with biological systems mentioned above (Section VI, Chapters 19 and 20). Nanotoxicological effects, indeed, should be treated with special caution until we exactly understand the effects of particles of this size range. On the other hand, the unique properties of NPs could also be used for our advantage. Since the lung is the main portal of entry and offers a very large surface area and a very thin air–blood tissue barrier for systemic drug delivery, an extensive knowledge on the interaction between the NPs and the lung will be invaluable for designing optimal NP-mediated drug delivery (Chapters 13 through 15). For instance, the alternative passive transmembrane pathway across the air–blood barrier, found in recent nanotoxicological studies, would provide an important opportunity for NP-based drug delivery.

A direct drug (gene) delivery from the nose to the brain (e.g., Oberdörster et al. 2004, 2009; Matsui et al. 2009) without involving the bloodstream would provide another exciting opportunity for the treatment of commonly occurring diseases of the central nervous system, including dementia (e.g., Alzheimer's disease). It should be noted that this pathway can be utilized only for drug particles in the nano-size range (Chapter 18).

Recent *in vitro* and *in vivo* data demonstrating that macrophages and dendritic cells (both sentinels of the immune system) collaborate in handling deposited NPs represent another exciting area of research (Gehr et al. 2006; Blank et al. 2007; Fiole et al. 2014) (Chapters 3, 9, and 10). Furthermore, since NPs can quickly appear in the draining lymph nodes of the lung (Chapter 11), they can be used to deliver antigens in the form of NP vaccines into the secondary lymphoid organs where B and T cells are activated and interact.

In summary, by improving the basic knowledge on the interaction of the lungs and NPs, we could better estimate the risk and potential pathology caused by environmental NP, and at the same time, we could deliver by inhalation and manipulate the uptake of NP with therapeutic effects to exploit their special properties. As nanotechnology is rapidly developing, future studies on NPs in the lung promise many new opportunities.

References

Blank, F., B. Rothen-Rutishauser, and P. Gehr. 2007. Dendritic cells and macrophages form a transepithelial network against foreign particulate antigens. *Am. J. Respir. Cell Mol. Biol.* 36: 669–677.

Chichova, M., M. Shkodrova, P. Vasileva, K. Kirilova, and D. Doncheva-Stoimenova. 2014. Influence of silver nanoparticles on the activity of rat liver mitochondrial ATPase. *J. Nanopart. Res.* 16: 2243: 1–14.

Chithrani, B.D., A.A. Ghazani, and W.C. Chan. 2006. Determining the size and shape dependence of gold nanoparticle uptake into mammalian cells. *Nano Lett.* 6(4): 662–668.

Choi, H.S., Y. Ashitate, J.H. Lee et al. 2010. Rapid translocation of nanoparticles from the lung airspaces to the body. *Nat. Biotechnol.* 28: 1300–1303.

Conner, S.D., and S.L. Schmid. 2003. Regulated portals of entry into the cell. *Nature* 422: 37–44.

Costa, C.S., J.V. Ronconi, J.F. Daufenbach et al. 2010. *In vitro* effects of silver nanoparticles on the mitochondrial respiratory chain. *Mol. Cell Biochem.* 342: 52–56.

Fiole, D., P. Deman, Y. Trescos et al. 2014. Two-photon intravital imaging of lungs during anthrax infection reveals long-lasting macrophage-dendritic cell contacts. *Infect. Immun.* 82: 864–872.

Gehr, P., F. Blank, and B.M. Rothen-Rutishauser. 2006. Fate of inhaled particles after interaction with the lung surface *Paediatr. Respir. Rev.* 7 Suppl. 1: S73–S75.

Geiser, M., B. Rothen-Rutishauser, N. Kapp et al. 2005. Ultrafine particles cross cellular membranes by nonphagocytic mechanisms in lungs and in cultured cells. *Environ. Health Perspect.* 113: 1555–1560.

Jiang, X., C. Röcker, M. Hafner, S. Brandholt, R.M. Dörlich, and G.U. Nienhaus. 2010. Endo- and exocytosis of zwitterionic quantum dot nanoparticles by live HeLa cells. *ACS Nano* 4: 6787–6797.

Kreyling, W.G., M. Semmler, F. Erbe et al. 2002. Translocation of ultrafine insoluble iridium particles from lung epithelium to extrapulmonary organs is size dependent but very low. *J. Toxicol. Environ. Health A* 65: 1513–1530.

Lesniak, W., A.U. Bielinska, K. Sun et al. 2005. Silver/dendrimer nanocomposites as biomarkers: Fabrication, characterization, *in vitro* toxicity, and intracellular detection. *Nano Lett.* 5: 2123–2130.

Magdolenova, Z., M. Drlickova, K. Henjum et al. 2013. Coating-dependent induction of cytotoxicity and genotoxicity of iron oxide nanoparticles. *Nanotoxicology* November 14, doi: 10.3109/17435390.2013.847505.

Matsui, Y., N. Sakai, A. Tsuda et al. 2009. Tracking the pathway of diesel exhaust particles from the nose to the brain by X-ray florescence analysis. *Spectrochim. Acta Part B.* 64: 796–801.

Monopoli, M.P., D. Walczyk, A. Campbell et al. 2011. Physical–chemical aspects of protein corona: Relevance to *in vitro* and *in vivo* biological impacts of nanoparticles. *J. Am. Chem. Soc.* 133: 2525–2534.

Mu, Q., N.S. Hondow, L. Ski, A.P. Brown, L.J. Jeuken, and M.N. Routledge. 2012. Mechanism of cellular uptake of genotoxic silica nanoparticles. *Part. Fibre Toxicol.* 9: 29.

Oberdörster, G., A. Elder, and A. Rinderknecht. 2009. Nanoparticles and the brain: Cause for concern? *J. Nanosci. Nanotechnol.* 9: 4996–5007.

Oberdörster, G., E. Oberdörster, and J. Oberdörster. 2005. Nanotoxicology: An emerging discipline evolving from studies of ultrafine particles. *Environ. Health Perspect.* 113: 823–839.

Oberdorster, G., Z. Sharp, V. Atudorei et al. 2004. Translocation of inhaled ultrafine particles to the brain. *Inhal. Toxicol.* 16: 437–445.

Rim, K.T., S.W. Song, and H.Y. Kim. 2013. Oxidative DNA damage from nanoparticle exposure and its application to workers' health: A literature review. *Saf. Health Work* 4: 177–186.

Rothen-Rutishauser, B., S. Schurch, and P. Gehr. 2007. Interaction of particles with membranes. In: *The Toxicology of Particles*, eds. K. Donaldson, and P.J. Borm, 139–160. Taylor & Francis Group, LLC, CRC Press, Boca Raton, FL.

Rothen-Rutishauser, B.M., S. Schurch, B. Haenni, N. Kapp, and P. Gehr. 2006. Interaction of fine particles and nanoparticles with red blood cells visualized with advanced microscopic techniques. *Environ. Sci. Technol.* 40: 4353–4359.

Shang, L., K. Nienhaus, and G.U. Nienhaus. 2014. Engineered nanoparticles interacting with cells: Size matters. *J. Nanobiotechnol.* 12: 5.

Teodoro, J.S., A.M. Simoes, F.V. Duarte et al. 2011. Assessment of the toxicity of silver nanoparticles *in vitro*: A mitochondrial perspective. *Toxicol. In Vitro* 25: 664–670.

Wang, T., J. Bai, X. Jiang, and G.U. Nienhaus. 2012. Cellular uptake of nanoparticles by membrane penetration: A study combining confocal microscopy with RTIR spectroelectrochemistry. *ACS Nano* 6: 1251–1259.

Index

Page numbers followed by f and t indicate figures and tables, respectively.

B

C

F

G

H